AF606297

5 043502 01

Statistical Dynamics of Sampled Data Systems

Statistical Dynamics of Sampled Data Systems

P. D. KRUT'KO

Translated by Scripta Technica Ltd

LONDON ILIFFE BOOKS LTD

ILIFFE BOOKS LTD
42 RUSSELL SQUARE
LONDON, W.C.1

Originally published in Moscow by
Soviet Radio Press
under the title
Statisticheskaya dinamika impul'snykh sistem

Translated and prepared for press by Scripta Technica Ltd.,
139a New Bond Street, London W.1

English edition first published in 1969

592 03936 6

Printed and bound in Great Britain by
J. W. Arrowsmith Ltd,
Bristol

Contents

PREFACE

Sampled data automatic control systems are extensively used in many branches of technology. This is responsible for the great interest in sampled data systems which has arisen in recent years, and for the many extensions and developments of the theory which are described in the literature. The work of Tsypkin [1] and Perov is of fundamental importance in this field.

Nevertheless a number of problems of this theory, particularly the statistical dynamics of linear and non-linear systems, have not as yet been comprehensively treated. For example, analytical methods are already available for investigating sampled data systems in the case of given and random disturbances. But in the statistical investigation of sampled data systems, and especially non-stationary systems (those with time-dependent parameters), the application of such analytical methods encounters serious difficulties, and one must employ methods of mathematical simulation based on the available computer facilities. There is therefore a current need for the development of effective practical methods for the statistical investigation of sampled data systems using computerised techniques.

The theory of statistical synthesis of sampled data systems now has methods at its disposal which allow optimal systems to be determined for stationary random disturbances, but methods are lacking for determining optimal sampled data systems for non-stationary random disturbances. An attempt is made in this monograph to fill this gap to some extent. The mathematical techniques and fundamental characteristics of linear sampled data systems are discussed in the first chapter. The techniques are those of the theory of difference equations, which describe sampled data systems in the same way as differential equations describe continuous systems. At the same time it is found that the discrete analogue of the Dirac δ-function, which is widely used in the theory of differential equations and in studies of continuous systems, is useful in the theory of sampled data systems. This analogue of the δ-function is called the σ-function.

The fundamental dynamic characteristic of a sampled data system in the time domain is the weighting function, which is defined as the system response to a σ-function disturbance. This definition has the advantage that it leads to a far-reaching formal analogy between sampled data and continuous systems, and this considerably simplifies the solution of a number of problems in the theory of sampled data systems. The simplification results from the fact that the highly developed procedures and methods of investigating continuous systems are found to be applicable in their entirety to sampled data systems. The first chapter concludes with a discussion of the basic dynamic characteristics of sampled data systems in the transform domain; this discussion forms the basis for the remainder of the book. A method is also given for determining the difference equations for the transfer function of a non-stationary sampled data system. This method is based on a formula obtained at the beginning of the first chapter; it is the discrete analogue of the Leibnitz formula used in the theory of differential operators.

Adjoint systems play an important role in the theory of continuous systems. A similar state of affairs prevails in the theory of sampled data systems, and the theory of adjoint sampled data systems is introduced in the second chapter. In this connection a number of problems of the theory of difference equations with variable parameters must first be considered, notably difference boundary value problems, which are the generalisation of the well known Gel'fond problem. Existence theorems are proved for homogeneous difference boundary value problems, and a function is introduced which is the discrete analogue of Green's function. The existence and uniqueness of this function are established by proving appropriate theorems. A theorem is also proved relating to the solution of a non-homogeneous difference equation defined by a summation operator whose kernel is the Green function of the corresponding difference boundary value problem.

An equation which is the discrete analogue of the Lagrange identity is obtained for linear difference operators, and an adjoint difference equation is introduced which determines the adjoint difference boundary value problem. A correspondence is also established between Green's functions of the original and the corresponding adjoint boundary value problem.

Using the theorems and properties relating to weighting functions which were proved in the first chapter, together with the properties of the Green's function, a connection is established between the weighting function and Green's function of the corresponding difference boundary value problem. In addition, an adjoint sampled data system is introduced, its basic properties are determined, and rules are formulated for its simulation using digital and analogue computers.

The third chapter is concerned with the properties of discrete random processes and the transformation of random functions by non-stationary sampled data systems. In particular, it is proved that the correlation function for the random process at the output of a sampled data system, when discrete white noise is present at the input of this system, satisfies the appropriate homogeneous and non-homogeneous difference equations and is the Green function of the corresponding self-adjoint difference boundary value problem. The results obtained enable a theory to be developed which allows discrete stationary and non-stationary random processes with given correlation functions to be set up on digital computers.

In the same chapter methods are given which enable the statistical characteristics of the output coordinates of stationary and non-stationary sampled data systems to be simulated mathematically. The procedures developed are based on the correlation theory of random processes, and enable the exact characteristics of the systems to be determined for both steady and transient states. The most effective methods of the statistical analysis of sampled data systems are based on discrete forming filters and adjoint sampled data systems.

All the methods developed for the statistical analysis of

sampled data systems presuppose the use of modern computational facilities: analogue and digital computers.

In the fourth chapter the methods which are found to be most effective in the statistical analysis of continuous systems are developed for application to non-linear sampled data systems. These are the approximate methods which use the correlation theory of random processes. Examples are the method of direct linearisation and the method of statistical linearisation of non-linear relationships.

Besides the direct development of the method of statistical linearisation in the analysis of non-linear sampled data systems given in this chapter, a general method is also given from which the method of statistical linearisation arises as a first approximation. This more general method results from the replacement of the non-linear sampled data system by an infinite number of appropriate linear systems. It enables the statistical characteristics of the system under consideration to be determined with any desired accuracy.

All the methods given can be used in statistical studies of non-linear sampled data systems. The choice of one method of analysis rather than another must be based in each case on the nature of the problem and the computational facilities available to the investigator.

Methods for determining optimal sampled data systems are presented in the last three chapters. These methods are based on the ideas of the general theory of optimal continuous systems developed by Pugachev. In Chapters 5 and 6 new methods are given for solving the discrete analogues of Wiener's problem and the Zadeh-Ragazzini problem. These methods employ the theory of adjoint sampled data systems and the theory of discrete shaping filters (Chapters 2 and 3). Chapter 6 gives the solution of a problem of great importance in practical applications: the discrete analogue of Semenov's problem. Here also the theory of adjoint sampled data systems and discrete shaping filters is used to obtain the solution.

In Chapter 7 methods are given for determining optimal sampled data systems in the case of non-stationary random disturbances. These methods presuppose that the correlation functions can be represented by polynomials of finite degree in the time variable; they determine the weighting functions of the required optimal systems by means of recurrence relations and thus obviate the need for solving systems of high-order algebraic equations.

The algorithms for determining optical non-stationary systems can easily be programmed for a digital computer. The problem of determining optimal characteristics by analogue computer is not considered in the present text.

The author is greatly indebted to Ya. S. Itskhoki, Ya. Z. Tsypkin, A.P. Grishin and V.M. Semenov for their continued interest in the present work and for their assistance in bringing it to fruition. Thanks are also due to P.G. Burlakov, V.N. Gotesman, L.A. Osipov and, A.A. Krasovskii and his colleagues, who reviewed the first three chapters of the book in manuscript.

The author would also like to express his gratitude to V.P. Perov and S.F. Bavarov whose suggestions resulted in many improvements.

Chapter 1

MATHEMATICAL THEORY: LINEAR SYSTEMS

1.1 DEFINITION OF A SAMPLED DATA SYSTEM

A system is called a sampled data system if its operation is characterised by discrete action. In the simplest case, a sampled data system can be represented by the series connection of a sampler (S) converting the input disturbance $x(t)$ of the system into a sequence of pulses $x_1(t)$, and a continuous part characterised by the transfer function W.

The sampled data system illustrated in Fig. 1.1 is the

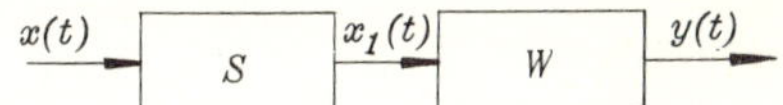

Fig. 1.1 The simplest open-loop sampled data system

simplest open-loop system. A diagram of the simplest closed-loop system, to which any other system with a single sampler may be reduced, is shown in Fig. 1.2. In this case the sampler, instead of converting the input disturbance $x(t)$ of the system into a sequence of pulses, converts the error $\mathcal{E}(t)$ defined by

$$\mathcal{E}(t) = x(t) - y(t)$$

where $y(t)$ is the output variable of the system.

If the sampler parameters are constant, and the transfer

function of the continuous part of the system is independent of time, i.e.

$$W \equiv W(p), \ p \equiv \frac{d}{dt}$$

then the sampled data system is called stationary. If the sampler parameters do depend on time, or if the transfer function of the continuous part is

$$W \equiv W(p, t), \ p \equiv \frac{d}{dt}$$

then the system is called non-stationary.

Three types of sampled data system may be distinguished, depending on the form of the sampler [1]:

1. sampled data systems of the first type, where the sampler carries out pulse-amplitude modulation of the input signal;
2. sampled data systems of the second type, where the sampler carries out pulse-width modulation;
3. sampled data systems of the third type, where the sampler carries out pulse-time modulation.

We shall first consider non-stationary sampled data

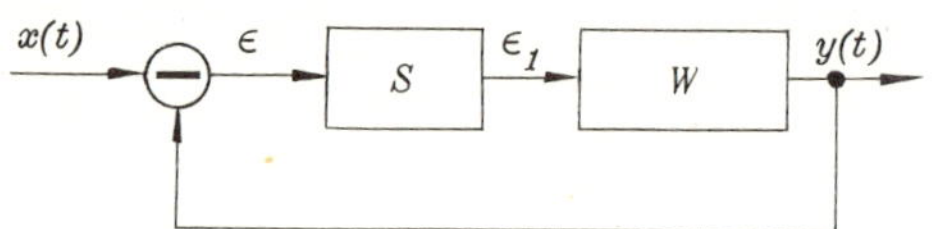

Fig. 1.2 The simplest closed-loop sampled data system

systems of the first type, where the sampler converts the input into a sequence of rectangular pulses of duration

$$\tau = \gamma T_0$$

where T_0 is the sampling period (time the sampler key is closed), and $\gamma = \frac{\tau}{T_0}$ is the relative pulse duration at the output of the sampler. Sampled data systems with samplers of the second or third type are non-linear systems. Methods of analysing such systems will be considered in Chapter 4.

The relative pulse duration γ can vary within the range $0 < \gamma \leqslant 1$. When $\gamma < 1$, the input of the sampler is converted into a sequence of equally spaced pulses whose height is proportional (or equal) to the values of the input at the instants of time immediately preceding the closure of the key. When $\gamma < 1$, the sampler is sometimes called a sampler with spread. When $\gamma = 1$, the input of the sampler is converted into a step function whose values are proportional (or equal) to the values of the input at the instants of time immediately preceding key closure. When $\gamma \ll 1$ and the input of the sampler remains substantially constant during the time τ, we may assume that the sampler converts the input into a sequence of instantaneous pulses whose area is equal (or proportional) to the values of the input at the instants when the key is closed. Such samplers are called δ-pulse samplers, and it is assumed that they generate δ-functions having weights equal (or proportional) to the values of the input at the discrete time instants corresponding to the instants when the sampler key is closed.

We shall therefore consider sampled data systems in which the sampler 'generates' both rectangular pulses and δ-pulses.

The above will also apply to those sampled data systems in which the sampler can be represented by the series connection of a sampler generating δ-pulses, and an appropriate shaping filter.

Examples of sampled data systems and their application in technology may be found in the books of Tsypkin [1] and Perov [14]. We shall always assume that a sampled data system is formed by suitably connecting a sampler and a continuous part, although in more complex cases a sampled data system can also be formed by including a digital computer operating according to a definite programme in the loop of a continuous system.

1.2 EQUATIONS FOR SAMPLED DATA SYSTEMS

1.2.1 Difference equations for sampled data systems

Every sampled data system transforms the input, and, as the result of the transformation, produces a function which characterises the state of the system at each instant of time. For example, the input for a sampled data system which is intended for the measurement of the kinematic parameters of a certain object in terms of its coordinates will be the coordinates measured at the discrete instants of time nT_0, $n=0, 1, \ldots$ The output variable of such a system is the measured quantity observable at the instants of time t. Another example is the control of the motion of an aircraft with respect to a given course by discrete commands from the ground. This reduces to the transformation of the aircraft's coordinates into deviations from the given course, and to the elimination of these deviations. The input for such a sampled data control system is given by the coordinates of the aircraft which are measured by radar at the times nT_0, $n=0, 1, \ldots$ The output of the system will be the position of the aircraft relative to the given course at any time t.

The nature of the sampled data system which transforms the input does not matter from the mathematical viewpoint. Only the set of mathematical operations and the law according to which this transformation is carried out are important.

A set of mathematical operations which relates given functions to some other functions is called an operator. In this way every sampled data system is specified as a transforming system by its operator.

Denoting the operator by the symbol A we can express the relationship between the functions of time $x(t)$ and $y(t)$ in the form

$$y=Ax \tag{2.1}$$

This means that the function $y(t)$ is the result of applying the operator A to the function $x(t)$. An operator is called linear if it obeys the principle of superposition.

$$A\sum_{k=1}^{N} c_k x_k = \sum_{k=1}^{N} c_k A x_k \tag{2.2}$$

for any N, c_k and x_k [13]. If (2.2) holds only for a particular choice of c_k and x_k, then the operator is called nonlinear. We shall give a few examples of the simplest operators for sampled data systems.

In the relation

$$y[n]=\Delta_n x[n] \tag{2.3}$$

where $y[n]$ and $x[n]$ are functions of discrete arguments (indicated by square brackets), the operator Δ_n places the first difference of the function $x[n]$ into correspondence with the function $y[n]$.

In the relation

$$y[n]=\sum_{m=-\infty}^{\infty} W[n;\, m]\, x[m] \tag{2.4}$$

the summation operator establishes a correspondence between the function $y[n]$ and the sum of the products of the two functions $W[n;\, m]$ and $x[n]$, the first of which depends on the two arguments n and m.

Let us now define the function $W[n;\, m]$ in (2.4) in the following way:

$$W[n;\, m]=\begin{cases} 1 \text{ when } n=m \\ 0 \text{ when } n\neq m \end{cases} \tag{2.5}$$

Then from (2.4) we have

$$y[n]\equiv x[n]$$

Consequently, the summation operator whose function $W[n;\, m]$ is defined by (2.5) corresponds to a system which carries out the identity transformation of the input function $x[n]$, i.e. at any instant n the output variable of such a system is equal to the input action.

Let us define the function $W[n;\, m]$ in (2.4) by

$$W[n;\, m]=\begin{cases} -1 \text{ when } n=m \\ 1 \text{ when } n=m+1 \\ 0 \text{ when } n\neq m,\ n\neq m+1 \end{cases} \tag{2.6}$$

Substituting (2.6) into (2.4) and summing we obtain

$$y[n]=\Delta_n x[n] \tag{2.7}$$

Consequently, a system whose operator is defined by (2.4) and (2.6) takes the exact difference of the input action.

The operator for any dynamical system can be specified by equations defining physical processes in the sampled data system and finally establishing the relation between the output variables of the system and the input actions.

Processes in a linear sampled data system are defined by difference equations. Using relative time, which is conventional in the theory of sampled data systems [1], the difference equations describing processes in a sampled data system can be written

$$\sum_{i=0}^{k} a_i[n]\,\Delta_n^i y[n,\varepsilon]=\sum_{j=0}^{h_1} b_j[n,\varepsilon]\,\Delta_n^j x[n,0],\quad 0\leqslant\varepsilon\leqslant\gamma$$
$$\sum_{i=0}^{k} c_i[n]\,\Delta_n^i y[n,\varepsilon]=\sum_{j=0}^{h_2} d_j[n,\varepsilon]\,\Delta_n^j x[n,0],\quad \gamma\leqslant\varepsilon\leqslant 1 \tag{2.8}$$

where $x[n, 0]$ is the input action of the system ($\varepsilon=0$);
$y[n,\varepsilon]=y[\bar{t}]$ is the output variable of the system at the time $\bar{t}=\frac{t}{T_0}=n+\varepsilon$, where $n=0, 1, \ldots, 0\leqslant\varepsilon\leqslant 1$;
T_0 is the sampling period (time for which the sampler key is closed);
γ is the relative duration of pulses generated by the sampler;
$\Delta_n^i y[n,\varepsilon]$ is the i-th order difference obtained from $y[\bar{t}]$ with respect to the argument n;
$\Delta_n^j x[n, 0]$ is the j-th order difference obtained from $x[n, 0]$ with respect to the argument n when $\varepsilon=0$;
$a_i[n]$, $b_j[n,\varepsilon]$, $c_i[n]$ and $d_j[n,\varepsilon]$ are coefficients, depending on time for a non-stationary system, of which $b_j[n,\varepsilon]$ and $d_j[n,\varepsilon]$ are continuous functions of ε;
k is the order of the difference equation, if $a_i\neq 0$ and $c_i\neq 0\,(i=0, 1, 2, \ldots\ k)$.

The first of the two equations in (2.8) defines processes in the sampled data system during the time the pulse from the sampler acts on the continuous part, whilst the second defines processes during the interval between the pulses.

For the sake of brevity we write (2.8) as follows:

$$L_n(\Delta, n)\,y[n,\varepsilon]=M_n(\Delta,\bar{t})\,x[n, 0],\quad 0\leqslant\varepsilon\leqslant\gamma$$
$$S_n(\Delta, n)\,y[n,\varepsilon]=G_n(\Delta,\bar{t})\,x[n, 0],\quad \gamma\leqslant\varepsilon\leqslant 1 \tag{2.9}$$

where the difference operators $L_n(\Delta, n)$, $M_n(\Delta,\bar{t})$, $S_n(\Delta, n)$ and $G_n(\Delta,\bar{t})$ are defined by

$$L_n(\Delta, n)\equiv\sum_{i=0}^{k} a_i[n]\,\Delta_n^i \tag{2.10}$$

$$M_n(\Delta, \overline{t}) \equiv \sum_{j=0}^{h_1} b_j [n, \varepsilon] \Delta_n^j \quad . \tag{2.11}$$

$$S_n(\Delta, n) \equiv \sum_{i=0}^{k} c_i [n] \Delta_n^i \tag{2.12}$$

$$G_n(\Delta, \overline{t}) \equiv \sum_{j=0}^{h_2} d_j [n, \varepsilon] \Delta_n^j \tag{2.13}$$

Processes in a sampled data system are defined by difference equations of the form given by (2.8) when the sampler generates rectangular pulses of finite duration γT_0 $(0 < \gamma < 1)$. But when $\gamma = 1$, the processes in a sampled data system are defined by equations of the form

$$K_n(\Delta, n) y [n, \varepsilon] = H_n(\Delta, \overline{t}) x [n, 0], \quad 0 \leqslant \varepsilon \leqslant 1 \tag{2.14}$$

the difference operators $K_n(\Delta, n)$ and $H_n(\Delta, t)$ are defined as follows:

$$\left. \begin{aligned} K_n(\Delta, n) &\equiv \sum_{i=0}^{k} f_i [n] \Delta_n^i \\ H_n(\Delta, \overline{t}) &\equiv \sum_{j=0}^{h} g_j [n, \varepsilon] \Delta_n^j \end{aligned} \right\} \tag{2.15}$$

Processes in those sampled data systems whose samplers generate δ-functions are defined by equations of the form given by (2.14). It was pointed out above that systems for which $\gamma \ll 1$ are of this kind.

There is yet another way of writing difference equations [18], which differs from (2.8) or (2.14). Using the expression for the i-th difference $\Delta^i \varphi[n, \varepsilon]$ in terms of the displaced functions $\varphi[n, \varepsilon]$, $\varphi[n+1 \ \varepsilon]$, . . ., i.e. using formula

$$\Delta^i \varphi [n, \varepsilon] = \sum_{\nu=0}^{i} (-1)^{i-\nu} C_i^\nu \varphi [n + \nu, \varepsilon] \tag{2.16}$$

we transform (2.8) so that (the coefficients having the previous meaning)

$$\left.\begin{aligned} &\sum_{i=0}^{k} a_i[n]\, y[n+i, \varepsilon] = \sum_{j=0}^{h_1} b_j[n, \varepsilon]\, x[n+j, 0], \ 0 \leqslant \varepsilon \leqslant \gamma \\ &\sum_{i=0}^{k} c_i[n]\, y[n+i, \varepsilon] = \sum_{j=0}^{h_2} d_j[n, \varepsilon]\, x[n+j, 0], \ \gamma \leqslant \varepsilon \leqslant 1 \end{aligned}\right\} \tag{2.17}$$

The form of difference equations given by (2.17) is preferable for the numerical solution of the equations; for analytical purposes it will be convenient to use (2.8) and (2.14). This form of difference equations has the advantage that they formally resemble differential equations, if we recall that in a difference equation the operator $\Delta_n\{. . .\}$, which is defined by the relation

$$\Delta_n y[n, \varepsilon] = y[n+1, \varepsilon] - y[n, \varepsilon]$$

is analogous to the operator $d_t\{\ldots\}$ in a differential equation.

1.2.2 Replacement of k*-th-order difference equation by a system of* k *first order equations*

When sampled data systems are investigated it is occasionally convenient to replace the difference equation which defines processes in the system by a system of first-order equations. Let the processes in the system be defined by the k-th order difference equation

$$\sum_{i=0}^{k} f_i[n]\, \Delta_n^i y[n, \varepsilon] = \sum_{j=0}^{h} g_j[n, \varepsilon]\, \Delta_n^i x[n, 0], \ 0 \leqslant \varepsilon \leqslant 1 \tag{2.18}$$

Without any loss of generality we may assume that $f_k[n] = 1$. We shall prove that (2.18) can be reduced to a system of k first-order difference equations of the form

$$y[n, \varepsilon]=y_1[n, \varepsilon]+F_0[n, \varepsilon]x[n, 0]$$
$$\Delta y_1[n, \varepsilon]=y_2[n, \varepsilon]+F_1[n, \varepsilon]x[n, 0]$$
$$\cdots\cdots\cdots\cdots\cdots\cdots$$
$$\Delta y_{k-1}[n, \varepsilon]=y_k[n, \varepsilon]+F_{k-1}[n, \varepsilon]x[n, 0] \qquad (2.19)$$
$$\Delta y_k[n, \varepsilon]=-f_{k-1}[n]y_k[n, \varepsilon]-f_{k-2}[n]y_{k-1}[n, \varepsilon]$$
$$-\ldots-f_0[n]y_1[n, \varepsilon]+F_k[n, \varepsilon]x[n, 0]$$
$$0\leqslant\varepsilon\leqslant1$$

To determine the functions $F_l[n, \varepsilon]$, which are combinations of $f[n]$ and $g_j[n, \varepsilon]$, all the $y_1[n, \varepsilon]$, $y_2[n, \varepsilon]$, . . ., $y_k[n, \varepsilon]$ must in turn be eliminated from (2.19), and the resulting difference equation be stipulated as equivalent to the initial equation (2.18). Proceeding in this way we obtain

$$\Delta^k\{F_0[n, \varepsilon]x[n, 0]\}+\ldots+\Delta\{F_{k-1}[n, \varepsilon]x[n, 0]\}$$
$$+F_k[n, \varepsilon]x[n, 0]+f_{k-1}[n]\left[\Delta^{k-1}\{F_0[n, \varepsilon]x[n, 0]\}\right.$$
$$\left.+\ldots+\Delta\{F_{k-2}[n, \varepsilon]x[n, 0]\}+F_{k-1}[n, \varepsilon]x[n, 0]\right]$$
$$+f_{k-2}[n]\left[\Delta^{k-2}\{F_0[n, \varepsilon]x[n, 0]\}\right.$$
$$\left.+\ldots+\Delta\{F_{k-3}[n, \varepsilon]x[n, 0]\}+F_{k-2}[n, \varepsilon]x[n, 0]\right]$$
$$\cdots\cdots\cdots\cdots\cdots\cdots$$
$$+f_1[n]\left[\Delta\{F_0[n, \varepsilon]x[n, 0]\}+F_1[n, \varepsilon]x[n, 0]\right]$$
$$+f_0[n]F_0[n, \varepsilon]x[n, 0]=\sum_{j=0}^{h}g_j[n, \varepsilon]\Delta_n^i x[n, 0]$$
$$0\leqslant\varepsilon\leqslant1 \qquad (2.20)$$

This equation can be regarded as an identity for the function $x[n, 0]$ and its differences $\Delta^j x[n, 0]$, $j=1,2,\ldots,k$. Putting $j=0, 1,\ldots, k$, in (2.20) we obtain a system of $k+1$ algebraic equations for the required functions $F_l[n, \varepsilon]$, $l=0, 1, \ldots, k$. Since the $g_j[n, \varepsilon]$ are not all zero (since otherwise we would have a homogeneous difference equation), the $F_l[n, \varepsilon]$ can be uniquely determined by solving the system of $k+1$ algebraic equations for $F_l[n, \varepsilon]$. This proves that a k-th order difference equation can be represented by a system of k first-order difference equations.

Following the terminology of the theory of differential equations, the system (2.19) will be called the normal form of difference equations.

Methods of establishing the difference equations for processes in sampled data systems with variable parameters have been considered by Tartakovskii [20, 21]. We shall use difference equations and their theory as a mathematical tool for investigating sampled data systems, but will not concern ourselves with the derivation of these equations.

1.2.3 Properties of linear difference operators

We shall establish certain properties of linear difference operators which will be required in subsequent discussions. As was indicated in Section 1.2.1, a linear difference operator is defined by

$$L_n(\Delta, \bar{t}) = \sum_{i=0}^{k} a_i[n, \varepsilon]\,\Delta_n^i, \quad 0 \leqslant \varepsilon \leqslant 1 \tag{2.21}$$

where the highest power of the operator Δ_n specifies the order of the difference operator $L_n(\Delta, \bar{t})$. An operator of this form, whose coefficients are functions of time $\bar{t} = n + \varepsilon$, i.e. $a_i = a_i[n, \varepsilon]$, will be called a variable difference operator, or a difference operator with variable coefficients. An operator of the form

$$L_n(\Delta, \varepsilon) = \sum_{i=0}^{k} a_i(\varepsilon)\,\Delta_n^i \tag{2.22}$$

whose coefficients are functions of the parameter ε, i.e., $a_i = a_i(\varepsilon)$, will be called a constant difference operator, or a difference operator with constant coefficients.

If a particular case, the coefficients of a variable operator depend only on n $a_i = a_i[n]$, that is, they are functions of an integral argument, then according to the definition given by (2.21), we have

$$L_n(\Delta, n) = \sum_{i=0}^{k} a_i[n]\,\Delta_n^i \tag{2.23}$$

In the same way, a particular form of the constant difference operator is furnished by an operator whose coefficients are constants independent of ε, that is,

$$a_i = \text{const}$$

In this instance, from the definition of an operator with

constant coefficients, we have

$$L_n(\Delta)=\sum_{i=0}^{k} a_i\Delta_n^i \tag{2.24}$$

If $k=0$ and $a_0=1$, then

$$L_n(\Delta, \overline{t})\equiv 1, \quad L_n(\Delta, \varepsilon)\equiv 1, \quad L_n(\Delta)\equiv 1 \tag{2.25}$$

Operators of this form will be called unity or identity operators.

We shall denote by

$$L_n(\Delta, \overline{t})\, x[n, \varepsilon] \tag{2.26}$$

the action of the difference operator $L_n(\Delta, \overline{t})$ on the function $x[n, \varepsilon]$. In a more general case, (2.26) can be written

$$L_n(\Delta, \overline{t})\, G[n, \varepsilon; m] \tag{2.27}$$

where the index n of the difference operator symbol L denotes the fact that this symbol operates on the function $G[n, \varepsilon; m]$ with respect to the variable n.

For both operators with variable coefficients and those with constant coefficients the symbol denoting the effect of an operator on a function is not commutative. For example, the expression $L_n(\Delta, \overline{t})\, x_1[n, \varepsilon]\, x_2[n, \varepsilon]$ denotes

$$a_k[n, \varepsilon]\,\Delta_n^k\{x_1[n, \varepsilon]\,x_2[n, \varepsilon]\}$$
$$+a_{k-1}[n, \varepsilon]\,\Delta_n^{k-1}\{x_1[n, \varepsilon]\,x_2[n, \varepsilon]\}+\ldots$$
$$\ldots+a_0[n, \varepsilon]\,x_1[n, \varepsilon]\,x_2[n, \varepsilon]$$

On the other hand, the expression $x_1[n, \varepsilon]\, L_n(\Delta, \overline{t})\, x_2[n, \varepsilon]$ denotes

$$x_1[n, \varepsilon]\{a_k[n, \varepsilon]\,\Delta_n^k\, x_2[n, \varepsilon]$$
$$+a_{k-1}[n, \varepsilon]\,\Delta_n^{k-1}\, x_2[n, \varepsilon]+\ldots+a_0[n, \varepsilon]\,x_2[n, \varepsilon]\}$$

Clearly, since the difference operators are linear, we have

$$\begin{aligned} &L_n(\Delta, \overline{t})\{x_1[n, \varepsilon]+x_2[n, \varepsilon]\} \\ &=L_n(\Delta, \overline{t})\,x_1[n, \varepsilon]+L_n(\Delta, \overline{t})\,x_2[n, \varepsilon] \end{aligned} \tag{2.28}$$

$$L_n^{\mathrm{I}}(\Delta, \bar{t})\, x_1[n, \varepsilon] + L_n^{\mathrm{II}}(\Delta, \bar{t})\, x_1[n, \varepsilon] = \left(L_n^{\mathrm{I}}(\Delta, \bar{t}) + L_n^{\mathrm{II}}(\Delta, \bar{t})\right) x_1[n, \varepsilon] \tag{2.29}$$

$$aL_n(\Delta, \bar{t})\, x_1[n, \varepsilon] + bL_n(\Delta, \bar{t})\, x_2[n, \varepsilon] = L_n(\Delta, \bar{t})\,(ax_1[n, \varepsilon] + bx_2[n, \varepsilon]) \tag{2.30}$$

where $a =$ const and $b =$ const.

Suppose the following two relations are given:

$$L_n^{\mathrm{I}}(\Delta, \bar{t})\, y[n, \varepsilon] = z[n, \varepsilon] \tag{2.31}$$

$$L_n^{\mathrm{II}}(\Delta, \bar{t})\, z[n, \varepsilon] = x[n, \varepsilon] \tag{2.32}$$

We shall determine the difference operator connecting the functions $y[n, \varepsilon]$ and $x[n, \varepsilon]$.

Substituting the value of $z[n, \varepsilon]$ from (2.31) into (2.32) we find that

$$[L_n^{\mathrm{II}}(\Delta, \bar{t}) * L_n^{\mathrm{I}}(\Delta, \bar{t})]\, y[n, \varepsilon] = x[n, \varepsilon] \tag{2.33}$$

or more concisely

$$L_n(\Delta, \bar{t})\, y[n, \varepsilon] = x[n, \varepsilon] \tag{2.34}$$

where

$$L_n(\Delta, \bar{t}) = L_n^{\mathrm{II}}(\Delta, \bar{t}) * L_n^{\mathrm{I}}(\Delta, \bar{t}) \tag{2.35}$$

The asterisk symbolises the operation of the difference operator $L_n^{\mathrm{II}}(\Delta, \bar{t})$ on the operator $L_n^{\mathrm{I}}(\Delta, \bar{t})$. It follows from (2.35) that the operation of one operator on another leads to a new operator. Let us establish the rules for the operation of one operator on another.

Suppose two linear difference operators are defined by

$$L_n(\Delta, \bar{t}) = \sum_{i=0}^{k} a_i[n, \varepsilon]\, \Delta_n^i \tag{2.36}$$

$$M_n(\Delta, \bar{t}) = \sum_{i=0}^{h} b_i[n, \varepsilon]\, \Delta_n^i \tag{2.37}$$

We find the operator $S_n(\Delta, \bar{t})$, by applying $L_n(\Delta, \bar{t})$ to $M_n(\Delta, \bar{t})$

in accordance with (2.35), i.e.

$$S_n(\Delta, \bar{t}) = L_n(\Delta, \bar{t}) * M_n(\Delta, \bar{t}) \tag{2.38}$$

Let us suppose first that

$$L_n(\Delta, \bar{t}) = a_1[n, \varepsilon]\Delta_n + a_0[n, \varepsilon] \tag{2.39}$$

$$M_n(\Delta, \bar{t}) = b_1[n, \varepsilon]\Delta_n + b_0[n, \varepsilon] \tag{2.40}$$

so that from (2.38) we have

$$S_n(\Delta, \bar{t}) = (a_1[n, \varepsilon]\Delta_n + a_0[n, \varepsilon]) * (b_1[n, \varepsilon]\Delta_n + b_0[n, \varepsilon]) \tag{2.41}$$

In view of the linearity property (2.29) of the difference operators we can write (2.41) in the form

$$\begin{aligned} S_n(\Delta, \bar{t}) = {} & a_1[n, \varepsilon]\Delta_n * (b_1[n, \varepsilon]\Delta_n + b_0[n, \varepsilon]) \\ & + a_0[n, \varepsilon] * (b_1[n, \varepsilon]\Delta_n + b_0[n, \varepsilon]) \end{aligned} \tag{2.42}$$

It can readily be shown that Δ_n:

$$a_i[n, \varepsilon] * b_j[n, \varepsilon] = a_i[n, \varepsilon]\, b_j[n, \varepsilon] \tag{2.43}$$

$$a_i[n, \varepsilon] * b_j[n, \varepsilon]\Delta_n = a_i[n, \varepsilon]\, b_j[n, \varepsilon]\Delta_n \tag{2.44}$$

$$\begin{aligned} & a_i[n, \varepsilon]\Delta_n * b_j[n, \varepsilon]\cdot 1 \\ & = a_i[n, \varepsilon](\Delta_n b_j[n, \varepsilon] + b_j[n+1, \varepsilon]\Delta_n) \end{aligned} \tag{2.45}$$

$$\begin{aligned} & a_i[n, \varepsilon]\Delta_n * b_j[n, \varepsilon]\Delta_n \\ & = a_i[n, \varepsilon](\Delta_n b_j[n, \varepsilon]\Delta_n + b_j[n+1, \varepsilon]\Delta_n^2) \end{aligned} \tag{2.46}$$

Using (2.42) to (2.46)

$$\begin{aligned} S_n(\Delta, \bar{t}) = {} & (a_1[n, \varepsilon]\Delta_n + a_0[n, \varepsilon])(b_1[n, \varepsilon]\Delta_n + b_0[n, \varepsilon]) \\ & + a_1[n, \varepsilon](b_1[n+1, \varepsilon]\Delta_n + b_0[n+1, \varepsilon])\Delta_n \end{aligned} \tag{2.47}$$

Since

$$a_1[n, \varepsilon] = \frac{\partial L_n(\Delta, \overline{t})}{\partial \Delta_n} \tag{2.48}$$

and

$$b_1[n+1, \varepsilon]\,\Delta_n + b_0[n+1, \varepsilon] = M_n(\Delta, \overline{t}+1) \tag{2.49}$$

we finally have

$$\begin{aligned} S_n(\Delta, \overline{t}) &= L_n(\Delta, \overline{t}) * M_n(\Delta, \overline{t}) \\ &= L_n(\Delta, \overline{t})\,M_n(\Delta, \overline{t}) + \frac{\partial L_n(\Delta, \overline{t})}{\partial \Delta_n} M_n(\Delta, \overline{t}+1)\,\Delta_n \end{aligned} \tag{2.50}$$

where only the products are formed of the difference operators, the operation of the operator Δ_n being extended to the function immediately after it.

We can show by induction that in general

$$\begin{aligned} S_n(\Delta, \overline{t}) &= L_n(\Delta, \overline{t}) * M_n(\Delta, \overline{t}) \\ &= L_n(\Delta, \overline{t})\,M_n(\Delta, \overline{t}) + \frac{\partial L_n(\Delta, \overline{t})}{\partial \Delta_n} M_n(\Delta, \overline{t}+1)\,\Delta_n \\ &\quad + \frac{1}{2!}\frac{\partial^2 L_n(\Delta, \overline{t})}{\partial \Delta_n^2} M_n(\Delta, \overline{t}+2)\,\Delta_n^2 \\ &\quad + \ldots + \frac{1}{k!}\frac{\partial^k L_n(\Delta, \overline{t})}{\partial \Delta_n^k} M_n(\Delta, \overline{t}+k)\,\Delta_n^k \end{aligned} \tag{2.51}$$

where in accordance with (2.37)

$$M_n(\Delta, \overline{t}+k) = \sum_{i=0}^{h} b_i[n+k, \varepsilon]\,\Delta_n^i \tag{2.52}$$

Now let the difference operators be defined by

$$L_n(\Delta, \overline{t}) = \sum_{i=0}^{k} a_i[n, \varepsilon]\,\Delta_n^i \tag{2.53}$$

$$M_n(\Delta, \varepsilon) = \sum_{i=0}^{h} b_i(\varepsilon)\,\Delta_n^i \tag{2.54}$$

We shall determine the operator defined by the application of $L_n(\Delta, \overline{t})$ the operator which is variable to the constant operator $M_n(\Delta, \varepsilon)$:

$$G_n(\Delta, \bar{t}) = L_n(\Delta, \bar{t}) * M_n(\Delta, \varepsilon) \tag{2.55}$$

Suppose first that

$$L_n(\Delta, \bar{t}) = a_1[n, \varepsilon]\,\Delta_n + a_0[n, \varepsilon] \tag{2.56}$$

$$M_n(\Delta, \varepsilon) = b_1(\varepsilon)\,\Delta_n + b_0(\varepsilon) \tag{2.57}$$

From (2.55) we now have

$$G_n(\Delta, \bar{t}) = (a_1[n, \varepsilon]\,\Delta_n + a_0[n, \varepsilon]) * \big(b_1(\varepsilon)\,\Delta_n + b_0(\varepsilon)\big) \tag{2.58}$$

Since the difference operators are linear, we can write (2.58) in the form

$$\begin{aligned} G_n(\Delta, \bar{t}) = {} & a_1[n, \varepsilon]\,\Delta_n * \big(b_1(\varepsilon)\,\Delta_n + b_0(\varepsilon)\big) \\ & + a_0[n, \varepsilon] * \big(b_1(\varepsilon)\,\Delta_n + b_0(\varepsilon)\big) \end{aligned} \tag{2.59}$$

We now use the properties of the Δ_n defined by (2.43) to (2.46). For the case under consideration we can write

$$a_i[n, \varepsilon] * b_j(\varepsilon) = a_i[n, \varepsilon]\,b_j(\varepsilon) \tag{2.60}$$

$$a_i[n, \varepsilon] * b_j(\varepsilon)\,\Delta_n = a_i[n, \varepsilon]\,b_j(\varepsilon)\,\Delta_n \tag{2.61}$$

$$a_i[n, \varepsilon]\,\Delta_n * b_j(\varepsilon)\cdot 1 = a_i[n, \varepsilon]\,b_j(\varepsilon)\,\Delta_n \tag{2.62}$$

$$a_i[n, \varepsilon]\,\Delta_n * b_j(\varepsilon)\,\Delta_n = a_i[n, \varepsilon]\,b_j(\varepsilon)\,\Delta_n^2 \tag{2.63}$$

Consequently, when (2.60) to (2.63) are taken into account, Equation (2.59) can be represented in the following form:

$$\begin{aligned} G_n(\Delta, \bar{t}) = {} & a_1[n, \varepsilon]\big(b_1(\varepsilon)\,\Delta_n + b_0(\varepsilon)\big)\,\Delta_n \\ & + a_0[n, \varepsilon]\big(b_1(\varepsilon)\,\Delta_n + b_0(\varepsilon)\big) \end{aligned} \tag{2.64}$$

From (2.56) and (2.57) we finally obtain

$$\begin{aligned} G_n(\Delta, \bar{t}) & = L_n(\Delta, \bar{t}) * M_n(\Delta, \varepsilon) \\ & = a_1[n, \varepsilon]\,M_n(\Delta, \varepsilon)\,\Delta_n + a_0[n, \varepsilon]\,M_n(\Delta, \varepsilon) \end{aligned} \tag{2.65}$$

This can also be written in another way. Transforming Equation (2.64) so that

$$G_n(\Delta, \bar{t}) = b_1(\varepsilon)(a_1[n, \varepsilon]\Delta_n + a_0[n, \varepsilon])\Delta_n + b_0(\varepsilon)(a_1[n, \varepsilon]\Delta_n + a_0[n, \varepsilon])$$

we find that

$$\begin{aligned} G_n(\Delta, \bar{t}) &= L_n(\Delta, \bar{t}) * M_n(\Delta, \varepsilon) \\ &= b_1(\varepsilon) L_n(\Delta, \bar{t})\Delta_n + b_0(\varepsilon) L_n(\Delta, \bar{t}) \end{aligned} \qquad (2.66)$$

We can show by induction that in general

$$\begin{aligned} G_n(\Delta, \bar{t}) &= L_n(\Delta, \bar{t}) * M_n(\Delta, \varepsilon) \\ &= a_k[n, \varepsilon] M_n(\Delta, \varepsilon)\Delta_n^k + a_{k-1}[n, \varepsilon] M_n(\Delta, \varepsilon)\Delta_n^{k-1} \\ &+ \ldots + a_0[n, \varepsilon] M_n(\Delta, \varepsilon) \end{aligned} \qquad (2.67)$$

where $a_i[n, \varepsilon]$ are the coefficients of the operator $L_n(\Delta, \bar{t})$;
k is the order of the operator $L_n(\Delta, \bar{t})$.

Similarly

$$\begin{aligned} G_n(\Delta, \bar{t}) &= L_n(\Delta, \bar{t}) * M_n(\Delta) = b_h(\varepsilon) L_n(\Delta, \bar{t})\Delta_n^h \\ &+ b_{h-1}(\varepsilon) L_n(\Delta, \bar{t})\Delta_n^{h-1} + \ldots + b_0(\varepsilon) L_n(\Delta, \bar{t}) \end{aligned} \qquad (2.68)$$

where $b_i(\varepsilon)$ are the coefficients of the operator $M_n(\Delta)$;
h is the order of the operator $M_n(\Delta)$.

It is clear that a constant operator operating on another gives

$$C_n(\Delta) = L_n(\Delta) * M_n(\Delta) = L_n(\Delta) M_n(\Delta) = M_n(\Delta) L_n(\Delta) \qquad (2.69)$$

In other words, the commutative property holds for constant operators.

It is worth noting that the following is true for any operator:

$$L_n^{(k)}(\Delta, \bar{t}) * M_n^{(h)}(\Delta, \bar{t}) = S_n^{(N)}(\Delta, \bar{t}) \qquad (2.70)$$

where the symbols k, h and N denote the order of the difference operators. The order of the operator formed by applying one operator to another is equal to the sum of the orders of these operators.

The above properties of difference operators and rules for operations on them have direct applications to sampled data systems with variable parameters, which are described by linear difference equations with variable coefficients.

To conclude, we shall prove an important formula which will be used below. Consider the difference operator

$$D(\Delta, \bar{t}) = \sum_{i=0}^{k} a_i[n, \varepsilon]\Delta^i \tag{2.71}$$

where for the sake of brevity we have omitted the index indicating the variable with respect to which the operator operates. The application of the operator to the product of $u[n, \varepsilon]$ and $v[n, \varepsilon]$ is defined by

$$\begin{aligned} D(\Delta, \bar{t})u[n, \varepsilon]v[n, \varepsilon] &= u[n, \varepsilon]D(\Delta, \bar{t})v[n, \varepsilon] \\ &+ \Delta u[n, \varepsilon]\frac{\partial D(\Delta, \bar{t})}{\partial \Delta}v[n+1, \varepsilon] + \dots \\ &+ \frac{1}{k!}\Delta^k u[n, \varepsilon]\frac{\partial^k D(\Delta, \bar{t})}{\partial \Delta^k}v[n+k, \varepsilon] \end{aligned} \tag{2.72}$$

It can readily be shown that

$$\Delta^l\{u[n, \varepsilon]v[n, \varepsilon]\} = \sum_{i=0}^{l} C_l^{l-i}\Delta^{l-i}u[n+i, \varepsilon]\Delta^i v[n, \varepsilon] \tag{2.73}$$

Applying (2.73) to $D(\Delta, \bar{t})$ which operates on the product of the functions, we find that

$$\begin{aligned} D(\Delta, \bar{t})u[n, \varepsilon]v[n, \varepsilon] &= \sum_{i=0}^{k} a_i[n, \varepsilon]\Delta^i\{u[n, \varepsilon]v[n, \varepsilon]\} \\ &= \sum_{i=0}^{k} a_i[n, \varepsilon]\sum_{j=0}^{i} C_i^{i-j}\Delta^{i-j}u[n+j, \varepsilon]\Delta^j v[n, \varepsilon] \end{aligned} \tag{2.74}$$

If we change the order of summation in (2.74) we find that

$$\begin{aligned} & D(\Delta, \bar{t})u[n, \varepsilon]v[n, \varepsilon] \\ &= \sum_{j=0}^{k} \Delta^j v[n, \varepsilon]\sum_{i=j}^{k} C_i^{i-j}a_i[n, \varepsilon]\Delta^{i-j}u[n+j, \varepsilon] \end{aligned} \tag{2.75}$$

Since

$$\sum_{i=j}^{k} C_i^{i-j}a_i[n, \varepsilon]\Delta^{i-j} = \frac{\partial^j D(\Delta, \bar{t})}{\partial \Delta^j}, \quad j = 1, 2, \dots, k-1 \tag{2.76}$$

and

$$C_k^0 a_k[n, \varepsilon] = \frac{1}{k!} \frac{\partial^k D(\Delta, \bar{t})}{\partial \Delta^k} \tag{2.77}$$

we finally obtain a formula from (2.74) which is the same as (2.72). The latter is analogous to the generalised Leibnitz formula in the theory of differential equations, which gives the result of the application of a differential operator to the product of two functions.

Since

$$\Delta^l \{u[n, \varepsilon] v[n, \varepsilon]\} = \sum_{i=0}^{l} C_l^{l-i} \Delta^{l-i} v[n+i, \varepsilon] \Delta^i u[n, \varepsilon] \tag{2.78}$$

we have

$$\begin{aligned} D(\Delta, \bar{t}) u[n, \varepsilon] v[n, \varepsilon] &= v[n, \varepsilon] D(\Delta, \bar{t}) u[n, \varepsilon] \\ &+ \Delta v[n, \varepsilon] \frac{\partial D(\Delta, \bar{t})}{\partial \Delta} u[n+1, \varepsilon] + \dots \\ &+ \frac{1}{k!} \Delta^k v[n, \varepsilon] \frac{\partial^k D(\Delta, \bar{t})}{\partial \Delta^k} u[n+k, \varepsilon] \end{aligned} \tag{2.79}$$

The formulae given by (2.72) and (2.79) are completely equivalent.

1.2.4 Definition of an inverse sampled data system

By analogy with continuous systems we can define an inverse sampled data system with respect to an original. A system is called inverse if its connection in series with the original system produces an identity transformation of the input, i.e. at every instant of time the output variable of the series connection of the original and inverse systems is equal to the input action.

This property of an inverse sampled data system can be expressed mathematically as follows. If A is the operator for the original system, then A^{-1} will denote the operator of the inverse sampled data system. This property can be expressed in the form

$$A^{-1}\{Ax\} = x \tag{2.80}$$

where x is the input action of the system, transformed by

the series connection of the direct and inverse systems.

We shall now establish the difference equation for an inverse sampled data system. Consider two sampled data systems described by difference equations of the form

$$\begin{aligned} L_n(\Delta, n)\, y[n] &= M_n(\Delta, n)\, x[n] \\ M_n(\Delta, n)\, z[n] &= L_n(\Delta, n)\, u[n] \end{aligned} \tag{2.81}$$

For the sake of simplicity we consider lattice functions.

Let these systems be connected in series so that the output variable of the first is the input of the second, i.e., $y[n]=u[n]$. The operator $L_n^{-1}(\Delta, n)$, which is the inverse of $L_n(\Delta, n)$, is then applied to the second equation (2.81), and by the property of mutually inverse operators we have

$$L_n^{-1}(\Delta, n)\, M_n(\Delta, n)\, z[n] = L_n^{-1}(\Delta, n)\, L_n(\Delta, n)\, u[n] = u[n]$$

Since in the given case $u[n]=y[n]$,

$$y[n] = L_n^{-1}(\Delta, n)\, M_n(\Delta, n)\, z[n] \tag{2.82}$$

Substituting $y[n]$ from (2.82) into the first equation in (2.81) we find that

$$L_n(\Delta, n)\, L_n^{-1}(\Delta, n)\, M_n(\Delta, n)\, z[n] = M_n(\Delta, n)\, x[n]$$

Since

$$L_n(\Delta, n)\, L_n^{-1}(\Delta, n) = L_n^{-1}(\Delta, n)\, L_n(\Delta, n) = 1$$

then

$$M_n(\Delta, n)\, z[n] = M_n(\Delta, n)\, x[n]$$

It follows that the output variable $z[n]$ of the two systems connected in series is identical to the input of the combined systems.

The rule for obtaining the equation of an inverse sampled data system can now be formulated as follows: the equation is derived from the equation for the original system by formally interchanging the roles of the input action and output variable. Thus, if the input of the original system is

$$L_n(\Delta, n)\, y[n, \varepsilon] = M_n(\Delta, \bar{t})\, x[n, 0], \quad 0 \leqslant \varepsilon \leqslant 1$$

then the equation of the corresponding inverse system is

$$M_n(\Delta, \bar{t})\, y[n, \varepsilon] = L_n(\Delta, n)\, x[n, 0], \quad 0 \leqslant \varepsilon \leqslant 1$$

where, as before, x is the input action and y is the output variable.

1.3 MODELS OF SAMPLED DATA SYSTEMS

Methods based on the use of analogue and digital computers have an important place in the investigation of sampled data systems, and require the system under consideration to be represented by either an analogue or a digital model.

Since an open-loop sampled data system consists of the sampler and the continuous part connected in series, it is clear that the analogue of a sampled data system consists of a model of the sampler and a model of the continuous part connected in series. The analogue of a closed-loop system is formed by adding a closing link to the analogue of the corresponding open-loop system. We confine ourselves to this remark about analogues of sampled data systems, since diagrams of such analogues are considered in [19], whilst the simulation of continuous linear systems is treated quite adequately in the literature.

We shall now develop digital models of sampled data systems. Consider the simplest first-order difference equation

$$\Delta y[n] = \varphi[n] \tag{3.1}$$

where $\varphi[n]$ is a given function of a discrete argument, which is defined for a set of values $n = 0, 1, \ldots$ To obtain a particular solution of (3.1), i.e. to find a function $y[n]$ which when substituted into (3.1) will transform it into an identity, we proceed as follows. We represent (3.1) in the form

$$y[n+1] = y[n] + \varphi[n]$$

or

$$y[k] = y[k-1] + \varphi[k-1] \tag{3.2}$$

where $k = n+1$. Without loss of generality we can put $\varphi[l] = 0$ when $l < 0$. Then, giving k the values $0, 1, \ldots, n$, we obtain

$$y[0]=0$$
$$y[1]=y[0]+\varphi[0]$$
$$y[2]=y[1]+\varphi[1]$$
$$\cdots\cdots\cdots\cdots$$
$$y[n]=y[n-1]+\varphi[n-1]$$

Grouping the terms, we find that

$$\sum_{k=0}^{n} y[k]=\sum_{k=0}^{n-1} y[k]+\sum_{k=0}^{n-1} \varphi[k]$$

or

$$\sum_{k=0}^{n} y[k]-\sum_{k=0}^{n-1} y[k]=\sum_{k=0}^{n-1} \varphi[k]$$

It follows that

$$y[n]=\sum_{k=0}^{n-1} \varphi[k] \tag{3.3}$$

This is in fact the required particular solution of (3.1). Indeed, since

$$\Delta y[n]=y[n+1]-y[n]$$

and since from (3.3)

$$y[n+1]=\sum_{k=0}^{n} \varphi[k]$$

then

$$\Delta y[n]=\sum_{k=0}^{n} \varphi[k]-\sum_{k=0}^{n-1} \varphi[k]\equiv\varphi[n] \tag{3.4}$$

Consequently, the substitution of (3.3) into (3.1) transforms it into an identity.

Comparison of (3.3) and (3.4) shows that, by taking the difference of a discrete function, we obtain the following function after the summation sign:

$$\Delta\left\{\sum_{k=0}^{n-1} \varphi[k]\right\}=\varphi[n] \tag{3.5}$$

i.e. the symbols Δ and Σ, cancel each other out when the first stands in front of the second.

Returning now to Equation (3.1), written

$$y[n]=\frac{1}{\Delta}\varphi[n] \tag{3.6}$$

and comparing (3.6) and (3.3) we conclude that the symbol $1/\Delta$ is a sum operator. The sum and difference elements are shown diagramatically in Fig. 1.3.

Now suppose that processes in a sampled data system are

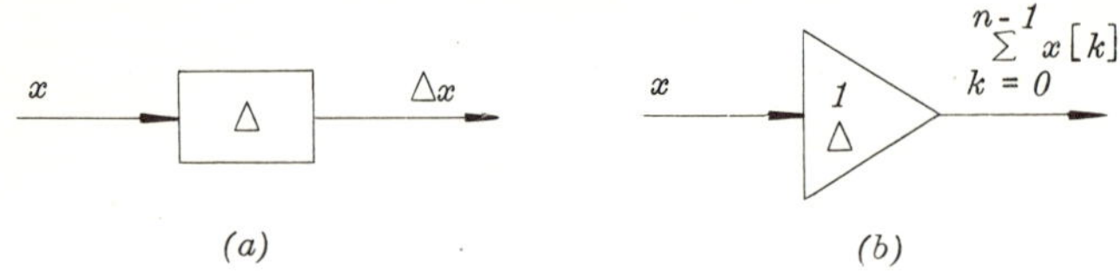

Fig. 1.3 System elements: (a) - difference element; (b) - summing element

defined by difference equations of the form

$$\sum_{i=0}^{k} a_i[n]\,\Delta_n^i y[n,\ \varepsilon]=b_0[n,\ \varepsilon]\,x[n,\ 0],\quad 0\leqslant\varepsilon\leqslant 1 \tag{3.7}$$

from which we can determine the highest difference of the output variable $\Delta_n^k y[n,\ \varepsilon]$:

$$\Delta_n^k y[n,\ \varepsilon]=\frac{1}{a_k[n]}\left\{b_0[n,\ \varepsilon]\,x[n,\ 0]-\sum_{i=0}^{k-1} a_i[n]\,\Delta_n^i y[n,\ \varepsilon]\right\}$$

$$0\leqslant\varepsilon\leqslant 1 \tag{3.8}$$

Summing both sides of (3.8) k times, we obtain in accordance with the definition of a sum (3.3)

$$y[n,\ \varepsilon]=\sum\nolimits^{(k)}\frac{1}{a_k[n]}\left\{b_0[n,\ \varepsilon]\,x[n,\ 0]-\sum_{i=0}^{k-1} a_i[n]\,\Delta_n^i y[n,\ \varepsilon]\right\}$$

$$0\leqslant\varepsilon\leqslant 1 \tag{3.9}$$

where the symbol Σ^k denotes k-ple summation. Equation (3.9) allows a digital model of the system under consideration to be constructed, as shown in Fig. 1.4.

Figure 1.5 shows a digital model of a system with

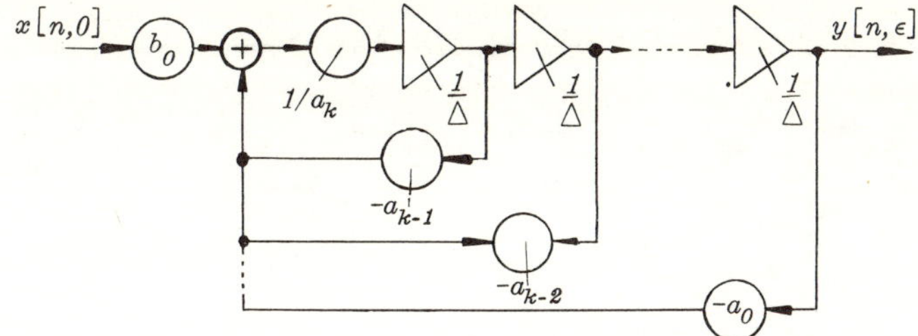

Fig. 1.4 A digital model of the system given by (3.9)

processes defined by a k-th order difference equation of the form

$$\sum_{i=0}^{k} a_i[n]\Delta_n^i y[n,\ \omega] = \sum_{j=0}^{h} b_j[n,\ \varepsilon]\Delta_n^j x[n,\ 0] \qquad (3.10)$$

$$0 \leqslant \varepsilon \leqslant 1$$

In the same way models can be constructed for a general system, processes which are defined by two difference equations for time intervals $0 \leqslant \varepsilon \leqslant \gamma$ and $\gamma \leqslant \varepsilon \leqslant 1$ respectively.

Let us now construct the diagrams of the digital models of the corresponding inverse sampled data systems. If the original system is described by a difference equation of the form (3.7), then processes in the corresponding inverse system are defined by

$$b_0[n,\ \varepsilon]y[n,\ \varepsilon] = \sum_{i=0}^{k} a_i[n]\Delta_n^i x[n,\ 0],\ 0 \leqslant \varepsilon \leqslant 1 \qquad (3.11)$$

where $y[n,\ \varepsilon]$ is the output variable of the system;
$x[n,\ 0]$ is the input action.

The diagram of a digital model corresponding to (3.11) is shown in Fig. 1.6.

If the original system is described by (3.10), then processes in the inverse system are defined by

$$\sum_{j=0}^{h} b_j[n,\ \varepsilon]\Delta_n^j y[n,\ \varepsilon] = \sum_{i=0}^{k} a_i[n]\Delta_n^i x[n,\ 0] \qquad (3.12)$$

$$0 \leqslant \varepsilon \leqslant 1$$

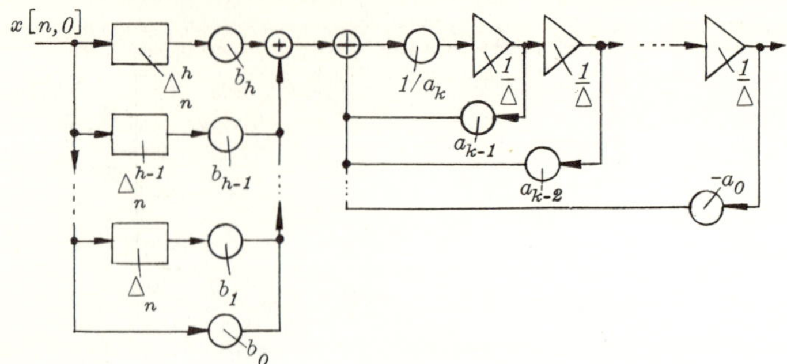

Fig. 1.5 A digital model of the system given by (3.10)

where $y[n, \varepsilon]$ is the output variable of the inverse sampled data system;

$x[n, 0]$ is the input action.

The diagram of the digital model corresponding to (3.12) is shown in Fig. 1.7.

A comparison of the system models shown in Figs. 1.4 and 1.6, and again in Figs. 1.5 and 1.7 shows that the original and inverse systems connected in series form a system which identically transforms the input action.

The diagrams of the models of sampled data systems constructed from (3.9) and (3.10) and shown in Figs. 1.4 to 1.7 emphasise the deep-rooted analogy between continuous and sampled data systems. Such models, generally speaking, are of considerable theoretical importance. In practice it is advisable to construct different diagrams of the digital

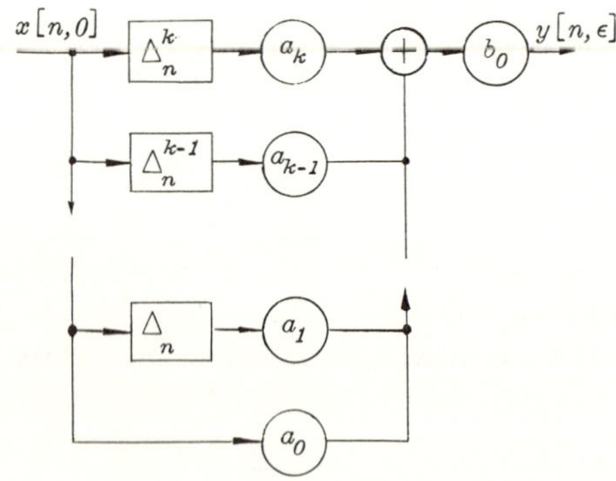

Fig. 1.6 A digital model of the system inverse to the system given by (3.9)

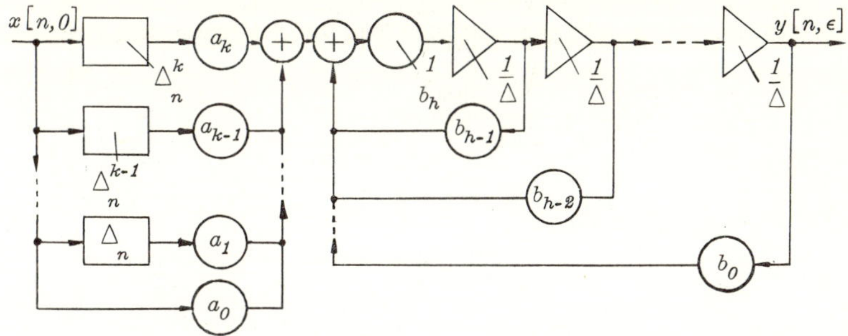

Fig. 1.7 A digital model of the system inverse to that given by (3.10)

models of the systems; the equations should be written not in terms of finite differences but in terms of displaced functions.

If we replace the differences in (3.10) by the corresponding displaced functions, we obtain, with the coefficients having the same meaning as before,

$$\sum_{i=0}^{k} a_i[n]\,y[n+i,\ \varepsilon]=\sum_{j=0}^{h} b_j[n,\ \varepsilon]\,x[n+j,\ 0],\ 0\leqslant\varepsilon\leqslant 1 \tag{3.13}$$

We determine $y[n+k, \varepsilon]$ from (3.13)

$$y[n+k,\ \varepsilon]$$

$$=\frac{1}{a_k[n]}\times\left\{\sum_{j=0}^{h} b_j[n,\ \varepsilon]\,x[n+j,\ 0]-\sum_{i=0}^{k-1} a_i[n]\,y[n+i,\ \varepsilon]\right\}$$

$$0\leqslant\varepsilon\leqslant 1$$

Introducing the new variable $n+k=l$, into the last equation we obtain

$$y[l,\ \varepsilon]=\frac{1}{a_k[l-k]}\left\{\sum_{j=0}^{h} b_j[l-k,\ \varepsilon]\,x[l+j-k,\ 0]\right.$$

$$\left.-\sum_{i=0}^{k-1} a_i[l-k]\,y[l+i-k,\ \varepsilon]\right\} \tag{3.14}$$

$$0\leqslant\varepsilon\leqslant 1$$

The diagram of the digital model corresponding to (3.14)

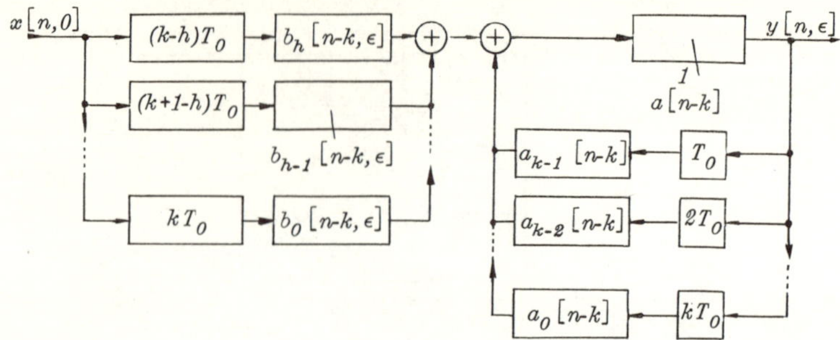

Fig. 1.8 A digital model of the system given by (3.14)

is shown in Fig. 1.8. The symbol mT_0 here denotes the unit of pure delay of the input by m closings of the sampler key of the system.

Let us construct the diagram of the model of the sampled data system which is inverse to the system described by (3.13). In accordance with (3.14) the difference equation of the inverse system has the form

$$y[l, \varepsilon]=\frac{1}{b_h[l-h, \varepsilon]}\left\{\sum_{i=0}^{k} a_i[l-h]x[l+i-h, 0]\right.$$

$$\left.-\sum_{j=0}^{h-1} b_j[l-h, \varepsilon]y[l+j-h, \varepsilon]\right\} \tag{3.15}$$

$$0 \leqslant \varepsilon \leqslant 1$$

where $y[n,\varepsilon]$ is the output variable of the sampled data system;

$x[n,0]$ is the input action.

The diagram of the model of a system corresponding to (3.15) is shown in Fig. 1.9.

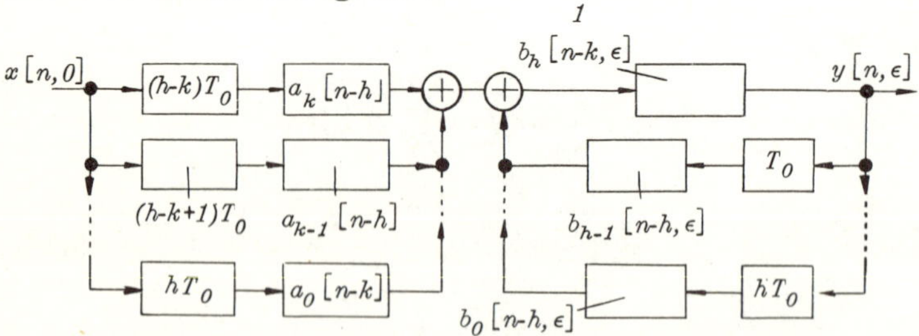

Fig. 1.9 A digital model of the system inverse to that described by (3.14)

The diagrams of digital models of sampled data systems may be regarded as diagrams of algorithms for calculating the output variable of a system from the input action. On the other hand, a physical model may be constructed from the digital model; its elements are digital adders, multiplying devices, etc. The problems involved in constructing such diagrams have also been considered in [22].

1.4 DEFINITION AND PROPERTIES OF σ-FUNCTIONS

The Dirac δ-function is widely used in the theory of differential equations and in the analysis of continuous automatic control systems. The σ-function plays a similar part in the theory of difference equations and the analysis of sampled data automatic systems. By a σ-function is meant a function defined by

$$\sigma[\bar{t}]=\begin{cases}1 & \text{when} \quad 0\leqslant\bar{t}<1 \\ 0 & \text{when} \quad 0>\bar{t},\ \bar{t}\geqslant 1\end{cases} \tag{4.1}$$

By definition it follows that the σ-function is the forward difference of the unit step function

$$\sigma[\bar{t}]=\mathbf{1}[\bar{t}]-\mathbf{1}[\bar{t}-1] \tag{4.2}$$

where $1[\bar{t}]$ is the unit step function of a discrete argument. As before, the discrete argument $\bar{t}$ is defined by

$$\left.\begin{aligned}\bar{t}=\frac{t}{T_0}=n+\varepsilon,\quad n=0,\ 1,\ 2,\dots \\ 0\leqslant\varepsilon\leqslant 1\end{aligned}\right\} \tag{4.3}$$

Using (4.3) we write (4.2) as

$$\begin{gathered}\sigma[n,\ \varepsilon]=\mathbf{1}[n,\ \varepsilon]-\mathbf{1}[n-1,\ \varepsilon] \\ 0\leqslant\varepsilon\leqslant 1\end{gathered} \tag{4.4}$$

By definition, the σ-function is a single pulse of unit height whose area is equal to unity.

The displaced σ-function is defined by

$$\begin{gathered}\sigma[n-m,\ \varepsilon]=\mathbf{1}[n-m,\ \varepsilon]-\mathbf{1}[n-(m+1),\ \varepsilon] \\ 0\leqslant\varepsilon\leqslant 1\end{gathered} \tag{4.5}$$

$$\left.\begin{aligned}\sigma[n-m,\ \varepsilon]&=\begin{cases}1 & \text{when } \bar{\tau}\leqslant\bar{t}<\bar{\tau}+1\\ 0 & \text{when } \bar{\tau}>\bar{t},\ \bar{t}\geqslant\bar{\tau}+1\end{cases}\\ \bar{\tau}&=m+\varepsilon,\ m=0,\ 1,\ldots,\ 0\leqslant\varepsilon\leqslant 1\end{aligned}\right\}\quad(4.6)$$

and is thus a single pulse displaced to the right along the time axis by $\bar{\tau}=m+\varepsilon$. It follows from (4.5) that the displaced σ-function can take place for any $n=0,\ 1,\ldots$ and $0\leqslant\varepsilon\leqslant 1$(Fig. 1.10).

We shall now prove certain important properties of the σ-function. According to the definition [18], we have for the first difference of the displaced σ-function with respect to the discrete argument n

$$\Delta_n\sigma[n-m,\ \varepsilon]=\sigma[n-(m-1),\ \varepsilon]-\sigma[n-m,\ \varepsilon]\quad(4.7)$$

We point out that in (4.7) the operation of taking the difference when $\varepsilon\neq 0$ should be regarded as calculating the difference for any fixed ε within the limits of the closed

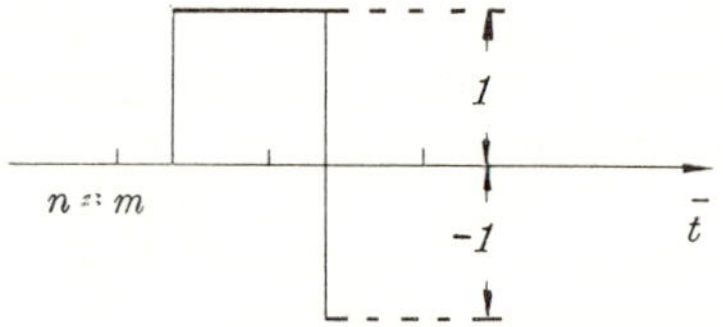

Fig. 1.10 Graphical representation of the displaced σ-function

interval [0; 1] when $n=0,\ 1,\ldots$ and $m=0,\ 1,\ldots$. By the definition (4.5) we have from (4.7)

$$\begin{aligned}\Delta_n\sigma[n-m,\ \varepsilon]&=\mathbf{1}[n-(m-1),\ \varepsilon]\\&-2\cdot\mathbf{1}[n-m,\ \varepsilon]+\mathbf{1}[n-(m+1),\ \varepsilon]\end{aligned}\quad(4.8)$$

Applying (4.8) in succession, for an s-th order difference defined by

$$\Delta^s_n\sigma[n-m,\ \varepsilon]=\Delta^{s-1}_n\sigma[n-(m-1),\ \varepsilon]-\Delta^{s-1}_n\sigma[n-m,\ \varepsilon]$$

we find

$$\Delta^s_n\sigma[n-m,\ \varepsilon]=\sum_{k=0}^{s+1}(-1)^k C^k_{s+1}\cdot\mathbf{1}[n-(m-s+k),\ \varepsilon]\quad(4.9)$$

where

$$C^k_{s+1} = \frac{(s+1)!}{k!\,(s+1-k)!}$$

It follows from (4.9) that

$$\Delta^s_n \sigma[n-m, \varepsilon] = 0 \text{ when } m+1+\varepsilon \leqslant n+\varepsilon,\ n+\varepsilon < m-s+\varepsilon$$

$$\Delta^s_n \sigma[n-m, \varepsilon] \neq 0 \text{ when } m-s+\varepsilon \leqslant n+\varepsilon < m+1+\varepsilon$$

In the interval $m-s+\varepsilon \leqslant n+\varepsilon < m+1+\varepsilon$ the s-th order difference of the displaced σ-function with respect to the argument n, is a step function defined by

$$\left.\begin{array}{c}\Delta^s_n \sigma[n-m, \varepsilon] = (-1)^k C^k_s \cdot \mathbf{1}[n-(m-s+k), \varepsilon] \\ k = 0, 1, 2, \ldots, s\end{array}\right\} \quad (4.10)$$

Figure 1.11 is a graphical representation of the displaced σ-function and its differences with respect to the argument n for $\varepsilon = 0$

$$\left.\begin{array}{c}\Delta^s_n \sigma[n-m, 0] = (-1)^k C^k_s \cdot \mathbf{1}[n-(m-s+k), 0] \\ k = 0, 1, \ldots, s\end{array}\right\} \quad (4.11)$$

when $s = 1$. 2 and 3.

We shall now find the difference of the displaced σ-function with respect to its second argument m. By (4.7) we have for the first-order difference

$$\Delta_m \sigma[n-m, \varepsilon] = \sigma[n-(m+1), \varepsilon] - \sigma[n-m, \varepsilon] \quad (4.12)$$

By the definition (4.5) and from Formula (4.12) we have

$$\begin{array}{c}\Delta_m \sigma[n-m, \varepsilon] = -\mathbf{1}[n-m, \varepsilon] \\ +2\cdot\mathbf{1}[n-(m+1), \varepsilon] - \mathbf{1}[n-(m+2), \varepsilon]\end{array} \quad (4.13)$$

The s-th order difference of the displaced σ-function with respect to the argument m is defined by

$$\Delta^s_m \sigma[n-m, \varepsilon] = \Delta^{s-1}_m \sigma[n-(m+1), \varepsilon] - \Delta^{s-1}_m \sigma[n-m, \varepsilon] \quad (4.14)$$

Consequently, applying (4.13) successively, we obtain the

expression for the s-th order difference as

$$\Delta^{s}_{m}\sigma[n-m,\ \varepsilon]=\sum_{k=0}^{s+1}(-1)^{s-k}C^{k}_{s+1}\cdot\mathbf{1}[n-(m+k),\ \varepsilon] \quad (4.15)$$

It follows from (4.15) that

$$\Delta^{s}_{m}\sigma[n-m,\ \varepsilon]=0 \text{ when } m+s+\varepsilon\leqslant n+\varepsilon,\ n+\varepsilon<m+\varepsilon$$
$$\Delta^{s}_{m}\sigma[n-m,\ \varepsilon]\neq 0 \text{ when } m+s+\varepsilon>n+\varepsilon\geqslant m+\varepsilon \quad (4.16)$$

In the interval $m+s+\varepsilon>n+\varepsilon\geqslant m+\varepsilon$ the s-th order difference is a step function defined by

$$\left.\begin{array}{c}\Delta^{s}_{m}\sigma[n-m,\ \varepsilon]=(-1)^{s-k}C^{k}_{s}\cdot\mathbf{1}[n-(m+k),\ \varepsilon]\\ k=0,\ 1,\ 2,\dots,\ s\end{array}\right\} \quad (4.17)$$

Figure 1.12 is a graphical representation of the displaced σ-function and its differences with respect to the argument m for $\varepsilon=0$

$$\left.\begin{array}{c}\Delta^{s}_{m}\sigma[n-m,\ 0]=(-1)^{s-k}C^{k}_{s}\cdot\mathbf{1}[n-(m+k),\ 0]\\ k=0,\ 1,\ 2,\dots,\ s\end{array}\right\} \quad (4.18)$$

when $s=1$, 2 and 3.

We point out that the σ-function can be regarded not only as a step function but also as a lattice function. By definition the lattice σ-function is

$$\sigma[n]=\begin{cases}1 & \text{when } n=0\\ 0 & \text{when } n\neq 0\end{cases} \quad (4.19)$$

Analogously the displaced lattice σ-function is defined as

$$\sigma[n-m]=\begin{cases}1 & \text{when } n=m\\ 0 & \text{when } n\neq m\end{cases} \quad (4.20)$$

It follows from (4.11) that the s-th order difference of the displaced lattice σ-function with respect to the argument n is

$$\Delta^{s}_{n}\sigma[n-m]=\begin{cases}(-1)^{k}C^{k}_{s}\cdot\mathbf{1}[n-(m-s+k)], & n=m-s+k\\ 0, & n\neq m-s+k\end{cases}$$
$$k=0,\ 1,\ 2,\dots,\ s \quad (4.21)$$

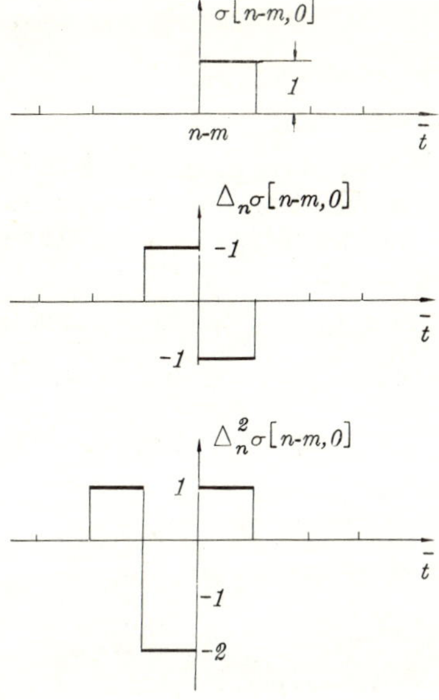

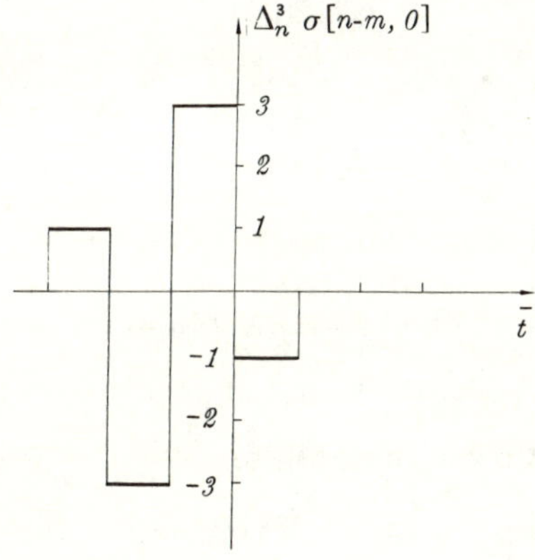

Fig. 1.11 Graphical representation of the displaced σ-function and its differences with respect to the argument n

Analogously the s-th order difference with respect to the argument m is

$$\Delta_m^s \sigma[n-m] = \begin{cases} (-1)^{s-k} C_s^k \cdot 1[n-(m+k)], & n=m+k \\ 0 & , \; n \neq m+k \end{cases}$$

$$k=0,\ 1,\ 2, \ldots, s \qquad (4.22)$$

Figures 1.13 and 1.14 show graphically the differences of the displaced lattice σ-function with respect to the arguments n and m. Since by the definition of the difference of a function of a discrete argument

$$\Delta_n f[n-m,\ \varepsilon] = f[n-(m-1),\ \varepsilon] - f[n-m,\ \varepsilon] \qquad (4.23)$$

and

$$\Delta_m f[n-m,\ \varepsilon] = f[n-(m+1),\ \varepsilon] - f[n-m,\ \varepsilon] \qquad (4.24)$$

then

$$\Delta_n f[n-m,\ \varepsilon] = -\Delta_m f[n-(m-1),\ \varepsilon] \qquad (4.25)$$

Using (4.23) to (4.25) in succession we find the general relation

$$\left.\begin{array}{c} \Delta_n^s f[n-m,\ \varepsilon] = (-1)^s \Delta_m^s f[n-(m-s),\ \varepsilon] \\ s=0,\ 1, \ldots \end{array}\right\} \qquad (4.26)$$

where

$$\Delta_n^0 f[n-m,\ \varepsilon] = \Delta_m^0 f[n-m,\ \varepsilon] = f[n-m,\ \varepsilon]$$

When $\varepsilon=0$, (4.26) becomes

$$\left.\begin{array}{c} \Delta_n^s f[n-m,\ 0] = (-1)^s \Delta_m^s f[n-(m-s),\ 0] \\ s=0,\ 1, \ldots \end{array}\right\} \qquad (4.27)$$

whence it follows that (4.26) holds for lattice functions

$$\left.\begin{array}{c} \Delta_n^s f[n-m] = (-1)^s \Delta_m^s f[n-(m-s)] \\ s=0,\ 1, \ldots \end{array}\right\} \qquad (4.28)$$

where, as before,

$$\Delta_n^0 f[n-m] = \Delta_m^0 f[n-m] = f[n-m]$$

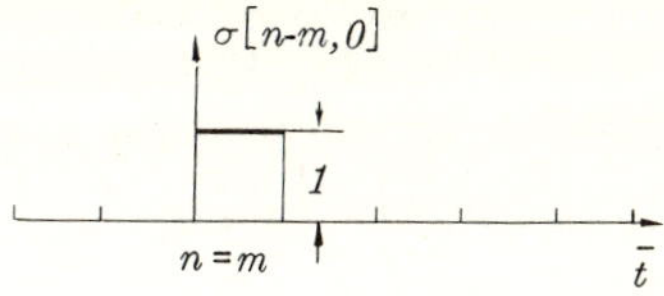

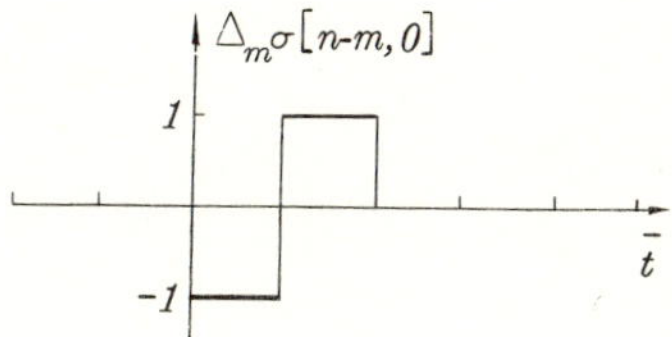

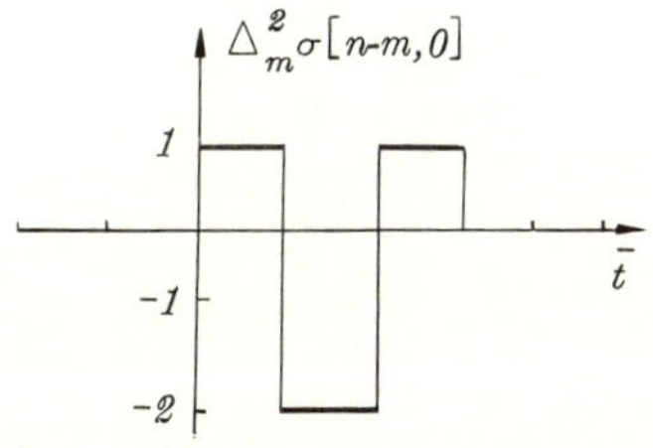

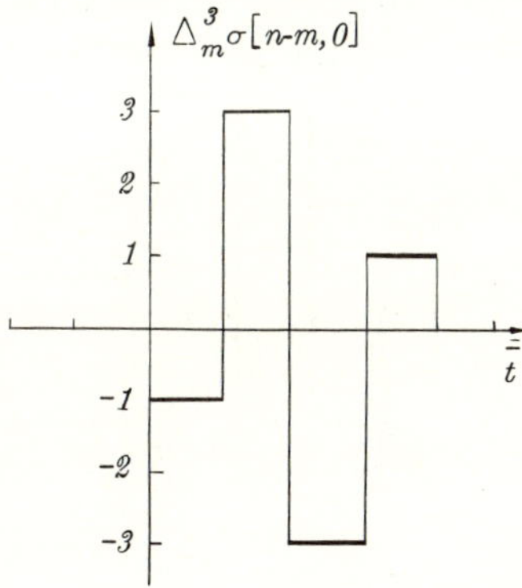

Fig. 1.12 Graphical representation of the displaced σ-function and its differences with respect to the argument m

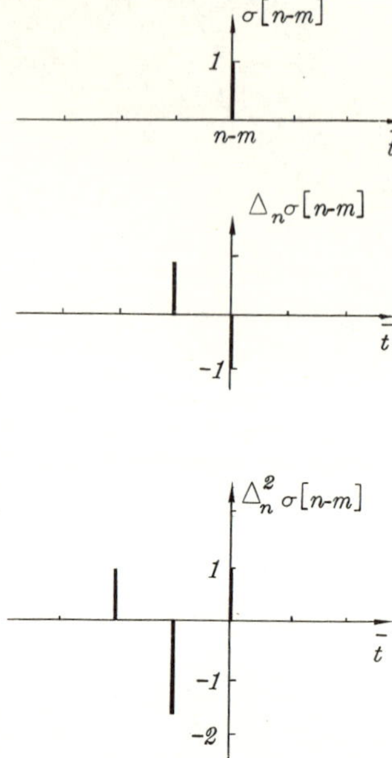

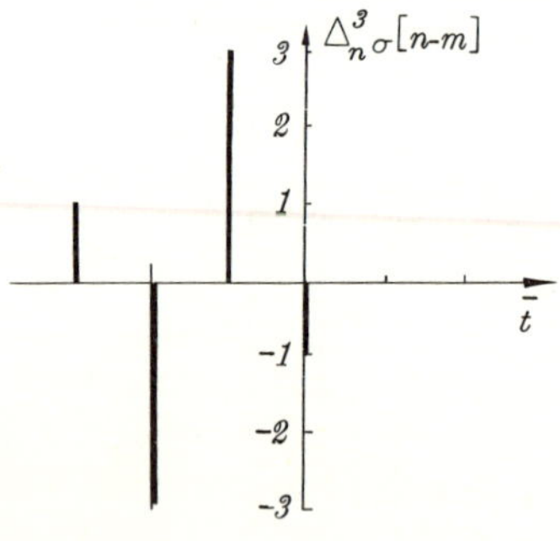

Fig. 1.13 Graphical representation of the displaced lattice σ-function and its differences with respect to the argument n

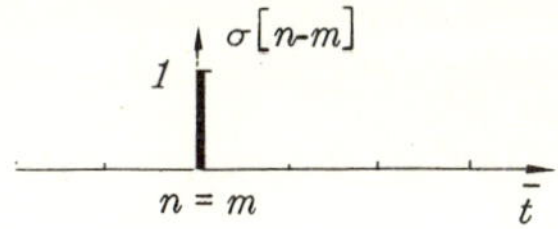

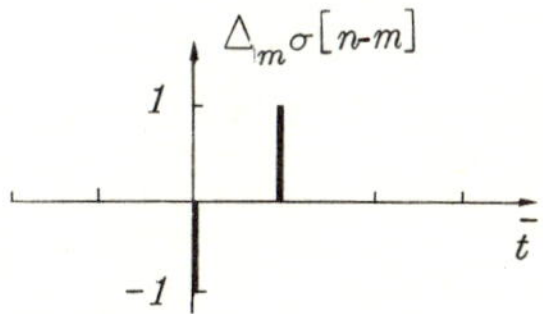

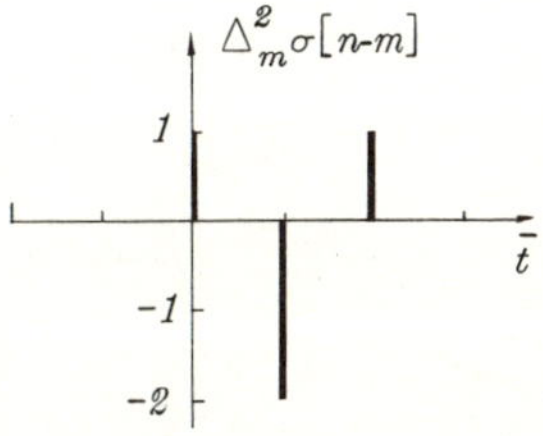

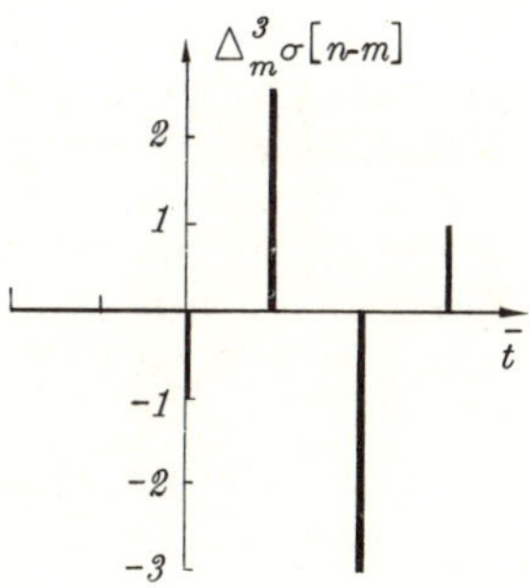

Fig. 1.14 Graphical representation of the displaced lattice σ-function and its differences with respect to the argument m

If we now put

$$f[n-m,\ \varepsilon]=\sigma[n-m,\ \varepsilon]$$

in (4.25) and (4.26), then from (4.25) we obtain

$$\Delta_n\sigma[n-m,\ \varepsilon]=-\Delta_m\sigma[n-(m-1),\ \varepsilon] \tag{4.29}$$

Analogously (4.26) gives

$$\left.\begin{array}{c}\Delta^s_n\sigma[n-m,\ \varepsilon]=(-1)^s\Delta^s_m\sigma[n-(m-s),\ \varepsilon]\\ s=0,\ 1,\ 2,\ldots\end{array}\right\} \tag{4.30}$$

where again we must put

$$\Delta^0_n\sigma[n-m,\ \varepsilon]=\Delta^0_m\sigma[n-m,\ \varepsilon]=\sigma[n-m,\ \varepsilon]$$

Putting $\varepsilon=0$, in (4.30) we obtain

$$\left.\begin{array}{c}\Delta^s_n\sigma[n-m,0]=(-1)^s\Delta^s_m\sigma[n-(m-s),0]\\ s=0,\ 1,\ldots\end{array}\right\} \tag{4.31}$$

whence it follows that (4.30) remains valid also for the lattice σ-function, i.e.

$$\left.\begin{array}{c}\Delta^s_n\sigma[n-m]=(-1)^s\Delta^s_m\sigma[n-(m-s)]\\ s=0,\ 1,\ldots\end{array}\right\} \tag{4.32}$$

The above properties proved for the σ-function play an important part in the analysis of sampled data systems, and will be used repeatedly in subsequent discussions.

From the definition of the σ-function, the multiplication of a two-sided function of the argument $\bar{t}=n+\varepsilon$ by the displaced σ-function excludes a part of the right-hand part of this function

$$\begin{array}{c}f[n,\ \varepsilon]\sigma[n-m,\ \varepsilon]=f[m,\ \varepsilon]\\ n+1+\varepsilon>m+\varepsilon\geqslant n+\varepsilon\end{array} \tag{4.33}$$

Now let

$$f[n,\ \varepsilon]=x[n,\ \varepsilon]\cdot\mathbf{1}[n-m,\ \varepsilon] \tag{4.34}$$

Then

$$\Delta_n \{x[n, \varepsilon]\cdot \mathbf{1}[n-m, \varepsilon]\}$$
$$=\Delta_n x[n, \varepsilon]\cdot \mathbf{1}[n-m, \varepsilon]+x[n+1, \varepsilon]\,\sigma[n-(m-1), \varepsilon]$$

and with (4.33) taken into account, the last relation can be written in the form

$$\begin{aligned}\Delta_n \{x[n, \varepsilon]\cdot \mathbf{1}[n-m, \varepsilon]\}=\Delta_n x[n, \varepsilon]\cdot \mathbf{1}[n-m, \varepsilon]\\ +x[m, \varepsilon]\,\sigma[n-(m-1), \varepsilon]\end{aligned} \tag{4.35}$$

The first term of (4.35) is the usual difference of the two-sided function of which only the right-hand part is used. The second term indicates the presence of discontinuities of the first difference of the function (4.34) if $x[m, \varepsilon]\neq 0$.

Thus (4.35) is an expression for the generalised first-order difference of a discontinuous function. From this follows the validity of the equation

$$x[n, \varepsilon]\,\sigma[n-m, \varepsilon]=x[m, \varepsilon]\,\sigma[n-m, \varepsilon] \tag{4.36}$$

which characterises the filtering properties of the σ-function.

Assuming $x[m,\varepsilon]=\text{const}$ in (4.35) we determine the expression for the second difference

$$\Delta^2_n \{x[n, \varepsilon]\cdot \mathbf{1}[n-m, \varepsilon]\}=\Delta^2_n x[n, \varepsilon]\cdot \mathbf{1}[n-m, \varepsilon]$$
$$+\Delta_n x[m, \varepsilon]\,\sigma[n-(m-1), \varepsilon]+x[m, \varepsilon]\,\Delta_n \sigma[n-(m-1), \varepsilon]$$

Calculating the differences s times in succession, we find the expression for the s-th order difference

$$\begin{aligned}\Delta^s_n \{x[n, \varepsilon]\cdot \mathbf{1}[n-m, \varepsilon]\}=\Delta^s_n x[n, \varepsilon]\cdot \mathbf{1}[n-m, \varepsilon]\\ +\sum_{i=0}^{s-1} \Delta^i_n x[m, \varepsilon]\,\Delta^{s-i-1}_n \sigma[n-(m-1), \varepsilon]\end{aligned} \tag{4.37}$$

which is valid also when $\bar{\tau}=m+\varepsilon=0$

$$\begin{aligned}\Delta^s_n \{x[n, \varepsilon]\cdot \mathbf{1}[n, \varepsilon]\}=\Delta^s_n x[n, \varepsilon]\cdot \mathbf{1}[n, \varepsilon]\\ +\sum_{i=0}^{s-1} \Delta^i_n x[0, 0]\,\Delta^{s-i-1}_n \sigma[n+1, \varepsilon]\end{aligned} \tag{4.38}$$

The formula (4.37) is the expression for the generalised s-th order difference of a discontinuous function of a discrete argument. Using (4.25) and (4.26), we can find from (4.37) the generalised s-th order difference taken with respect to the second argument, provided $x[n, \varepsilon]$ is a displaced function, i.e. $\bar{t}=\bar{\tau}-\bar{\lambda}$. Putting $\varepsilon=0$ in (4.37) we find

$$\Delta_n^s \{x[n]\cdot \mathbf{1}[n-m, 0]\} = \Delta_n^s x[n, 0]\cdot \mathbf{1}[n-m, 0] + \sum_{i=0}^{s-1} \Delta_n^i x[m, 0]\, \Delta_n^{s-i-1} \sigma[n-(m-1), 0] \tag{4.39}$$

whence it follows that (4.37) is in fact valid for the σ-function, i.e.

$$\Delta_n^s \{x[n, 0]\cdot \mathbf{1}[n-m]\} = \Delta_n^s x[n]\cdot \mathbf{1}[n-m] + \sum_{i=0}^{s-1} \Delta_n^i x[m]\, \Delta_n^{s-i-1} \sigma[n-(m-1)] \tag{4.40}$$

We have demonstrated the validity of the equation

$$x[n, \varepsilon]\,\sigma[n-m, \varepsilon] = x[m, \varepsilon]\,\sigma[n-m, \varepsilon]$$

Determining now the value of the product of a function and the first difference of the σ-function with respect to the argument n, we have, taking into account (4.10),

$$x[n, \varepsilon]\,\Delta_n \sigma[n-m, \varepsilon] = x[m-1, \varepsilon] - x[m, \varepsilon] \tag{4.41}$$

Analogously, bearing in mind (4.17), we obtain

$$x[m, \varepsilon]\,\Delta_m \sigma[n-m, \varepsilon] = x[m+1, \varepsilon] - x[m, \varepsilon] \tag{4.42}$$

Comparing (4.41) and (4.42) it is clear that

$$x[n, \varepsilon]\,\Delta_n \sigma[n-m, \varepsilon] = -x[m, \varepsilon]\,\Delta_m \sigma[n-(m-1), \varepsilon] \tag{4.43}$$

Calculating the differences of both sides of (4.43) in succession, we can show that

$$x[n, \varepsilon]\,\Delta_n^s \sigma[n-m, \varepsilon] = (-1)^s x[m, \varepsilon]\,\Delta_m^s \sigma[n-(m-s), \varepsilon] \tag{4.44}$$

$$s=0, 1, 2, \ldots$$

where

$$x[n, \varepsilon]\,\Delta_n^0\,\sigma[n-m, \varepsilon] = x[n, \varepsilon]\,\sigma[n-m, \varepsilon]$$

$$x[m, \varepsilon]\,\Delta_m^0\,\sigma[n-m, \varepsilon] = x[m, \varepsilon]\,\sigma[n-m, \varepsilon]$$

Equation (4.44) is valid also for lattice functions, i.e.

$$x[n]\,\Delta_n^s\,\sigma[n-m] = (-1)^s x[m]\,\Delta_m^s\,\sigma[n-(m-s)] \qquad s=0, 1, \ldots \tag{4.45}$$

We shall now consider (4.36)

$$x[n, \varepsilon]\,\sigma[n-m, \varepsilon] = x[m, \varepsilon]\,\sigma[n-m, \varepsilon]$$

Calculating the first difference of both sides with respect to m, we obtain

$$x[n, \varepsilon]\,\Delta_m\sigma[n-m, \varepsilon] = \Delta_m\{x[m, \varepsilon]\,\sigma[n-m, \varepsilon]\} \tag{4.46}$$

Since the relation

$$\Delta_n\sigma[n-m, \varepsilon] = -\Delta_m\sigma[n-(m-1), \varepsilon]$$

holds, we can write (4.46) in the form

$$x[n, \varepsilon]\,\Delta_n\sigma[n-(m+1), \varepsilon] = -\Delta_m\{x[m, \varepsilon]\,\sigma[n-m, \varepsilon]\}$$

Introducing the new variable $m'+1=m$, we finally obtain, with the previous meaning of the arguments,

$$x[n, \varepsilon]\,\Delta_n\sigma[n-m, \varepsilon] = -\Delta_m\{x[m-1, \varepsilon]\,\sigma[n-(m-1), \varepsilon]\} \tag{4.47}$$

In general it is possible to show that the following formula is valid:

$$x[n, \varepsilon]\,\Delta_n^s\,\sigma[n-m, \varepsilon] = (-1)^s\,\Delta_m^s\{x[m-s, \varepsilon]\,\sigma[n-(m-s), \varepsilon]\} \qquad s=0, 1, \ldots \tag{4.48}$$

This relation is also valid for lattice functions, i.e.

$$x[n]\,\Delta_n^s\,\sigma[n-m] = (-1)^s\Delta_m^s\{x[m-s]\,\sigma[n-(m-s)]\} \qquad s=0, 1, \ldots \tag{4.49}$$

We have now proved a group of properties of the σ-function which play an important part in the analysis of sampled data systems, and the definite analogy which exists between the σ-function and Dirac's δ-function is demonstrated below.

1. The δ-function is the derivative of the unit step function

$$\delta(t)=\frac{d}{dt}\mathbf{1}(t)$$

$$\mathbf{1}(t)=\begin{cases}1 \text{ when } t\geqslant 0,\\ 0 \text{ when } t<0\end{cases}$$

The σ-function is the difference of the unit step function

$$\sigma[n, \varepsilon]=\mathbf{1}[n, \varepsilon]-\mathbf{1}[n-1, \varepsilon]$$

It is easy to see that δ-function is obtained from the σ-function as the result of proceeding to the limit

$$\delta(t)=\lim_{T_0\to 0}\left[\frac{1}{T_0}\sigma[n, \varepsilon]\right]_{(n+\varepsilon)T_0=t}$$

2. The property of the δ-function

$$\delta_t^{(s)}(t-\tau)=(-1)^s\,\delta_\tau^{(s)}(t-\tau), \quad s=0, 1, \ldots$$

is parallelled by the property of σ-function

$$\Delta_n^s\,\sigma[n-m, \varepsilon]=(-1)^s\,\Delta_m^s\,\sigma[n-(m-s), \varepsilon]$$

$$s=0, 1, \ldots$$

The above property of the δ-function can then be obtained from the analogous property of the σ-function, taking into consideration that

$$\delta^{(s)}(t)=\lim_{T_0\to 0}\left[\frac{1}{T_0^{s+1}}\Delta^s\sigma[n, \varepsilon]\right]_{(n+\varepsilon)T_0=t}$$

3. To the generalised s-th order derivative of the discontinuous function

$$\frac{d^s}{dt^s}[x(t)\cdot\mathbf{1}(t-\tau)]=x_t^{(s)}(t)\cdot\mathbf{1}(t-\tau)$$

$$+\sum_{i=0}^{s-1}x_t^{(i)}(\tau)\,\delta_t^{(s-i-1)}(t-\tau)$$

corresponds the generalised s-th order difference of the discontinuous function of a discrete argument

$$\Delta_n^s \left[x[n, \varepsilon]\cdot 1\,[n-m, \varepsilon]\right] = \Delta_n^s x[n, \varepsilon]\cdot 1\,[n-m, \varepsilon] + \sum_{i=0}^{s-1} \Delta_n^i x[m, \varepsilon]\,\Delta_n^{s-i-1} \sigma[n-(m-1), \varepsilon]$$

4. The property of the δ-function

$$x(t)\,\delta_t^{(s)}(t-\tau) = (-1)^s \frac{\partial^s}{\partial\tau^s}\left[x(\tau)\,\delta(t-\tau)\right]$$

is parallelled by the property of the σ-function

$$x[n, \varepsilon]\,\Delta_n^s \sigma[n-m, \varepsilon] = (-1)^s \Delta_m^s \left[x[m-s, \varepsilon]\,\sigma[n-(m-s), \varepsilon]\right]$$

The fundamental properties of the σ-function and the analogous properties of the δ-function are listed in Table 1.1.

We shall now proceed to prove some other important properties of the σ-function. From the definition of the σ-function the following holds:

$$\sum_{m=-\infty}^{\infty} \sigma[n-m, \varepsilon] = 1 \tag{4.50}$$

The summation with respect to m when $\varepsilon \neq 0$ should be understood as taking the sum of the values of the function for any fixed value $\varepsilon=\varepsilon_0=$ const $(0 \leqslant \varepsilon_0 \leqslant 1)$.

When $\varepsilon = 0$ (4.50) assumes the form

$$\sum_{m=-\infty}^{\infty} \sigma[n-m, 0] = 1 \tag{4.51}$$

whence it follows that (4.50) is valid for the lattice σ-function, i.e.

$$\sum_{m=-\infty}^{\infty} \sigma[n-m] = 1 \tag{4.52}$$

In view of the selective properties of the σ-function it is clear that

$$\sum_{m=-\infty}^{\infty} x[m, \varepsilon]\,\sigma[n-m, \varepsilon] = x[n, \varepsilon] \tag{4.53}$$

Putting $\varepsilon=0$ in (4.53) we obtain

$$\sum_{m=-\infty}^{\infty} x[m, 0]\,\sigma[n-m, 0]=x[n, 0] \tag{4.54}$$

whence it follows that for lattice functions the following also holds:

$$\sum_{m=-\infty}^{\infty} x[m]\,\sigma[n-m]=x[n] \tag{4.55}$$

Now consider

$$x[n, \varepsilon]\,\Delta_n\sigma[n-m, \varepsilon]=-\Delta_m\{x[m-1, \varepsilon]\,\sigma[n-(m-1), \varepsilon]\} \tag{4.56}$$

Summing both sides with respect to n we have

$$\begin{aligned}\sum_{n=-\infty}^{\infty} x[n, \varepsilon]\,\Delta_n\sigma[n-m, \varepsilon]\\ =-\sum_{n=-\infty}^{\infty} \Delta_m\{x[m-1, \varepsilon]\,\sigma[n-(m-1), \varepsilon]\}\end{aligned} \tag{4.57}$$

Since

$$\begin{aligned}&\Delta_m\{x[m-1, \varepsilon]\,\sigma[n-(m-1), \varepsilon]\}\\ &=\Delta_m x[m-1, \varepsilon]\,\sigma[n-(m-1), \varepsilon]\\ &\quad+x[m, \varepsilon]\,\Delta_m\sigma[n-(m-1), \varepsilon]\end{aligned}$$

(4.57) can be written

$$\begin{aligned}&\sum_{n=-\infty}^{\infty} x[n, \varepsilon]\,\Delta_n\sigma[n-m, \varepsilon]\\ &=-\sum_{n=-\infty}^{\infty} \Delta_m x[m-1, \varepsilon]\,\sigma[n-(m-1), \varepsilon]\\ &\quad-\sum_{n=-\infty}^{\infty} x[m, \varepsilon]\,\Delta_m\sigma[n-(m-1), \varepsilon]\end{aligned}$$

It is easy to show that

$$\sum_{n=-\infty}^{\infty} \Delta_m\sigma[n-(m-1), \varepsilon]=0$$

Consequently, the last relation can be written

$$\sum_{n=-\infty}^{\infty} x[n, \varepsilon]\, \Delta_n \sigma[n-m, \varepsilon]$$

$$= -\sum_{n=-\infty}^{n} \Delta_m x[m-1, \varepsilon]\, \sigma[n-(m-1), \varepsilon]$$

Since by (4.50)

$$\sum_{n=-\infty}^{\infty} \Delta_m x[m-1, \varepsilon]\, \sigma[n-(m-1), \varepsilon]$$

$$= \Delta_m x[m-1, \varepsilon] \sum_{n=-\infty}^{\infty} \sigma[n-(m-1), \varepsilon] = \Delta_m x[m-1, \varepsilon]$$

we finally have

$$\sum_{n=-\infty}^{\infty} x[n, \varepsilon]\, \Delta_n \sigma[n-m, \varepsilon] = -\Delta_m x[m-1, \varepsilon] \tag{4.58}$$

For lattice functions the following is also valid:

$$\sum_{n=-\infty}^{n} x[n]\, \Delta_n \sigma[n-m] = -\Delta_m x[m-1] \tag{4.59}$$

Now consider (4.48)

$$x[n, \varepsilon]\, \Delta_n^s \sigma[n-m, \varepsilon]$$

$$= (-1)^s \Delta_m^s \{x[m-s, \varepsilon]\, \sigma[n-(m-s), \varepsilon]\}$$

Summing both sides with respect to n between the limits $-\infty$ and ∞, we have

$$\sum_{n=-\infty}^{\infty} x[n, \varepsilon]\, \Delta_n^s \sigma[n-m, \varepsilon]$$

$$= (-1)^s \sum_{n=-\infty}^{\infty} \Delta_m^s \{x[m-s, \varepsilon]\, \sigma[n-(m-s), \varepsilon]\} \tag{4.60}$$

Since

$$\Delta_m^s \{x[m-s,\ \varepsilon]\,\sigma[n-(m-s),\ \varepsilon]\}$$
$$=\Delta_m^s x[m-s,\ \varepsilon]\,\sigma[n-(m-s),\ \varepsilon]$$
$$+x[m-s+1,\ \varepsilon]\,\Delta_m^s\sigma[n-(m-s),\ \varepsilon]$$

(4.60) can be written

$$\sum_{n=-\infty}^{\infty} x[n,\ \varepsilon]\,\Delta_n^s\,\sigma[n-m,\ \varepsilon]$$
$$=(-1)^s\sum_{n=-\infty}^{\infty}\Delta_m^s x[m-s,\ \varepsilon]\,\sigma[n-[m-s),\ \varepsilon] \qquad (4.61)$$
$$+(-1)^s\sum_{n=-\infty}^{\infty} x[m-s+1,\ \varepsilon]\,\Delta_m^s\sigma[n-(m-s),\ \varepsilon]$$

Since the s-th order difference of the displaced σ-function with respect to the argument m is defined by

$$\Delta_m^s\sigma[n-(m-s),\ \varepsilon]$$
$$=(-1)^{s-k}C_s^k\cdot 1\,[n-(m-s+k),\ \varepsilon] \qquad (4.62)$$
$$k=0,\ 1,\ 2,\ldots,\ s$$

we can show that

$$\sum_{n=-\infty}^{\infty}\Delta_m^s\sigma[n-(m-s),\ \varepsilon]=0 \qquad (4.63)$$

Consequently, when (4.63) is taken into account (4.61) assumes the form

$$\sum_{n=-\infty}^{\infty} x[n,\ \varepsilon]\,\Delta_n^s\,\sigma[n-m,\ \varepsilon]$$
$$=(-1)^s\sum_{n=-\infty}^{\infty}\Delta_m^s x[m-s,\ \varepsilon]\,\sigma[n-(m-s),\ \varepsilon] \qquad (4.64)$$

It is clear that

$$\sum_{n=-\infty}^{\infty}\Delta_m^s x[m-s,\ \varepsilon]\,\sigma[n-(m-s),\ \varepsilon]=\Delta_m^s x[m-s,\ \varepsilon]$$

so that we finally have

$$\sum_{n=-\infty}^{\infty} x[n, \varepsilon]\, \Delta_n^s\, \sigma[n-m, \varepsilon] = (-1)^s\, \Delta_m^s x[m-s, \varepsilon] \qquad s=0,\ 1,\ 2,\ldots \tag{4.65}$$

where

$$\sum_{n=-\infty}^{\infty} x[n, \varepsilon]\, \Delta_n^0\, \sigma[n-m, \varepsilon] = \sum_{n=-\infty}^{\infty} x[n, \varepsilon]\, \sigma[n-m, \varepsilon] = x[m, \varepsilon]$$

i.e. the relation (4.53) obtained earlier in a different way is a particular case of (4.65).

In exactly the same way it is possible to show that the other general relation also holds:

$$\sum_{m=-\infty}^{\infty} x[m, \varepsilon]\, \Delta_n^s\, \sigma[n-m, \varepsilon] = \Delta_n^s\, \dot{x}[n, \varepsilon] \qquad s=0,\ 1,\ 2,\ldots \tag{4.66}$$

Relations analogous to (4.65) and (4.66) also hold for lattice functions:

$$\sum_{n=-\infty}^{\infty} x[n]\, \Delta_n^s\, \sigma[n-m] = (-1)^s\, \Delta_m^s x[m-s] \qquad s=0,\ 1,\ 2,\ldots \tag{4.67}$$

$$\sum_{m=-\infty}^{\infty} x[m]\, \Delta_n^s\, \sigma[n-m] = \Delta_n^s\, x[n] \qquad s=0,\ 1,\ 2,\ldots \tag{4.68}$$

We omit the proofs, remarking only that it is also easy to obtain relations analogous to (4.66) and (4.67):

$$\sum_{m=-\infty}^{\infty} x[m, \varepsilon]\, \Delta_m^s\, \sigma[n-m, \varepsilon] = (-1)^s\, \Delta_n^s\, x[n-s, \varepsilon] \qquad s=0,\ 1,\ 2,\ldots \tag{4.69}$$

$$\sum_{n=-\infty}^{\infty} x[n, \varepsilon]\, \Delta_m^s\, \sigma[n-m, \varepsilon] = \Delta_m^s\, x[m, \varepsilon] \qquad s=0,\ 1,\ 2,\ldots \tag{4.70}$$

Table 1

No.	Properties of δ-function	Properties of σ-function
1	$\delta(t) = \frac{d}{dt}\mathbf{1}(t)$ $\delta(t) = \begin{cases} 0, & t \neq 0 \\ \infty, & t = 0 \end{cases}$	$\sigma[n, \varepsilon] = \Delta_n \mathbf{1}[n, \varepsilon]$ $\sigma[n, \varepsilon]$ $= \begin{cases} 0, & 1 \leqslant n + \varepsilon,\ n + \varepsilon < 0 \\ 1, & 1 > n + \varepsilon \geqslant 0 \end{cases}$
2	$\delta_t^{(p)}(t - \tau)$ $= (-1)^p \delta_\tau^{(p)}(t - \tau),\ p = 0, 1, \ldots$	$\Delta_n^p \sigma[n - m, \varepsilon]$ $= (-1)^p \Delta_m^p \sigma[n - (m - p), \varepsilon]$ $p = 0, 1, \ldots$
3	$\frac{\partial^p}{\partial t^p}[x(t) \cdot \mathbf{1}(t - \tau)]$ $= x_t^{(p)}(t) \cdot \mathbf{1}(t - \tau)$ $+ \sum_{i=0}^{p-1} x^{(i)}(\tau)\, \delta_t^{(p-i-1)}(t - \tau)$	$\Delta_n^p \{x[n, \varepsilon] \cdot \mathbf{1}[n - m, \varepsilon]\}$ $= \Delta_n^p \{x[n, \varepsilon] \cdot \mathbf{1}[n - m, \varepsilon]\}$ $+ \sum_{i=0}^{p-1} \Delta_n^i x[m, \varepsilon]\, \Delta_n^{p-i-1} \sigma$ $[n - (m - 1), \varepsilon]$
4	$x(t)\, \delta_t^{(p)}(t - \tau)$ $= (-1)^p \frac{\partial^p}{\partial \tau^p}[x(\tau)\, \delta(t - \tau)]$ $p = 0, 1, \ldots$	$x[n, \varepsilon]\, \Delta_n^p \sigma[n - m, \varepsilon]$ $= (-1)^p \Delta_m^p \{x[m - p, \varepsilon]\, \sigma$ $[n - (m - p), \varepsilon]\},\ p = 0, 1, \ldots$

For lattice functions the following relations also hold:

$$\sum_{m=-\infty}^{\infty} x[m]\, \Delta_m^s \sigma[n-m] = (-1)^s \Delta_n^s x[n-s] \qquad s = 0, 1, 2, \ldots \tag{4.71}$$

$$\sum_{n=-\infty}^{\infty} x[n]\, \Delta_m^s \sigma[n - m] = \Delta_m^s x[m] \qquad s = 0, 1, 2, \ldots \tag{4.72}$$

(4.69) and (4.70) can be obtained in exactly the same way as (4.65). Their validity can easily be verified, putting

$$\Delta^{s}_{m}\sigma[n-m,\ \varepsilon]=\Delta^{s-1}_{m}\sigma[n+1-m,\ \varepsilon]-\Delta^{s-1}_{m}\sigma[n-m,\ \varepsilon]$$

$$s=1,\ 2,\ldots$$

and using the properties of the σ-function proved above.

We have now proved a second group of properties of the σ-function, which has considerable importance in the analysis of sampled data systems. They are given in Table 1.2, together with the analogous properties of the δ-function.

The σ-function introduced in this section plays the same part in the theory of difference equations as the δ-function of Dirac does in the theory of differential equations. It will be shown later that the analysis of sampled data systems using the σ-function is the same, from a mathematical viewpoint, as the analysis of continuous systems. This is of considerable importance, since the methods of analysis and synthesis worked out for continuous systems can thus be extended to sampled data systems.

We now determine the basic characteristics of linear sampled data systems.

1.5 WEIGHTING FUNCTION OF A LINEAR SYSTEM

As is known [13], the principle of superposition applied to continuous linear systems enables the response to an arbitrary action to be represented as the sum of the responses to the elementary actions which make up the input of the system. This statement applies in full to sampled data automatic systems [1, 14, 19]. Thus, if the response of a sampled data system to an elementary action is known, we can determine its response to an arbitrary action. It is important, however, to choose the standard elementary action such that actions of arbitrary form can be expressed fairly simply by it.

It was shown in Section 1.4 that any function of a discrete argument is fairly simply expressed in terms of the σ-function. Indeed, using the σ-function we may represent a function $x[n,\ \varepsilon]$ of any arbitrary form by means of

$$x[n,\ \varepsilon]=\sum_{m=-\infty}^{\infty}x[m,\ \varepsilon]\,\sigma[n-m,\ \varepsilon] \tag{5.1}$$

The operator defined by (5.1) will be called the summation

Table 2

No.	σ-function and its properties	δ-function and its properties
1	$\sum_{n=-\infty}^{\infty} \sigma[n, \varepsilon]=1$	$\int_{-\infty}^{\infty} \delta(t)\, dt=1$
2	$\sum_{m=-\infty}^{\infty} x[m, \varepsilon]\, \sigma[n-m, \varepsilon] = x[n, \varepsilon]$	$\int_{-\infty}^{\infty} x(\tau)\, \delta(t-\tau)\, d\tau = x(t)$
3	$\sum_{n=-\infty}^{\infty} x[n, \varepsilon]\, \Delta_n^s \sigma[n-m, \varepsilon] = (-1)^s \Delta_m^s x[m-s, \varepsilon]$	$\int_{-\infty}^{\infty} x(t)\, \delta_t^{(s)}(t-\tau)\, dt = (-1)^s x_\tau^{(s)}(\tau)$
4	$\sum_{m=-\infty}^{\infty} x[m, \varepsilon]\, \Delta_n^s \sigma[n-m, \varepsilon] = \Delta_n^s x[n, \varepsilon]$	$\int_{-\infty}^{\infty} x(\tau)\, \delta_t^{(s)}(t-\tau)\, d\tau = x_t^{(s)}(t)$
5	$\sum_{m=-\infty}^{\infty} x[m, \varepsilon]\, \Delta_m^s \sigma[n-m, \varepsilon] = (-1)^s \Delta_n^s x[n-s, \varepsilon]$	$\int_{-\infty}^{\infty} x(\tau)\, \delta_\tau^{(s)}(t-\tau)\, d\tau = (-1)^s x_t^{(s)}(t)$
6	$\sum_{n=-\infty}^{\infty} x[n, \varepsilon]\, \Delta_m^s \sigma[n-m, \varepsilon] = \Delta_m^s x[m, \varepsilon]$	$\int_{-\infty}^{\infty} x(t)\, \delta_\tau^{(s)}(t-\tau)\, dt = x_\tau^{(s)}(\tau)$

operator. According to (5.1) the σ-function is the kernel of this operator.

Let A be the operator of a non-stationary sampled data system. Applying this operator with respect to the argument n to both sides of (5.1) we have

$$y[n, \varepsilon]=\sum_{m=-\infty}^{\infty} x[m, \varepsilon]\, A_n \sigma[n-m, \varepsilon] \tag{5.2}$$

where $y[n, \varepsilon]$ is the result of applying the operator A_n to the function $x[n, \varepsilon]$

$$y[n, \varepsilon] = A_n x[n, \varepsilon]$$

In the sampled data systems considered, which by Tsypkin's classification [1] are systems of the first type, the sampler generates rectangular pulses whose heights are proportional to the values of the disturbing function $x[n, \varepsilon]$ at the instants of closing the key ($\varepsilon=0$). Accordingly the following relation must be used for systems of the first type

$$y[n, \varepsilon] = \sum_{m=-\infty}^{\infty} x[m, 0] A_n \sigma[n-m, 0] \tag{5.3}$$

where

$$y[n, \varepsilon] = A_n x[n, 0]$$

According to (2.1) the kernel of the summation operator $A_n\sigma[n-m, 0]$ in (5.3) is merely the response of the system to the σ-function applied to the input at the instant $\bar{t}=m$. Consequently, (5.3) determines the response of a system with operator A_n to an arbitrary action $x[n, \varepsilon]$ in terms of the reaction of the system to the σ-function.

Writing

$$A_n\sigma[n-m, 0] = W[n, \varepsilon; m] \tag{5.4}$$

where $\bar{t}=n+\varepsilon$ is the instant at which the output variable is observed;

m is the instant at which the σ-function is applied to the input,

we make the following definition: the weighting function of a sampled data system is the response of this system to the σ-function.

Since no physical system whatever can respond at a given instant $\bar{t}$ to a disturbance applied afterwards, the weighting function $W[n, \varepsilon; m]$ satisfied the condition of physical realisability; mathematically this is expressed as

$$W[n, \varepsilon; m] = \begin{cases} 0 & \text{when } m > n \\ W[n, \varepsilon; m] & \text{when } m \leqslant n \end{cases} \tag{5.5}$$

From the definition (5.4) the weighting function $W[n, \varepsilon; m]$ of a non-stationary sampled data system is a function of two arguments: $\bar{t}=n+\varepsilon$ is the instant at which the output variable is observed, and m is the instant at which the σ-function is applied. The weighting function of a stationary sampled data system (system with constant parameters) depends only on the interval between the instant at which the action is applied and the instant at which the signal is observed at the output of the system. This follows, in particular, from the definition of the weighting function (5.4) if we take into account that the operator A_n has constant coefficients. Thus for a stationary sampled data system the weighting function satisfies the condition

$$W[n,\ \varepsilon;\ m]=W[n-m,\ \varepsilon]=W[l,\ \varepsilon] \tag{5.6}$$

where $n-m=l$.

The output variable $y[n,\ \varepsilon]$ of a sampled data system is determined by (5.3) and (5.4) in terms of its weighting function as follows:

$$y[n,\ \varepsilon]=\sum_{m=-\infty}^{\infty} W[n,\ \varepsilon;\ m]\,x[m,\ 0] \tag{5.7}$$

where $x[m,\ 0]$ are the values of the input action at the instants the key is closed.

Since the weighting function satisfies the condition of physical realisability, (5.7) can be written in the form

$$y[n,\ \varepsilon]=\sum_{m=-\infty}^{n} W[n,\ \varepsilon;\ m]\,x[m,\ 0] \tag{5.8}$$

since, from (5.5),

$$W[n,\ \varepsilon;\ m]\equiv 0 \text{ when } n<m$$

If the system is at rest up to the instant $\bar{t}=0$, and at this instant the disturbance $x[n,\ \varepsilon]$ begins to act on it, then the output variable observed at time $\bar{t}$, is determined by

$$y[n,\ \varepsilon]=\sum_{m=0}^{n} W[n,\ \varepsilon;\ m]\,x[m,\ 0] \tag{5.9}$$

For a stationary system (5.7) to (5.9) assume the forms:

$$y[n,\ \varepsilon]=\sum_{m=-\infty}^{\infty} W[n-m,\ \varepsilon]\,x[m,\ 0] \tag{5.10}$$

$$y[n,\ \varepsilon]=\sum_{m=-\infty}^{n} W[n-m,\ \varepsilon]\, x[m,\ 0] \tag{5.11}$$

$$y[n,\ \varepsilon]=\sum_{m=0}^{n} W[n-m,\ \varepsilon]\, x[m,\ 0] \tag{5.12}$$

When the new variable $n-l=m$ is introduced (5.11) and (5.12) are reduced to

$$y[n,\ \varepsilon]=\sum_{l=0}^{\infty} W[l,\ \varepsilon]\, x[n-l,\ 0] \tag{5.13}$$

$$y[n,\ \varepsilon]=\sum_{l=0}^{n} W[l,\ \varepsilon]\, x[n-l,\ 0] \tag{5.14}$$

Very often when sampled data systems are investigated, the calculation of the output variable is confined to discrete instants at which the key of the sampler is closed ($\varepsilon=0$). Accordingly, putting $\varepsilon=0$ in (5.8) and (5.14) we obtain

$$y[n,\ 0]=\sum_{l=0}^{\infty} W[l,\ 0]\, x[n-l,\ 0] \tag{5.15}$$

$$y[n,\ 0]=\sum_{l=0}^{n} W[l,\ 0]\, x[n-l,\ 0] \tag{5.16}$$

where all the functions should be regarded as lattice functions. The expressions (5.15) and (5.16) determine the values of the output variable of non-stationary and stationary systems at the discrete instants of time at which the key is closed.

It is important to note that the above relations determining the output variable hold for the case where the input disturbance $x\ [n,\ \varepsilon]$ is continuous. In the general case, when $x\ [n,\ \varepsilon]$ has discontinuities, we must bear in mind that a pulsed disturbance acting on the continuous part of the system at the instant $\bar{t}=m$, is interpreted by the sampler in accordance with its value at the instant immediately preceding that at which the key is closed, i.e., at the instant

$$\bar{t}=[m-1,\ \varepsilon]_{\varepsilon\to 1}$$

This fact was first observed by Perov [14], and by it the fundamental relations assume the form

$$y[n, \varepsilon]=\sum_{m=-\infty}^{\infty} W[n, \varepsilon; m]\, x[m-1,1] \tag{5.17}$$

$$y[n, \varepsilon]=\sum_{m=-\infty}^{\infty} W[n-m, \varepsilon]\, x[m-1,1] \tag{5.18}$$

For physically realisable systems (5.17) and (5.18) are written

$$y[n, \varepsilon]=\sum_{m=-\infty}^{n} W[n, \varepsilon; m]\, x[m-1,1] \tag{5.19}$$

$$y[n, \varepsilon]=\sum_{m=-\infty}^{n} W[n-m, \varepsilon]\, x[m-1,1] \tag{5.20}$$

Equations (5.17) to (5.20) are of a general nature, since where the input disturbance is continuous

$$x[m-1,1]=x[m, 0]$$

and (5.7), (5.8), (5.10) and (5.11) are obtained from (5.17) to (5.20) respectively.

Thus the weighting function is an exhaustive characteristic of a sampled data dynamical system. Knowing it we can by appropriate formulae determine the output variable for an arbitrary input disturbance. An analogy between sampled data and continuous systems can be pointed out. As is known [13] the weighting function $W(t;\ \tau)$ of a continuous system with operator A_t is the response of this system to the δ-function, i.e.

$$A_t\delta(t-\tau)=W(t;\ \tau) \tag{5.21}$$

The weighting function $W[n, \varepsilon; m]$ of a sampled data system with operator A_n, is the response of this system to the σ-function, i.e.

$$A_n\sigma[n-m, 0]=W[n, \varepsilon; m] \tag{5.22}$$

Comparing (5.21) and (5.22) establishes the appropriate analogy between continuous and sampled data systems.

Owing to the fact that

$$\lim_{T_0 \to 0} \left[\frac{1}{T_0} \sigma[n, \varepsilon] \right]_{(n+\varepsilon) T_0 = t} = \delta(t)$$

we have

$$\lim_{T_0 \to 0} \left[\frac{1}{T_0} W[n, \varepsilon; m] \right]_{\substack{(n+\varepsilon) T_0 = t \\ m T_0 = \tau}} = W(t; \tau) \tag{5.23}$$

whence it follows that the weighting function of a continuous system is obtained from the weighting function of the corresponding sampled data system by going to the limit (5.23).

In the literature [1, 19] the weighting function is defined differently. In fact, the weighting function of an open-loop sampled data system is defined as the response of the corresponding normalised continuous part of the system to the δ-function. We briefly recall that the normalised continuous part of a sampled data system consists of the continuous part and a shaping filter connected in series, the transfer function of the filter corresponding to the sampler (Fig. 1.15).

Thus we have two definitions of the weighting function

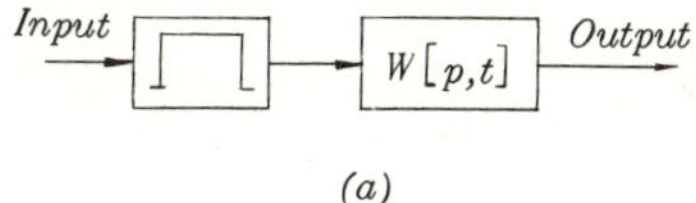

(a)

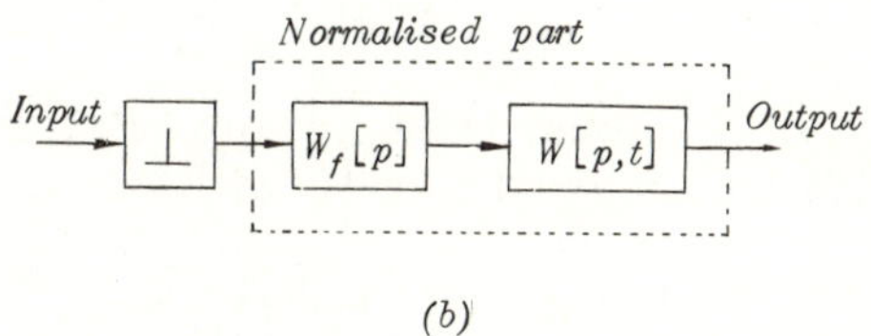

(b)

Fig. 1.15 Definition of the normalised continuous part of a sampled data system

which are not incompatible. We point out, however, that the definition of the weighting function as the response of the normalised part to the δ-function, as given in the theory of sampled data systems [1, 19], is of great practical importance. Following it we can find the difference equations of a given system, making use of the decomposition theorem. On the other hand, the definition given here as the response of this system to the σ-function has great theoretical importance. Using it as the fundamental dynamical characteristic of a system we can build up the theory of sampled data systems in the same form as it is developed for continuous systems. The development of certain problems of this theory forms in fact the content of the following chapters.

We shall now consider the weighting functions of certain elementary sampled data systems. Let the operator of the system be

$$A_n \equiv k[n]$$

Then, from (5.4), the weighting function of this system is the result of applying this operator to the σ-function

$$W[n;\ m] = A_n \sigma[n-m] = k[n]\sigma[n-m]$$

When $k[n]=1$ we find that the inertia-free amplifier is an element which identically transforms the input disturbance.

If the operator $A_n \equiv \Delta_n$, the weighting function of the system is

$$W[n;\ m] = \Delta_n \sigma[n-m]$$

Similarly, for $A_n \equiv \Delta^s_n$,

$$W[n;\ m] = \Delta^s_n \sigma[n-m] \qquad (5.24)$$

Consequently, a sampled data system with the weighting function (5.23) realises the s-th difference of the input disturbance.

Indeed on the basis of (5.15) we have

$$y[n] = \sum_{m=-\infty}^{\infty} W[n;\ m]\, x[m]$$

$$= \sum_{m=-\infty}^{\infty} \Delta^s_n \sigma[n-m]\, x[m]$$

but since

$$\sum_{m=-\infty}^{\infty} \Delta_n^s \sigma[n-m]\, x[m] = \Delta_n^s x[n]$$

we finally obtain

$$y[n] = \Delta_n^s x[n]$$

Furthermore, since

$$\sum_{m=-\infty}^{\infty} x[m]\, \sigma[n-(m+l)] = x[n-l] \tag{5.25}$$

a system with the weighting function

$$W[n;\ m] = \sigma[n-(m+l)]$$

displaces the input signal by l sampling periods.

In order to emphasise the analogy between continuous and sampled data systems we point out that a sampled data system realising the calculated difference of the input signal and having the weighting function

$$W[n;\ m] = \Delta_n \sigma[n-m]$$

corresponds to a continuous system of exact differentiation with the weighting function

$$W(t;\ \tau) = \delta_t'(t-\tau)$$

Similarly a sampled data system with the weighting function

$$W[n;\ m] = \Delta_n^s \sigma[n-m]$$

corresponds to a continuous system of s-ple differentiation with the weighting function

$$W(t;\ \tau) = \delta_t^{(s)}(t-\tau)$$

Finally, a system with the weighting function

$$W[n;\ m] = \sigma[n-(m+k)]$$

corresponds to an ideal delay element with the weighting function

$$W(t;\ \tau) = \delta[t-(\tau+\lambda)]$$

1.6 WEIGHTING FUNCTIONS OF PARALLEL AND SERIES SYSTEMS

In practice it is often necessary to determine the weighting functions of complex sampled data systems consisting of individual sampled data and continuous systems with known weighting functions connected in various ways.

Methods of determining the weighting functions of complex stationary and non-stationary systems are discussed below.

These methods are based on determining the weighting function of a system as the response of the system to the σ-function.

1.6.1 Systems connected in parallel

Let a sampled data system consist of two systems (Fig. 1.16) connected in parallel and having the weighting functions $W_1[n, \varepsilon; m]$ and $W_2[n, \varepsilon; m]$. We assume that the samplers of both systems are synchronised and in phase. In this case

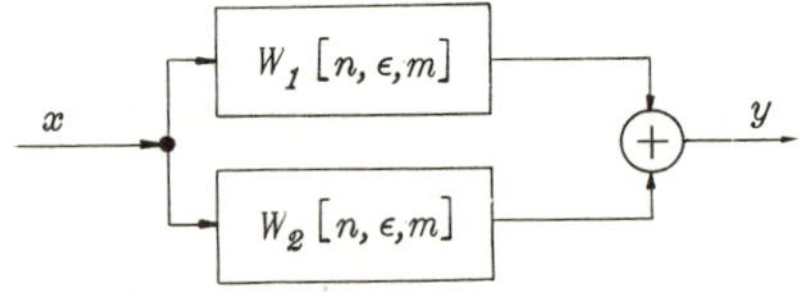

Fig. 1.16 Systems connected in parallel

the output variables of the systems with these weighting functions are determined by (5.17) as follows:

$$y_1[n, \varepsilon] = \sum_{m=-\infty}^{\infty} W_1[n, \varepsilon; m]\, x[m, 0] \tag{6.1}$$

$$y_2[n, \varepsilon] = \sum_{m=-\infty}^{\infty} W_2[n, \varepsilon; m]\, x[m, 0] \tag{6.2}$$

Since the output variable of the combined system equals

$$y[n, \varepsilon] = y_1[n, \varepsilon] + y_2[n, \varepsilon]$$

then using (6.1) and (6.2) we have

$$y[n, \varepsilon] = \sum_{m=-\infty}^{\infty} W_1[n, \varepsilon; m]\, x[m, 0] + \sum_{m=-\infty}^{\infty} W_2[n, \varepsilon; m]\, x[m, 0] \tag{6.3}$$

or

$$y[n, \varepsilon] = \sum_{m=-\infty}^{\infty} (W_1[n, \varepsilon; m] + W_2[n, \varepsilon; m])\, x[m, 0] \tag{6.4}$$

Equation (6.4) defines the output variable of the parallel connection of the two systems in terms of the weighting functions W_1 and W_2 for the input $x[n, \varepsilon]$. To determine the weighting function of the systems connected in parallel we assume that the input action x is the σ-function applied to the system at the instant $\bar{l} = m$. Consequently, substituting the arbitrary action in (6.4) by the σ-function, we have

$$y[n, \varepsilon] = W[n, \varepsilon; m] = \sum_{\nu=-\infty}^{\infty} (W_1[n, \varepsilon; \nu] + W_2[n, \varepsilon; \nu])\, \sigma[m-\nu, 0] = W_1[n, \varepsilon; m] + W_2[n, \varepsilon; m]$$

whence it follows that the weighting function $W[n, \varepsilon; m]$ of the parallel connection is the sum of the two weighting functions, i.e.

$$W[n, \varepsilon; m] = W_1[n, \varepsilon; m] + W_2[n, \varepsilon; m] \tag{6.5}$$

We point out that (6.5) also follows directly from the comparison of (6.4) and (5.17).

The weighting function of k systems connected in parallel, each having weighting functions $W_i[n, \varepsilon; m]$, $i = 1, 2, \ldots, k$, equals

$$W[n, \varepsilon; m] = \sum_{i=1}^{k} W_i[n, \varepsilon; m] \tag{6.6}$$

Equations (6.5) and (6.6) occur in the literature on the theory of sampled data systems [1, 14, 19].

1.6.2 Systems connected in series

Let a sampled data system consist of two systems connected in series and having weighting functions $W_1[n, \varepsilon; m]$ and $W_2[n, \varepsilon; m]$. We assume that the samplers of both systems

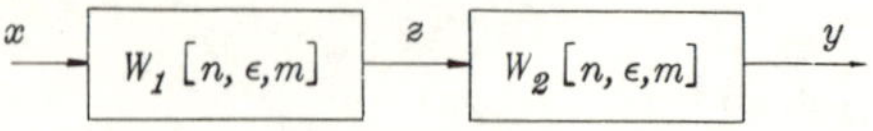

$x \to \boxed{W_1[n, \epsilon, m]} \xrightarrow{z} \boxed{W_2[n, \epsilon, m]} \xrightarrow{y}$

Fig. 1.17 Systems connected in series

are synchronised and in phase. In this case (Fig. 1.17) the processes are determined thus (for simplicity it is assumed that x and z are continuous):

$$z[n, \varepsilon] = \sum_{m=-\infty}^{\infty} W_1[n, \varepsilon; m]\, x[m, 0] \tag{6.7}$$

$$y[n, \varepsilon] = \sum_{l=-\infty}^{\infty} W_2[n, \varepsilon; l]\, z[l, 0] \tag{6.8}$$

where $x[n, \varepsilon]$ is the input action of the combined system;
$z[n, \varepsilon]$ is the output variable of the system with the weighting function W_1 and the input action of the second system with the weighting function W_2;
$y[n, \varepsilon]$ is the output variable of the combined system.

Since

$$z[l, 0] = \sum_{m=-\infty}^{\infty} W_1[l, 0; m]\, x[m, 0] \tag{6.9}$$

(6.8) can be written in the form

$$y[n, \varepsilon] = \sum_{l=-\infty}^{\infty} W_2[n, \varepsilon; l] \sum_{m=-\infty}^{\infty} W_1[l, 0; m]\, x[m, 0] \tag{6.10}$$

Thus the output variable of two non-stationary sampled data systems connected in series is determined by (6.10) for the input disturbance $x[n, \varepsilon]$. Replacing the input action

x by the σ-function in (6.10) we have

$$W[n, \varepsilon; m] = \sum_{l=-\infty}^{\infty} W_2[n, \varepsilon; l] \sum_{\nu=-\infty}^{\infty} W_1[l, 0; \nu]\,\sigma[\nu - m, 0]$$

$$= \sum_{l=-\infty}^{\infty} W_2[n, \varepsilon; l]\, W_1[l, 0; m]$$

Thus the weighting function $W[n, \varepsilon; m]$ of two non-stationary systems in series is determined by:

$$W[n, \varepsilon; m] = \sum_{l=-\infty}^{\infty} W_2[n, \varepsilon; l]\, W_1[l, 0; m] \qquad (6.11)$$

When the condition of physical realisability of the system is taken into account (6.11) can be written:

$$W[n, \varepsilon; m] = \sum_{l=m}^{n} W_2[n, \varepsilon; l]\, W_1[l, 0; m] \qquad (6.12)$$

where n assumes the values $n = m,\ m+1, \ldots$ Equation (6.12) has the condition that the weighting function $W_1[n, \varepsilon; m]$ has no discontinuities. It is easy to show that if W_1 is not a continuous function of the argument $\bar{t}$, then $W[n, \varepsilon; m]$ is determined by

$$W[n, \varepsilon; m] = \sum_{l=m}^{n} W_2[n, \varepsilon; l]\, W_1[l-1, 1; m] \qquad (6.13)$$

If the equation below

$$W_1[l-1, 1; m] = W_1[l, 0; m]$$

then (6.13) and (6.12) are equivalent.

The following equations hold for two non-stationary mutually inverse sampled data systems connected in series:

$$\sum_{l=m}^{n} W[n, \varepsilon; l]\, W^{-}[l-1, 1; m] = \sigma[n-m, 0]$$

$$\sum_{l=m}^{n} W^{-}[n, \varepsilon; l]\, W[l-1, 1; m] = \sigma[n-m, 0] \qquad (6.14)$$

Equations (6.12) and (6.11) are analogous to the corresponding formulae for continuous systems. Let a continuous system consist of two non-stationary systems with weighting functions $W_1(t;\tau)$ and $W_2(t;\tau)$ connected in series. If $x(t)$ is the input of the system, and $y(t)$ its output, we have

$$z(t)=\int_{-\infty}^{\infty} W_1(t;\tau)\,x(\tau)\,d\tau \tag{6.15}$$

$$y(t)=\int_{-\infty}^{\infty} W_2(t;\alpha)\,z(\alpha)\,d\alpha \tag{6.16}$$

where $z(t)$ is the output variable of the first system and the input of the second system.

Substituting (6.15) into (6.16) we obtain

$$y(t)=\int_{-\infty}^{\infty} W_2(t;\alpha)\int_{-\infty}^{\infty} W_1(\alpha;\tau)\,x(\tau)\,d\tau\,d\alpha \tag{6.17}$$

If now the input action in (6.17) is replaced by the δ-function applied at $t=\tau$, we have

$$\begin{aligned} W_1(t;\tau) &= \int_{-\infty}^{\infty} W_2(t;\alpha)\int_{-\infty}^{\infty} W_1(\lambda;\tau)\,\delta(\lambda-\alpha)\,d\lambda\,d\alpha \\ &= \int_{-\infty}^{\infty} W_2(t;\alpha)\,W_1(\alpha;\ \tau)\,d\alpha \end{aligned} \tag{6.18}$$

where $W(t;\ \tau)$ is the weighting function of the combined system. When the physical realisability of the system is taken into consideration, (6.18) can be written

$$W(t;\tau)=\int_{\tau}^{t} W_2(t;\alpha)\,W_1(\alpha;\tau)\,d\alpha \tag{6.19}$$

By comparing (6.19) and (6.18) with (6.11) and (6.12) the analogy between them can be seen.

From (6.12) it is seen that change in the order in which the systems are connected results in a new system being formed.

If continuous systems with weighting functions W_1 and W_2 are mutually inverse systems, that is, $W_2=W_1^-$, then from (6.18) and (6.17) we obtain

$$\int_{-\infty}^{\infty} W_1^-(t;\alpha)\,W_1(\alpha;\tau)\,d\alpha=\delta(t-\tau)$$

It is clear that the following relation also holds

$$\int_{-\infty}^{\infty} W_1(t;\alpha)\, W_1^{-}(\alpha;\tau)\, d\alpha = \delta(t-\tau)$$

These relations are analogous to the corresponding relations for sampled data systems.

From (6.12) and (6.13) we can easily obtain the corresponding relations for stationary sampled data systems connected in series. These have the form

$$W[n,\varepsilon] = \sum_{l=0}^{n} W_2[l,\varepsilon]\, W_1[n-l,0] \qquad (6.20)$$

in the case when $W_1[n, \varepsilon]$ has no discontinuities, and

$$W[n,\varepsilon] = \sum_{l=0}^{n} W_2[l,\varepsilon]\, W_1[n-l-1,1] \qquad (6.21)$$

when $W_1[n, \varepsilon]$ has discontinuities.

Using (6.12) and (6.13) or (6.20) and (6.21) repeatedly we can determine the weighting function of a system consisting of any number of systems connected in series.

In the general case, when a sampled data system is connected in series with a system of arbitrary form (continuous or sampled data; see Fig. 1.18), the following rule for determining the weighting function of the combined system applies: the weighting function $W_H[n, \varepsilon; m]$ of the system formed by connecting in series a sampled data system with weighting function $W[n, \varepsilon; m]$ and a system with the operator $H_{\bar{t}}$ (differential or difference) is the result of applying the operator $H_{\bar{t}}$ to the weighting function $W[n, \varepsilon; m]$, i.e.

$$W_H[n, \varepsilon; m] = H_{\bar{t}} W[n, \varepsilon; m] \qquad (6.22)$$

To show this, apply an operator $H_{\bar{t}}^{-1}$, which is inverse to the

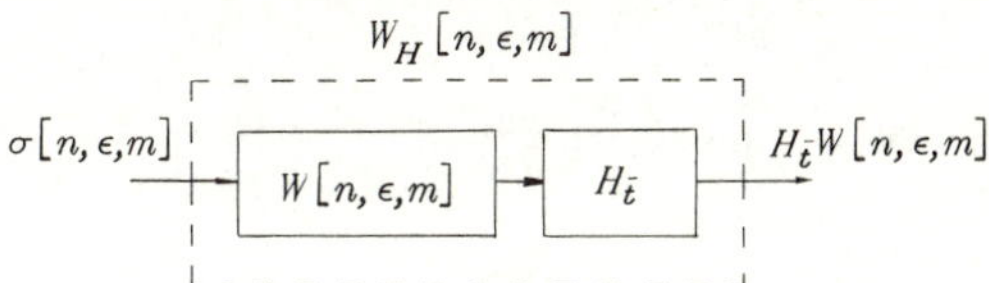

Fig. 1.18 A system connected in series with a system of arbitrary form

operator $H_{\bar{t}}$, to both sides of (6.22); we obtain

$$H_{\bar{t}}^{-1} W_{\mathrm{H}}[n, \varepsilon; m] = H_{\bar{t}}^{-1} \{H_{\bar{t}} W[n, \varepsilon; m]\} \equiv W[n, \varepsilon; m]$$

which proves the above rule.

1.7 WORKED EXAMPLES

We shall apply the results obtained in Sections 1.5 and 1.6 to some concrete examples.

Example 1. Determine the response of a sampled data system whose weighting function is

$$W[n, \varepsilon] = \begin{cases} 1+\varepsilon, & n=0 \\ -\varepsilon, & n=1 \\ 0, & n \geqslant 2 \end{cases} \tag{7.1}$$

to an action of the form

$$x[n, \varepsilon] = n+\varepsilon \tag{7.2}$$

$$x[n, \varepsilon] = (n+\varepsilon)^2 \tag{7.3}$$

Let us find the response of the system to the action (7.2). Since $x[n-1,1]=x[n,0]$ we can use the formula

$$y[n,\varepsilon] = \sum_{l=0}^{n} W[l, \varepsilon]\, x[n-l, 0] \tag{7.4}$$

to determine the output variable of the system. Putting $n=0,1,\ldots,$ in (7.4) we have when (7.1) and (7.2) are taken into account:

$$y[0, \varepsilon] = W[0,\varepsilon]\, x[0, 0] = 0$$
$$y[1, \varepsilon] = W[0,\varepsilon]\, x[1, 0] + W[1,\varepsilon] x[0,0] = 1+\varepsilon$$
$$y[2,\varepsilon] = W[0,\varepsilon]\, x[2,0] + W[1,\varepsilon]\, x[1,0] = 2+\varepsilon$$
$$\cdots\cdots\cdots\cdots\cdots\cdots\cdots\cdots$$
$$y[n,\varepsilon] = W[0,\varepsilon]\, x[n,0] + W[1,\varepsilon]\, x[n-1, 0] = n+\varepsilon$$

Thus

$$y[n, \varepsilon] = \begin{cases} 0, & n \leqslant 0 \\ n+\varepsilon, & n \geqslant 1 \end{cases} \tag{7.5}$$

It follows from (7.5) that for $n \geqslant 1$ the sampled data system accurately reproduces linear action. The error of the system $\mathcal{E}[n, \varepsilon]$ is non-zero only for $n=0$. Then

$$\mathcal{E}[n, \varepsilon] = \begin{cases} \varepsilon \text{ when } n=0 \\ 0 \text{ when } n \neq 0 \end{cases} \tag{7.6}$$

We shall now determine the response of a system with the weighting function (7.1) to a quadratic function of the form (7.3). Since in this case we have $x[n-1,1]=x[n,0]$, we can use (7.4) to determine the response. Thus, putting $n=0,1,\ldots$, in (7.4) we have:

$$y[0,\varepsilon] = W[0,\varepsilon]\, x[0,0] = 0$$

$$y[1,\varepsilon] = W[0,\varepsilon]\, x[1,0] + W[1,\varepsilon]\, x[0,0] = 1+\varepsilon$$

$$y[2,\varepsilon] = W[0,\varepsilon]\, x[2,0] + W[1,\varepsilon]\, x[1,0] = 2^2\cdot(1+\varepsilon) - \varepsilon\cdot 1$$

$$\cdots\cdots\cdots\cdots\cdots\cdots\cdots\cdots$$

$$y[n,\varepsilon] = W[0,\varepsilon]\, x[n,0] + W[1,\varepsilon]\, x[n-1,0] = (1+\varepsilon)\, n^2 - \varepsilon\,(n-1)^2$$

Consequently, we have

$$y[n,\varepsilon] = \begin{cases} 0 & \text{when } n=0 \\ 1+\varepsilon & \text{when } n=1 \\ (1+\varepsilon)\, n^2 - \varepsilon\,(n-1)^2 & \text{when } n \geqslant 2 \end{cases} \tag{7.7}$$

It follows from (7.7) that the system accurately reproduces the quadratic function only at discrete instants of time. When $\varepsilon \neq 0$ the error of the system, $\mathcal{E}[n, \varepsilon]$ is determined by

$$\mathcal{E}[n,\varepsilon] = \begin{cases} \varepsilon^2 & \text{when } n=0 \\ \varepsilon^2+\varepsilon & \text{when } n \geqslant 1 \end{cases} \tag{7.8}$$

A comparison of the results obtained shows that the sampled data system considered is a follow-up system which accurately reproduces the action $x[n, \varepsilon]$=const and a linear action. In addition, transient processes in this system cease after a single sampling period.

Example 2. Find the response of a sampled data system with the weighting function

$$W[n,\varepsilon] = \begin{cases} 1 \text{ when } n=0 \\ -1 \text{ when } n=1 \\ 0 \text{ when } n \geqslant 2 \end{cases} \tag{7.9}$$

to actions of the form (7.2) and (7.3).

Applying (7.4), we have

$$y[0,\varepsilon]=W[0,\varepsilon]\,x[0,0]=0$$

$$y[1,\varepsilon]=W[0,\varepsilon]\,x[1,0]+W[1,\varepsilon]\,x[0,0]=1$$

$$y[2,\varepsilon]=W[0,\varepsilon]\,x[2,0]+W[1,\varepsilon]\,x[1,0]=1$$

....

$$y[n,\varepsilon]=\sum_{m=0}^{n} W[m,\varepsilon]\,x[n-m,0]=1$$

Thus we obtain

$$y[n,\varepsilon]=\begin{cases} 0 \text{ when } n\leqslant 0 \\ 1 \text{ when } n\geqslant 1 \end{cases} \tag{7.10}$$

We shall now determine the response of the same system to a quadratic function of the form (7.3). Applying (7.4) we obtain

$$y[n,\varepsilon]=\begin{cases} 0 & \text{when } n\leqslant 0 \\ 1 & \text{when } n=1 \\ 2n-1 & \text{when } n\geqslant 2 \end{cases} \tag{7.11}$$

It is easy to see that the system is one which computes the first difference of the input action with a delay of one sampling period. Indeed, the weighting function $W[n,\varepsilon]$, determined by (7.9) is nothing else but the first difference of the displaced σ-function with respect to the second argument, taken with the negative sign, i.e.

$$W[n-m,\varepsilon]=-\Delta_m\sigma[n-m,0]$$

As a result of this we have

$$y[n,\varepsilon]=-\sum_{l=0}^{n}\Delta_l\sigma[n-l,0]\,x[l,0]$$

or, when the property of the σ-function defined by the formula

$$\sum_{l=-\infty}^{\infty} x[l,0]\,\Delta_l^s\,\sigma[n-l,0]=(-1)^s\Delta_n^s\,x[n-s,0]$$

is taken into account,

$$y[n,\varepsilon]=\Delta x[n-1,0] \tag{7.12}$$

Determining the first differences of step functions corresponding to (7.2) and (7.3) we find the relations

$$y[n, \varepsilon]=\Delta x[n-1, 0]=1 \tag{7.13}$$

$$y[n, \varepsilon]=\Delta x[n-1, 0]=2n-1 \tag{7.14}$$

which coincide with (7.10) and (7.11) when $n\geqslant 1$.

Example 3. Find the response of a sampled data system whose weighting function is determined by

$$W[n,\varepsilon]=\begin{cases} 1 \text{ when } n=0 \\ -2 \text{ when } n=1 \\ 1 \text{ when } n=2 \\ 0 \text{ when } n\geqslant 3 \end{cases} \tag{7.15}$$

to an action of the form (7.3).

Following (7.4), we have

$$y[n,\varepsilon]=\begin{cases} 0 \text{ when } n=0 \\ 1 \text{ when } n=1 \\ 2 \text{ when } n\geqslant 2 \end{cases} \tag{7. 16}$$

It is obvious that the weighting function of this system is the second difference of the σ-function, i.e.

$$W[n-m,\varepsilon]=\Delta^2_m \sigma[n-m,0] \tag{7.17}$$

Accordingly a system with the weighting function (7.17) calculates the second difference of the input action with a delay of two sampling periods. Thus, the output variable of the system is determined by

$$y[n,\varepsilon]=\sum_{m=0}^{n} \Delta^2_m \sigma[n-m,0]\, x[m,0]$$

or, taking into account the property of the σ-function expressed by (4.66), we finally obtain

$$y[n,\varepsilon]=\Delta^2_n x[n-2,0] \tag{7.18}$$

Calculating the second difference of the input action $x[n, 0]=n^2$, we have from (7.18)

$$\Delta^2_n x[n-2, 0]=2 \tag{7.19}$$

which coincides with (7.16) for $n \geqslant 2$. From (7.16) it follows, in particular, that transient processes in the system considered cease after a time corresponding to two sampling periods.

It is easy to see that for the calculation of the second-order difference of the input action without delay we must have a sampled data system with a weighting function defined by

$$W[n,\varepsilon]=\begin{cases} 1 & \text{when } n=-2 \\ -2 & \text{when } n=-1 \\ 1 & \text{when } n=0 \\ 0 & \text{when for other } n \end{cases} \tag{7.20}$$

Comparing (7.20) and (4.21) we note that the weighting function (7.20) is the second difference of the displaced σ-function with respect to the first argument, that is, we have

$$W[n-m,\varepsilon]=\Delta_n^2\,\sigma[n-m,0] \tag{7.21}$$

Making use of the property of the σ-function defined by

$$\sum_{m=-\infty}^{\infty} x[m,0]\,\Delta_n^s\,\sigma\,[n-m,0]=\Delta_n^s\,x[n,0]$$

we determine the output variable of the system under consideration with the weighting function (7.21)

$$y[n,\varepsilon]=\sum_{m=-\infty}^{\infty}\Delta_n^2\,\sigma[n-m,0]\,x[m,0]=\Delta_n^2\,x[n,0] \tag{7.22}$$

whence it follows that the system with the weighting function (7.2) is indeed capable of computing the second difference of the input action without delay. We point out, however, that such a system is physically not realisable, since according to (7.20) $W[n,\ \varepsilon]\neq 0$ when $n<0$.

Example 4. Find the weighting function of two sampled data systems connected in parallel, the samplers of which have the same repetition periods and which work in unison and in phase, whilst the weighting functions of the systems are defined by

$$W_1[n,\varepsilon]=\begin{cases}1+\varepsilon \text{ when } n=0\\ -\varepsilon \text{ when } n=1\\ 0 \text{ when } n\geqslant 2\end{cases} \tag{7.23}$$

$$W_2[n,\varepsilon]=\begin{cases}1 \text{ when } n=0\\ -1 \text{ when } n=1\\ 0 \text{ when } n\geqslant 2\end{cases} \tag{7.24}$$

Find also the response of these systems to linear and quadratic actions of the form of (7.2) and (7.3).

To determine the weighting function of the combination we use (6.6) which, when (7.23) and (7.24) are taken into account, gives the following expression for $W[n, \varepsilon]$:

$$W[n,\varepsilon]=\begin{cases}2+\varepsilon \text{ when } n=0\\ -(1+\varepsilon) \text{ when } n=1\\ 0 \text{ when } n\geqslant 2\end{cases} \tag{7.25}$$

We now find the response of a system with the weighting function (7.25) to a linear disturbance. By (7.4) we have

$$y[n,\varepsilon]=\begin{cases}0 \text{ when } n\leqslant 0\\ 2+\varepsilon \text{ when } n=1\\ 1+(n+\varepsilon) \text{ when } n\geqslant 2\end{cases} \tag{7.26}$$

The response to a linear disturbance plotted from the equations above is shown in Fig. 1.19. It follows from the figure that the system under consideration accurately predicts the input signal by one sampling period. Physically this is explained by the circumstance that, as was shown in Examples 1 and 2, the system with the weighting function defined by (7.23) exactly copies linear input, i.e. is an accurate follow-up system, whilst a system with the weighting function (7.24) computes the difference of the signal, i.e. determines the rate of its change over one period. Accordingly it is a natural consequence that a parallel combination of these systems forms a system which predicts the input signal by one period. In particular, from Fig. 1.19 it follows that transient processes in the system are terminated after a single sampling period.

We now find the response of this system to an action

given by the quadratic function

$$x[n, \varepsilon] = (n+\varepsilon)^2$$

Applying (7.4) we have

$$y[n,\varepsilon] = \begin{cases} 0 & \text{when } n \leqslant 0 \\ 2+\varepsilon & \text{when } n = 1 \\ n^2 + (2n-1)(1+\varepsilon) & \text{when } n \geqslant 2 \end{cases} \tag{7.27}$$

We determine the error which occurs when the quadratic signal is predicted. Then

$$\mathcal{E}[n,\varepsilon] = \begin{cases} \varepsilon^2 & \text{when } n = 0 \\ 2+3\varepsilon+\varepsilon^2 & \text{when } n \geqslant 1 \end{cases} \tag{7.28}$$

whence it is seen that the error has a constant component $\mathcal{E}_1[n,\varepsilon] = 2$ and a periodic component determined by

$$\mathcal{E}_2[n,\varepsilon] = 3\varepsilon + \varepsilon^2$$

The constant component of the prediction is due to the fact that the signal is predicted without taking the acceleration into account. Consequently, if a system determining the second difference is connected in parallel with a system with the weighting function (7.25), we shall have a sampled data system which exactly predicts a quadratic signal by one period.

It was shown in Example 2 that a system computing the

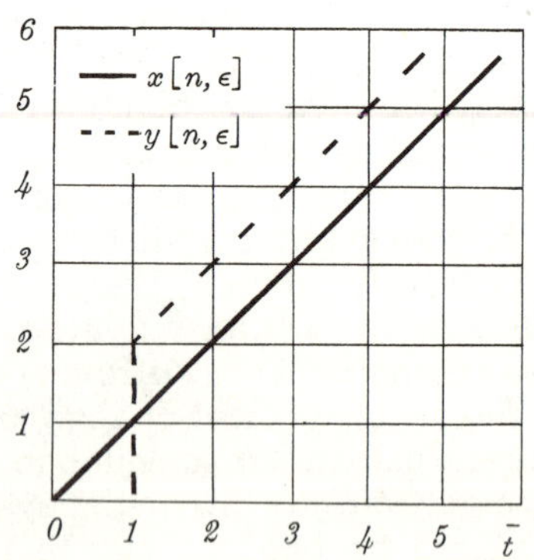

Fig. 1.19 The response curve of a system with the weighting function (7.25) to a linear action

second-order difference of the input has the weighting function

$$W_3[n-m,\varepsilon]=\Delta_m^2\,\sigma[n-m,0] \tag{7.29}$$

Using (6.21) we determine the weighting function $W[n,\varepsilon]$ of three systems connected in parallel and having weighting functions W_1, W_2, and W_3. We have, taking into account (7.23), (7.24) and (7.29):

$$W[n,\varepsilon]=\begin{cases} 3+\varepsilon & \text{when } n=0 \\ -(3+\varepsilon) & \text{when } n=1 \\ 1 & \text{when } n=2 \\ 0 & \text{when } n\geqslant 3 \end{cases} \tag{7.30}$$

The response of a system with the weighting function (7.30) to a quadratic action equals

$$y[n,\varepsilon]=\begin{cases} 0 & \text{when } n=0 \\ (3+\varepsilon)(2n-1)+(n-2)^2 & \text{when } n\geqslant 1 \end{cases} \tag{7.31}$$

The error in predicting the quadratic signal by a system with the weighting function (7.30) is

$$\mathcal{E}[n,\varepsilon]=\begin{cases} \varepsilon^2 & \text{when } n=0 \\ 3\varepsilon+\varepsilon^2 & \text{when } n\geqslant 1 \end{cases} \tag{7.32}$$

as follows from (7.31). Consequently, a system with the weighting function (7.3) exactly predicts a quadratic signal only at discrete instants of time ($\varepsilon=0$).

Example 5. Find the weighting function of two sampled data systems connected in series and having the weighting functions

$$W_1[n,\varepsilon]=\begin{cases} 1 & \text{when } n\geqslant 0 \\ 0 & \text{when } n<0 \end{cases} \tag{7.33}$$

$$W_2[n,\varepsilon]=\begin{cases} 1 & \text{when } n\geqslant 0 \\ 0 & \text{when } n<0 \end{cases} \tag{7.34}$$

Since the weighting functions are discontinuous we must use the following formula to determine the weighting function $W[n,\varepsilon]$ of the systems connected in series:

$$W[n,\varepsilon]=\sum_{m=0}^{n} W_2[m,\varepsilon]\,W_1[n-m-1,1] \tag{7.35}$$

Putting $n=0, 1, \ldots$ in (7.35) we find

$$W[0,\varepsilon]=W_2[0,\varepsilon]\,W_1[-1,1]=0$$

$$W[1,\varepsilon]=W_2[0,\varepsilon]\,W_1[0,1]+W_2[1,\varepsilon]\,W_1[-1,1]=1$$

$$W[2,\varepsilon]=W_2[0,\varepsilon]\,W_1[1,1]+W_2[1,\varepsilon]\,W_1[0,1]=2$$

$$\cdots\cdots\cdots\cdots\cdots\cdots\cdots$$

$$W[n,\varepsilon]=W_2[0,\varepsilon]\,W_1[n-1,1]+W_2[1,\varepsilon]\,W_1[n-2,1]=n$$

Thus, we have

$$W[n,\varepsilon]=\begin{cases} n \text{ when } n\geqslant 1 \\ 0 \text{ when } n\leqslant 0 \end{cases} \tag{7.36}$$

Let now

$$W_1[n,\varepsilon]=\begin{cases} 1 \text{ when } n\geqslant 0 \\ 0 \text{ when } n<0 \end{cases} \tag{7.37}$$

$$W_2[n,\varepsilon]=\begin{cases} 1 \text{ when } n=0 \\ -1 \text{ when } n=1 \\ 0 \text{ when } n\neq 0,\ n\neq 1 \end{cases} \tag{7.38}$$

In this case from (7.35) we have

$$W[n,\varepsilon]=\begin{cases} 1 \text{ when } n=1 \\ 0 \text{ when } n\neq 1 \end{cases} \tag{7.39}$$

i.e. the weighting function $W[n, \varepsilon]$ is the displaced σ-function (in the direction of delay)

$$W[n, \varepsilon]=\sigma[n-1, 0] \tag{7.40}$$

It can be shown that by altering the order in which systems with the weighting functions (7.37) and (7.38) are connected in series we do not alter the weighting function of the combined system.

It follows from (7.40) that the output variable of a system with such a weighting function will repeat any input disturbance, but with a delay of one period.

If now

$$W_1[n,\varepsilon]=\begin{cases} 1 \text{ when } n=-1, 0, 1, \ldots \\ 0 \text{ when } n\leqslant -2 \end{cases} \tag{7.41}$$

$$W_2[n,\varepsilon] = \begin{cases} 1 \text{ when } n=0 \\ -1 \text{ when } n=1 \\ 0 \text{ when } n \neq 0,\ n \neq 1 \end{cases} \tag{7.42}$$

then, applying (7.45), we find

$$W[n,\varepsilon] = \begin{cases} 1 \text{ when } n=0 \\ 0 \text{ when } n \neq 0 \end{cases} \tag{7.43}$$

i.e. in this case we have

$$W[n,\varepsilon] = \sum_{l=-1}^{n} W_2[l,\varepsilon]\, W_1[n-l-1,1] = \sigma[n,0] \tag{7.44}$$

Changing the order in which the systems with the weighting functions (7.41) and (7.42) are connected, we obtain

$$W[n,\varepsilon] = \sum_{l=-1}^{n} W_1[l,\varepsilon]\, W_2[n-l-1,1] = \sigma[n,0] \tag{7.45}$$

Consequently, a system formed by connecting in series two sampled data systems with the weighting functions (7.41) and (7.42) identically transforms the input action, and thus systems with such weighting functions are mutually inverse systems. We point out, however, that a system with the weighting function (7.41) is physically not realisable, since $W_1[n,\varepsilon] \neq 0$ for $n<0$.

1.8 WEIGHTING FUNCTIONS OF CLOSED-LOOP SYSTEMS

1.8.1 Stationary systems

The simplest closed-loop system is shown in Fig. 1.2. In a closed-loop system the input signal of the sampler, $\mathcal{E}[n,\varepsilon]$, is determined by the difference of the input action $x[n,\varepsilon]$ and the output variable of the system $y[n,\varepsilon]$ at the instants when the sampler key is closed:

$$\mathcal{E}[n,0] = \big[x[n,\varepsilon] - y[n,\varepsilon]\big]_{\varepsilon=0} \tag{8.1}$$

We denote the weighting function of the closed-loop system by $W_c[n,\varepsilon]$. Its output variable $y[n,\varepsilon]$, when the input signal x acts on the system at the instant $\bar{t}=0$, is determined as follows:

$$y[n,\varepsilon] = \sum_{m=0}^{n} W_c[m,\varepsilon]\, x[n-m,0] \tag{8.2}$$

On the other hand, the output variable of the system can be defined in terms of the weighting function of the open-loop system, $W_o[n,\varepsilon]$ and the error $\mathcal{E}[n,0]$:

$$y[n,\varepsilon] = \sum_{l=0}^{n} W_o[l,\varepsilon]\, \mathcal{E}[n-l,0] \tag{8.3}$$

where the error $\mathcal{E}$ of the system is determined from (8.1). Substituting (8.1) for the error into (8.3), we find

$$y[n,\varepsilon] = \sum_{l=0}^{n} W_o[l,\varepsilon]\, x[n-l,0] - \sum_{l=0}^{n} W_o[l,\varepsilon]\, y[n-l,0] \tag{8.4}$$

Since

$$y[n-l,0] = \left[\sum_{m=0}^{n} W_c[m,\varepsilon]\, x[n-m,0]\right]_{\bar{t}=n-l+0}$$

the expression for the output variable of the closed-loop system can be written as

$$y[n,\varepsilon] = \sum_{l=0}^{n} W_o[l,\varepsilon]\, x[n-l,0] - \sum_{l=0}^{n} W_o[l,\varepsilon]\left[\sum_{m=0}^{n} W_c[m,\varepsilon]\, x[n-m,0]\right]_{\bar{t}=n-l+0} \tag{8.5}$$

Equation (8.5) establishes a connection between the output variable $y[n,\varepsilon]$ of a closed-loop system and the input action $x[n,\varepsilon]$ in terms of the weighting functions of the open-loop and corresponding closed-loop systems.

To determine the weighting function of the closed-loop system in question we must apply an action in the form of the σ-function to its input. Accordingly, replacing the input action x by the σ-function, applied at the instant $\bar{t}=m$, in (8.2) we obtain

$$y[n,\varepsilon] = \sum_{m=0}^{n} W_c[m,\varepsilon]\, \sigma[n-m,0] = W_c[n,\varepsilon] \tag{8.6}$$

whence it follows that

$$y[n-l, 0]=W_c[n-l, 0] \tag{8.7}$$

Replacing now $x[n, 0]$ by the σ-function in (8.4) and using (8.7) we find

$$W_c[n,\varepsilon]=W_o[n,\varepsilon]-\sum_{l=0}^{n} W_o[l,\varepsilon]\, W_c[n-l,0] \tag{8.8}$$

where the upper limit of summation must not be included in the interval of summation, since the comparison of the output variable of the closed-loop system with the input action is carried out one sampling period after the instant at which the input is applied. Consequently, we finally have

$$W_c[n,\varepsilon]=W_o[n,\varepsilon]-\sum_{l=0}^{n-1} W_o[l,\varepsilon]\, W_c[n-l,0] \tag{8.9}$$

where the limit points of summation must be included in the interval of summation.

Equation (8.9) establishes a connection between the weighting function of the closed- and the corresponding open-loop sampled data systems. It holds for the case where the weighting function of the closed-loop system $W_c[n, \varepsilon]$ is a discontinuous function. In the case where $W_c[n, \varepsilon]$ has discontinuities at the instants $n=0,1, \ldots$, the following applies:

$$W_c[n,\varepsilon]=W_o[n,\varepsilon]-\sum_{l=0}^{n-1} W_o[l,\varepsilon]\, W_o[n-l-1,1] \tag{8.10}$$

Formulae (8.9) and (8.10) are essentially recurrence relations by which, in successive computations, we can determine the values of $W_c[n, \varepsilon]$ for $n=0, 1, \ldots$ and $0 \leqslant \varepsilon \leqslant 1$ from the given values of the weighting function of the open-loop system $W_o[n, \varepsilon]$, $n=0,1, \ldots, 0 \leqslant \varepsilon \leqslant 1$.

We point out that (8.10) can be obtained from physical considerations. (This is how the relation analogous to (8.10) was obtained by Perov [14].) Thus, if at the instant $\bar{t}=0$ a unit pulse acts on the closed-loop system, the weighting function over the time interval 0 to $(0+\varepsilon)$ is

$$W_c[0, \varepsilon]=W_o[0, \varepsilon]$$

In the following sampling period the process $W_o[1, \varepsilon]$ takes place at the output of the system. To this is added the component $W_c[0, 1]W_o[0, \varepsilon]$ which arises as the result of the action of the system error at the instant $\bar{t}=1$ defined by $\mathscr{E}[1, 0]=W[0, 1]$. Thus for the time interval 1 to $(1+\varepsilon)$ the following relation holds:

$$W_c[1, \varepsilon]=W_o[1, \varepsilon]-W_c[0, 1]\cdot W_o[0, \varepsilon]$$

Reasoning in the same way we obtain for the next interval

$$W_c[2, \varepsilon]=W_o[2, \varepsilon]-W_c[0, 1]W_o[1, \varepsilon]-W_c[1, 1]W_o[0, \varepsilon]$$

Finally, for an arbitrary sampling period we obtain

$$W_c[n, \varepsilon]=W_o[n, \varepsilon]-\sum_{l=0}^{n-1} W_c[l, 1]W_o[n-l-1, \varepsilon] \quad (8.10')$$

Equations (8.10) and (8.10') differ only in the order of summation and one can be transformed into the other by changing the variable. Thus, putting $n-l-1=k$, we obtain a formula coinciding with (8.10).

Using (8.10) we can easily solve the inverse problem, i.e. we can determine the expression for the weighting function of the open-loop system:

$$W_o[n,\varepsilon]=W_c[n,\varepsilon]+\sum_{l=0}^{n-1} W_c[l,\varepsilon]W_o[n-l-1,1] \quad (8.11)$$

Equation (8.11) is easily verified if we put $n=0,1, \ldots$ in (8.10) and determine in succession $W_o[n, \varepsilon]$, $n=0,1, \ldots$

We now consider a stationary system whose feedback loop contains also a stationary system with the operator H (Fig. 1.20) and weighting function $W_C[n, \varepsilon]$. Assuming that the samplers of both systems work in unison and in phase, we find the weighting function of this closed-loop system.

When the disturbance $x[n, \varepsilon]$ acts at the input of the

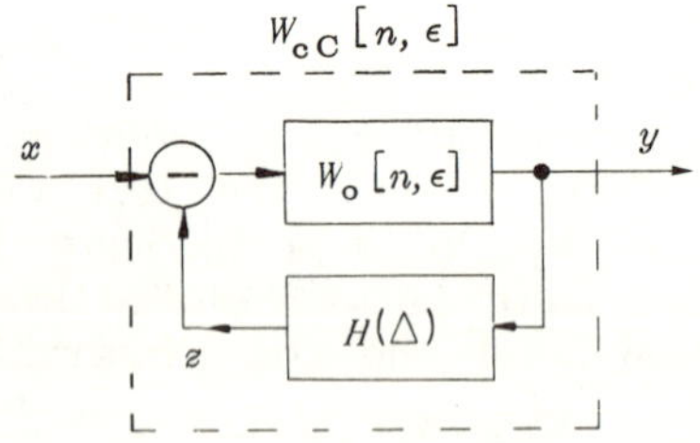

Fig. 1.20 A closed-loop sampled data system

system its output $y[n, \varepsilon]$ is determined by

$$y[n,\varepsilon] = \sum_{m=-\infty}^{\infty} W_{cC}[m,\varepsilon]\, x[n-m,0] \tag{8.12}$$

where $W_{cC}[n, \varepsilon]$ is the weighting function of the closed-loop system whose feedback loop contains a system with the operator H. On the other hand, the output variable can be determined in terms of the weighting function $W_o[n, \varepsilon]$ of the open-loop system and the error of the closed-loop system

$$y[n,\varepsilon] = \sum_{l=-\infty}^{\infty} W_o[l,\varepsilon]\, \mathcal{E}[n-l,0] \tag{8.13}$$

where

$$\mathcal{E}[n-l,0] = x[n-l,0] - z[n-l,0] \tag{8.14}$$

and $z[n, \varepsilon]$ is the output variable of the system included in the feedback loop. Since $y[n, \varepsilon]$ is the input disturbance of the system in the feedback loop,

$$z[n,\varepsilon] = \sum_{\nu=-\infty}^{\infty} W_C[\nu,\varepsilon]\, y[n-\nu,0] \tag{8.15}$$

Substituting the value of the system error defined by (8.14) into (8.15) we have

$$y[n,\varepsilon] = \sum_{l=-\infty}^{\infty} W_o[l,\varepsilon]\, x[n-l,0] - \sum_{l=-\infty}^{\infty} W_o[l,\varepsilon]\, z[n-l,0] \tag{8.16}$$

From (8.13) we have

$$y[n-\nu,0] = \left[\sum_{m=-\infty}^{\infty} W_{cC}[m,\varepsilon]\, x[n-m,0]\right]_{\bar{t}=n-\nu+0} \tag{8.17}$$

Consequently (8.15) can be written

$$z[n,\varepsilon] = \sum_{\nu=-\infty}^{\infty} W_C[\nu,\varepsilon] \left[\sum_{m=-\infty}^{\infty} W_{cC}[m,\varepsilon]\, x[n-m,0]\right]_{\bar{t}=n-\nu+0} \tag{8.18}$$

Equations (8.16) and (8.18) define processes in the stationary systems under consideration.

Replacing x in (8.12) by the σ-function we obtain

$$y[n,\varepsilon]=\sum_{m=-\infty}^{\infty} W_{cC}[m,\varepsilon]\,\sigma[n-m,0]=W_{cC}[n,\varepsilon] \tag{8.19}$$

whence

$$y[n-\nu,0]=W_{cC}[n-\nu,0] \tag{8.20}$$

Consequently, substituting (8.20) into (8.15) we find

$$z[n,\varepsilon]=\sum_{\nu=-\infty}^{\infty} W_C[\nu,\varepsilon]\,W_{cC}[n-\nu,0] \tag{8.21}$$

whence we obtain

$$z[n-l,0]=\left[\sum_{\nu=-\infty}^{\infty} W_C[\nu,\varepsilon]\,W_{cC}[n-\nu,0]\right]_{\bar{t}=n-l+0} \tag{8.22}$$

Thus, when (8.19) to (8.22) are taken into account, (8.12) gives

$$\begin{aligned} W_{cC}[n,\varepsilon]=W_o[n,\varepsilon]-\sum_{l=-\infty}^{\infty} W_o[l,\varepsilon] \\ \times\left[\sum_{\nu=-\infty}^{\infty} W_C[\nu,\varepsilon]\,W_{cC}[n-\nu,0]\right]_{\bar{t}=n-l+0} \end{aligned} \tag{8.23}$$

It is not difficult to see that (8.21) is the weighting function of two systems with the weighting functions W_C and W_{cC} connected in series. Denoting the weighting function of this combination by

$$W_{oCC}[n,\varepsilon]=\sum_{\nu=-\infty}^{\infty} W_C[\nu,\varepsilon]\,W_{oC}[n-\nu,0] \tag{8.24}$$

we finally obtain from (8.23)

$$W_{cC}[n,\varepsilon]=W_o[n,\varepsilon]-\sum_{l=-\infty}^{\infty} W_o[l,\varepsilon]\,W_{cCC}[n-l,0] \tag{8.25}$$

Equation (8.25) establishes a connection between the weighting functions $W_o[n,\varepsilon]$ and $W_C[n,\varepsilon]$ on one side and the weighting function of the closed-loop system $W_{cC}[n,\varepsilon]$ on the other. This relation is valid for the case where W_o and W_C have no breaks in the continuity for $n=0,1,\ldots$ It is easy to show that if any of the weighting functions W_o and W_C has discontinuities at the instants the sampler key is closed, then (8.25) assumes the form

$$W_{cC}[n, \varepsilon] = W_o[n, \varepsilon] - \sum_{l=-\infty}^{\infty} W_o[l, \varepsilon] W_{cCC}[n-l-1,1] \tag{8.26}$$

where the weighting function W_{cCC} is determined by the relation

$$W_{cCC}[n, \varepsilon] = \sum_{\nu=-\infty}^{\infty} W_C[\nu, \varepsilon] W_{cC}[n-\nu-1,1] \tag{8.27}$$

From (8.26) we can easily determine

$$W_o[n, \varepsilon] = W_{cC}[n, \varepsilon] + \sum_{l=-\infty}^{\infty} W_{cCC}[l, \varepsilon] W_o[n-l,0] \tag{8.28}$$

Equation (8.28) enables the inverse problem to be solved, i.e. the weighting function $W_o[n, \varepsilon]$ of the open-loop system can be determined from the given weighting functions W_{cC} and W_{cCC}. Similarly, from (8.27) we can obtain

$$W_o[n, \varepsilon] = W_{cC}[n, \varepsilon] + \sum_{l=-\infty}^{\infty} W_{cCC}[l, \varepsilon] W_o[n-l-1,1] \tag{8.29}$$

For a system that is physically realisable, (8.25) and (8.26) assume the form

$$W_{cC}[n, \varepsilon] = W_o[n, \varepsilon] - \sum_{l=0}^{n-2} W_o[l, \varepsilon] W_{cCC}[n-l, 0] \tag{8.30}$$

$$W_{cC}[n, \varepsilon] = W_o[n, \varepsilon] - \sum_{l=0}^{n-2} W_o[l, \varepsilon] W_{cCC}[n-l-1,1] \tag{8.31}$$

since the signal from the output of the feedback loop enters the input of the system with the weighting function $W_o(n, \varepsilon)$, beginning with the instant $\bar{t}=2$. In view of this, the expressions determining the weighting function $W_o(n, \varepsilon)$ of the open-loop system can be written

$$W_o[n, \varepsilon] = W_{cC}[n, \varepsilon] + \sum_{l=0}^{n-2} W_{cCC}[l, \varepsilon] W_o[n-l, 0] \tag{8.32}$$

$$W_o[n, \varepsilon] = W_{oC}[n, \varepsilon] + \sum_{l=0}^{n-2} W_{oCC}[l, \varepsilon] W_o[n-l-1, 1] \tag{8.33}$$

1.8.2 Non-stationary systems

We shall consider a non-stationary sampled data system closed by a negative feedback loop with an operator equal to unity. When a disturbance $x[n, \varepsilon]$ acts at the input its output variable $y[n, \varepsilon]$ is determined by

$$y[n, \varepsilon] = \sum_{m=-\infty}^{\infty} W_c[n, \varepsilon; m] x[m, 0] \tag{8.34}$$

where $W_c[n, \varepsilon; m]$ is the weighting function of the closed-loop non-stationary system. On the other hand, the output variable $y[n, \varepsilon]$ can be determined also in terms of the weighting function $W_o[n, \varepsilon; m]$ of the corresponding open-loop system and the error $\mathscr{E}$ of the closed-loop system:

$$y[n, \varepsilon] = \sum_{l=-\infty}^{\infty} W_o[n, \varepsilon; l] \mathscr{E}[l, 0] \tag{8.35}$$

where the error of the system is determined by

$$\mathscr{E}[l, 0] = [x[l, \varepsilon] - y[l, \varepsilon]]_{\varepsilon=0}$$

Substituting the expression for the error into (8.35) we find

$$y[n, \varepsilon] = \sum_{l=-\infty}^{\infty} W_o[n, \varepsilon; l] x[l, 0] - \sum_{l=-\infty}^{\infty} W_o[n, \varepsilon; l] y[l, 0] \tag{8.36}$$

It is clear that after carrying out the summation in (8.35) we obtain a function of the argument $\bar{t} = n + \varepsilon$. Consequently, from (8.35) we have

$$y[l, 0] = \Big[\sum_{m=-\infty}^{\infty} W_c[n, \varepsilon; m] x[m, 0]\Big]_{\bar{t}=l+0} \tag{8.37}$$

Substituting (8.37) into (8.36) we obtain

$$y[n,\varepsilon]=\sum_{l=-\infty}^{\infty} W_{o}[n,\varepsilon;l]\,x[l,0]-\sum_{l=-\infty}^{\infty} W_{o}[n,\varepsilon;l]\times\Big[\sum_{m=-\infty}^{\infty} W_{c}[n,\varepsilon;m]\,x[m,0]\Big]_{\bar{t}=l+0} \tag{8.38}$$

which establishes a connection between the output variable and the input disturbance in terms of the weighting functions of the open-loop system and the corresponding closed-loop non-stationary system.

Substituting the input disturbance $x[n,0]$ in (8.38) by the σ-function applied at the instant $\bar{t}=m$, we obtain

$$W_{c}[n,\varepsilon;m]=\sum_{\nu=-\infty}^{\infty} W_{o}[n,\varepsilon;\nu]\,\sigma[\nu-m,0]-\sum_{l=-\infty}^{\infty} W_{o}[n,\varepsilon;l]\sum_{\nu=-\infty}^{\infty} W_{c}[l,0;\nu]\,\sigma[\nu-m,0] \tag{8.39}$$

whence, taking into consideration the properties of σ-functions, we finally obtain

$$W_{c}[n,\varepsilon;m]=W_{o}[n,\varepsilon;m]-\sum_{l=-\infty}^{\infty} W_{o}[n,\varepsilon;l]\,W_{c}[l,0;m] \tag{8.40}$$

Equation (8.40) establishes a connection between the weighting functions of the closed-loop system and the corresponding open-loop sampled data system. It is valid for the case where the weighting functions of the closed-loop system $W_{c}[n,\varepsilon;m]$ has no discontinuities at the instants when the sampler key is closed. In the contrary case it is easy to show that the following relation holds:

$$W_{c}[n,\varepsilon;m]=W_{o}[n,\varepsilon;m]-\sum_{l=-\infty}^{\infty} W_{o}[n,\varepsilon;l]\,W_{c}[l-1,1;m] \tag{8.41}$$

Since the weighting function of a non-stationary sampled data system satisfies the condition of physical realisability

$$W[n,\varepsilon;m]=\begin{cases}0 & \text{when } n<m\\ W[n,\varepsilon;m] & \text{when } n\geqslant m\end{cases} \tag{8.42}$$

Equations (8.40) and (8.41) can be written in the form

$$W_c[n, \varepsilon; m] = W_o[n, \varepsilon; m] - \sum_{l=m+1}^{n} W_o[n, \varepsilon; l]\, W_c[l, 0; m] \tag{8.43}$$

$$W_c[n, \varepsilon; m] = W_o[n, \varepsilon; m] - \sum_{l=m+1}^{n} W_o[n, \varepsilon; l]\, W_c[l-1, 1; m] \tag{8.44}$$

$$n = m + 1 + k, \quad k = 0, 1, \ldots$$

The lower limits of summation in (8.43) and (8.44) are not m, but $m+1$, since in the system under consideration the signal from the feedback loop enters the input of the system with the weighting function $W_o[n, \varepsilon; m]$, by one period of closure of the sampler key after the instant at which the input action is applied.

The relations obtained determine the weighting function of the simplest closed-loop system. Using similar methods we can find the corresponding relation determining the weighting function of a non-stationary closed-loop system whose feedback loop contains a sampled data system with the weighting function $W_C[n, \varepsilon; m]$.

The method described for finding the weighting functions of complex systems is based on determining the weighting function of a sampled data system as a response of this system to the σ-function. The advantage of this method is its simplicity and mathematical rigour. Furthermore, the procedure of obtaining some relation or other for sampled data systems remains the same as for continuous systems, methods of analysis of which are by now fairly well worked out. Besides the theoretical importance of these considerations a practical advantage is that the continuous and sampled data systems can be investigated by a unified procedure.

1.9 WEIGHTING FUNCTIONS OF MULTI-DIMENSIONAL SYSTEMS

In a general case a sampled data system can have l inputs and k outputs. Such a system is called a multi-dimensional sampled data system, and we shall now determine its weighting function.

We denote the input action entering the system inputs

$j=1, 2, \ldots, l$ by $x_1, x_2, \ldots, x_l$. The output variables corresponding to the k outputs $i=1, 2, \ldots, k$, are denoted by $y_1, y_2, \ldots, y_k$. Further values of the output variables corresponding to the action of only one disturbance x_j at the j-th input is denoted by y_{ij}, so that y_{ij} denotes the output variable at the i-th output when a disturbance acts on the j-th input. Hence, incidentally, it follows that the component systems forming the multi-dimensional system are mutually connected.

On the basis of the superposition principle we can write

$$y_i[n, \varepsilon] = \sum_{j=1}^{l} y_{ij}[n, \varepsilon], \quad i = 1, 2, \ldots, k \tag{9.1}$$

By definition $y_{ij}[n, \varepsilon]$ is the result of a transformation of the input disturbance j by a difference operator A_{ij}, establishing the connection between x_j and y_i, i.e.

$$\begin{gathered} y_{ij}[n, \varepsilon] = A_{ij} x_j \\ i = 1, 2, \ldots, k_j;\ j = 1, 2, \ldots, l \end{gathered} \tag{9.2}$$

To determine the output variables of a multi-dimensional system we represent the input disturbances $x_j[n, 0]$ thus:

$$\begin{gathered} x_j[n, 0] = \sum_{m=-\infty}^{\infty} \sigma[n-m, 0]\, x_j[m, 0] \\ j = 1, 2, \ldots, l \end{gathered} \tag{9.3}$$

Applying the operators A_{ij} with respect to the variable n to both sides of (9.3) we find

$$\begin{gathered} A_{ij} x_j[n, 0] = \sum_{m=-\infty}^{\infty} A_{ij}\sigma[n-m, 0]\, x_j[m, 0] \\ i = 1, 2, \ldots, k;\ j = 1, 2, \ldots, l \end{gathered} \tag{9.4}$$

Since by (9.2)

$$A_{ij} x_j[n, 0] = y_{ij}[n, \varepsilon] \tag{9.5}$$

then Equation (9.4) can be written as

$$\begin{gathered} y_{ij}[n, \varepsilon] = \sum_{m=-\infty}^{\infty} A_{ij}\sigma[n-m, 0]\, x_j[m, 0] \\ i = 1, 2, \ldots, k;\ j = 1, 2, \ldots, l \end{gathered} \tag{9.6}$$

Equation (9.6) determines the response of the system at the

i-th output for the disturbance x_j acting at the j-th input in terms of the response to the σ-function, also acting on the j-th input. Consequently, we can write

$$y_{ij}[n, \varepsilon] = \sum_{m=-\infty}^{\infty} W_{ij}[n, \varepsilon; m] x_j[m, 0]$$

$$i = 1, 2, \ldots, k; \quad j = 1, 2, \ldots, l \tag{9.7}$$

where

$$W_{ij}[n, \varepsilon; m] = A_{ij}\sigma[n - m, 0] \tag{9.8}$$

are the weighting functions of the system, constituting the response at the i-th output for the σ-function acting at the j-th input at the instant $\bar{t} = m$

Substituting the values of the output variables y_{ij} from (9.7) into (9.1) we find

$$y_i[n, \varepsilon] = \sum_{j=1}^{l} \sum_{m=-\infty}^{\infty} W_{ij}[n, \varepsilon; m] x_j[m, 0]$$

$$i = 1, 2, \ldots, k \tag{9.9}$$

Equation (9.9) determines the variable of the i-th output of the system for the disturbances x_j acting at the j inputs in terms of the weighting functions W_{ij}.

Thus, a multi-dimensional sampled data system with l inputs and k outputs is characterised by the rectangular $l \times k$ matrix of weighting functions

$$W = \begin{Vmatrix} W_{11} & W_{12} & \ldots & W_{1l} \\ W_{21} & W_{22} & \ldots & W_{2l} \\ \ldots & \ldots & \ldots & \ldots \\ W_{k1} & W_{k2} & \ldots & W_{kl} \end{Vmatrix} \tag{9.10}$$

The weighting functions W_{ij} of a real system satisfy the condition

$$W_{ij}[n, \varepsilon; m] = \begin{cases} 0 & \text{when } n < m \\ W_{ij}[n, \varepsilon; m] & \text{when } n \geqslant m \end{cases} \tag{9.11}$$

which is the condition of physical realisability of a multi-dimensional sampled data system. When (9.11) is taken into account, (9.9) can be written as

$$y_i[n, \varepsilon] = \sum_{j=1}^{l} \sum_{m=h}^{\infty} W_{ij}[n, \varepsilon; m]\, x_j[m, 0] \qquad i = 1, 2, \ldots, k \tag{9.12}$$

where h is the instant at which the actions are applied to the input of the system.

In a particular case, putting in (9.12) the input disturbances equal to the σ-function applied at the instant $\bar{t} = m$, we find

$$y_i[n, \varepsilon] = \sum_{j=1}^{l} \sum_{\nu=h}^{n} W_{ij}[n, \varepsilon; \nu]\, \sigma_j[\nu - m, 0] \qquad i = 1, 2, \ldots, k$$

or, finally,

$$W_i[n, \varepsilon; m] = \sum_{j=1}^{l} W_{ij}[n, \varepsilon; m] \qquad i = 1, 2, \ldots, k \tag{9.13}$$

A multi-dimensional sampled data system with k inputs and k outputs is described by a system of simultaneous linear difference equations

$$\Delta_n y_i[n, \varepsilon] = \sum_{j=1}^{k} a_{ij}[n]\, y_i[n, \varepsilon] + \varphi_i[n, 0] \qquad i = 1, 2, \ldots, k \tag{9.14}$$

where y_i are the output variables;

φ_i are the input disturbances.

A particular case of the system (9.14) is furnished by

$$\begin{aligned} y[n, \varepsilon] &= y_1[n, \varepsilon] \\ \Delta y_1[n, \varepsilon] &= y_2[n, \varepsilon] \\ &\cdots\cdots \\ \Delta y_{k-1}[n, \varepsilon] &= y_k[n, \varepsilon] \\ \Delta y_k[n, \varepsilon] &= -a_{k-1}[n]\, y_k[n, \varepsilon] - \ldots - a_0[n]\, y_1[n, \varepsilon] + \varphi[n, 0] \end{aligned} \tag{9.15}$$

which can be obtained from a k-th order equation of the form

$$\sum_{i=0}^{k} a_i[n]\, \Delta^i y[n, \varepsilon] = \varphi[n, 0] \qquad a_k[n] = 1 \tag{9.16}$$

by introducing the new variables $y_1, y_2, \ldots, y_k$. The method of obtaining such systems of first-order difference equations was described in Section 1.2.

The system of difference equations (9.15) describes k holding circuits (sampled data follow-up systems with a single integrator when $\gamma=1$) connected with each other. If $x_j[n, 0]$ is the input action of the j-th holding circuit, and $y_i[n, \varepsilon]$ is the output variable of the i-th circuit, then (9.9) for $k=l$ assumes the form

$$y_i[n, \varepsilon]=\sum_{j=1}^{k}\sum_{m=-\infty}^{\infty} W_{ij}[n, \varepsilon; m]\, x_j[m, 0] \qquad i=1, 2, \ldots, k \tag{9.17}$$

If the σ-function is the input action of the j-th holding circuit, whilst all the remaining inputs are zero, then the weighting functions W_{ij} of this system are determined by

$$W_{ij}[m+1, \varepsilon; m]=\begin{cases} 0 \text{ when } i\neq j \\ 1 \text{ when } i=j \end{cases} \tag{9.18}$$

Equation (9.18) is valid, because at the output a holding circuit repeats the input quantity which directly precedes the instant at which the key is closed.

For the given system the matrix of weighting functions (9.10) reduces, for $n=m+1$, to the square unit matrix

$$W=\begin{Vmatrix} 1 & 0 & 0 \ldots & 0 \\ 0 & 1 & 0 \ldots & 0 \\ 0 & 0 & 1 \ldots & 0 \\ \cdot & \cdot & \cdot & \cdot \\ 0 & 0 & 0 \ldots & 1 \end{Vmatrix} \tag{9.19}$$

1.10 DETERMINATION OF THE WEIGHTING FUNCTION

In the preceding sections we considered methods of determining the weighting functions of sampled data systems formed by various combinations of individual systems. It was assumed that the weighting functions of the component systems were given either analytically or by tables of their values. We now determine the weighting functions directly from the given difference equations of sampled data systems.

1.10.1 Determination from the non-homogeneous difference equations

It was shown in Section 1.5 that the weighting function of a sampled data system is the response of this system to the σ-function. Mathematically this definition was expressed in the form

$$W[n, \varepsilon; m] = A_n \sigma[n-m, 0] \tag{10.1}$$

where A_n is the operator of the sampled data system.

Let the processes in the system be determined by the difference equations

$$\begin{aligned} L_n(\Delta, n)\, y[n, \varepsilon] &= x[n, 0], \quad 0 \leqslant \varepsilon \leqslant \gamma \\ S_n(\Delta, n)\, y[n, \varepsilon] &= x[n, 0], \quad \gamma \leqslant \varepsilon \leqslant 1 \end{aligned} \tag{10.2}$$

Then from (10.1) the weighting function of such a system is the solution of (10.2) with zero initial conditions, if the σ-function applied at the instant $\bar{t} = m$ is the right-hand side of these equations. To prove this statement it is sufficient to show that the output variable $y[n, \varepsilon]$ of the system, defined in terms of its weighting function by

$$y[n, \varepsilon] = \sum_{m=-\infty}^{\infty} W[n, \varepsilon; m]\, x[m, 0] \tag{10.3}$$

satisfies (10.2) with the condition that the weighting function $W[n, \varepsilon; m]$ is the solution of the équations

$$\begin{aligned} L_n(\Delta, n)\, W[n, \varepsilon; m] &= \sigma[n-m, 0], \quad 0 \leqslant \varepsilon \leqslant \gamma \\ S_n(\Delta, n)\, W[n, \varepsilon; m] &= \sigma[n-m, 0], \quad \gamma \leqslant \varepsilon \leqslant 1 \end{aligned} \tag{10.4}$$

where the difference operators are defined by the relations

$$L_n(\Delta, n) = \sum_{i=0}^{k} a_i[n]\, \Delta_n^i \tag{10.5}$$

$$S_n(\Delta, n) = \sum_{i=0}^{k} c_i[n]\, \Delta_n^i \tag{10.6}$$

For a proof we determine the differences of the output variable of the system under consideration $\Delta_n y[n, \varepsilon]$, $\Delta_n^2 y[n, \varepsilon], \ldots, \Delta_n^k y[n, \varepsilon]$, using (10.3). Taking into account that $y[n, \varepsilon]$ is determined by a sum with a constant upper limit, we obtain by the linearity property of the operator

$$\Delta_n y[n, \varepsilon] = \sum_{m=-\infty}^{\infty} \Delta_n W[n, \varepsilon; m] x[m, 0]$$

$$\Delta_n^2 y[n, \varepsilon] = \sum_{m=-\infty}^{\infty} \Delta_n^2 W[n, \varepsilon; m] x[m, 0] \tag{10.7}$$

$$\cdots\cdots\cdots\cdots\cdots\cdots$$

$$\Delta_n^k y[n, \varepsilon] = \sum_{m=-\infty}^{\infty} \Delta_n^k W[n, \varepsilon; m] x[m, 0]$$

Substituting now the values of the differences from (10.7) into (10.2), we obtain the following identities when (10.4) is taken into account

$$\sum_{m=-\infty}^{\infty} L_n(\Delta, n) W[n, \varepsilon; m] x[m, 0]$$

$$= \sum_{m=-\infty}^{\infty} x[m, 0] \sigma[n-m, 0] = x[n, 0], \quad 0 \leqslant \varepsilon \leqslant \gamma$$

$$\sum_{m=-\infty}^{\infty} S_n(\Delta, n) W[n, \varepsilon; m] x[m, 0]$$

$$= \sum_{m=-\infty}^{\infty} x[m, 0] \sigma[n-m, 0] = x[n, 0], \quad \gamma \leqslant \varepsilon \leqslant 1$$

which confirm that the output variable of the system written in the form (10.3) satisfies (10.2). Consequently, the weighting of a system described by the difference equations (10.2) is determined by solving the non-homogeneous difference equations (10.4) with zero initial conditions and the condition that the σ-function is the right-hand side of these equations.

We now consider a system whose processes are determined by the difference equations

$$\begin{aligned} L_n(\Delta, n) y[n, \varepsilon] &= M_n(\Delta, \bar{t}) x[n, 0], \quad 0 \leqslant \varepsilon \leqslant \gamma \\ S_n(\Delta, n) y[n, \varepsilon] &= G_n(\Delta, \bar{t}) x[n, 0], \quad \gamma \leqslant \varepsilon \leqslant 1 \end{aligned} \tag{10.8}$$

where the difference operators $L_n(\Delta, n)$ and $S_n(\Delta, n)$ are defined by (10.5) and (10.6) respectively, whilst the operators of the right-hand sides are

$$M_n(\Delta, \bar{t}) = \sum_{j=0}^{h_1} b_j[n, \varepsilon]\,\Delta_n^j \tag{10.9}$$

$$G_n(\Delta, \bar{t}) = \sum_{j=0}^{h_2} d_j[n, \varepsilon]\,\Delta_n^j \tag{10.10}$$

The weighting function of this system is denoted by the symbol $W_1[n, \varepsilon; m]$. In keeping with the definition (10.1) the weighting function $W_1[n, \varepsilon; m]$ is the solution of the non-homogeneous difference equations

$$\begin{aligned} L_n(\Delta, n)\,W_1[n, \varepsilon; m] &= M_n(\Delta, \bar{t})\,\sigma[n-m, 0], \quad 0 \leqslant \varepsilon \leqslant \gamma \\ S_n(\Delta, n)\,W_1[n, \varepsilon; m] &= G_n(\Delta, \bar{t})\,\sigma[n-m, 0], \quad \gamma \leqslant \varepsilon \leqslant 1 \end{aligned} \tag{10.11}$$

with zero initial conditions. We shall now prove this statement. The output variable of the system under consideration having the weighting function $W_1[n, \varepsilon; m]$, is determined by

$$y[n, \varepsilon] = \sum_{m=-\infty}^{\infty} W_1[n, \varepsilon; m]\,x[m, 0] \tag{10.12}$$

Determining the differences $\Delta_n^r y[n, \varepsilon]$, $r = 1, 2, \ldots, k$, we find, using (10.12):

$$\begin{aligned} \Delta_n y[n, \varepsilon] &= \sum_{m=-\infty}^{\infty} \Delta_n W_1[n, \varepsilon; m]\,x[m, 0] \\ \Delta_n^2 y[n, \varepsilon] &= \sum_{m=-\infty}^{\infty} \Delta_n^2 W_1[n, \varepsilon; m]\,x[m, 0] \\ &\ldots\ldots\ldots\ldots\ldots\ldots \\ \Delta_n^k y[n, \varepsilon] &= \sum_{m=-\infty}^{\infty} \Delta_n^k W_1[n, \varepsilon; m]\,x[m, 0] \end{aligned} \tag{10.13}$$

Multiplying $\Delta_n^i y[n, \varepsilon]$ by $a_i[n]$ and assuming the products thus obtained with respect to i, we find allowing for (10.5),

$$\sum_{i=0}^{k} a_i[n]\,\Delta_n^i\, y[n, \varepsilon] = \sum_{i=0}^{k} a_i[n] \sum_{m=-\infty}^{\infty} \Delta_n^i\, W_1[n, \varepsilon; m]\, x[m, 0]$$

$$= \sum_{m=-\infty}^{\infty} x[m, 0] \sum_{i=0}^{k} a_i[n]\,\Delta_n^i\, W_1[n, \varepsilon; m] \qquad (10.14)$$

$$= \sum_{m=-\infty}^{\infty} L_n(\Delta, n)\, W_1[n, \varepsilon; m]\, x[m, 0]$$

Multiplying now $\Delta_n^i\, y[n, \varepsilon]$ by $c_i[n]$ and summing these products with respect to i, we have analogously to (10.14), when (10.6) is taken into account,

$$\sum_{i=0}^{k} c_i[n]\,\Delta_n^i\, y[n, \varepsilon] = \sum_{i=0}^{k} c_i[n] \sum_{m=-\infty}^{\infty} \Delta_n^i\, W_1[n, \varepsilon; m]\, x[m, 0]$$

$$= \sum_{m=-\infty}^{\infty} x[m, 0] \sum_{i=0}^{k} c_i[n]\,\Delta_n^i\, W_1[n, \varepsilon; m] \qquad (10.15)$$

$$= \sum_{m=-\infty}^{\infty} S_n(\Delta, n)\, W_1[n, \varepsilon; m]\, x[m, 0]$$

Taking into consideration (10.11) we obtain from these last relations the identities

$$\sum_{m=-\infty}^{\infty} L_n(\Delta, n)\, W_1[n, \varepsilon; m]\, x[m, 0]$$

$$= \sum_{m=-\infty}^{\infty} M_n(\Delta, \bar{t})\, \sigma[n-m, 0]\, x[m, 0] \qquad (10.16)$$

$$= M_n(\Delta, \bar{t})\, x[n, 0], \quad 0 \leqslant \varepsilon \leqslant \gamma$$

$$\sum_{m=-\infty}^{\infty} S_n(\Delta, n)\, W_1[n, \varepsilon; m]\, x[m, 0]$$

$$= \sum_{m=-\infty}^{\infty} G_n(\Delta, \bar{t})\, \sigma[n-m, 0]\, x[m, 0] \qquad (10.17)$$

$$= G_n(\Delta, \bar{t})\, x[n, 0], \quad \gamma \leqslant \varepsilon \leqslant 1$$

which prove that the weighting function of a sampled data

system of general form is the solution of the corresponding non-homogeneous difference equations whose disturbing function is the σ-function.

If the weighting function $W[n, \varepsilon; m]$ of the system is known and the processes in it are determined by difference equations of the form

$$\begin{aligned} L_n(\Delta, n)\, W[n, \varepsilon; m] &= \sigma[n-m, 0], \quad 0 \leqslant \varepsilon \leqslant \gamma \\ S_n(\Delta, n)\, W[n, \varepsilon; m] &= \sigma[n-m, 0], \quad \gamma \leqslant \varepsilon \leqslant 1 \end{aligned} \tag{10.18}$$

then the weighting function can be determined by another method, without solving the non-homogeneous difference equations (10.11). Thus, the output variable of the system whose processes are given by (10.8) can be determined by

$$y_1[n, \varepsilon] = \sum_{m=-\infty}^{\infty} W_1[n, \varepsilon; m]\, x[m, 0] \tag{10.19}$$

The output variable of a system, the difference equations of which do not contain differences of the input disturbance, is determined in terms of its weighting function $W[n, \varepsilon; m]$ by

$$y[n, \varepsilon] = \sum_{m=-\infty}^{\infty} W[n, \varepsilon; m]\, x[m, 0] \tag{10.20}$$

Replacing now the disturbance $x[m, 0]$ (10.20) by the right-hand sides (10.8) we obtain for the variable $y_1[n, \varepsilon]$ the following expression:

$$\begin{aligned} y_1[n, \varepsilon] &= \sum_{m=-\infty}^{\infty} W[n, \varepsilon; m]\{M_m(\Delta, m, \varepsilon)\, x[m, 0]\}, \ 0 \leqslant \varepsilon \leqslant \gamma \\ y_1[n, \varepsilon] &= \sum_{m=-\infty}^{\infty} W[n, \varepsilon; m]\{G_m(\Delta, m, \varepsilon)\, x[m, 0]\}, \ 0 \leqslant \varepsilon \leqslant 1 \end{aligned} \tag{10.21}$$

Substituting in (10.21) the values of the difference operators $M_n(\Delta, \bar{t})$ and $G_n(\Delta, \bar{t})$ for $n = m$ from (10.9) and (10.10), we obtain in expanded form

$$y_1[n, \varepsilon] = \sum_{m=-\infty}^{\infty} W[n, \varepsilon; m] \sum_{j=0}^{h_1} b_j[m, \varepsilon]\, \Delta_m^j\, x[m, 0], 0 \leqslant \varepsilon \leqslant \gamma$$

$$y_1[n, \varepsilon] = \sum_{m=-\infty}^{\infty} W[n, \varepsilon; m] \sum_{j=0}^{h_2} d_j[m, \varepsilon] \Delta_m^j x[m, 0], \quad \gamma \leqslant \varepsilon \leqslant 1 \tag{10.22}$$

We shall now transform these relations. For this we successively sum by parts in (10.22), so that there are no terms under the summation sign whose factors are differences of the input disturbance $x[m, 0]$.

Applying the rule for summing by parts we have for (10.22)

$$\sum_{m=-\infty}^{\infty} W[n, \varepsilon; m] b_0[m, \varepsilon] x[m, 0] = \sum_{m=-\infty}^{\infty} W[n, \varepsilon; m] b_0[m, \varepsilon] x[m, 0]$$

$$\sum_{m=-\infty}^{\infty} W[n, \varepsilon; m] b_1[m, \varepsilon] \Delta_m x[m, 0] = \{W[n, \varepsilon; m] b_1[m, \varepsilon]\}_{m=-\infty}^{\infty} - \sum_{m=-\infty}^{\infty} \Delta_m \{W[n, \varepsilon; m] b_1[m, \varepsilon]\} x[m+1, 0] \tag{10.23}$$

Bearing in mind that for stable sampled data systems the following holds

$$W[n, \varepsilon, m]|_{m=\infty} = W[n, \varepsilon; m]|_{m=-\infty} = 0$$

and that the second term of the last relation can be written in the form

$$\sum_{m=-\infty}^{\infty} \Delta_m \{W[n, \varepsilon; m] b_1[m, \varepsilon]\} x[m+1, 0] = \sum_{m=-\infty}^{\infty} \Delta_m \{W[n, \varepsilon; m-1] b_1[m-1, \varepsilon]\} x[m, 0]$$

we finally have for (10.23)

$$\sum_{m=-\infty}^{\infty} W[n, \varepsilon; m] b_1[m, \varepsilon] \Delta_m x[m, 0] = - \sum_{m=-\infty}^{\infty} \Delta_m \{W[n, \varepsilon; m-1] b_1[m-1, \varepsilon]\} x[m, 0] \tag{10.24}$$

Similarly it can be shown that the following relations hold

$$\sum_{m=-\infty}^{\infty} W[n, \varepsilon; m]\, b_l[m, \varepsilon]\, \Delta_m^l x[m, 0]$$

$$= (-1)^l \sum_{m=-\infty}^{\infty} \Delta_m^l \{W[n, \varepsilon; m-l]\, b_l[m-l, \varepsilon]\}\, x[m, 0] \tag{10.25}$$

$$l = 0, 1, \ldots$$

In a similar way for the second equation (10.22) we obtain

$$\sum_{m=-\infty}^{\infty} W[n, \varepsilon; m]\, d_l[m, \varepsilon]\, \Delta_m^l x[m, 0]$$

$$= (-1)^l \sum_{m=-\infty}^{\infty} \Delta_m^l \{W[n, \varepsilon; m-l]\, d_l[m-l, \varepsilon]\}\, x[m, 0] \tag{10.26}$$

$$l = 0, 1, \ldots$$

Substituting (10.25) and (10.26) into (10.22) we finally obtain the following expression for the output variable of a sampled data system of general form

$$y_1[n, \varepsilon] = \sum_{m=-\infty}^{\infty} x[m, 0] \sum_{j=0}^{h_1} (-1)^j$$

$$\times \Delta_m^j \{W[n, \varepsilon; m-j]\, b_j[m-j, \varepsilon]\}, \quad 0 \leqslant \varepsilon \leqslant \gamma$$

$$y_1[n, \varepsilon] = \sum_{m=-\infty}^{\infty} x[m, 0] \sum_{j=0}^{h_2} (-1)^j \tag{10.27}$$

$$\times \Delta_m^j \{W[n, \varepsilon; m-j]\, d_j[m-j, \varepsilon]\}, \quad \gamma \leqslant \varepsilon \leqslant 1$$

Equating now the right-hand sides of (10.27) and (10.19) we have

$$W_1[n, \varepsilon; m] = \sum_{j=0}^{h_1} (-1)^j \Delta_m^j \{W[n, \varepsilon; m-j]\, b_j[m-j, \varepsilon]\}$$

$$0 \leqslant \varepsilon \leqslant \gamma$$

$$W_1[n, \varepsilon; m] = \sum_{j=0}^{h_2} (-1)^j \Delta_m^j \{W[n, \varepsilon; m-j]\, d_j[m-j, \varepsilon]\}$$

$$\gamma \leqslant \varepsilon \leqslant 1 \tag{10.28}$$

The relations (10.28) enable us to find the weighting function of a sampled data system of general form from the weighting function of the system whose difference equations do not contain differences of the input disturbance.

We note that the relations (10.28) can be obtained in another way. We replace the input disturbance x in (10.22) by the σ-function applied to the system at the instant $\bar{t}=m$. In this case by the definition of the weighting function of a sampled data system, we have $y_1[n, \varepsilon] = W_1[n, \varepsilon; m]$ and from (10.22) we obtain

$$W_1[n, \varepsilon; m] = \sum_{l=-\infty}^{\infty} W[n, \varepsilon; l] \sum_{j=0}^{h_1} b_j[l, \varepsilon] \Delta_l^j \sigma[l-m, 0]$$

$$0 \leqslant \varepsilon \leqslant \gamma$$

$$W_1[n, \varepsilon; m] = \sum_{l=-\infty}^{\infty} W[n, \varepsilon; l] \sum_{j=0}^{h_2} d_j[l, \varepsilon] \Delta_l^j \sigma[l-m, 0] \tag{10.29}$$

$$\gamma \leqslant \varepsilon \leqslant 1$$

Using the property of the σ-function expressed by

$$f[l,0]\Delta_l^j \sigma[l-m,0] = (-1)^j \Delta_m^j \{f[m-j,0]\sigma[l-(m-j),0\}$$

we write the right-hand sides of the last relations

$$\sum_{l=-\infty}^{\infty} W[n, \varepsilon; l] b_j[l, \varepsilon] \Delta_l^j \sigma[l-m, 0]$$
$$= (-1)^j \Delta_m^j \{b_j[m-j, \varepsilon] W[n, \varepsilon; m-j]\} \qquad j = 0, 1, \ldots, h_1$$
$$\sum_{l=-\infty}^{\infty} W[n, \varepsilon; l] d_j[l, \varepsilon] \Delta_l^j \sigma[l-m, 0] \tag{10.30}$$
$$= (-1)^j \Delta_m^j \{d_j[m-j, \varepsilon] W[n, \varepsilon; m-j]\} \qquad j = 0, 1, \ldots, h_2$$

Substituting (10.30) into (10.29) we thus obtain relations coinciding with (10.28).

We have now shown that the weighting function of a sampled data system whose sampler generates pulses of relative duration $\gamma<1$ is determined by solving the corresponding non-homogeneous difference equations with zero initial conditions.

Without giving proofs we shall now extend the results to systems whose samplers generate pulses of relative duration $\gamma=1$, and also to systems with δ-pulse elements. It was shown above that the difference equations of systems either with δ-pulse elements or with samplers with $\gamma=1$ have the form

$$K_n(\Delta, n)\, y[n, \varepsilon] = H_n(\Delta, \bar{t})\, x[n, 0], \quad 0 \leqslant \varepsilon \leqslant 1 \tag{10.31}$$

where the difference operators $K_n(\Delta, n)$ and $H_n(\Delta, \bar{t})$ are determined by the relations

$$\begin{aligned} K_n(\Delta, n) &= \sum_{i=0}^{k} f_i[n]\, \Delta_n^i \\ H_n(\Delta, \bar{t}) &= \sum_{j=0}^{h} g_j[n, \varepsilon]\, \Delta_n^j \end{aligned} \tag{10.32}$$

From what has been said above, the weighting function $W_1[n, \varepsilon; m]$ of a system whose processes are determined by difference equations of the form (10.31) is given by the solution of the difference equation

$$\begin{gathered} K_n(\Delta, n)\, W_1[n, \varepsilon; m] = H_n(\Delta, \bar{t})\, \sigma[n-m, 0] \\ 0 \leqslant \varepsilon \leqslant 1 \end{gathered} \tag{10.33}$$

for zero initial conditions.

If we know the weighting function $W[n, \varepsilon; m]$ of a system whose difference equation has the form

$$K_n(\Delta, n)\, y[n, \varepsilon] = x[n, 0], \quad 0 \leqslant \varepsilon \leqslant 1 \tag{10.34}$$

then the weighting function $W_1[n, \varepsilon; m]$ can also be determined from

$$\begin{gathered} W_1[n, \varepsilon; m] = \sum_{j=0}^{h} (-1)^j \Delta_m^j \{W[n, \varepsilon; m-j]\, g_j[m-j, \varepsilon]\} \\ 0 \leqslant \varepsilon \leqslant 1 \end{gathered} \tag{10.35}$$

The results of this section remain valid also for stationary systems, which therefore need not be considered separately. We point out only that (10.28) for a stationary system can be written

$$\begin{gathered} W_1[n-m, \varepsilon] = \sum_{j=0}^{h_1} (-1)^j b_j(\varepsilon)\, \Delta_m^j\, W[n-m+j, \varepsilon] \\ 0 \leqslant \varepsilon \leqslant \gamma \end{gathered}$$

$$W_1[n-m, \varepsilon] = \sum_{j=0}^{h_2} (-1)^j d_j(\varepsilon) \Delta_m^j W[n-m+j, \varepsilon]$$
$$\gamma \leqslant \varepsilon \leqslant 1 \tag{10.36}$$

Putting $n-m=l$ in (10.36) we finally obtain

$$W_1[l, \varepsilon] = \sum_{j=0}^{h_1} b_j(\varepsilon) \Delta_l^j W[l, \varepsilon], \quad 0 \leqslant \varepsilon \leqslant \gamma$$
$$W_1[l, \varepsilon] = \sum_{j=0}^{h_2} d_j(\varepsilon) \Delta_l^j W[l, \varepsilon], \quad \gamma \leqslant \varepsilon \leqslant 1 \tag{10.37}$$

It has now been shown that the weighting function of a sampled data system constitutes the solution of the corresponding non-homogeneous difference equations with zero initial conditions. It will be shown later that this method of determining the weighting function is in fact the most convenient one. However, it can also be found by solving the corresponding homogeneous equations with non-zero initial conditions. We shall discuss later the method of discovering the appropriate initial conditions for which the weighting function can be thus determined. We consider the discussion of this cumbersome method advisable for two reasons. In the first place, we shall have an opportunity to convince ourselves once more of the fact that the σ-function must be treated in exactly the same way as the δ-function is treated when dealing with differential equations. In the second place - and this is the main point - the results have importance for the subsequent development of the theory of sampled data systems. In particular, using these results we shall define Green's function for difference equations, and by means of this establish a number of important properties of sampled data systems.

1.10.2 Determination from the homogeneous difference equations

We shall now give the method of determining the initial conditions for which the weighting can be found from the homogeneous difference equations. For brevity this method is expounded for a system having a sampler with $\gamma=1$ (or having a δ-pulse element). The results can easily be extended to a general sampled data system.

The problem determining the corresponding initial conditions can be formulated as follows. Given the difference equation of the system of the order k

$$L_n(\Delta, n) W[n, \varepsilon; m] = M_n(\Delta, \bar{t}) \sigma[n-m, 0], \ 0 \leqslant \varepsilon \leqslant 1 \tag{10.38}$$

and the initial conditions

$$\Delta_n^i W[n, \varepsilon; m]|_{n=m} = 0, \ i = 0, 1, \ldots, k-1 \quad 0 \leqslant \varepsilon \leqslant 1 \tag{10.39}$$

find the initial conditions

$$\Delta_n^i \overline{W}[n, \varepsilon; m]|_{n=m}, \quad i = 0, 1, \ldots, k-1 \quad 0 \leqslant \varepsilon \leqslant 1 \tag{10.40}$$

for which the weighting function constitutes the solution of the homogeneous difference equation

$$L_n(\Delta, n) \overline{W}[n, \varepsilon; m] = 0, \ 0 \leqslant \varepsilon \leqslant 1 \tag{10.41}$$

where the symbol $\overline{W}$ denotes the weighting function with non-zero initial conditions.

Since the difference operator $L_n(\Delta, n)$ is defined by

$$L_n(\Delta, n) = \sum_{i=0}^{k} a_i[n] \Delta_n^i \tag{10.42}$$

then for the solution of the problem we must find the differences

$$\Delta_n^i \overline{W}[n, \varepsilon; m]$$
$$(i = 0, 1, \ldots, k)$$

which constitute the generalised differences of the weighting function $\overline{W}[n, \varepsilon; m]$. Using (4.37) we find the expression for the generalised difference of the weighting function

$$\Delta_n^i \overline{W}[n, \varepsilon; m] = \Delta_n^i W[n, \varepsilon; m] - \sum_{l=0}^{i-1} \Delta_n^l \overline{W}[m, \varepsilon; m] \Delta_n^{i-l-1} \sigma[n-(m-1), 0] \tag{10.43}$$

We note that (10.43) can be found in another way. Denoting

the discrete Laplace transformation of the weighting function $W[n, \varepsilon; m]$ by

$$W^*(q, \varepsilon, m) = D\{W[n, \varepsilon; m]\} = \sum_{n=0}^{\infty} e^{-qn} W[n, \varepsilon; m] \quad (10.44)$$

we can find the transform (in the sense of the Laplace transformation) of the i-th difference of $\overline{W}[n, \varepsilon; m]$:

$$D\{\Delta_n^i \overline{W}[n, \varepsilon; m]\} = (e^q - 1)^i W^*(q, \varepsilon; m) - e^q \cdot e^{-qm} \times \sum_{l=0}^{i-1} \Delta_n^{i-l-1} \overline{W}[m, \varepsilon; m] (e^q - 1)^l \quad (10.45)$$

Passing on to the originals in (10.45) we have

$$\Delta_n^i \overline{W}[n, \varepsilon; m] = \Delta_n^i W[n, \varepsilon; m] - \sum_{l=0}^{i-1} \Delta_n^{i-l-1} \overline{W}[m, \varepsilon; m] \Delta_n^l \sigma[n - (m-1), 0] \quad (10.46)$$

since the transform of the differences of the displaced σ-function is

$$D\{\Delta_n^l \sigma[n - m, 0]\} = e^{-qm} (e^q - 1)^l \qquad l = 0, 1, \ldots \quad (10.47)$$

Equations (10.43) and (10.46) differ from each other only in the order of summation, and are transformed one into the other by changing the summation variable. Thus, denoting $i - l - 1 = \lambda$ (10.46) we obtain

$$\Delta_n^i \overline{W}[n, \varepsilon; m] = \Delta_n^i W[n, \varepsilon; m] - \sum_{\lambda=0}^{i-1} \Delta_n^\lambda \overline{W}[m, \varepsilon; m] \Delta_n^{i-\lambda-1} \sigma[n - (m-1), 0] \quad (10.48)$$

which coincides with (10.43).

Multiplying both sides of (10.46) by $a_i[n]$ and summing the products thus obtained with respect to i from 0 to k, we obtain, when (10.42) is taken into account,

$$L_n(\Delta, n) \overline{W}[n, \varepsilon; m] = L_n(\Delta, n) W[n, \varepsilon; m] - \sum_{i=1}^{k} \sum_{l=0}^{i-1} \Delta_n^{i-l-1} \overline{W}[m, \varepsilon; m] a_i[n] \Delta_n^l \sigma[n - (m-1), 0] \quad (10.49)$$

$$0 \leqslant \varepsilon \leqslant 1$$

Taking into consideration (10.38) and (10.41), we obtain from the last relation*

$$M_n(\Delta, \bar{t})\,\sigma[n-m, 0] = \sum_{i=1}^{k}\sum_{l=0}^{i-1} \Delta_n^{i-l-1} W[m, \varepsilon; m] \times a_i[n]\,\Delta_n^l\,\sigma[n-(m-1), 0], \quad 0 \leqslant \varepsilon \leqslant 1 \tag{10.50}$$

We shall now transform this equation.

We assume that the difference equation $M_n(\Delta, \bar{t})$ is defined by

$$M_n(\Delta, \bar{t}) = \sum_{j=0}^{h} b_j[n, \varepsilon]\,\Delta_n^j \tag{10.51}$$

where h is the order of the difference operator $M_n(\Delta, \bar{t})$, which can be equal to $h \leqslant k$. By the property of the σ-function expressed by

$$x[n, \varepsilon]\,\Delta_n^j\,\sigma[n-m, 0] = (-1)^j\,\Delta_m^j\{x[m-j, \varepsilon]\,\sigma[n-(m-j), 0]\} \tag{10.52}$$

we have for the left-hand side of (10.50)

$$\sum_{j=0}^{h} b_i[n, \varepsilon]\,\Delta_n^j\,\sigma[n-m, 0] = \sum_{j=0}^{h}(-1)^j\,\Delta_m^j\{b_j[m-j, \varepsilon]\,\sigma[n-(m-1), 0]\} \tag{10.53}$$

Applying the rule for taking the difference of the product of two functions we find for the right-hand side of (10.53):

$$\begin{aligned}
&\Delta_m^0\{b_0[m, \varepsilon]\,\sigma[n-m, 0]\} = b_0[m, \varepsilon]\,\sigma[n-m, 0] \\
&\Delta_m\{b_1[m-1, \varepsilon]\,\sigma[n-(m-1), 0]\} \\
&= b_1[m-1, \varepsilon]\,\Delta_m\sigma[n-(m-1), 0] \\
&+ \Delta_m b_1[m-1, \varepsilon]\,\sigma[n-(m-2), 0] \\
&\cdots\cdots\cdots\cdots\cdots\cdots \\
&\Delta_m^h\{b_h[m-h, \varepsilon]\,\sigma[n-(m-h), 0]\} \\
&= \sum_{\nu=0}^{h} C_h^{h-\nu}\Delta_m^\nu b_h[m-h, \varepsilon]\,\Delta_m^{h-\nu}\sigma[n-(m-h-\nu), 0]
\end{aligned} \tag{10.54}$$

* For convenience in writing we use $W[n, \varepsilon; m]$ instead of $\bar{W}[n, \varepsilon; m]$ in what follows.

Taking into consideration these relations, the right-hand side of (10.53) can be written

$$\sum_{j=0}^{h}(-1)^j \Delta_m^j \{b_j[m-j, \varepsilon]\, \sigma[n-(m-j), 0]\}$$

$$=\sum_{j=0}^{h}(-1)^j \sum_{\nu=0}^{j} C_j^{j-\nu} \Delta_m^\nu b_j[m-j, \varepsilon]\, \Delta_m^{j-\nu} \sigma[n-[m-j-\nu), 0] \tag{10.55}$$

Because of

$$\Delta_n^s \sigma[n-m, 0]=(-1)^s \Delta_m^s \sigma[n-(m-s), 0] \tag{10.56}$$

the right-hand side of (10.55) can be transformed thus:

$$\sum_{j=0}^{h}(-1)^j \sum_{\nu=0}^{j} C_j^{j-\nu} \Delta_m^\nu b_j[m-j, \varepsilon]\, \Delta_m^{j-\nu} \sigma[n-(m-j-\nu), 0]$$

$$=\sum_{j=0}^{h}(-1)^j \sum_{\nu=0}^{j} (-1)^{j-\nu} C_j^{j-\nu} \Delta_m^\nu b_j[m-j, \varepsilon]\, \Delta_m^{j-\nu} \sigma[n-m, 0] \tag{10.57}$$

Transforming the right-hand side of (10.57) we can write it in the following form:

$$\sum_{j=0}^{h}(-1)^j \sum_{\nu=0}^{j} C_j^{j-\nu} \Delta_m^\nu b_j[m-j, \varepsilon]\, \Delta_m^{j-\nu} \sigma[n-(m-j-\nu), 0]$$

$$=\sum_{j=0}^{h}\sum_{\nu=0}^{j}(-1)^\nu C_j^{j-\nu} \Delta_m^\nu b_j[m-j, \varepsilon]\, \Delta_n^{j-\nu} \sigma[n-m, 0]$$

When the last relation is taken into consideration the left-hand side of (10.50) can be written in the final form

$$\sum_{j=0}^{h} b_j[n, \varepsilon]\, \Delta_n^j \sigma[n-(m-1), 0]=\sum_{j=0}^{h}\sum_{\nu=0}^{j}(-1)^\nu C_j^{j-\nu} \times \Delta_m^\nu b_j[m-j, \varepsilon]\, \Delta_n^{j-\nu} \sigma[n-(m-1), 0] \tag{10.58}$$

We now transform the right-hand side of (10.50). Using (10.52) we can write

$$a_i[n]\, \Delta_n^l \sigma[n-(m-1), 0]$$

$$=(-1)^l \Delta_m^l \{a_i[m-l-1]\, \sigma[n-(m-l-1), 0]\} \tag{10.59}$$

$$l=0, 1, \ldots$$

Applying (10.59) to the right-hand side of (10.50) we obtain

$$\sum_{i=1}^{k}\sum_{l=0}^{i-1}\Delta_n^{i-l-1} W[m, \varepsilon; m]\, a_i[n]\, \Delta_n^{l}\, \sigma[n-(m-1), 0]$$

$$=\sum_{i=1}^{k}\sum_{l=0}^{i-1}\Delta_n^{i-l-1} W[m, \varepsilon; m](-1)^{l} \tag{10.60}$$

$$\times \Delta_m^{l}\{a_i[m-l-1)\sigma[n-(m-l-1), 0]\}$$

Using (10.54) we can easily establish that

$$\Delta_m^{l}\{a_i[m-l-1]\,\sigma[n-(m-l-1), 0]\}$$

$$=\sum_{\mu=0}^{l} C_l^{l-\mu}\Delta_m^{\mu}\, a_i[m-l-1]\, \Delta_m^{l-\mu}\sigma(n-(m-l-\mu-1), 0] \tag{10.61}$$

or since

$$\Delta_m^{l-\mu}\sigma[n-(m-l-\mu-1), 0]$$

$$=(-1)^{l-\mu}\Delta_n^{l-\mu}\sigma[n-(m-1), 0]$$

we have instead of (10.61)

$$\Delta_m^{l}\{a_i[m-l-1]\,\sigma[n-(m-l-1), 0]\}$$

$$=\sum_{\mu=0}^{l}(-1)^{l-\mu} C_l^{l-\mu}\Delta_m^{\mu}\, a_i[m-l-1]\, \Delta_n^{l-\mu}\sigma[n-(m-1), 0] \tag{10.62}$$

Applying then (10.62) to the right-hand side of (10.60), we find that

$$\sum_{i=1}^{k}\sum_{l=0}^{i-1}\Delta_n^{i-l-1} W[m, \varepsilon; m](-1)^{l}$$

$$\times \Delta_m^{l}\{a_i[m-l-1]\,\sigma[n-(m-l-1), 0]\}$$

$$=\sum_{i=1}^{k}\sum_{l=0}^{i-1}(-1)^{l}\Delta_n^{i-l-1} W[m, \varepsilon; m]\sum_{\mu=0}^{l}(-1)^{l-\mu} C_l^{l-\mu}$$

$$\times \Delta_m^{\mu}\, a_i[m-l-1]\, \Delta_n^{l-\mu}\sigma[n-(m-1), 0]$$

Transforming the right-hand side of the last relation we

finally write (10.60) in the form

$$\sum_{i=1}^{k}\sum_{l=0}^{i-1}\Delta_n^{i-l-1}W[m, \varepsilon; m]\,a_i[n]\,\Delta_n^l\sigma[n-(m-1), 0]$$

$$=\sum_{i=1}^{k}\sum_{l=0}^{i-1}\Delta_n^{i-l-1}W[m, \varepsilon; m]\sum_{\mu=0}^{l}(-1)^{\mu}C_l^{l-\mu} \qquad (10.63)$$

$$\times\Delta_m^{\mu}\,a_i[m-l-1]\,\Delta_n^{l-\mu}\sigma[n-(m-1), 0]$$

Thus when (10.58) and (10.63) are taken into account the original equation (10.50) can be written

$$\sum_{j=0}^{h}\sum_{\nu=0}^{j}(-1)^{\nu}C_j^{j-\nu}\Delta_m^{\nu}b_j[m-j, \varepsilon]\,\Delta_n^{j-\nu}\sigma[n-(m-1), 0]$$

$$=\sum_{i=1}^{k}\sum_{l=0}^{i-1}\Delta_n^{i-l-1}W[m, \varepsilon; m]\sum_{\mu=0}^{l}(-1)^{\mu}C_l^{l-\mu} \qquad (10.64)$$

$$\times\Delta_m^{\mu}\,a_i[m-l-1]\,\Delta_n^{l-\mu}\sigma[n-(m-1), 0]$$

$$0\leqslant\varepsilon\leqslant 1$$

Equation (10.64) enables us to obtain a system of algebraic equations for the sought-after values $\Delta_n^r W[n, \varepsilon; m]_{n=m}$, $r=0,1, \dots, k-1$ from the known coefficients of the difference equation of the system. This system of equations can be obtained by equating the terms with the same order differences of the σ-function in both sides of (10.64). As a rule, the process of obtaining the corresponding systems of equations in closed form is very laborious and cumbersome; this is due to the method of direct choice used to obtain the recurrence relations. We shall now give a much simpler method.

Since in (10.64) we must form the coefficients with the same order of differences of the σ-function, then, specifying this order for the left-hand side of (10.64) by introducing a new summation variable, $j-\nu=r$, $r=0, 1, \dots, h$, we obtain

$$\sum_{j=0}^{h}\sum_{\nu=0}^{j}(-1)^{\nu}C_j^{j-\nu}\Delta_m^{\nu}b_j[m-j, \varepsilon]\,\Delta_n^{j-\nu}\sigma[n-(m-1), 0]$$

$$=\sum_{j=0}^{h}\sum_{j-r=0}^{j}(-1)^{j-r}C_j^r\,\Delta_m^{j-r}b_j[m-j, \varepsilon]\,\Delta_n^r\,\sigma[n-(m-1), 0]$$

$$r=0, 1, \dots, h \qquad (10.65)$$

Since for a fixed value of r the relation

$$\sum_{j=0}^{h}\sum_{j-r=0}^{j}=\sum_{j=r}^{h},\quad r=0,\ 1,\ \dots,\ h \tag{10.66}$$

is valid, the right-hand side of (10.65) can be written in the form

$$\sum_{j=r}^{h}(-1)^{j-r}C_j^r\,\Delta_m^{j-r}b_j\,[m-j,\ \varepsilon]\,\Delta_n^r\,\sigma\,[n-(m-1),\ 0] \tag{10.67}$$

whence it follows that when $\Delta_n^r\,\sigma[n-(m-1),\ 0]$, we have the coefficient

$$\beta_r=\sum_{j=r}^{h}(-1)^{j-r}C_j^r\,\Delta_m^{j-r}b_j\,[m-j,\ \varepsilon] \tag{10.68}$$

We analogously transform the right-hand side of (10.64). Introducing a new summation variable $l-\mu=\lambda,\ \lambda=0,\ 1,\ \dots,\ k-1$, we have for the right-hand side, from (10.64),

$$\begin{aligned}&\sum_{i=1}^{k}\sum_{l=0}^{i-1}\sum_{l-\lambda=0}^{l}(-1)^{l-\lambda}\Delta_n^{i-l-1}W\,[m,\ \varepsilon;\ m]\,C_l^\lambda\\&\times\Delta_m^{l-\lambda}a_i\,[m-l-1]\,\Delta_n^\lambda\sigma[n-(m-1),\ 0]\end{aligned} \tag{10.69}$$

But since for a fixed value $\lambda=0,1,\dots,\ k-1$,

$$\sum_{i=1}^{k}\sum_{l=0}^{i-1}\sum_{l-\lambda=0}^{l}=\sum_{i=1}^{k}\sum_{l=\lambda}^{i-1} \tag{10.70}$$

we finally obtain for the right-hand side of (10.64)

$$\begin{aligned}&\sum_{i=1}^{k}\sum_{l=\lambda}^{i-1}(-1)^{l-\lambda}\Delta_n^{i-l-1}W\,[m,\ \varepsilon;\ m]\,C_l^\lambda\\&\times\Delta_m^{l-\lambda}a_i\,[m-l-1]\,\Delta_n^\lambda\sigma\,[n-(m-1),\ 0]\end{aligned} \tag{10.71}$$

It follows from (10.71) that the coefficients for $\Delta_n^\lambda\sigma\,[n-(m-1),\ 0]$ are determined by

$$\alpha_\lambda=\sum_{i=1}^{k}\sum_{l=\lambda}^{i-1}(-1)^{l-\lambda}\,\Delta_n^{i-l-1}W\,[m,\ \varepsilon;\ m]\,C_l^\lambda\Delta_m^{l-\lambda}a_i\,[m-l-1] \tag{10.72}$$

Thus (10.64) can finally be written as follows, when (10.67) and (10.71) are taken into account,

$$\sum_{j=r}^{h}(-1)^{j-r}C_j^r\Delta_m^{j-r}b_j[m-j,\varepsilon]\Delta_n^r\sigma[n-(m-1),0]$$
$$=\sum_{i=1}^{k}\sum_{l=\lambda}^{i-1}(-1)^{l-\lambda}\Delta_n^{i-l-1}W[m,\varepsilon;m]C_l^\lambda \tag{10.73}$$
$$\times\Delta_m^{l-\lambda}a_i[m-l-1]\Delta_n^\lambda\sigma[n-(m-1),0]$$
$$r=0,1,\dots,h;\ \lambda=0,1,\dots,k-1,\ 0\leqslant\varepsilon\leqslant 1$$

Now equating the coefficients β_r and α_λ with the same numbers in (10.73) we have

$$\sum_{i=1}^{k}\sum_{l=\lambda}^{i-1}(-1)^{l-\lambda}\Delta_n^{i-l-1}W[m,\varepsilon;m]C_l^\lambda\Delta_m^{l-\lambda}a_i[m-l-1]$$
$$=\sum_{j=\lambda}^{h}(-1)^{j-\lambda}C_j^\lambda\Delta_m^{j-\lambda}b_i[m-j,\varepsilon]$$
$$\lambda=0,1,\dots,h \tag{10.74}$$

$$\sum_{i=1}^{k}\sum_{l=\nu}^{i-1}(-1)^{l-\nu}\Delta_n^{i-l-1}W[m,\varepsilon;m]C_l^\nu\Delta_m^{l-\nu}a_i[m-l-1]=0$$
$$\nu=h+1,\ h+2,\dots,k-1,\ 0\leqslant\varepsilon\leqslant 1$$

The system of equations (10.74) thus obtained is the one required. The solution of this system determines the corresponding initial conditions $\Delta_n^r W[n,\varepsilon;m]|_{n=m}$ ($r=0,1,\dots,k-1$), for which the weighting function of the system in question can be found as the solution of the homogeneous difference equation

$$L_n(\Delta,n)W[n,\varepsilon;m]=0,\quad 0\leqslant\varepsilon\leqslant 1 \tag{10.75}$$

Putting $i=h+2$ and $\nu=h+1$ in the second group of equations (10.74) we find

$$i-l-1=0 \tag{10.76}$$

When $i=k$ and $\nu=h+1$ we have

$$i-l-1=k-h-2 \tag{10.77}$$

Thus from (10.74), (10.76) and (10.77) it follows that for the system in question

$$\Delta_n^\mu W[n,\varepsilon;m]|_{n=m}=0,\quad 0\leqslant\varepsilon\leqslant 1$$
$$\mu=0,1,\dots,k-h-2 \tag{10.78}$$

All the subsequent differences of higher orders of the weighting function are non-zero at the point $n=m$, whilst the difference of order $k-h-1$, the lowest of the non-zero differences, equals

$$\Delta_n^{k-h-1} W[n, \varepsilon; m]|_{n=m} = \frac{b_h[m-h, \varepsilon]}{a_k[m-h]}, \quad 0 \leqslant \varepsilon \leqslant 1 \quad (10.79)$$

All differences of the weighting function of order higher than $(k-h-1)$ can be determined from the system of equations (10.74).

The results so obtained can easily be extended to a system of general form, processes in which are determined by two difference equations for the two time intervals $0 \leqslant \varepsilon \leqslant \gamma$ and $\gamma \leqslant \varepsilon \leqslant 1$ respectively. Let the processes in a system be determined by the difference equations

$$\begin{aligned} \sum_{i=0}^{k} a_i[n] \Delta_n^i y[n, \varepsilon] &= \sum_{i=0}^{h_1} b_j[n, \varepsilon] \Delta_n^j x[n, 0], \quad 0 \leqslant \varepsilon \leqslant \gamma \\ \sum_{i=0}^{k} c_i[n] \Delta_n^i y[n, \varepsilon] &= \sum_{j=0}^{h_2} d_j[n, \varepsilon] \Delta_n^j x[n, 0], \quad \gamma \leqslant \varepsilon \leqslant 1 \end{aligned} \quad (10.80)$$

where $h_1 \leqslant k$, $h_2 \leqslant k$.

The weighting function $W_1[n, \varepsilon; m]$ of such a system is determined in accordance with (10.80) by the solution of the homogeneous difference equations

$$\begin{aligned} L_n(\Delta, n) W_1[n, \varepsilon; m] &= 0, \quad 0 \leqslant \varepsilon \leqslant \gamma \\ S_n(\Delta, n) W_1[n, \varepsilon; m] &= 0, \quad \gamma \leqslant \varepsilon \leqslant 1 \end{aligned} \quad (10.81)$$

(the operators $L_n(\Delta, n)$ and $S_n(\Delta, n)$ are determined by Equations (10.80) with initial conditions which can be found from the following system of equations for the time interval $0 \leqslant \varepsilon \leqslant \gamma$:

$$\begin{aligned} \sum_{i=1}^{k} \sum_{i=\lambda}^{i-1} (-1)^{l-\lambda} \Delta_n^{i-l-1} W_1[m, \varepsilon; m] C_l^\lambda \Delta_m^{l-\lambda} a_i[m-l-1] \\ = \sum_{j=\lambda}^{h_1} (-1)^{j-\lambda} C_j^\lambda \Delta_m^{j-\lambda} b_j[m-j, \varepsilon] \\ \lambda = 0, 1, 2, \ldots, h_1 \\ \Delta_n^r W_1[n, \varepsilon; m]|_{n=m} = 0, \quad r = 0, 1, \ldots k - h_1 - 2 \\ 0 \leqslant \varepsilon \leqslant \gamma \end{aligned} \quad (10.82)$$

For the interval $\gamma \leqslant \varepsilon \leqslant 1$ the initial conditions $\Delta_n^r W[n, \varepsilon; m]|_{n=m}$ $(r=0,1,\ldots,k-1)$ are determined from the system of equations

$$\sum_{i=1}^{k}\sum_{l=\lambda}^{i-1}(-1)^{l-\lambda}\Delta_n^{i-l-1}W[m, \varepsilon;\ m]C_l^\lambda\Delta_m^{l-\lambda}c_i[m-l-1]$$
$$=\sum_{j=\lambda}^{h_2}(-1)^{j-\lambda}C_j^\lambda\ \Delta_m^{j-\lambda}d_j[m-j,\ \varepsilon] \tag{10.83}$$
$$\lambda=0,\ 1,\ 2,\ \ldots,\ h_2$$
$$\Delta_n^r W[n, \varepsilon;\ m]|_{n=m}=0,\ \ r=0,\ 1,\ 2,\ \ldots,\ k-h-2$$
$$\gamma\leqslant\varepsilon\leqslant 1$$

In the cases considered above, the right-hand side of the difference equations contained the differences $\Delta_n^i x[n, 0]$ of the input disturbance.

We now consider a particular case when the sampled data system has an equation of the form

$$\mathrm{K}_n(\Delta, n)\,y[n, \varepsilon]=b[n,\ \varepsilon]\,x[n, 0],\quad 0\leqslant\varepsilon\leqslant 1$$
$$\mathrm{K}_n(\Delta,\ n)=\sum_{i=0}^{k}a_i[n]\,\Delta_n^i \tag{10.84}$$

In accordance with the foregoing, the weighting function of this system is defined from the homogeneous difference equation

$$\mathrm{K}_n(\Delta,\ n)\,W[n,\ \varepsilon;\ m]=0,\ \ 0\leqslant\varepsilon\leqslant 1 \tag{10.85}$$

with initial conditions defined by

$$\Delta_n^\mu W[n,\ \varepsilon;\ m]|_{n=m}=0,\quad \mu=0,1,\ \ldots,\ k-2$$
$$\Delta_n^{k-1}W[n,\ \varepsilon;\ m]|_{n=m}=\frac{b[m,\ \varepsilon]}{a_k[m]},\quad 0\leqslant\varepsilon\leqslant 1 \tag{10.86}$$

These results remain valid, of course, for stationary systems. Without giving proofs we formulate these results for a stationary system whose difference equations take the form

$$\sum_{i=0}^{k}a_i\Delta_n^i y[n,\ \varepsilon]=\sum_{j=0}^{h_1}b_j(\varepsilon)\,\Delta_n^j x[n,\ 0],\ \ 0\leqslant\varepsilon\leqslant\gamma$$
$$\sum_{i=0}^{k}c_i\Delta_n^i y[n,\ \varepsilon]=\sum_{j=0}^{h_2}d_j(\varepsilon)\,\Delta_n^j x[n, 0],\quad \gamma\leqslant\varepsilon\leqslant 1 \tag{10.87}$$

The weighting function of such a system can be determined from the corresponding homogeneous equations. The initial conditions of this problem are determined from the following system of algebraic equations for the time interval $0 \leqslant \varepsilon \leqslant \gamma$:

$$\sum_{i=1}^{k} a_i \Delta_n^{i-\lambda-1} W[n-m, \varepsilon]\,|_{n=m} = b_\lambda(\varepsilon)$$
$$\lambda = 0, 1, \ldots, h_1 \tag{10.88}$$
$$\Delta_n^r W[n-m, \varepsilon]\,|_{n=m} = 0, \quad r = 0, 1, \ldots, k-h_1-2$$
$$0 \leqslant \varepsilon \leqslant \gamma$$

For the interval $\gamma \leqslant \varepsilon \leqslant 1$ the initial conditions are determined from

$$\sum_{i=1}^{k} c_i \Delta_n^{i-\lambda-1} W[n-m, \varepsilon]\,|_{n=m} = d_\lambda(\varepsilon)$$
$$\lambda = 0, 1, \ldots, h_2 \tag{10.89}$$
$$\Delta_n^r W[n-m, \varepsilon]\,|_{n=m} = 0, \; r = 0, 1, \ldots, k-h_2-2$$
$$\gamma \leqslant \varepsilon \leqslant 1$$

1.10.3 Analogy between continuous and sampled data systems

From the definition of the weighting function of a sampled data system as the response of this system to the σ-function, we gave two methods for finding the weighting function from the difference equations of the system. We shall now summarise the results obtained, assuming that the system is described by a single difference equation of the form

$$L_n(\Delta, n)\, y[n, \varepsilon] = M_n(\Delta, \overline{t})\, x[n, 0], \; 0 \leqslant \varepsilon \leqslant 1 \tag{10.90}$$

where

$$L_n(\Delta, n) = \sum_{i=0}^{k} a_i[n]\, \Delta_n^i$$

$$M_n(\Delta, \overline{t}) = \sum_{j=0}^{h} b_j[n, \varepsilon]\, \Delta_n^j$$

As shown in Section 1.10.1 the weighting function of such a system can be found from the non-homogeneous difference equation (10.90), whose right-hand side contains the σ-function

$$L_n(\Delta, n) W[n, \varepsilon; m] = M_n(\Delta, \overline{t})\, \sigma[n-m, 0] \qquad 0 \leqslant \varepsilon \leqslant 1 \tag{10.91}$$

and whose initial conditions are

$$\Delta_n^i W[n, \varepsilon; m]\,|_{n=m} = 0, \; i = 0, 1, \ldots, k-1 \qquad 0 \leqslant \varepsilon \leqslant 1 \tag{10.92}$$

Here we have a complete analogy with continuous systems, processes which are determined by linear differential equations. Thus, let the differential equation of a continuous system be

$$L_t(p, t)\, y(t) = M_t(p, t)\, x(t), \quad p = \frac{d}{dt}$$

$$L_t(p, t) = \sum_{i=0}^{k} a_i(t) \frac{d^i}{dt^i} \tag{10.93}$$

$$M_t(p, t) = \sum_{j=0}^{h} b_j(t) \frac{d^j}{dt^j}$$

As is known [13, 23], the weighting function $W(t; \tau)$ of such a system is determined from the non-homogeneous differential equation (10.93), whose right-hand side contains the δ-function

$$L_t(p, t)\, W(t; \tau) = M_t(p, t)\, \delta(t-\tau) \tag{10.94}$$

and whose initial conditions are

$$\left. \frac{\partial^i W(t; \tau)}{\partial t^i} \right|_{t=\tau} = 0, \; i = 0, 1, \ldots, k-1 \tag{10.95}$$

Bearing in mind then that the weighting function of a sampled data system is the response of this system to the σ-function, while the weighting function of a continuous system is its response to the δ-function, we note the corresponding analogy between continuous and sampled data systems by comparing (10.90) to (10.92) with (10.93) to (10.95).

As was shown in Section 1.10.2, the weighting function of the system, determined by (10.90), can be found also from the corresponding homogeneous difference equation with non-zero initial conditions:

$$L_n(\Delta,\ n)\,W[n,\ \varepsilon;\ m]=0,\ 0\leqslant\varepsilon\leqslant 1 \tag{10.96}$$

The initial conditions of this problem are determined by solving a system of algebraic equations of the following form:

$$\sum_{i=1}^{k}\sum_{l=\lambda}^{i-1}(-1)^{l-\lambda}\,\Delta_n^{i-l-1}W[m,\ \varepsilon;\ m]\,C_l^{\lambda}\,\Delta_m^{l-\lambda}a_i[m-l-1]$$
$$=\sum_{j=\nu}^{h}(-1)^{j-\nu}C_j^{\nu}\Delta_m^{j-\nu}b_j[m-j,\ \varepsilon] \tag{10.97}$$
$$\lambda=0,\ 1,\ldots,\ h,\ \nu=0,\ 1,\ldots,\ h,\ 0\leqslant\varepsilon\leqslant 1.$$

The weighting function of such a system, as has been shown, has the properties

$$\Delta_n^r W[n,\ \varepsilon;\ m]\,|_{n=m}=0,\ r=0,\ 1,\ldots,\ k-h-2,\ \ 0\leqslant\varepsilon\leqslant 1 \tag{10.98}$$

while the lowest of the non-zero differences has a finite jump at $n=m$, which equals the ratio of the higher coefficients

$$\Delta_n^{k-h-1}W[n,\ \varepsilon;\ m]\,|_{n=m}=\frac{b_h[m-h,\ \varepsilon]}{a_k[m-h]},\ \ 0\leqslant\varepsilon\leqslant 1 \tag{10.99}$$

Here also an analogy can be seen with continuous systems. Thus, according to (10.93) the weighting function of a continuous system can be found as the solution of the homogeneous differential equation

$$L_t(p,\ t)\ W(t;\ \tau)=0 \tag{10.100}$$

with the initial conditions determined from the system of algebraic equations [13]

$$\sum_{r=0}^{k-\mu-1}W_t^{(r)}(\tau;\ \tau)\sum_{\lambda=0}^{k-\mu-r-1}(-1)^{\lambda}C_{\lambda+\mu}^{\mu}\,a_{\lambda+\mu+r+1}^{(\lambda)}(\tau)$$
$$=\sum_{\nu=\mu}^{h}(-1)^{\nu-\mu}C_{\nu}^{\mu}\,b_{\nu}^{(\nu-\mu)}(\tau),\ \mu=0,\ 1,\ldots,\ h \tag{10.101}$$

The weighting function $W(t;\ \tau)$ of a continuous system has the properties

$$\left.\frac{\partial^r W(t;\ \tau)}{\partial t^r}\right|_{t=\tau} = 0,\ r = 0,\ 1, \ldots,\ k-h-2 \quad (10.102)$$

whilst the lowest non-zero derivative for $t=\tau$ has a jump, equal to the ratio of the higher coefficients

$$\left.\frac{\partial^{k-h-1} W(t;\ \tau)}{\partial t^{k-h-1}}\right|_{t=\tau} = \frac{b_h(\tau)}{a_k(\tau)} \quad (10.103)$$

Comparing (10.96) to (10.99) with (10.100) to (10.103) we note the similarity in the properties of the weighting functions of continuous and sampled data systems.

Assuming that $M_n(\Delta, \bar{t}) = b[n, \varepsilon]$, in (10.90), i.e. the sampled data system is described by an equation of the form

$$L_n(\Delta,\ n)\, y[n,\ \varepsilon] = b[n,\ \varepsilon]\, x[n,\ 0],\ 0 \leqslant \varepsilon \leqslant 1 \quad (10.104)$$

we give the physical interpretation of the jump y of the difference of the weighting function $W[n,\ \varepsilon;\ m]$. Figure 1.21 shows a digital model of a system whose difference equation

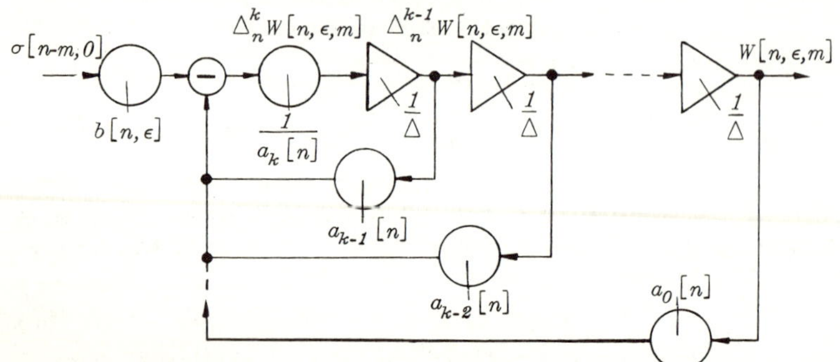

Fig. 1.21 The system described by (10.104)

is (10.104). If the system is at rest before the instant $t = m$, then the difference of the weighting function, when the σ-function acts at its input, is

$$\Delta_n^k W[n,\ \epsilon;\ m]\,|_{n=m} = \frac{b[n,\ \varepsilon]}{a_k[n]}\, \sigma[n-m,\ 0] \quad (10.105)$$

But by the definition of a sum

$$\Delta_n^{k-1} W[n, \varepsilon; m]|_{n=m} = \frac{b[m, \varepsilon]}{a_k[m]} \tag{10.106}$$

i.e. the difference of order $k-1$ of the weighting function undergoes a jump, as follows from (10.99) for $h=0$.

This shows that in determining the weighting function of a sampled data system with the help of a digital model, actions in the form of a σ-function and its differences can be replaced by the corresponding initial conditions in the summators.

1.10.4 Worked examples

We shall consider a number of examples of determining the weighting function of the simplest sampled data systems.

Example 1. Find the weighting function of a sampled data system whose difference equations have the form

$$\begin{aligned} \Delta_n y[n, \varepsilon] + (1 - e^{-\beta}) y[n, \varepsilon] &= (1 - e^{-\beta\varepsilon}) \Delta_n x[n, 0] \\ &+ \{(1 - e^{-\beta}) + e^{-\beta\varepsilon}(e^{-\beta(1-\gamma)} - 1)\} x[n, 0] \\ &0 \leqslant \varepsilon \leqslant \gamma \\ \Delta_n y[n, \varepsilon] + (1 - e^{-\beta}) y[n, \varepsilon] &= (e^{\beta\gamma} - 1) e^{-\beta\varepsilon} \Delta_n x[n, 0] \\ &+ (e^{\beta\gamma} - 1) e^{-\beta\varepsilon} x[n, 0] \\ &\gamma \leqslant \varepsilon \leqslant 1 \end{aligned} \tag{10.107}$$

where $y[n, \varepsilon]$ and $x[n, 0]$ are the output and input variables of the system respectively.

Equations (10.107) determine the processes in the system, which consists of an inertia element and a sampler generating pulses with duration $\gamma < 1$ and gain factor $k_s = 1$. The diagram of this system is shown in Fig. 1.22. The weighting function of the system in question is determined

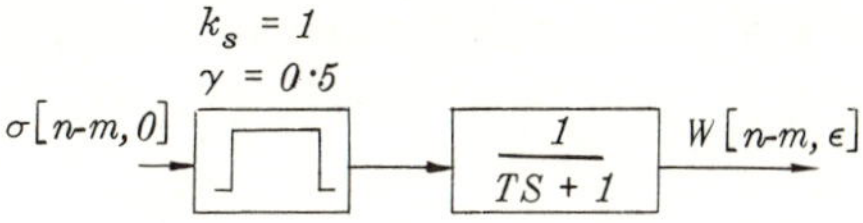

Fig. 1.22 The system described by (10.107)

by the solution of the non-homogeneous difference equations

$$\Delta_n W[n-m, \varepsilon] + a_0 W[n-m, \varepsilon] = b_1(\varepsilon)\Delta_n \sigma[n-m, 0] + b_0(\varepsilon)\sigma[n-m, 0], \quad 0 \leqslant \varepsilon \leqslant \gamma$$

$$\Delta_n W[n-m, \varepsilon] + c_0 W[n-m, \varepsilon] = d_1(\varepsilon)\Delta_n \sigma[n-m, 0] + d_0(\varepsilon)\sigma[n-m, 0], \quad \gamma \leqslant \varepsilon \leqslant 1 \tag{10.108}$$

with zero initial conditions. In (10.108)

$$\begin{gathered} a_0 = c_0 = (1 - e^{-\beta}), \; a_1 = c_1 = 1 \\ b_0(\varepsilon) = 1 + e^{-\beta} + e^{-\beta\varepsilon}(e^{-\beta(1-\gamma)} - 1) \\ b_1(\varepsilon) = 1 - e^{-\beta\varepsilon} \\ d_0(\varepsilon) = d_1(\varepsilon) = (e^{\beta\gamma} - 1)e^{-\beta\varepsilon}, \quad \beta = \frac{T_0}{T} \end{gathered} \tag{10.109}$$

where T_0 is the sampling period. Putting $n-m=l$ in (10.108) we write these equations in more convenient form (with the independent variable denoted as before)

$$\begin{gathered} W[n+1, \varepsilon] + (a_0 - 1)W[n, \varepsilon] = b_1(\varepsilon)\sigma[n+1, 0] + (b_0(\varepsilon) - b_1(\varepsilon))\sigma[n, 0], \; 0 \leqslant \varepsilon \leqslant \gamma \\ W[n+1, \varepsilon] + (c_0 - 1)W[n, \varepsilon] = d_1(\varepsilon)\sigma[n+1, 0] + (d_0(\varepsilon) - d_1(\varepsilon))\sigma[n, 0], \quad \gamma \leqslant \varepsilon \leqslant 1 \end{gathered} \tag{10.110}$$

Introducing a new variable $n+1=k$ in (10.110) we obtain the equations

$$\begin{gathered} W[k, \varepsilon] + (a_0 - 1)W[k-1, \varepsilon] = b_1(\varepsilon)\sigma[k, 0] + (b_0(\varepsilon) - b_1(\varepsilon))\sigma[k-1, 0], \quad 0 \leqslant \varepsilon \leqslant \gamma \\ W[k, \varepsilon] + (c_0 - 1)W[k-1, \varepsilon] = d_1(\varepsilon)\sigma[k, 0] + (d_0(\varepsilon) - d_1(\varepsilon))\sigma[k-1, 0], \quad \gamma \leqslant \varepsilon \leqslant 1 \end{gathered} \tag{10.111}$$

whence it follows that when $k=0$

$$W[-1, \varepsilon] = 0, \quad \sigma[-1, 0] = 0$$

therefore

$$W[0, \varepsilon] = \begin{cases} b_1(\varepsilon), & 0 \leqslant \varepsilon \leqslant \gamma \\ d_1(\varepsilon), & \gamma \leqslant \varepsilon \leqslant 1 \end{cases} \tag{10.112}$$

For the time interval in question ($k = 1$) we obtain

$$W[1, \varepsilon] = \begin{cases} (1 - a_0) W[0, \varepsilon] + [b_0(\varepsilon) - b_1(\varepsilon)] \\ \qquad 0 \leqslant \varepsilon \leqslant \gamma \\ (1 - c_0) W[0, \varepsilon] + [d_0(\varepsilon) - d_1(\varepsilon)] \\ \qquad \gamma \leqslant \varepsilon \leqslant 1 \end{cases} \tag{10.113}$$

since $\sigma[1,0] = 0$. The values of the weighting function for $k \geqslant 2$ are determined from the recurrent formula

$$W[k, \varepsilon] = a_0 W[k - 1, \varepsilon], \quad k \geqslant 2, \quad 0 \leqslant \varepsilon \leqslant 1 \tag{10.114}$$

Taking into account the values of the coefficient (10.109) we can finally write, on the basis of (10.112),

$$W[0, \varepsilon] = \begin{cases} 1 - e^{-\beta\varepsilon}, & 0 \leqslant \varepsilon \leqslant \gamma \\ (1 - e^{-\beta\gamma}) e^{-\beta\varepsilon}, & \gamma \leqslant \varepsilon \leqslant 1 \end{cases} \tag{10.115}$$

Analogously, from (10.113) we have

$$W[1, \varepsilon] = \begin{cases} e^{-\beta}(1 - e^{-\beta\varepsilon}) + e^{-\beta}(1 + e^{-\beta(\varepsilon - \gamma)}) \\ \qquad 0 \leqslant \varepsilon \leqslant \gamma \\ e^{-\beta}(1 - e^{-\beta\gamma}) e^{-\beta(\varepsilon - \gamma)} \\ \qquad \gamma \leqslant \varepsilon \leqslant 1 \end{cases} \tag{10.116}$$

Finally, from (10.114) we obtain

$$W[k, \varepsilon] = e^{-\beta} W[k - 1, \varepsilon], \; k \geqslant 2, \; 0 \leqslant \varepsilon \leqslant 1 \tag{10.117}$$

The weighting function of the system under consideration, calculated from (10.115) to (10.117), is shown in Fig. 1.23. It was assumed in the calculations that $\gamma = 0{,}5$, $\beta = 1$.

In the case in question the difference operators $L_n(\Delta)$ and $M_n(\Delta, \varepsilon)$, $S_n(\Delta)$ and $G_n(\Delta, \varepsilon)$ are of the same order, equal to unity ($k = h_1 = h_2 = 1$). Therefore in accordance with (10.82) and (10.83) the zero-order differences of the weighting function have jump discontinuities equal to

$$\begin{aligned} W[n - m, \varepsilon]|_{n=m} &= \frac{b_1(\varepsilon)}{1} = (1 - e^{-\beta\varepsilon}), \quad 0 \leqslant \varepsilon \leqslant \gamma \\ W[n - m, \varepsilon]|_{n=m} &= \frac{d_1(\varepsilon)}{1} = (1 - e^{-\beta\gamma}) e^{-\beta\varepsilon}, \quad \gamma \leqslant \varepsilon \leqslant 1 \end{aligned} \tag{10.118}$$

Comparing this with (10.115) we find that these are identical

Example 2. Find the weighting function of a sampled data system whose processes are determined by the difference equation

$$\Delta_n y[n, \varepsilon] + (1 - e^{-\beta}) y[n, \varepsilon] = (1 - e^{-\beta\varepsilon}) \Delta_n x[n, 0] + (1 - e^{-\beta}) x[n, 0], \quad 0 \leqslant \varepsilon \leqslant 1 \tag{10.119}$$

Equation (10.119) corresponds to a system consisting of a sampler with $\gamma = 1$ and an inertia element. The weighting

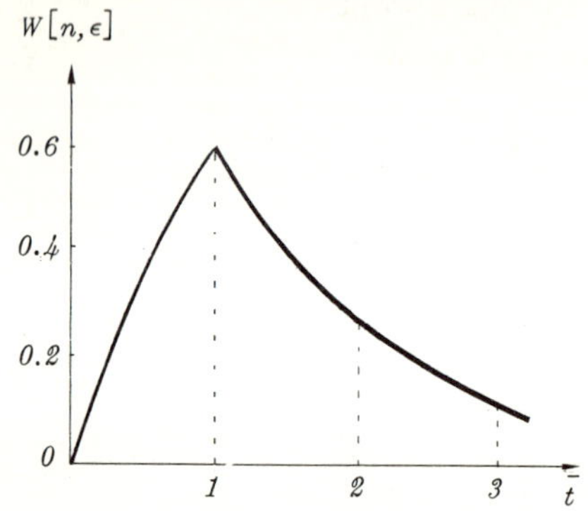

Fig. 1.23 The weighting function of the system shown in Fig. 1.22, for $\gamma = 0.5$ and $\beta = 1$

function of the system under consideration can be determined as the solution of (10.119) for $x[n, 0] = \sigma[n, 0]$ and zero initial conditions. Replacing the differences in (10.119) by the corresponding displaced functions, we obtain

$$W[n+1, \varepsilon] - e^{-\beta} W[n, \varepsilon] = (1 - e^{-\beta\varepsilon}) \sigma[n+1, 0] + (e^{-\beta\varepsilon} - e^{-\beta}) \sigma[n, 0], \quad 0 \leqslant \varepsilon \leqslant 1 \tag{10.120}$$

Introducing a new variable $n + 1 = k$ in (10.120) we have

$$W[k, \varepsilon] = e^{-\beta} W[k-1, \varepsilon] + (1 - e^{-\beta\varepsilon}) \sigma[k, 0] + (e^{-\beta\varepsilon} - e^{-\beta}) \sigma[k-1, 0], \quad 0 \leqslant \varepsilon \leqslant 1 \tag{10.121}$$

From (10.121) it follows directly, when the property of the σ-function is taken into account, that

$$W[k, \varepsilon] = \begin{cases} 1 - e^{-\beta\varepsilon}, & k = 0 \\ e^{-\beta}(1 - e^{-\beta\varepsilon}) + (e^{-\beta\varepsilon} - e^{-\beta}), & k = 1 \\ e^{-\beta} W[k-1, \varepsilon], & k \geqslant 2, \ 0 \leqslant \varepsilon \leqslant 1 \end{cases} \tag{10.122}$$

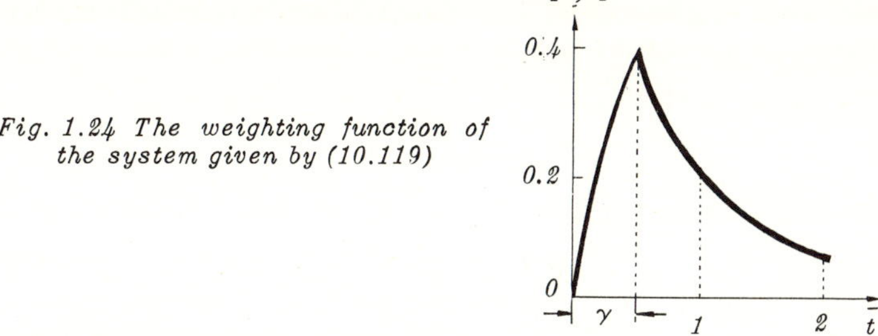

Fig. 1.24 The weighting function of the system given by (10.119)

The weighting function calculated from (10.122) for $\beta=1$ is shown in Fig. 1.24.

In the case in question the operators $L_n(\Delta)$ and $M_n(\Delta, \varepsilon)$ are of the same order, equal to unity ($k=h=1$); therefore in accordance with (10.79) the zero-order difference of the weighting function for $n=m$ has the jump

$$W[n-m, \varepsilon]\,|_{n=m}=\frac{b_1(\varepsilon)}{a_1}=\frac{1-e^{-\beta\varepsilon}}{1}=1-e^{-\beta\varepsilon}$$
$$0\leqslant\varepsilon\leqslant 1 \qquad (10.123)$$

which also coincides with (10.122).

1.11 PROPERTIES OF WEIGHTING FUNCTIONS

We shall now prove a number of theorems which characterise the properties of the weighting functions of sampled data systems. These theorems will be formulated for the case where the processes in a system are determined by a single difference equation.

Theorem 1. If a sampled data system is described by the difference equation of order k:

$$L_n(\Delta, n)\,y[n, \varepsilon]=b[n, \varepsilon], \quad 0\leqslant\varepsilon\leqslant 1$$
$$L_n(\Delta, n)=\sum_{i=0}^{k} a_i[n]\,\Delta_n^i \qquad (11.1)$$

then the weighting function $W[n, \varepsilon, m]$ of such a system can be represented in the form of a linear combination of linearly independent solutions $y_i[n, \varepsilon]$, $i=1, 2, \ldots, k$ of the homogeneous equation

$$L_n(\Delta, n)\, y[n, \varepsilon] = 0, \quad 0 \leqslant \varepsilon \leqslant 1 \tag{11.2}$$

that is, in the form

$$W[n, \varepsilon; m] = \sum_{i=1}^{k} y_i^*[m, \varepsilon]\, y_i[n, \varepsilon], \quad 0 \leqslant \varepsilon \leqslant 1 \tag{11.3}$$

Here $a_i[n]$ and $b[n, \varepsilon]$ are functions of a discrete argument which are defined and bounded for each n of the set $N\{-\infty < n < \infty\}$; furthermore, $a_k[n] \neq 0$ for any n belonging to N.

Proof. For the proof we shall find the general solution of the difference equation (11.1) with zero initial conditions, when the σ-function is on its right-hand side. In this case, as has been shown above, the weighting function of the system is the solution of the equation.

Thus, if the conditions of the theorem are satisfied there exists a fundamental set of solutions $y_i[n, \varepsilon]$ of the homogeneous equation (11.2). The general solution of the non-homogeneous equation can be written in the form [18]

$$W[n, \varepsilon; m] = \sum_{i=1}^{k} C_i y_i[n, \varepsilon], \quad 0 \leqslant \varepsilon \leqslant 1 \tag{11.4}$$

where C_i must be chosen so that (11.4) satisfies the equation

$$L_n(\Delta, n)\, W[n, \varepsilon; m] = b[n, \varepsilon]\, \sigma[n - m, 0], \quad 0 \leqslant \varepsilon \leqslant 1 \tag{11.5}$$

For the determination of C_i $(i = 1, 2, \ldots, k)$ we must have k equations. We shall obtain one equation as the result of substituting (11.4) into (11.5). The remaining $k-1$ equations will be obtained as follows. Using (11.4) we find the first difference of the function $W[n, \varepsilon; m]$ with respect to the variable n, assuming that C_i are functions of the time $\bar{t}$. We then have

$$\Delta_n W[n, \varepsilon; m] = \sum_{i=1}^{k} C_i[n, \varepsilon]\, \Delta y_i[n, \varepsilon] + \sum_{i=1}^{k} \Delta C_i[n, \varepsilon]\, y_i[n+1, \varepsilon] \tag{11.6}$$

We then stipulate the second term in (11.6) as zero, i.e.

$$\sum_{i=1}^{k} \Delta C_i[n, \varepsilon]\, y_i[n+1, \varepsilon] = 0 \tag{11.6-1}$$

In this case (11.6) gives

$$\Delta_n W[n, \varepsilon; m] = \sum_{i=1}^{k} C_i[n, \varepsilon]\, \Delta y_i[n, \varepsilon] \qquad (11.7\text{-}1)$$

Determining $\Delta^2 n\, W[n, \varepsilon; m]$ from (11.7-1) we find

$$\Delta_n^2 W[n, \varepsilon; m] = \sum_{i=1}^{k} C_i[n, \varepsilon]\, \Delta^2 y_i[n, \varepsilon]$$

$$+ \sum_{i=1}^{k} \Delta C_i[n, \varepsilon] \Delta y_i[n+1, \varepsilon]$$

Putting

$$\sum_{i=1}^{k} \Delta C_i[n, \varepsilon]\, \Delta y_i[n+1, \varepsilon] = 0 \qquad (11.6\text{-}2)$$

in this equation, we have for the second-order difference of the weighting function

$$\Delta_n^2 W[n, \varepsilon; m] = \sum_{i=1}^{k} C_i[n, \varepsilon]\, \Delta^2 y_i[n, \varepsilon] \qquad (11.7\text{-}2)$$

Similarly, for the $(k-1)$-th difference we obtain

$$\Delta_n^{k-1} W[n, \varepsilon; m] = \sum_{i=1}^{k} C_i[n, \varepsilon]\, \Delta^{k-1} y_i[n, \varepsilon]$$

$$+ \sum_{i=1}^{k} \Delta C_i[n, \varepsilon]\, \Delta^{k-2} y_i[n+1, \varepsilon]$$

Stipulating that

$$\sum_{i=1}^{k} \Delta C_i[n, \varepsilon]\, \Delta^{k-2} y_i[n+1, \varepsilon] = 0 \qquad (11.6\text{-}k\text{-}1)$$

we obtain

$$\Delta_n^{k-1} W[n, \varepsilon; m] = \sum_{i=1}^{k} C_i[n, \varepsilon]\, \Delta^{k-1} y_i[n, \varepsilon] \qquad (11.7\text{-}k\text{-}1)$$

Finally, the k-th difference of the weighting function, on the basis of (11.7-k-1), is

$$\Delta_n^k W[n, \varepsilon; m] = \sum_{i=1}^{k} C_i[n, \varepsilon]\, \Delta_n^k y_i[n, \varepsilon] + \sum_{i=1}^{k} \Delta C_i[n, \varepsilon] \Delta^{k-1} y_i[n, \varepsilon] \qquad (11.7\text{-}k)$$

Substituting (11.7-1) into (11.7-k) and (11.4) into (11.5) we have

$$\sum_{i=1}^{k} C_i[n, \varepsilon] \sum_{j=0}^{k} a_j[n] \Delta^j y_i[n, \varepsilon] + \sum_{i=1}^{k} \Delta C_i[n, \varepsilon]\, \Delta^{k-1} y_i[n+1, \varepsilon] = b[n, \varepsilon]\, \sigma[n-m, 0]$$

Since $y_i[n, \varepsilon]$ is the solution of (11.2), then from the last equation we have

$$\sum_{i=1}^{k} \Delta C_i[n, \varepsilon]\, \Delta^{k-1} y_i[n+1, \varepsilon] = b[n, \varepsilon]\, \sigma[n-m, 0] \qquad (11.6\text{-}k)$$

Thus we have a system of k algebraic equations for the determination of $C_i[n, \varepsilon]$:

$$\begin{aligned} &\sum_{i=1}^{k} \Delta C_i[n, \varepsilon]\, y_i[n+1, \varepsilon] = 0 \\ &\sum_{i=1}^{k} \Delta C_i[n, \varepsilon]\, \Delta y_i[n+1, \varepsilon] = 0 \\ &\quad \cdots\cdots\cdots\cdots\cdots\cdots \\ &\sum_{i=1}^{k} \Delta C_i[n, \varepsilon]\, \Delta^{k-2} y_i[n+1, \varepsilon] = 0 \\ &\sum_{i=1}^{k} \Delta C_i[n, \varepsilon]\, \Delta^{k-1} y_i[n+1, \varepsilon] = b[n, \varepsilon]\, \sigma[n-m, 0] \end{aligned} \qquad (11.8)$$

The determinant $D[y_1[n+1, \varepsilon], \ldots, y_k[n+1, \varepsilon]]$ of the system of equations (11.8) is non-zero (for the fundamental set of solutions $y_i[n, \varepsilon]$) for any n belonging to N. Consequently, the system of (11.8) has a unique solution for

$\Delta C_i[n, \varepsilon]$. Solving this system we then have by the definition of a sum,

$$C_i[n, \varepsilon] = \sum_{\lambda=0}^{n-1} (-1)^{k+i-1} b[\lambda, \varepsilon]\,\sigma[\lambda - m, 0]\,\frac{D_i[\lambda, \varepsilon]}{D[\lambda, \varepsilon]} \tag{11.9}$$

$$i = 1, 2, \ldots, k;\ 0 \leqslant \varepsilon \leqslant 1$$

where $D[\lambda,\varepsilon]$ is the determinant of the system of equations (11.8), defined as follows:

$$D[\lambda,\varepsilon] = \begin{vmatrix} y_1[\lambda+1, \varepsilon] & y_2[\lambda+1, \varepsilon] & \ldots & y_k[\lambda+1, \varepsilon] \\ \Delta y_1[\lambda+1, \varepsilon] & \Delta y_2[\lambda+1, \varepsilon] & \ldots & \Delta y_k[\lambda+1, \varepsilon] \\ \cdots & \cdots & \cdots & \cdots \\ \Delta^{k-1} y_1[\lambda+1, \varepsilon] & \Delta^{k-1} y_2[\lambda+1, \varepsilon] & \ldots & \Delta^{k-1} y_k[\lambda+1, \varepsilon] \end{vmatrix}$$

while $D_i[\lambda, \varepsilon]$ is obtained from $D[\lambda, \varepsilon]$ by the Cramer rule. Taking into account the property of the σ-function, we obtain from (11.9)

$$C_i[n, \varepsilon] = (-1)^{k+i-1} \frac{D_i[m, \varepsilon]}{D[m, \varepsilon]} b[m, \varepsilon] = C_i[m, \varepsilon] \tag{11.10}$$

$$i = 1, 2, \ldots, k,\ 0 \leqslant \varepsilon \leqslant 1$$

or, introducing the symbols

$$\overset{*}{y}_i[m, \varepsilon] = C_i[m, \varepsilon],\ i = 1, 2, \ldots, k$$

we have from (11.4)

$$W[n, \varepsilon; m] = \sum_{i=1}^{k} \overset{*}{y}_i[m, \varepsilon]\, y_i[n, \varepsilon],\ 0 \leqslant \varepsilon \leqslant 1 \tag{11.11}$$

which proves Theorem 1.

Using the results of the preceding section we can prove Theorem 1 in another way. Thus, if

$$y_i[n\ \varepsilon],\quad i = 1, 2, \ldots, k,\quad 0 \leqslant \varepsilon \leqslant 1 \tag{11.12}$$

is the fundamental set of solutions of the difference equations

$$L_n(\Delta, n)\, y[n, \varepsilon] = 0,\ 0 \leqslant \varepsilon \leqslant 1 \tag{11.13}$$

then the weighting function $W[n, \varepsilon; m]$, i.e. the general solution of the equation

$$L_n(\Delta, n)\, W[n, \varepsilon; m] = b[n, \varepsilon]\sigma[n - m, 0], 0 \leqslant \varepsilon \leqslant 1 \tag{11.14}$$

will be

$$W[n, \varepsilon; m] = \sum_{i=1}^{k} C_i y_i [n, \varepsilon] \tag{11.15}$$

where C_i are determined from the initial conditions ((10.98) and (10.99) when $h = 0$)

$$\begin{aligned} &\Delta_n^r W[n, \varepsilon; m]|_{n=m} = 0, \; r = 0, 1, \ldots, k-2 \\ &\Delta_n^{k-1} W[n, \varepsilon; m]|_{n=m} = \frac{b[m, \varepsilon]}{a_k[m]}, \; 0 \leqslant \varepsilon \leqslant 1 \end{aligned} \tag{11.16}$$

Substituting (11.15) into (11.16) and assuming C_i to be constant we find

$$\begin{aligned} &\sum_{i=1}^{k} C_i \Delta^r y_i [n, \varepsilon]|_{n=m} = 0, \; r = 0, 1, \ldots, k-2 \\ &\sum_{i=1}^{k} C_i \Delta^{k-1} y_i [n, \varepsilon]|_{n=m} = \frac{b[m, \varepsilon]}{a_k[m]}, \; 0 \leqslant \varepsilon \leqslant 1 \end{aligned} \tag{11.17}$$

Thus we have obtained a system of k algebraic equations for the unknowns $C_i, i = 1, 2, \ldots, k$. Solving this system we find

$$W[n, \varepsilon; m] = \sum_{i=1}^{k} (-1)^{k+i} \frac{D_i[m, \varepsilon]}{a_k[m] D[m, \varepsilon]} y_i[n, \varepsilon] \tag{11.18}$$

$$0 \leqslant \varepsilon \leqslant 1$$

where

$$D[m, \varepsilon] = D[\lambda, \varepsilon]|_{\lambda=m}$$

$$D_i[m, \varepsilon] = D_i[\lambda, \varepsilon]|_{\lambda=m}$$

Using in (11.18) the following notation

$$(-1)^{k+i} \frac{D_i[m, \varepsilon]}{a_k[m] D[m, \varepsilon]} = y_i^*[m, \varepsilon] \tag{11.19}$$

we finally obtain from (11.18) the equation below, which also prove Theorem 1.

$$W[n, \varepsilon; m] = \sum_{i=1}^{k} y_i^*[m, \varepsilon] y_i[n, \varepsilon], \; 0 \leqslant \varepsilon \leqslant 1 \tag{11.20}$$

Theorem 1 thus gives the property of the weighting function of a system whose difference equations do not contain differences of the input action. We shall now prove a similar theorem for a more general case.

Theorem 2. If a sampled data system is described by the k-th order difference equation:

$$L_n(\Delta, n)\, y[n, \varepsilon] = M_n(\Delta, \bar{t})\, x[n, 0], \quad 0 \leqslant \varepsilon \leqslant 1$$

$$L_n(\Delta, n) = \sum_{i=0}^{k} a_i[n]\, \Delta_n^i \tag{11.21}$$

$$M_n(\Delta, \bar{t}) = \sum_{j=0}^{h} b_j[n, \varepsilon]\, \Delta_n^j$$

then the weighting function $W[n, \varepsilon; m]$ of this system can be represented in the form of a linear combination of the linearly independent solutions $y_i[n, \varepsilon]$ of the homogeneous equation

$$L_n(\Delta, n)\, y[n, \varepsilon] = 0, \quad 0 \leqslant \varepsilon \leqslant 1 \tag{11.22}$$

that is, in the form

$$W[n, \varepsilon; m] = \sum_{i=1}^{k} y_i^*[m, \varepsilon]\, y_i[n, \varepsilon], \quad 0 \leqslant \varepsilon \leqslant 1 \tag{11.23}$$

Here $a_i[n]$ and $b_j[n, \varepsilon]$ are functions of a discrete argument, defined for each n belonging to $N\{-\infty < n < \infty\}$; in addition, $a_k[n] \neq 0$ for any n belonging to N.

Proof. We use $W_0[n, \varepsilon; m]$ to denote the weighting function of a system whose difference equation is

$$L_n(\Delta, n)\, W_0[n, \varepsilon; m] = \sigma[n - m, 0], \quad 0 \leqslant \varepsilon \leqslant 1 \tag{11.24}$$

It has been shown above that the weighting function $W[n, \varepsilon; m]$ can be defined in terms of the weighting functions $W_0[n, \varepsilon; m]$ with the aid of the relation

$$W[n, \varepsilon; m] = \sum_{j=0}^{h} (-1)^j \Delta_m^j \{W_0[n, \varepsilon; m - j]\, b_j[m - j, \varepsilon]\} \tag{11.25}$$

$$0 \leqslant \varepsilon \leqslant 1$$

Since by Theorem 1

$$W_0[n, \varepsilon; m] = \sum_{j=1}^{k} y_{0i}^{*}[m, \varepsilon]\, y_{0i}[n, \varepsilon], \quad 0 \leqslant \varepsilon \leqslant 1 \tag{11.26}$$

where $y_{0i}[n, \varepsilon]$ is the fundamental set of solutions of a homogeneous equation corresponding to (11.21), then for $W[n, \varepsilon; m]$ we have in accordance with (11.25)

$$W[n, \varepsilon; m] = \sum_{j=0}^{h} (-1)^j \Delta_m^j \Big\{ b_j[m-j, \varepsilon] \times \sum_{i=1}^{k} y_{0i}^{*}[m, \varepsilon]\, y_{0i}[n, \varepsilon] \Big\} \tag{11.27}$$

$$0 \leqslant \varepsilon \leqslant 1$$

Changing the order of summation in this expression we have

$$W[n, \varepsilon; m] = \sum_{i=1}^{k} y_{0i}[n, \varepsilon] \times \sum_{j=0}^{h} (-1)^j \Delta_m^j \{ b_j[m-j, \varepsilon]\, y_{0i}^{*}[m, \varepsilon] \} \tag{11.28}$$

$$0 \leqslant \varepsilon \leqslant 1$$

Denoting now

$$y_i[n, \varepsilon] = y_{0i}[n, \varepsilon]$$
$$y_i^{*}[m, \varepsilon] = \sum_{j=0}^{h} (-1)^j \Delta_m^j \{ b_j[m-j, \varepsilon]\, y_{0i}^{*}[m, \varepsilon] \} \tag{11.29}$$

$$i = 1, 2, \ldots, k, \quad 0 \leqslant \varepsilon \leqslant 1$$

in (11.28), we finally obtain

$$W[n, \varepsilon; m] = \sum_{i=1}^{k} y_i^{*}[m, \varepsilon]\, y_i[n, \varepsilon], \quad 0 \leqslant \varepsilon \leqslant 1 \tag{11.30}$$

which proves Theorem 2.

For a sampled data system of a general form, the processes in which are determined by two difference equations for the time intervals $0 \leqslant \varepsilon \leqslant \gamma$ and $\gamma \leqslant \varepsilon \leqslant 1$, it is easy to prove the following theorem.

Theorem 3. If a sampled data system is described by the k-th order difference equations

$$\begin{aligned} L_n(\Delta, n)\, y_1[n, \varepsilon] &= M_n(\Delta, \bar{t})\, x[n, 0], \quad 0 \leqslant \varepsilon \leqslant \gamma \\ S_n(\Delta, n)\, y_2[n, \varepsilon] &= G_n(\Delta, \bar{t})\, x[n, 0], \quad \gamma \leqslant \varepsilon \leqslant 1 \end{aligned} \tag{11.31}$$

the coefficients of which satisfy the conditions of Theorems 1 and 2, then the weighting function $W[n, \varepsilon; m]$ of this system can be represented for the time intervals $0 \leqslant \varepsilon \leqslant \gamma$ and $\gamma \leqslant \varepsilon \leqslant 1$ as follows

$$\begin{aligned} W[n, \varepsilon; m] &= \sum_{i=1}^{k} y_{1i}^{*}[m, \varepsilon]\, y_{1i}[n, \varepsilon], \quad 0 \leqslant \varepsilon \leqslant \gamma \\ W[n, \varepsilon; m] &= \sum_{i=1}^{k} y_{2i}^{*}[m, \varepsilon]\, y_{2i}[n, \varepsilon], \quad \gamma \leqslant \varepsilon \leqslant 1 \end{aligned} \tag{11.32}$$

where $y_{1i}[n, \varepsilon]$ are the fundamental set of solutions of the homogeneous equation

$$L_n(\Delta, n)\, y_1[n, \varepsilon] = 0, \quad 0 \leqslant \varepsilon \leqslant \gamma \tag{11.33}$$

and $y_{2i}[n, \varepsilon]$ are the fundamental set of solutions of the homogeneous equation

$$S_n(\Delta, n)\, y_2[n, \varepsilon] = 0, \quad \gamma \leqslant \varepsilon \leqslant 1 \tag{11.34}$$

The proof of Theorem 3 can be carried out in the same way as the proofs of Theorems 1 and 2.

We shall proceed now to prove a theorem on the piecewise continuity of the weighting function. We shall prove this theorem for a system whose processes are determined by a single difference equation.

Theorem 4. If a sampled data system is described by the difference equation

$$L_n(\Delta, n)\, y[n, \varepsilon] = M_n(\Delta, \bar{t})\, x[n, 0], \quad 0 \leqslant \varepsilon \leqslant 1$$

$$L_n(\Delta, n) = \sum_{i=1}^{k} a_i[n]\, \Delta_n^i, \quad M_n(\Delta, \bar{t}) = \sum_{j=0}^{h} b_j[n, \varepsilon]\, \Delta_n^j \tag{11.35}$$

then the weighting function $W[n, \varepsilon; m]$ of this system is a piecewise-continuous function of the argument $\bar{t} = n + \varepsilon$.

Here $a_i[n]$ are defined and, furthermore, $a_k[n] \neq 0$ for each n belonging to the set $N\{-\infty < n < \infty\}$, whilst $b_j[n, \varepsilon]$ are defined and continuous (or piecewise-continuous) on the set of values of N and $0 \leqslant \varepsilon \leqslant 1$ for any n belonging to N.

Proof. The weighting function $W_0[n, \varepsilon; m]$ of a system whose difference equation is

$$L_n(\Delta, n) W_0[n, \varepsilon; m] = b_0[n, \varepsilon]\,\sigma[n-m, 0], \quad 0 \leqslant \varepsilon \leqslant 1 \tag{11.36}$$

are given by the relation

$$W_0[m+k+l, \varepsilon; m] = \frac{1}{a_k^*[m+l]} \sum_{i=0}^{k-1} a_i^*[m+l]\, W_0[m+k+l-1, \varepsilon; m] \tag{11.37}$$

$$0 \leqslant \varepsilon \leqslant 1$$

where the initial values of the weighting function are

$$\begin{gathered} W_0[m+\lambda, \varepsilon; m] = 0, \quad \lambda = 0, 1, \ldots, k-2 \\ W_0[m+k-1, \varepsilon; m] = \frac{b_0[m, \varepsilon]}{a_k[m]}, \quad 0 \leqslant \varepsilon \leqslant 1 \end{gathered} \tag{11.38}$$

and the coefficients a_i^* are combinations of the coefficients a_i. The piecewise continuity of the weighting function $W_0[n, \varepsilon; m]$ follows directly from (11.37) and (11.38).

Further, the weighting function $W[n, \varepsilon; m]$ is connected with $W_0[n, \varepsilon; m]$ by the relation

$$W[n, \varepsilon; m] = \sum_{j=0}^{h} (-1)^j \Delta_m^j \{W_0[n, \varepsilon; m-j]\, b_j[m-j, \varepsilon]\} \tag{11.39}$$

$$0 \leqslant \varepsilon \leqslant 1$$

But since $W_0[n, \varepsilon; m)$ and $b_j[m-j, \varepsilon]$ $(j = 0, 1, \ldots, h)$ are piecewise-continuous functions of the argument $\overline{t}$, then their product is also a piecewise-continuous function of $\overline{t}$. Consequently, from the theorem on the continuity of a sum of a finite number of continuous functions, we have the piecewise continuity of $W[n, \varepsilon; m]$, determined by (11.39). This proves Theorem 4.

The properties of weighting functions just proved in the form of theorems are of great importance, and will be used in what follows.

1.12 OTHER CHARACTERISTICS OF SAMPLED DATA SYSTEMS

The weighting function is an all-embracing characteristic of a linear sampled data system in the domain of a real variable (in the time domain). But very often, when solving certain problems of sampled data systems, it is convenient to consider the equations of the systems in the domain of a complex variable. We then have to deal with the transfer functions and frequency characteristics of these systems.

This method of investigating sampled data systems with constant parameters has been treated in numerous publications by Tsypkin [1-12]. By the use of the basic dynamical characteristic, in the domain of the complex variable of the transfer function of a sampled data system, methods can be found for the realisation of sampled data systems (mainly stationary systems). These methods are discussed by Perov [14]. A number of problems in the theory of systems with variable parameters have been considered in the domain of a complex variable by Tartakovskii [20, 21, 22] and others.

In this section we shall present only the necessary information on the characteristics of sampled data systems in the domain of a complex variable. We shall also discuss a method for determining the equation for the transfer function of a non-stationary system, based on the formula obtained in Section 1.2, which is analogous to the Leibnitz formula in the theory of differential equations.

1.12.1 Determination of the transfer function and the frequency characteristic

The output variable $y[n, \varepsilon]$ of a non-stationary sampled data system with the weighting function $W[n, \varepsilon; m]$ is determined, as was shown above, by

$$y[n, \varepsilon] = \sum_{m=-\infty}^{n} W[n, \varepsilon; m]\, x[m-1,1] \qquad (12.1)$$

where x is the input disturbance of the system, which can be either a continuous or a discontinuous function. If $Y^*(q, \varepsilon)$ is the transform of the function $y[n, \varepsilon]$ (in the sense of the discrete Laplace transformation), we have

$$y[n, \varepsilon] = \frac{1}{2\pi j} \int_{c-j\pi}^{c+j\pi} Y^*(q, \varepsilon) e^{qn} dq \qquad (12.2)$$

In turn

$$x[m-1, 1] = \frac{1}{2\pi j} \int_{c-j\pi}^{c+j\pi} X^*(q, 1) e^{q(m-1)} dq \qquad (12.3)$$

where $X^*(q, 1)$ is the transform of the function $x[n, \varepsilon]$ when $\varepsilon = 1$. Substituting (12.3) into (12.1) we have

$$y[n, \varepsilon] = \sum_{m=-\infty}^{n} W[n, \varepsilon; m] \frac{1}{2\pi j} \int_{c-j\pi}^{c+j\pi} X^*(q, 1) e^{q(m-1)} dq \qquad (12.4)$$

Changing the order of summation and integration in (12.4) we obtain

$$y[n, \varepsilon] = \frac{1}{2\pi j} \int_{c-j\pi}^{c+j\pi} e^{-q} X^*(q, 1) \sum_{m=-\infty}^{n} W[n, \varepsilon; m] e^{qm} dq$$

$$= \frac{1}{2\pi j} \int_{c-j\pi}^{c+j\pi} e^{-q} X^*(q, 1) e^{qn} \sum_{m=-\infty}^{n} W[n, \varepsilon; m] e^{-q(n-m)} dq \qquad (12.5)$$

If in (12.5) we put

$$\sum_{m=-\infty}^{n} W[n, \varepsilon; m] e^{-q(n-m)} = W^*(q, n, \varepsilon) \qquad (12.6)$$

then

$$y[n, \varepsilon] = \frac{1}{2\pi j} \int_{c-j\pi}^{c+j\pi} e^{-q} X^*(q, 1) W^*(q, n, \varepsilon) e^{qn} dq \qquad (12.7)$$

whence on the basis of (12.2) we can write

$$Y^*(q, \varepsilon) = W^*(q, n, \varepsilon) X^*(q, 1) e^{-q} \qquad (12.8)$$

The function $W^*(q, n, \varepsilon)$ of the complex variable q, determined by (12.6), is called the transfer function of a non-stationary sampled data system. Then introducing the new variable $n-m=r$ in (12.6) we have

$$W^*(q, n, \varepsilon)=\sum_{r=0}^{\infty} W[n, \varepsilon; n-r]\, e^{-qr} \tag{12.9}$$

whence it follows that the transfer function of a sampled data system is the discrete Laplace transformation of its weighting function, i.e.

$$W^*(q, n, \varepsilon)= D\{W[n, \varepsilon; m]\} \tag{12.10}$$

Equation (12.8) is called the equation of a sampled data system in the domain of transforms. If the input disturbance of the system $x[\bar{t}]$ is a continuous function, then the equation of the sampled data system assumes the form

$$Y^*(q, \varepsilon)=W^*(q, n, \varepsilon)\, X^*(q, 0) \tag{12.11}$$

since in this case

$$X^*(q, 1)\, e^{-q}=X^*(q, 0) \tag{12.12}$$

As follows from the definition, the weighting function of a sampled data system with variable parameters depends on the time $\bar{t}=n+\varepsilon$. For a stationary system $W[n, \varepsilon; m]=W[n-m, \varepsilon]$ and (12.9) assumes the form

$$W^*(q, \varepsilon)= \sum_{r=0}^{\infty} W[r, \varepsilon]\, e^{-qr} \tag{12.13}$$

whence it follows that the weighting function of a stationary system depends on the parameter ε, and is not constant. When (12.13) is taken into account the equation of a stationary system is written as follows:

$$Y^*(q, \varepsilon)=W^*(q, \varepsilon)\, X^*(q, 1)\, e^{-q} \tag{12.14}$$

or

$$Y^*(q, \varepsilon) = W^*(q, \varepsilon)\, X^*(q, 0) \tag{12.15}$$

if $x[\bar{t}]$ is a continuous function.

Equations (12.8) and (12.14) have been obtained by Perov [14]. Equations (12.11) and (12.15) arise from these equations as particular cases.

If the transformation parameter q is a purely imaginary number

$$q = j\bar{\omega} \tag{12.16}$$

where $\bar{\omega} = \omega T_0$ is the relative frequency, then from (12.9) we obtain

$$W^*(j\bar{\omega}, n, \varepsilon) = \sum_{r=0}^{\infty} W[n, \varepsilon; n-r]\, e^{-j\bar{\omega}r} \tag{12.17}$$

The function $W^*(j\bar{\omega}, n, \varepsilon)$ determined by (12.17) is called the frequency characteristic of a sampled data system with variable parameters. The frequency characteristic of a stationary sampled data system is determined in accordance with (12.13) as follows:

$$W^*(j\bar{\omega}, \varepsilon) = \sum_{r=0}^{\infty} W[r, \varepsilon]\, e^{-j\bar{\omega}r} \tag{12.18}$$

We shall now give a method for determining the transfer function of a system from the given difference equations.

1.12.2 Determination of the transfer function from the given difference equation

The transfer function of a stationary sampled data system can be found immediately if the difference equations of the system are given. Since the commutative property applies to difference equations with constant coefficients, then the transfer function of a stationary system defined by difference equations with constant coefficients is obtained directly by replacing the operator Δ_n^l by $(e^q - 1)^l$. Consequently, if a system is described by the difference equation

$$L_n(\Delta)\, y[n, \varepsilon] = M_n(\Delta, \varepsilon)\, x[n, 0], \quad 0 \leqslant \varepsilon \leqslant 1 \tag{12.19}$$

then, in accordance with the rule just given and (12.14), the transfer function of this system is given by

$$\begin{aligned} W^*(q, \varepsilon) &= \frac{Y^*(q, \varepsilon)}{X^*(q, 1)\, e^{-q}} \\ &= \frac{M(e^q - 1, \varepsilon)}{L(e^q - 1)}, \quad 0 \leqslant \varepsilon \leqslant 1 \end{aligned} \tag{12.20}$$

or

$$W^*(q,\ \varepsilon)=\frac{\sum_{l=0}^{h} b_l(\varepsilon)(e^q-1)^l}{\sum_{i=0}^{k} a_i(e^q-1)^i},\ 0\leqslant\varepsilon\leqslant 1 \qquad (12.21)$$

where a_i and $b_l(\varepsilon)$ are the coefficients of the difference operators $L_n(\Delta)$ and $M_n(\Delta,\ \varepsilon)$ respectively. The transfer function of a stationary system described by two difference equations is determined in exactly the same way.

By definition of the frequency characteristic of a sampled data system we have from (12.20)

$$W^*(j\bar{\omega},\ \varepsilon)=\frac{M(e^{j\bar{\omega}}-1,\ \varepsilon)}{L(e^{j\bar{\omega}}-1)},\ 0\leqslant\varepsilon\leqslant 1 \qquad (12.22)$$

or

$$W^*(j\bar{\omega},\ \varepsilon)=\frac{\sum_{l=0}^{h} b_l(\varepsilon)(e^{j\bar{\omega}}-1)^l}{\sum_{i=0}^{k} a_i(e^{j\bar{\omega}}-1)^i},\ 0\leqslant\varepsilon\leqslant 1 \qquad (12.23)$$

Equations (12.22) and (12.23) define the frequency characteristic of a system described by the single difference equation (12.19).

Owing to the properties of linear difference operators with variable coefficients which define non-stationary sampled data systems, the transfer functions of the latter cannot be obtained directly by replacing the operators Δ_n^l by $(e^q-1)^l$. Consequently, from the difference equation of a system we can only determine another difference equation whose solution will be the transfer function of the non-stationary system.

We now discuss the determination of the difference equation for the transfer function of a non-stationary system. Let the system be given by the single difference equation of order k:

$$L_n(\Delta,\ n)\,y[n,\ \varepsilon]=M_n(\Delta,\ \bar{t})\,x[n,\ 0],\ 0\leqslant\varepsilon\leqslant 1$$
$$L_n(\Delta,\ n)=\sum_{i=0}^{k} a_i[n]\,\Delta_n^i,\ M_n(\Delta,\ \bar{t})=\sum_{j=0}^{h} b_j[n,\ \varepsilon]\,\Delta_n^j \qquad (12.24)$$

The weighting function $W[n, \varepsilon; m]$ of such a system satisfies the non-homogeneous equation (12.24) with zero initial conditions if the disturbance x is replaced by the σ-function:

$$L_n(\Delta, n) W[n, \varepsilon; m] = M_n(\Delta, \bar{t}) \sigma[n-m, 0] \quad 0 \leqslant \varepsilon \leqslant 1 \qquad (12.25)$$

Subjecting both sides of (12.25) to the discrete Laplace transformation with respect to the variable m, we have, when (12.7) is taken into account,

$$L_n(\Delta, n) W^*(q, n, \varepsilon) e^{qn} = M_n(\Delta, \bar{t}) e^{qn}, \; 0 \leqslant \varepsilon \leqslant 1 \qquad (12.26)$$

We now use the analogy to the Leibnitz formula obtained for difference operators:

$$\begin{aligned} D(\Delta, \bar{t}) u[n, \varepsilon] v[n, \varepsilon] &= u[n, \varepsilon] D(\Delta, \bar{t}) v[n, \varepsilon] \\ &+ \Delta u[n, \varepsilon] \frac{\partial D(\Delta, \bar{t})}{\partial \Delta} v[n+1, \varepsilon] + \dots \\ &+ \frac{1}{k!} \Delta^k u[n, \varepsilon] \frac{\partial^k D(\Delta, \bar{t})}{\partial \Delta^k} v[n+k, \varepsilon] \end{aligned} \qquad (12.27)$$

We apply (12.27) to the right-hand side of (12.26). Then

$$\begin{aligned} M_n(\Delta, \bar{t}) e^{qn} &= \sum_{j=0}^{h} b_j[n, \varepsilon] \Delta^j e^{qn} \\ &= \sum_{j=0}^{h} b_j[n, \varepsilon] (e^q - 1)^j e^{qn} \\ &= e^{qn} \sum_{j=0}^{h} b_j[n, \varepsilon] (e^q - 1)^j = e^{qn} M(q, \bar{t}) \end{aligned} \qquad (12.28)$$

where

$$M(q, \bar{t}) = \sum_{j=0}^{h} b_j[n, \varepsilon] (e^q - 1)^j \qquad (12.29)$$

Thus when the difference operator $M_n(\Delta, \bar{t})$ given by (12.24) is applied to the function e^{qn} we have

$$M_n(\Delta, \bar{t}) e^{qn} = e^{qn} M(q, \bar{t}) \qquad (12.30)$$

Let us now transform the left-hand side of (12.26). Putting

$$W^*(q, n, \varepsilon) = u[n, \varepsilon] \tag{12.31}$$

$$e^{qn} = v[n] \tag{12.32}$$

we obtain, when (12.27) is taken into account,

$$L_n(\Delta, n) W^*(q, n, \varepsilon) e^{qn}$$

$$= \sum_{i=0}^{k} a_i[n] \Delta_n^i \{W^*(q, n, \varepsilon) e^{qn}\}$$

$$= W^*(q, n, \varepsilon) L_n(\Delta, n) e^{qn} + \Delta_n W^*(q, n, \varepsilon) \frac{\partial L_n(\Delta, n)}{\partial \Delta} e^{q(n+1)}$$

$$+ \ldots + \frac{1}{k!} \Delta_n^k W^*(q, n, \varepsilon) \frac{\partial^k L_n(\Delta, n)}{\partial \Delta^k} e^{q(n+k)} \tag{12.33}$$

But since by (12.30)

$$\frac{\partial^\nu L_n(\Delta, n)}{\partial \Delta^\nu} e^{q(n+\nu)} = e^{q(n+\nu)} \left. \frac{\partial^\nu L_n(\Delta, n)}{\partial \Delta^\nu} \right|_{\Delta = e^q - 1} \tag{12.34}$$

(12.36) can be written in the form

$$L_n(\Delta, n) W^*(q, n, \varepsilon) e^{qn}$$

$$= \Big\{ W^*(q, n, \varepsilon) L(q, n) + \Delta_n W^*(q, n, \varepsilon) L'(q, n) \tag{12.35}$$

$$+ \ldots + \frac{1}{k!} \Delta_n^k W^*(q, n, \varepsilon) L^{(k)}(q, n) \Big\} e^{qn}$$

where

$$L^{(\nu)}(q, n) = \frac{\partial^\nu L(q, n)}{\partial (e^q - 1)^\nu} e^{q\nu}, \quad \nu = 1, 2, \ldots, k \tag{12.36}$$

$$L(q, n) = \sum_{i=0}^{k} a_i[n] (e^q - 1)^i \tag{12.37}$$

Thus, when (12.30) and (12.35) are taken into account, (12.26) can be written as follows:

$$\sum_{\mu=0}^{k} \frac{1}{\mu!} \Delta_n^\mu W^*(q, n, \varepsilon) L^{(\mu)}(q, n) = M(q, \bar{t}), \; 0 \leqslant \varepsilon \leqslant 1 \tag{12.38}$$

Denoting in (12.38)

$$\frac{1}{\mu!} L^{(\mu)}(q, n) = \frac{1}{\mu!} \frac{\partial^{\mu} L(q, n)}{\partial (e^q - 1)^{\mu}} = \overset{*}{a}_{\mu}[n] \qquad (12.39)$$

we write the required equation in its final form:

$$\sum_{\mu=0}^{k} \overset{*}{a}_{\mu}[n] \Delta_n^{\mu} W^*(q, n, \varepsilon) = M(q, \bar{t}), 0 \leqslant \varepsilon \leqslant 1 \qquad (12.40)$$

Equation (12.40) defines the transfer function of a non-stationary sampled data system. Since the weighting function $W[n, \varepsilon; m]$ is the solution of the corresponding non-homogeneous equation with zero initial conditions, (12.40) must also be solved with zero initial conditions in order to determine the transfer function $W^*[q, n, \varepsilon]$.

Equation (12.40) remains valid also for a stationary sampled data system. Then

$$\overset{*}{a}_{\mu}[n] = \frac{1}{\mu!} \frac{\partial^{\mu} L(q)}{\partial (e^q - 1)^{\mu}} = \overset{*}{a}_{\mu} \qquad (12.41)$$

$$\Delta_n^{\mu} W^*(q, n, \varepsilon) = \Delta_n^{\mu} W^*(q, \varepsilon) = 0, \ \mu = 1, 2, \ldots \qquad (12.42)$$

$$\overset{*}{a}_0[n] = \overset{*}{a}_0 = L(q) = \sum_{i=0}^{k} a_i (e^q - 1)^i \qquad (12.43)$$

$$M(q, \bar{t}) = \sum_{j=0}^{h} b_j(\varepsilon)(e^q - 1)^j \qquad (12.44)$$

When (12.41) to (12.44) are taken into account, (11.40) assumes the form

$$\sum_{i=0}^{k} a_i (e^q - 1)^i W^*(q, \varepsilon) = \sum_{j=0}^{h} b_j(\varepsilon)(e^q - 1)^j, \ 0 \leqslant \varepsilon \leqslant 1$$

or, in final form

$$W^*(q, \varepsilon) = \frac{\sum_{j=0}^{h} b_j(\varepsilon)(e^q - 1)^j}{\sum_{i=0}^{k} a_i(e^q - 1)^i}, \quad 0 \leqslant \varepsilon \leqslant 1 \qquad (12.45)$$

which coincides with the expression for the transfer function of a stationary system (12.21), obtained in another way.

If a sampled data system is described by two difference equations, then in accordance with the foregoing, which relates to the determination of the transfer function of such a system, we have two equations analogous to (12.40) for the time intervals $0 \leqslant \varepsilon \leqslant \gamma$ and $\gamma \leqslant \varepsilon \leqslant 1$.

Let us now find the difference equation for the transfer function of a closed-loop system, with the operator of the feedback loop equal to unity. Let the equation of the open-loop system be

$$\begin{gathered} \mathrm{H}_n(\Delta, n)\, y[n, \varepsilon] = S_n(\Delta, \overline{t})\, x[n, 0], \quad 0 \leqslant \varepsilon \leqslant 1 \\ \mathrm{H}_n(\Delta, n) = \sum_{i=0}^{k} a_i[n]\, \Delta_n^i \\ S_n(\Delta, \overline{t}) = \sum_{j=0}^{h} b_j[n, \varepsilon]\, \Delta_n^j \end{gathered} \qquad (12.46)$$

The equation of this closed-loop system has the form

$$\mathrm{H}_n(\Delta, n)\, y[n, \varepsilon] = S_n(\Delta, \overline{t})\{x[n, 0] - y[n, 0]\}, \quad 0 \leqslant \varepsilon \leqslant 1 \qquad (12.47)$$

If the input disturbance of the system is the σ-function, then for (12.47) we have

$$\mathrm{H}_n(\Delta, n)\, W[n, \varepsilon; m] = S_n(\Delta, \overline{t})\{\sigma[n - m, 0] - W[n, 0; m]\} \quad 0 \leqslant \varepsilon \leqslant 1 \qquad (12.48)$$

where $W[n, \varepsilon; m]$ is the weighting function of the closed-loop system. To find the equation for the transfer function we subject both sides of (12.48) to the discrete Laplace transformation with respect to the variable m. We then obtain

$$\mathbf{H}_n(\Delta, n)\, W^*(q, n, \varepsilon)\, e^{qn} = S_n(\Delta, \bar{t})\, e^{qn} - S_n(\Delta, \bar{t})\, W^*(q, n, 0)\, e^{qn}, \quad 0 \leqslant \varepsilon \leqslant 1 \tag{12.49}$$

This equation will now be transformed. Applying (12.27) to the first term of the right-hand side of Equation (12.49) we find

$$S_n(\Delta, \bar{t})\, e^{qn} = e^{qn} S(q, \bar{t}) \tag{12.50}$$

where according to (12.30)

$$S(q, \bar{t}) = S_n(\Delta, \bar{t})\big|_{\Delta = e^q - 1} = \sum_{j=0}^{h} b_j[n, \varepsilon]\,(e^q - 1)^j \tag{12.51}$$

Applying (12.27) to the second term of the right-hand side of Equation (12.49) we have

$$S_n(\Delta, \bar{t})\, W^*(q, n, 0)\, e^{qn} = W^*(q, n, 0)\, S_n(\Delta, \bar{t})\, e^{qn} + \Delta_n W^*(q, n, 0)\, \frac{\partial S_n(\Delta, \bar{t})}{\partial \Delta}\, e^{q(n+1)} + \ldots + \frac{1}{h!}\, \Delta_n^h W^*(q, n, 0)\, \frac{\partial^h S_n(\Delta, \bar{t})}{\partial \Delta^h}\, e^{q(n+h)} \tag{12.52}$$

Since

$$\frac{\partial^\nu S_n(\Delta, \bar{t})}{\partial \Delta^\nu}\, e^{q(n+\nu)} = e^{q(n+\nu)}\, \frac{\partial^\nu S_n(\Delta, \bar{t})}{\partial \Delta^\nu}\bigg|_{\Delta = e^q - 1} \tag{12.53}$$

Equation (12.52) can be written

$$S_n(\Delta, \bar{t})\, W^*(q, n, 0)\, e^{qn} = e^{qn} \sum_{\nu=0}^{h} \frac{1}{\nu!}\, \Delta_n^\nu W^*(q, n, 0)\, S^{(\nu)}(q, \bar{t}) \tag{12.54}$$

where

$$S^{(\nu)}(q, \bar{t}) = e^{q\nu}\, \frac{\partial^\nu S(q, \bar{t})}{\partial (e^q - 1)^\nu} \tag{12.55}$$

$$S(q, \bar{t}) = \sum_{j=0}^{h} b_j[n, \varepsilon]\,(e^q - 1)^j \tag{12.56}$$

Analogously, for the left-hand side of (12.49) we obtain

$$\begin{aligned} &\mathrm{H}_n(\Delta, n) W^*(q, n, \varepsilon) e^{qn} \\ &= e^{qn} \sum_{\mu=0}^{k} \frac{1}{\mu!} \Delta_n^{\mu} W^*(q, n, \varepsilon) \mathrm{H}^{(\mu)}(q, n) \end{aligned} \tag{12.57}$$

where

$$\mathrm{H}^{(\mu)}(q, n) = e^{q\mu} \frac{\partial^{\mu} \mathrm{H}(q, n)}{\partial (e^q - 1)^{\mu}} \tag{12.58}$$

$$\mathrm{H}(q, n) = \sum_{i=0}^{k} a_i[n](e^q - 1)^i \tag{12.59}$$

When (12.50), (12.54) and (12.57) are taken into account, (12.49) can be written as follows:

$$\begin{aligned} &\sum_{\mu=0}^{k} \frac{1}{\mu!} \Delta_n^{\mu} W^*(q, n, \varepsilon) \mathrm{H}^{(\mu)}(q, n) \\ &+ \sum_{\nu=0}^{h} \frac{1}{\nu!} \Delta_n^{\nu} W^*(q, n, 0) S^{(\nu)}(q, \overline{t}) = S(q, \overline{t}), \quad 0 \leqslant \varepsilon \leqslant 1 \end{aligned} \tag{12.60}$$

Equation (12.60) enables the transfer function $W^*(q, n, \varepsilon)$ of the closed-loop system in question to be found in terms of the difference equation of the open-loop system.

When $\varepsilon = 0$ transfer function $W^*(q, n, 0)$ can be found from the equation that is a particular case of (12.60):

$$\begin{aligned} &\sum_{\mu=0}^{h} \frac{1}{\mu!} \{\mathrm{H}^{(\mu)}(q, n) + S^{(\mu)}(q, n)\} \Delta_n^{\mu} W^*(q, n, 0) \\ &+ \sum_{\nu=h+1}^{k} \frac{1}{\nu!} \mathrm{H}^{(\nu)}(q, n) \Delta_n^{\nu} W^*(q, n, 0) = S(q, n), \quad 0 \leqslant \varepsilon \leqslant 1 \end{aligned} \tag{12.61}$$

Equation (12.61) coincides with that obtained in another way by Tartakovskii [20]. The above relations remain valid also for stationary sampled data systems.

We have now considered the most important characteristics of sampled data systems both in the time domain and in the domain of transforms. In analysing these systems we also have to use the concept of system error coefficients.

The determination of the error coefficients and methods for calculating them can be found in the work of Tsypkin and of Perov.

We shall now analyse an important class of sampled data systems, which we shall call adjoint sampled data systems. First, however, it will be necessary to consider certain problems of the theory of difference equations with variable coefficients.

Chapter 2

ADJOINT SYSTEMS

Adjoint systems play an important part in the theory of continuous automatic control systems. The same is true also for the theory of sampled data systems. In this chapter we shall define the adjoint sampled data system and study its fundamental properties, which are very useful in statistical analysis and synthesis of sampled data systems. Before defining the adjoint sampled data system we shall consider certain problems of the theory of linear difference equations with variable coefficients.

2.1 FORMULATION OF A DIFFERENCE BOUNDARY-VALUE PROBLEM

In the theory of finite differences [18] we have a problem concerning the solution of a homogeneous difference equation, which is analogous to the Cauchy problem in the theory of differential equations. Given the k-th order linear difference equation

$$L_n(\Delta, n)\, y[n] = \sum_{i=0}^{k} a_i[n]\, \Delta^i y[n] = 0 \tag{1.1}$$

(where $a_i[n]$ are given functions of the discrete argument $n = 0, 1, \ldots,$) and the initial conditions

$$\Delta^i y[0] = y_{0i},\ i = 0, 1, \ldots, k-1 \tag{1.2}$$

determine the function $y[n]$ which transforms (1.1) into an identity for a given set of values of n. In what follows this problem is called a linear difference problem with initial conditions.

The geometrical meaning of the solution is as follows: it is required to find such a curve $y[n]$ which, in accordance with (1.2), will pass through the point $(0; y_{00})$ and have differences at this point equal to y_{0i} $(i=1,2, \ldots, k-1)$.

We shall now give, without proof, a theorem on the solutions of a homogeneous difference equation with initial conditions, which will be required later. If the solutions of such an equation with the initial conditions $y_1[n]$, $y_2[n]$, ..., $y_n[n]$ are linearly dependent, i.e. for the set of values $n=0, 1, \ldots$ the following relation applies

$$\lambda_1 y_1[n]+\lambda_2 y_2[n]+\ldots+\lambda_k y_k[n]=0 \tag{1.3}$$

where λ_i are constants, which are not all zero at the same time, then the determinant

$$D\{y_1[n], \ldots, y_k[n]\}=\begin{vmatrix} y_1[n] & y_2[n] & \ldots & y_k[n] \\ \Delta y_1[n] & \Delta y_2[n] & \ldots & \Delta y_k[n] \\ \ldots & \ldots & \ldots & \ldots \\ \Delta^{k-1} y_1[n] & \Delta^{k-1} y_2[n] & \ldots & \Delta^{k-1} y_k[n] \end{vmatrix} \tag{1.4}$$

is zero for the entire set $N=\{n\}$, on which the linearly dependent solutions $y_i[n]$ are defined. Conversely, if $y_i[n]$ are linearly independent, then

$$D\{y_1[n], \ldots, y_k[n]\} \neq 0 \tag{1.5}$$

for any value of n belonging to the set N. The proof of this theorem is given by Gel'fond [18]. Without giving the proof we only remark that this theorem is generalised for the case where the domain of definition of the solutions is given by the set of values $N=\{-\infty<n<\infty\}$ and $0 \leqslant \varepsilon \leqslant 1$ for each n belonging to N, i.e. the following theorem is valid.

Theorem. If $y_1[n, \varepsilon]$, $y_2[n, \varepsilon], \ldots$, $y_k[n, \varepsilon]$ are found to be the linearly dependent solutions of the homogeneous difference problem with initial conditions

$$L_n(\Delta, n)\, y[n, \varepsilon] = \sum_{i=0}^{k} a_i[n]\, \Delta^i y[n, \varepsilon] = 0, \ 0 \leqslant \varepsilon \leqslant 1$$
$$\Delta^i y[0, \varepsilon] = y_{0i}(\varepsilon), \ i = 0, 1, \ldots, k-1 \tag{1.6}$$

where $a_i[n]$ are functions of n defined on the set N;
$y_{0i}(\varepsilon)$ are functions of ε defined on the set $0 \leqslant \varepsilon \leqslant 1$, then the determinant

$$D\{y_1[n, \varepsilon], \ldots, y_k[n, \varepsilon]\} = \begin{vmatrix} y_1[n, \varepsilon] & y_2[n, \varepsilon] & \ldots & y_k[n, \varepsilon] \\ \Delta y_1[n, \varepsilon] & \Delta y_2[n, \varepsilon] & \ldots & \Delta y_k[n, \varepsilon] \\ \cdots & \cdots & \cdots & \cdots \\ \Delta^{k-1} y_1[n, \varepsilon] & \Delta^{k-1} y_2[n, \varepsilon] & \ldots & \Delta^{k-1} y_k[n, \varepsilon] \end{vmatrix} \tag{1.7}$$

is zero for the entire set of values n belonging to N and $0 \leqslant \varepsilon \leqslant 1$, on which the $y_i[n, \varepsilon]$ are defined and are linearly dependent. Conversely, if $y_i[n, \varepsilon]$ are linearly independent, then

$$D\{y_1[n, \varepsilon], \ldots, y_k[n, \varepsilon]\} \neq 0 \tag{1.8}$$

for all values $\bar{t} = n + \varepsilon$ belonging to the set just mentioned.

A corollary of this theorem is that it is valid for any subset $N' = \{\nu \leqslant n \leqslant \mu\}$, $0 \leqslant \varepsilon \leqslant 1$ of the set N.

Let us now generalise the homogeneous difference problem with initial conditions. For this we consider the k-th order equation

$$L_n(\Delta, n)\, y[n, \varepsilon] = 0, \ 0 \leqslant \varepsilon \leqslant 1 \tag{1.9}$$

the coefficients of which, $a_i[n]$ $(i = 0, 1, \ldots, k)$, are defined on the finite interval $[\nu; \mu]$. In what follows we shall assume that this condition is satisfied in all cases.

In determing the solution of (1.9) we require that at the ends of the interval $[\nu, \varepsilon; \mu, \varepsilon]$ this solution satisfies the conditions expressed by

$$V_l(y[\nu, \varepsilon], \Delta y[\nu, \varepsilon], \ldots, \Delta^{s_1} y[\nu, \varepsilon], \ y[\mu, \varepsilon], \Delta y[\mu, \varepsilon], \ldots$$
$$\Delta^{s_2} y[\mu, \varepsilon]) = 0, \tag{1.10}$$
$$l = 1, 2, \ldots, m, \ s_1, \ s_2 \leqslant k-1$$

which, in general, can be non-linear functions of its arguments.

The relations (1.10) impose definite conditions on the solution $y[n, \varepsilon]$ of (1.9) and its differences $\Delta^i y[n, \varepsilon]$ at the boundaries of the interval $[\nu, \varepsilon; \mu, \varepsilon]$.These conditions are called boundary conditions. The boundary conditions (1.10) are assumed to be linearly independent, since if some of V_l, $l=1,2, \ldots, m$ are linear combinations of the rest, they can be discarded.

The problem of determining the solution of (1.9) that satisfies the boundary value problem (1.10) is called the difference boundary value problem. Geometrically the solution of a difference boundary value problem constitutes a curve $y[n, \varepsilon]$, which at the ends of the interval $[\nu, \varepsilon; \mu, \varepsilon]$ satisfied the conditions (1.10).

2.2 LINEAR DIFFERENCE BOUNDARY-VALUE PROBLEMS

We shall consider a particular case of the difference boundary value problem which is of practical importance, where the difference equation and the functions V_l, $l=1, 2, \ldots, m$ which define the difference boundary-value problem are linear. Difference boundary-value problems of this class are called linear difference boundary-value problems. We introduce the following definitions.

Definition 1. A non-homogeneous linear difference boundary-value problem is the term given to the problem of solving the linear difference equation

$$L_n(\Delta, n)\, y[n, \varepsilon] = 0, \quad 0 \leqslant \varepsilon \leqslant 1 \tag{2.1}$$

satisfying at the ends of the interval $[\nu, \varepsilon; \mu, \varepsilon]$ the following boundary conditions

$$\sum_{i=0}^{k-1} \{\alpha_{il} \Delta^i y[\nu, \varepsilon] + \beta_{il} \Delta^i y[\mu, \varepsilon]\} = \gamma_l \qquad l = 1, 2, \ldots, m \tag{2.2}$$

where α_{il}, β_{il} and γ_l are given constants, with

$$\sum_{i=0}^{k-1} \{|\alpha_{il}| + |\beta_{il}|\} \neq 0, \quad l = 1, 2, \ldots, m \tag{2.3}$$

for example all α_{il} and β_{il} are not zero at the same time.

Definition 2. A homogeneous linear difference boundary-value problem is the term given to the problem of solving the linear difference equation (2.1) which at the ends of the interval $[\nu, \varepsilon; \mu, \varepsilon]$ satisfies the boundary conditions

$$R_l(y) \equiv \sum_{i=0}^{k-1} \{\alpha_{il}\Delta^i y[\nu, \varepsilon] + \beta_{il}\Delta^i y[\mu, \varepsilon]\} = 0$$
$$l = 1, 2, \ldots, m \tag{2.4}$$

where (2.3) applies to α_{il} and β_{il}.

It is easy to see that if the boundary conditions are defined only for one end of the interval which, in particular, can be zero, then the homogeneous linear difference boundary-value problem degenerates into the homogeneous difference problem with initial conditions, considered by Gel'fond[18].

In the following we shall be interested mainly in the homogeneous linear difference boundary-value problem. For the sake of brevity we shall call it simply the difference boundary-value problem.

We now pass on to prove a number of theorems concerning the solutions of a difference boundary-value problem. These theorems will be required later.

Theorem 1. Every difference boundary-value problem has a fundamental set of solutions.*

Proof. We choose two sets of k^2 functions $\lambda_{ij}(\varepsilon)$ and $\gamma_{ij}(\varepsilon)$ $(i, j = 1, 2, \ldots, k, 0 \leqslant \varepsilon \leqslant 1)$ so that the determinants

$$\det \| \lambda_{ij}(\varepsilon) \|_{i, j = 1, 2, \ldots, k} \neq 0, \quad 0 \leqslant \varepsilon \leqslant 1 \tag{2.5}$$

$$\det \| \gamma_{ij}(\varepsilon) \|_{i, j = 1, 2, \ldots, k} \neq 0, \quad 0 \leqslant \varepsilon \leqslant 1 \tag{2.6}$$

Let $y_1[n, \varepsilon], \ldots, y_k[n, \varepsilon]$ be the particular solutions of the boundary-value problem under consideration, which has the boundary conditions

$$y_i[\nu, \varepsilon] = \lambda_{1i}(\varepsilon), \ \Delta y_i[\nu, \varepsilon] = \lambda_{2i}(\varepsilon), \ldots, \ \Delta^{k-1} y_i[\nu, \varepsilon] = \lambda_{ki}(\varepsilon)$$

$$y_i[\mu, \varepsilon] = \gamma_{1i}(\varepsilon), \ \Delta y_i[\mu, \varepsilon] = \gamma_{2i}(\varepsilon), \ldots, \ \Delta^{k-1} y_i[\mu, \varepsilon] = \gamma_{ki}(\varepsilon)$$

$$i = 1, 2, \ldots, k$$

* We shall prove the theorem for the case where $m = k$.

Then the determinants (2.5) and (2.6) are determinants of the form (1.7) for $n=\nu$ and $n=\mu$ respectively, and are non-zero at the points $\bar{t}=\nu+\varepsilon$, $\bar{t}=\mu+\varepsilon$ $(0\leqslant\varepsilon\leqslant1)$. Therefore, by the theorem of Section 2.1, $y_i[n, \varepsilon]$ are linearly independent, i.e. they form the fundamental set of solutions of the difference boundary-value problem in question. This proves Theorem 1.

Theorem 2. If $y_1[n, \varepsilon]$, $y_2[n, \varepsilon], \ldots,$ $y_k[n,\varepsilon]$ form the fundamental set of solutions of a difference boundary-value problem, then the general solution of the problem is represented in the form

$$y[n, \varepsilon]=c_1 y_1 [n, \varepsilon]+c_2y_2[n, \varepsilon]+\ldots+c_ky_k[n, \varepsilon] \qquad (2.7)$$

where $c_1, c_2, \ldots, c_k$ are arbitrary constants.

Proof. Substituting the general solution (2.7) into the boundary conditions (1.4) we obtain the system of linear homogeneous equations for the arbitrary constants c_j

$$\sum_{j=1}^{k} c_j \sum_{i=0}^{k-1} \{\alpha_{il}\Delta^i y_j[\nu, \varepsilon]+\beta_{il}\Delta^i y_j[\mu, \varepsilon]\} \qquad (2.8)$$

or

$$\left.\begin{aligned} &\sum_{j=1}^{k} c_jR_l(y_j)=0 \\ &l=1, 2, \ldots, m \end{aligned}\right\} \qquad (2.9)$$

As $R_l(y)=0$, the rank of the matrix formed from the coefficients of (2.9),

$$R=\| R_l(y_j) \|_{\substack{l=1, 2, \ldots, m\\ j=1, 2, \ldots, k}} \qquad (2.10)$$

is zero. Consequently, (2.9) admits only one single-valued set of solutions $c_1, c_2, \ldots, c_k$, for which the expression (2.7) satisfies the boundary conditions (1.4). This proves the theorem.

We have the following corollaries of Theorem 2.

Corollary 1. If the rank of matrix R is h, the difference boundary-value problem has $k-h$ independent solutions.

Corollary 2. If $m=k$ and the rank of the matrix R is $\imath=k$, the difference boundary-value problem has only the trivial solution $y[n, \varepsilon]\equiv 0$. As a consequence of this, for any fundamental set of solutions $y_i[n, \varepsilon]$ $(i=1, 2, \ldots, k)$ the determinant

$$\det \| R_l (y_j) \|_{j,\, l=1,\, 2, \ldots, k} \neq 0 \tag{2.11}$$

2.3 GREEN'S FUNCTION

Let a difference boundary-value problem be given by the k-th order difference equation

$$L_n (\Delta,\ n) y [n,\ \varepsilon] \equiv \sum_{i=0}^{k} a_i [n] \Delta^i y [n,\ \varepsilon] = 0,\ 0 \leqslant \varepsilon \leqslant 1 \tag{3.1}$$

and the boundary conditions

$$R_l (y) = 0,\ l = 1,\ 2, \ldots,\ k \tag{3.2}$$

Similarly to the proof for differential operators [24], it can be shown that the operator $L_n^{-1} (\Delta,\ n)$, which is the inverse of an operator $L_n (\Delta,\ n)$, exists if, and only if, the equation

$$L_n (\Delta,\ n) y [n,\ \varepsilon] = 0,\ 0 \leqslant \varepsilon \leqslant 1 \tag{3.3}$$

has the one and only trivial solution $y=0$. Accordingly, by Corollary 2 of Theorem 2 of the preceding section, the rank of the matrix

$$\left\| \begin{matrix} R_1 (y_1) & R_1 (y_2) \ldots & R_1 (y_k) \\ R_2 (y_1) & R_2 (y_2) \ldots & R_2 (y_k) \\ \cdots & \cdots & \cdots \\ R_k (y_1) & R_k (y_2) \ldots & R_k (y_k) \end{matrix} \right\|$$

where $y_i[n,\varepsilon]$ $(i=1, 2, \ldots, k)$, the fundamental set of solutions of the equation $L_n (\Delta,\ n) y=0$, is equal to k. Consequently,

$$\det \| R_l (y_j) \|_{l,\, j=1,\, 2, \ldots,\, k} \neq 0 \tag{3.4}$$

An example of $L_n^{-1} (\Delta,\ n)$ which is inverse to the operator

$L_n(\Delta, n)$ is given by the summation operator. The kernel of the summation operator is the Green's function of a difference boundary-value problem. We shall now define Green's function.

Definition. Green's function of a homogeneous linear difference boundary-value problem is the term given to a function $G[n, \varepsilon; m]$ which satisfies the following conditions:

1. $G[n, \varepsilon; m]$ is piecewise-continuous and has piecewise-continuous differences with respect to n up to and including the $(k-2)$-th order for all n and m belonging to the interval $[\nu, \varepsilon; \mu, \varepsilon]$.
2. For any fixed value of m belonging to the interval $[\nu, \varepsilon; \mu, \varepsilon]$ the function $G[n, \varepsilon; m]$ has for $n=m$ continuous differences with respect to n $(0<\varepsilon<1)$ up to and including the $(k-2)$-th order in each of the intervals $[\nu, \varepsilon; m]$ and $(m; \mu; \varepsilon]$, with the $(k-1)$-th order difference having for $n=m$ a jump equal to

$$\Delta_n^{k-1} G[n, \varepsilon; m]\,|_{n=m} = \frac{1}{a_k[m]} \tag{3.5}$$

3. Over each of the intervals $[\nu, \varepsilon; m)$ and $(m; \mu, \varepsilon]$ the function $G[n, \varepsilon; m]$ satisfies the homogeneous difference equation

$$L_n(\Delta, n)\, G[n, \varepsilon; m] = 0, \quad 0 \leqslant \varepsilon \leqslant 1$$

and the boundary conditions

$$R_l(G) = 0, \quad l = 1, 2, \ldots, k$$

Theorem 1. If a difference boundary-value problem is given by the difference equation

$$L_n(\Delta, n)\, y[n, \varepsilon] = 0, \quad 0 \leqslant \varepsilon \leqslant 1 \tag{3.6}$$

and the boundary conditions

$$R_l(y) = 0, \quad l = 1, 2, \ldots, k \tag{3.7}$$

and if it has only a trivial solution, then this boundary-value problem has only one Green's function.

Proof Let $y_1[n, \varepsilon], y_2[n, \varepsilon], \ldots, y_k[n, \varepsilon]$ be linearly independent solutions of the homogeneous equation (3.6). Then, by the third property of Green's function and Theorem 2 of

Section 2.2 the following is true:

$$G[n, \varepsilon; m] = \sum_{i=1}^{k} \alpha_i y_i[n, \varepsilon], \quad \nu \leqslant n < m, \quad 0 \leqslant \varepsilon \leqslant 1 \tag{3.8}$$

where α_i are functions of m and the parameter ε. Analogously for the second interval

$$G[n, \varepsilon; m] = \sum_{i=1}^{k} \beta_i y_i[n, \varepsilon], \quad m < n \leqslant \mu, \quad 0 \leqslant \varepsilon \leqslant 1 \tag{3.9}$$

where β_i are also functions of m and ε.

From the continuity condition for the differences of the function $G[n, \varepsilon; m]$ with respect to the argument n up to and including the $(k-2)$-th order, when $n=m$, we can write

$$\begin{gathered}
\sum_{i=1}^{k} \alpha_i y_i[m, \varepsilon] - \sum_{i=1}^{k} \beta_i y_i[m, \varepsilon] = 0 \\
\sum_{i=1}^{k} \alpha_i \Delta y_i[m, \varepsilon] - \sum_{i=1}^{k} \beta_i \Delta y_i[m, \varepsilon] = 0 \\
\cdots\cdots\cdots\cdots\cdots\cdots \\
\sum_{i=1}^{k} \alpha_i \Delta^{k-2} y_i[m, \varepsilon] - \sum_{i=1}^{k} \beta_i \Delta^{k-2} y_i[m, \varepsilon] = 0
\end{gathered} \tag{3.10}$$

In addition, (3.5) gives

$$\sum_{i=1}^{k} \alpha_i \Delta^{k-1} y_i[m, \varepsilon] - \sum_{i=1}^{k} \beta_i \Delta^{k-1} y_i[m, \varepsilon] = \frac{1}{a_k[m]} \tag{3.11}$$

Putting

$$c_i = \beta_i - \alpha_i, \quad i = 1, 2, \ldots, k$$

we write (3.10) and (3.11) in the form of a system of k equations:

$$\begin{gathered}
\sum_{i=1}^{k} c_i y_i[m, \varepsilon] = 0, \qquad \sum_{i=1}^{k} c_i \Delta y_i[m, \varepsilon] = 0 \\
\cdots\cdots\cdots\cdots \quad \sum_{i=1}^{k} c_i \Delta^{k-1} y_i[m, \varepsilon] = \frac{1}{a_k[m]}
\end{gathered} \tag{3.12}$$

The determinant of the system (3.12) is a determinant of

the type (1.7) of the fundamental set of solutions $y_1, y_2 \ldots, y_k$, being non-zero, for $n=m$, because of (3.4). Consequently, (3.12) uniquely defines c_i.

To determine the functions α_i and β_i we use the boundary conditions (3.7), written in the form

$$R_l(y)=R_{l_\nu}(y)+R_{l\mu}(y) \tag{3.13}$$

where $R_{l_\nu}(y)$ includes all the terms containing $\Delta^i y[\nu, \varepsilon]$, $i=0, 1, \ldots, k-1$, and $R_{l\mu}(y)$ includes all the terms containing analogous differences at the point $n=\mu$.

Taking into account (3.8), (3.9) and (3.13) we write the boundary conditions in the form

$$R_l(G)=\sum_{i=1}^{k}\alpha_i R_{l_\nu}(y_i)+\sum_{i=1}^{k}\beta_i R_{l\mu}(y_i)=0 \tag{3.14}$$

But since

$$\alpha_i=\beta_i-c_i$$

we obtain from (3.14)

$$\sum_{i=1}^{k}\beta_i R_{l\mu}(y_i)+\sum_{i=1}^{k}(\beta_i-c_i)R_{l_\nu}(y_i)=0 \tag{3.15}$$

Equation (3.15) can be rewritten in the form

$$\sum_{i=1}^{k}\beta_i[R_{l\mu}(y_i)+R_{l_\nu}(y_i)]-\sum_{i=1}^{k}c_i R_{l_\nu}(y_i)=0$$

or, when (3.13) is used, in the final form

$$\sum_{i=1}^{k}\beta_i R_l(y_i)=\sum_{i=1}^{k}c_i R_{l_\nu}(y_i), \quad l=1, 2, \ldots, k \tag{3.16}$$

Thus, for the determination of β_i we have (3.16), whose determinant is non-zero because of (3.4). Consequently, (3.16) has only one solution for β_i. But since $\alpha_i=\beta_i-c_i$, the function α_i is determined at the same time. This proves the existence and uniqueness of Green's functions for a difference boundary-value problem.

Theorem 2. If a difference boundary-value problem is given by the difference equation

$$L_n(\Delta, n)\,y[n, \varepsilon]=0, \quad 0\leqslant\varepsilon\leqslant 1$$
$$L_n(\Delta, n)=\sum_{i=0}^{k} a_i[n]\,\Delta_n^i \tag{3.17}$$

and the boundary conditions

$$R_l(y)=0, \quad l=1, 2, \ldots, k \tag{3.18}$$

and if it has only a trivial solution, then for any function $x[n, \varepsilon]$ which is continuous in the interval $[\nu, \varepsilon;\ \mu, \varepsilon]$ there is a solution for the equation

$$L_n(\Delta, n)\,y[n, \varepsilon]=x[n, 0], \quad 0\leqslant\varepsilon\leqslant 1 \tag{3.19}$$

This solution is given by

$$y[n, \varepsilon]=\sum_{m=\nu}^{\mu} G[n, \varepsilon;\ m]\,x[m, 0] \tag{3.20}$$

where $G[n, \varepsilon;\ m]$ is Green's function of the difference boundary-value problem under consideration.

Proof. We must show that the function $y[n, \varepsilon]$ determined from (3.20) satisfies (3.19) and the boundary conditions (3.18). We shall find the differences of order $i=1, 2, \ldots, k$ of the function $y[n, \varepsilon]$ determined from (3.20). Since according to the definition of Green's function the differences

$$\Delta_n^i G[n, \varepsilon;\ m], \quad i=1, 2, \ldots, k-2$$

are piecewise-continuous for any values of n and m belonging to the interval $[\nu, \varepsilon;\ \mu, \varepsilon]$, we have from (3.20)

$$\Delta_n^i y[n, \varepsilon]=\sum_{m=\nu}^{\mu} \Delta_n^i G[n, \varepsilon;\ m]\,x[m, 0]$$
$$i=1, 2, \ldots, k-2 \tag{3.21}$$

Here the function $y[n, \varepsilon]$ and the differences $\Delta_n^i y[n, \varepsilon]$ up to and including the $(k-2)$-th order are piecewise-continuous on the interval $[\nu, \varepsilon;\ \mu, \varepsilon]$. The $(k-1)$-th order difference of the Green's function has a jump for $n=m$. Hence, to determine the differences $\Delta_n^{k-1} y[n, \varepsilon]$ and $\Delta_n^k y[n, \varepsilon]$ we use the formula for a generalised difference obtained in Chapter 1:

$$\Delta_n \{f[n, \varepsilon]\cdot 1[n-m, 0]\} = \Delta f[n, \varepsilon]\cdot 1[n-m, 0] + f[m, \varepsilon]\,\sigma[n-(m-1), 0] \tag{3.22}$$

Using (3.21) we have

$$\Delta_n^{k-2} y[n, \varepsilon] = \sum_{m=\nu}^{\mu} \Delta_n^{k-2} G[n, \varepsilon; m]\, x[m, 0] \tag{3.23}$$

To determine $\Delta_n^{k-1} y[n, \varepsilon]$ we write (3.23) in the form

$$\Delta_n^{k-2} y[n, \varepsilon] = \sum_{m=\nu}^{\mu} \Delta_n^{k-2} G[n, \varepsilon; m]\cdot 1[n-m, 0]\, x[m, 0] \tag{3.24}$$

Applying (3.22) to (3.24) we have

$$\Delta_n^{k-1} y[n, \varepsilon] = \sum_{m=\nu}^{\mu} \{\Delta_n^{k-1} G[n, \varepsilon; m]\cdot 1[n-m, 0]\, x[m, 0] + \Delta_n^{k-2} G[m, \varepsilon; m]\,\sigma[n-(m-1), 0]\, x[m, 0]\}$$

But since

$$\Delta_n^{k-2} G[n, \varepsilon; m]|_{n=m} = 0$$

we finally obtain

$$\Delta_n^{k-1} y[n, \varepsilon] = \sum_{m=\nu}^{\mu} \Delta_n^{k-1} G[n, \varepsilon; m]\cdot 1[n-m, 0]\, x[m, 0] \tag{3.25}$$

Applying (3.22) to (3.25) we obtain for the k-th difference

$$\Delta_n^{k} y[n, \varepsilon] = \sum_{m=\nu}^{\mu} \Delta_n^{k} G[n, \varepsilon; m]\, x[m, 0] + \frac{1}{a_k[n]}\, x[n, 0] \tag{3.26}$$

Substituting now (3.20), (3.21), (3.25) and (3.26) into (3.19), we have

$$L_n(\Delta, n)\, y[n, \varepsilon] = \sum_{i=0}^{k} a_i[n] \sum_{m=\nu}^{\mu} \Delta_n^{i} G[n, \varepsilon; m]\, x[m, 0] + x[n, 0]$$

or

$$L_n(\Delta, n)\, y[n, \varepsilon] = \sum_{m=\nu}^{\mu} L_n(\Delta, n)\, G[n, \varepsilon; m]\, x[m, 0] + x[n, 0]$$

Since Green's function satisfies the homogeneous difference equation

$$L_n(\Delta, n)G[n, \varepsilon; m]=0$$

we obtain from the last relation

$$L_n(\Delta, n)y[n, \varepsilon]=x[n, 0]$$

which coincides with (3.19). Consequently, the function $y[n, \varepsilon]$ given by (3.20) satisfies the non-homogeneous difference equation (3.19).

We shall now prove that $y[n, \varepsilon]$ satisfies also the boundary conditions (3.18). Since

$$\left.\begin{array}{c} R_l(y)=\sum_{i=0}^{k-1}\alpha_{il}\Delta_n^i y[\nu, \varepsilon]+\sum_{i=0}^{k-1}\beta_{il}\Delta_n^i y[\mu, \varepsilon] \\ l=1, 2, \ldots, k \end{array}\right\} \quad (3.27)$$

then, substituting $\Delta_n^i y[n, \varepsilon]|_{n=\nu}$ and $\Delta_n^i y[n, \varepsilon]|_{n=\mu}$ into (3.27) we obtain

$$\begin{aligned} R_l(y)=&\sum_{i=0}^{k-1}\alpha_{il}\sum_{m=\nu}^{\mu}\Delta_n^i G[\nu, \varepsilon; m]\,x[m, 0] \\ &+\sum_{i=0}^{k-1}\beta_{il}\sum_{m=\nu}^{\mu}\Delta_n^i G[\mu, \varepsilon; m]\,x[m, 0] \end{aligned} \quad (3.28)$$

Changing the order of summation in (3.28) we have

$$R_l(y)=\sum_{m=\nu}^{\mu}R_{l\nu}(G)\,x[m, 0]+\sum_{m=\nu}^{\mu}R_{i\mu}(G)\,x[m, 0]$$

$$=\sum_{m=\nu}^{\mu}[R_{l\nu}(G)+R_{l\mu}(G)]\,x[m, 0]=\sum_{m=\nu}^{\mu}R_l(G)\,x[m, 0]$$

Since Green's function satisfies the boundary conditions, then

$$R_l(G)=0, \ l=1, 2, \ldots, k$$

and from the last relation we obtain

$$R_l(y)\equiv 0, \ l=1, 2, \ldots, k$$

which coincides with (3.18). Thus, the function $y[n, \varepsilon]$ given by (3.20) satisfies the difference equation (3.17) and the boundary conditions (3.18). This proves Theorem 2.

It follows from Theorem 1 of the present section that the Green's function $G[n, \varepsilon; m]$ of the difference boundary-value problem (3.1) and (3.2) is its fundamental solution over the interval $[\nu, \varepsilon; \mu, \varepsilon]$.

We shall now show that the Green's function $G[n, \varepsilon; m]$ of the difference boundary-value problem, as a function of n, can be represented in the form

$$G[n, \varepsilon; m] = \sum_{i=1}^{k} y_i^*[m, \varepsilon]\, y_i[n, \varepsilon] \tag{3.29}$$

where $y_i[n, \varepsilon]$, $i = 1, 2, \ldots, k$ are the fundamental set of solutions of (3.1); $y_i^*[m, \varepsilon]$ are functions of m and ε determined by the boundary conditions (3.2).

We give the proof of this property of the Green's function for the case where (3.2) has the form

$$\left.\begin{aligned} \Delta_n^i y[n, \varepsilon]\,|_{n=m} &= 0, \quad i = 0, 1, \ldots, k-2 \\ \Delta_n^{k-1} y[n, \varepsilon]\,|_{n=m} &= \frac{1}{a_k[m]}, \quad 0 \leqslant \varepsilon \leqslant 1 \end{aligned}\right\} \tag{3.30}$$

Suppose the fundamental set of solutions

$$y_i[n, \varepsilon], \; i = 1, 2, \ldots, k$$

of the difference equation (3.1) is known. Then, by Theorem 1 of this section, the Green's function is expressed by

$$G[n, \varepsilon; m] = \sum_{i=1}^{k} \eta_i y_i[n, \varepsilon] \tag{3.31}$$

where η_i are certain functions determined by the boundary conditions (3.30). Substituting (3.31) into (3.30) we obtain the system of equations for η_i:

$$\left.\begin{aligned} \sum_{i=1}^{k} \eta_i \Delta^l y_i[m, \varepsilon] &= 0, \quad l = 0, 1, \ldots, k-2 \\ \sum_{i=1}^{k} \eta_i \Delta^{k-1} y_i[m, \varepsilon] &= \frac{1}{a_k[m]}, \quad 0 \leqslant \varepsilon \leqslant 1 \end{aligned}\right\} \tag{3.32}$$

$D\{y_1[m, \varepsilon], \ldots, y_k[m, \varepsilon]\}$ the determinant of the system (3.22), is non-zero; consequently, the system has a single-valued solution for η_i. Solving this system and substituting the solution into (3.31) we have

$$G[n, \varepsilon; m] = \sum_{i=1}^{k} (-1)^{k+i} \frac{1}{a_k[m]} \frac{D_i[m, \varepsilon]}{D[m, \varepsilon]} y_i[n, \varepsilon] \quad (3.33)$$

where for the sake of brevity

$$D[m, \varepsilon] = D\{y_1[m, \varepsilon], \ldots, y_k[m, \varepsilon]\}$$

while $D_i[m, \varepsilon]$ is obtained from $D[m, \varepsilon]$ using the Cramer method.

Putting

$$(-1)^{k+i} \frac{1}{a_k[m]} \frac{D_i[m, \varepsilon]}{D[m, \varepsilon]} = y_i^*[m, \varepsilon]$$

in (3.33), we finally obtain

$$G[n, \varepsilon; m] = \sum_{i=1}^{k} y_i^*[m, \varepsilon] y_i[n, \varepsilon] \quad (3.34)$$

which coincides with (3.29).

We point out that if (3.33) is taken into account, the expression can be written in the form

$$G[n, \varepsilon; m] = \frac{1}{a_k[m]} \frac{\begin{vmatrix} y_1[m, \varepsilon] & y_2[m, \varepsilon] & \ldots & y_k[m, \varepsilon] \\ \Delta y_1[m, \varepsilon] & \Delta y_2[m, \varepsilon] & \ldots & \Delta y_k[m, \varepsilon] \\ \ldots & \ldots & \ldots & \ldots \\ \Delta^{k-2} y_1[m, \varepsilon] & \Delta^{k-2} y_2[m, \varepsilon] & \ldots & \Delta^{k-2} y_k[m, \varepsilon] \\ y_1[n, \varepsilon] & y_2[n, \varepsilon] & \ldots & y_k[n, \varepsilon] \end{vmatrix}}{\begin{vmatrix} y_1[m, \varepsilon] & y_2[m, \varepsilon] & \ldots & y_k[m, \varepsilon] \\ \Delta y_1[m, \varepsilon] & \Delta y_2[m, \varepsilon] & \ldots & \Delta y_k[m, \varepsilon] \\ \ldots & \ldots & \ldots & \ldots \\ \Delta^{k-1} y [m, \varepsilon] & \Delta^{k-1} y_2[m, \varepsilon] & \ldots & \Delta^{k-1} y_k[m, \varepsilon] \end{vmatrix}} \quad (3.35)$$

2.4 ADJOINT DIFFERENCE EQUATIONS

Suppose we are given the linear difference equation with variable coefficients (for the sake of simplicity we consider lattice functions)

$$L_n(\Delta, n)\, y[n] \equiv a_0[n]\, y[n] + a_1[n]\, \Delta y[n] + \ldots + a_k[n]\Delta^k y[n] \tag{4.1}$$

where the coefficients $a_i[n]$ $(i=0, 1, \ldots, k)$ are defined on the interval $[\nu, \mu]$. We multiply the difference operator $L_n(\Delta, n)$ applied to the function y by the function $z[n]$ and sum the expression so obtained by parts between the limits $[\nu; \mu]$. We carry out the summation, reducing the order of the difference $\Delta^i y[n]$ until the factor $y[n]$ alone is left behind the summation sign. Applying the formula of summing by parts

$$\sum_{n=\nu}^{\mu} u[n]\, \Delta v[n] = u[n]\, v[n] \Big|_{\nu}^{\mu-1} - \sum_{n=\nu}^{\mu} v[n+1]\, \Delta u[n] \tag{4.2}$$

we have

$$\sum_{n=\nu}^{\mu} a_0[n]\, z[n]\, y[n] = \sum_{n=\nu}^{\mu} a_0[n]\, z[n]\, y[n]$$

$$\sum_{n=\nu}^{\mu} a_1[n]\, z[n]\, \Delta y[n] = a_1[n]\, z[n]\, y[n] \Big|_{\nu}^{\mu-1} - \sum_{n=\nu}^{\mu} \Delta\{a_1[n]\, z[n]\}\, y[n+1]$$

$$\sum_{n=\nu}^{\mu} a_2[n]\, z[n]\, \Delta^2 y[n] = a_2[n]\, z[n]\, \Delta y[n] \Big|_{\nu}^{\mu-1}$$

$$-\sum_{n=\nu}^{\mu} \Delta\{a_2[n]\, z[n]\}\, \Delta y[n+1] = a_2[n]\, z[n]\, \Delta y[n] \Big|_{\nu}^{\mu-1}$$

$$-\Delta\{a_2[n]\, z[n]\}\, y[n+1] \Big|_{\nu}^{\mu-1} + \sum_{n=\nu}^{\mu} \Delta^2\{a_2[n]\, z[n]\}\, y[n+2]$$

$$\cdots\cdots\cdots\cdots\cdots$$

$$\sum_{n=\nu}^{\mu} a_k[n]\, z[n]\, \Delta^k y[n] = [a_k[n]\, z[n]\, \Delta^{k-1} y[n] - \Delta\{a_k[n]\, z[n]\}\, \Delta^{k-2} y[n+1] + \Delta^2\{a_k[n]\, z[n]\}$$

$$\times \Delta^{k-3} y[n+2] - \ldots + (-1)^{k-1}\Delta^{k-1}\{a_k[n]\, z[n]\}$$

$$\times y[n+k-1]]_{\nu}^{\mu-1} + (-1)^k \sum_{n=\nu}^{\mu} \Delta^k\{a_k[n]\, z[n]\}\, y[n+k]$$

We now transform the right-hand sides of these equations into a form such that the function $y[n]$ is outside the summation sign instead of $y[n+i]$ $(i=1, \ldots, k)$. For this we introduce the new variable $n+i=n'$; then, with the previous notation for the independent variable, we obtain

$$\sum_{n=\nu}^{\mu} \Delta^i \{a_i[n]\, z[n]\}\, y[n+i] = \sum_{n=\nu+i}^{\mu} \Delta^i \{a_i[n-i]\, z[n-i]\}\, y[n] \tag{4.3}$$

In addition, we note that the right-hand side of (4.3) can be written

$$\sum_{n=\nu+i}^{\mu} \Delta^i \{a_i[n-i]\, z[n-i\}\, y[n] = \sum_{n=\nu}^{\mu} \Delta^i \{a_i[n-i]\, z[n-i]\}\, y[n] - \sum_{n=\nu}^{\nu+i-1} \Delta^i \{a_i[n-i]\, z[n-i]\}\, y[n] \tag{4.4}$$

Adding now the relations thus obtained we have from (4.3) and (4.4)

$$\begin{aligned}
\sum_{n=\nu}^{\mu} z[n]\, L_n(\Delta, n)\, y[n] &= \sum_{n=\nu}^{\mu} [a_0[n]\, z[n] \\
&- \Delta \{a_1[n-1]\, z[n-1]\} + \Delta^2 \{a_2[n-2]\, z\,]n-2]\} \\
&- \ldots + (-1)^k \Delta^k \{a_k[n-k]\, z[n-k]\}]\, y[n] \\
&+ [a_1[n]\, z[n]\, y[n] - \Delta \{a_2[n]\, z[n]\}\, y[n+1] + \ldots \\
&+ (-1)^{k-1} \Delta^{k-1} \{a_k[n]\, z[n]\}\, y[n+k-1] \\
&+ a_2[n]\, z[n]\, \Delta y[n] - \Delta \{a_3[n]\, z[n]\}\, \Delta y[n+1] + \ldots \\
&+ (-1)^{k-2} \Delta^{k-2} \{a_k[n]\, z[n]\}\, \Delta y[n+k-2] \\
&\cdots\cdots\cdots\cdots\cdots\cdots \\
&+ a_{k-1}[n]\, z[n]\, \Delta^{k-2} y[n] - \Delta \{a_k[n]\, z[n]\} \\
&\times \Delta^{k-2} y[n+1] + a_k[n]\, z[n]\, \Delta^{k-1} y[n]]_{\nu}^{\mu-1} \\
&+ \sum_{i=1}^{k} (-1)^{i-1} \sum_{n=\nu}^{\nu+i-1} \Delta^i \{a_i[n-i]\, z[n-i]\}\, y[n]
\end{aligned} \tag{4.5}$$

We write this expression more concisely as

$$\sum_{n=\nu}^{\mu}\{z[n]L_n(\Delta,n)y[n]-y[n]L_n^*(\Delta,n)z[n]\} \\ =\Psi^*(y,z)+\Phi^*(y,z) \tag{4.6}$$

where

$$L_n^*(\Delta,n)z[n]=a_0[n]z[n]-\Delta\{a_1[n-1]z[n-1]\} \\ +\Delta^2\{a_2[n-2]z[n-2]\}-\dots \\ +(-1)^k\Delta^k\{a_k[n-k]z[n-k]\} \tag{4.7}$$

$$\Psi^*(y,z)=\Big[\sum_{j=0}^{k-1}\sum_{i=j+1}^{k}(-1)^{i-j-1} \\ \times\Delta^{i-j-1}\{a_i[n]z[n]\}\Delta^j y[n+i-j-1]\Big]_{\nu}^{\mu-1} \tag{4.8}$$

$$\Phi^*(y,z)=\sum_{i=1}^{k}(-1)^{i-1}\sum_{n=\nu}^{\nu+i-1}\Delta^i\{a_i[n-i]z[n-i]\}y[n] \tag{4.9}$$

The difference operator $L_n^*(\Delta, n)$ defined by the relation

$$L_n^*(\Delta,n)z[n]\equiv\sum_{i=0}^{k}(-1)^i\Delta^i\{a_i[n-i]z[n-i]\} \tag{4.10}$$

is called the adjoint difference operator with respect to $L_n(\Delta,n)$ (henceforth the asterisk is used to denote adjoint operators).

In conformity with the definition of an adjoint difference operator, the equation

$$L_n^*(\Delta,n)z[n]=0 \tag{4.11}$$

is called the adjoint difference equation with respect to the equation $L_n(\Delta, n)y[n]=0$.

Relation (4.6) is analogous to Lagrange's formula, which is well known in the theory of differential equations.

We now consider the sum

$$\sum_{n=\nu}^{\mu}y[n]L_n^*(\Delta,n)z[n]=\sum_{n=\nu}^{\mu}y[n]\sum_{i=0}^{k}(-1)^i \\ \times\Delta_n^i\{a_i[n-i]z[n-i]\} \tag{4.12}$$

Using the summation by parts formula, (4.2), we obtain

$$\sum_{n=\nu}^{\mu} y[n]\,a_0[n]\,z[n] = \sum_{n=\nu}^{\mu} z[n]\,a_0[n]\,y[n]$$

$$-\sum_{n=\nu}^{\mu} \Delta\{a_1[n-1]\,z[n-1]\}\,y[n]$$

$$= -y[n]\,a_1[n-1]\,z[n-1]\Big|_{\nu}^{\mu-1} + \sum_{n=\nu}^{\mu} z[n]\,a_1[n]\,\Delta y[n]$$

$$\sum_{n=\nu}^{\mu} \Delta^2\{a_2[n-2]\,z[n-2]\}\,y[n]$$

$$= \Delta\{a_2[n-2]\,z[n-2]\}\,y[n]\Big|_{\nu}^{\mu-1}$$

$$-a_2[n-1]\,z[n-1]\,\Delta y[n]\Big|_{\nu}^{\mu-1} + \sum_{n=\nu}^{\mu} a_2[n]\,z[n]\,\Delta^2 y[n]$$

. . . .

$$(-1)^k \sum_{n=\nu}^{\mu} \Delta^k\{a_k[n-k]\,z[n-k]\}\,y[n]$$

$$= (-1)^k\,[\Delta^{k-1}\{a_k[n-k]\,z[n-k]\}\,y[n]$$

$$-\Delta^{k-2}\{a_k[n-k+1]\,z[n-k+1]\}\,\Delta y[n] + \dots$$

$$+(-1)^{k-1} a_k[n-1]\,z[n-1]\,\Delta^{k-1} y[n]]_{\nu}^{\mu-1}$$

$$+(-1)^k \sum_{n=\nu}^{\mu} a_k[n]\,z[n]\,\Delta^k y[n]$$

Summing the above relations and carrying out certain transformations we obtain

$$\sum_{n=\nu}^{\mu} \{y[n] L_n^*(\Delta, n)\,z[n] - z[n]\,L_n(\Delta, n)\,y[n]\} = \Psi(y, z) + \Phi(y, z) \tag{4.13}$$

where Ψ and Φ are functions which, relative to $\Delta^i y[\nu]$ and $\Delta^i y[\mu]$, are analogous to the functions Ψ^* and Φ^*.

It follows from (4.13) that the difference expression $L_n(\Delta, n)\, y[n]$ is adjoint to the difference expression $L^*_n(\Delta, n)\, z[n]$, i.e. the following relation applies

$$L^{**}_n(\Delta, n)\, y[n] = L_n(\Delta, n)\, y[n] \tag{4.14}$$

The property (4.14) of difference operators can be formulated as follows: the difference operators $L_n(\Delta, n)$ and $L^*_n(\Delta, n)$ are mutually adjoint.

Putting now

$$L_n(\Delta, n) = L_n(\Delta, n)_1 + L_n(\Delta, n)_2$$

where

$$L_n(\Delta, n)_1 = \sum^{k_1} a_{1i}[n]\, \Delta^i \qquad L_n(\Delta, n)_2 = \sum_{i=0}^{k_2} a_{2i}[n]\, \Delta^i$$

we obtain from (4.10)

$$L^*_n(\Delta, n)\, z[n] = \sum_{i=0}^{k_1} (-1)^i \Delta^i \{a_{1i}[n-i]\, z[n-i]\}$$

$$+ \sum_{i=0}^{k_2} (-1)^i \Delta^i \{a_{2i}[n-i]\, z[n-i]\}$$

whence it follows that

$$(L_n(\Delta, n)_1 + L_n(\Delta, n)_2)^* = L^*_n(\Delta, n)_1 + L^*_n(\Delta, n)_2 \tag{4.15}$$

Similarly we can obtain the equation

$$(\lambda L_n(\Delta, n))^* = \lambda L^*_n(\Delta, n) \tag{4.16}$$

where λ is a real number.

The difference operator $L_n(\Delta, n)$ will be called self-adjoint if

$$L^*_n(\Delta, n) = L_n(\Delta, n) \tag{4.17}$$

2.5 THE ADJOINT BOUNDARY-VALUE PROBLEM

The boundary-value problem defined by the difference equation

$$L^*_n(\Delta, n)\, y[n, \varepsilon] = 0, \quad 0 \leqslant \varepsilon \leqslant 1 \tag{5.1}$$

which is adjoint to (3.1), and the boundary conditions

$$R_l^*(y)=0, \quad l=1, 2, \ldots, k \tag{5.2}$$

which are adjoint to the boundary conditions (3.2), is called the adjoint difference homogeneous boundary-value problem. For brevity this will subsequently be called simply the adjoint boundary-value problem.

Theorems corresponding to those proved in Sections 2.2 and 2.3 for the difference boundary-value problem can be proved for the adjoint boundary-value problem. Without giving proofs we formulate the basic results for the case where the number of independent boundary conditions coincides with the order of the difference equation.

1. An adjoint boundary-value problem has only a trivial solution if the corresponding original boundary-value problem has only a trivial solution.
2. The adjoint difference operator $L_n^*(\Delta, n)$ has the inverse $(L_n^*(\Delta, n))^{-1}$. This latter is a summation operator whose kernel is the Green's operator of the adjoint boundary-value problem.
3. The Green's function of the adjoint boundary-value problem is the fundamental solution of the corresponding homogeneous adjoint difference equation.
4. The Green's function of the adjoint boundary-value problem $G^*[n, \varepsilon; m]$ is the Green's function $G[n, \varepsilon; m]$ of the corresponding original boundary-value problem with the roles of $n+\varepsilon$ and m reversed, i.e. with $n+\varepsilon$ as the parameter:

$$G^*[n, \varepsilon; m]=G[m; n, \varepsilon] \tag{5.3}$$

We can verify this by representing the Green's function $G[n, \varepsilon; m]$ as a function m in the form

$$G^*[n, \varepsilon; m]=\sum_{i=1}^{k} y_i^*[n, \varepsilon]\, y_i[m, \varepsilon] \tag{5.4}$$

where $y_i[m, \varepsilon]$, $i=1, 2, \ldots, k$ is the fundamental set of solutions of the adjoint equation (5.1) with the boundary conditions which are adjoint to those of (3.2):

$$\begin{aligned} &\Delta_m^i y[m, \varepsilon]\,|_{m=n}=0, \quad i=0, 1, \ldots, k-2 \\ &\Delta_m^{k-1} y[m, \varepsilon]\,|_{m=n}=(-1)^{k-1}\frac{1}{a_k[n]}, \quad 0\leqslant\varepsilon\leqslant 1 \end{aligned} \tag{5.5}$$

In fact, if the fundamental set of solutions of (5.1) is known, the Green's function as a function of m can be expressed by

$$G[m; n, \varepsilon] = \sum_{i=1}^{k} \eta_i^* y_i [m, \varepsilon] \tag{5.6}$$

where η_i^* are certain functions of $\bar{t} = n + \varepsilon$ determined by the boundary conditions (5.5). Substituting (5.6) into (5.5) we obtain a system of equations from which η_i^* $(i = 1, 2, \ldots, k)$ can be determined. Since the determinant of this system is non-zero, its solution yields

$$\eta_i^* = (-1)^{k+i} \frac{1}{a_k[n]} \frac{D_i^*[n, \varepsilon]}{D^*[n, \varepsilon]}, \quad i = 1, 2, \ldots, k \tag{5.7}$$

where $D_i^*[n, \varepsilon]$ is obtained from $D^*[n, \varepsilon]$ by the well-known rule.

Substituting (5.7) into (5.6) we have

$$G[m; n, \varepsilon] = \sum_{i=1}^{k} (-1)^{k+i} \frac{1}{a_k[n]} \frac{D_i^*[n, \varepsilon]}{D^*[n, \varepsilon]} y_i [m, \varepsilon] \tag{5.8}$$

Putting

$$(-1)^{k+i} \frac{1}{a_k[n]} \frac{D_i^*[n, \varepsilon]}{D^*[n, \varepsilon]} = y_i^*[n, \varepsilon]$$
$$i = 1, 2, \ldots, k$$

in (5.8) we finally obtain

$$G[m; n, \varepsilon] = \sum_{i=1}^{k} y_i^*[n, \varepsilon]\, y_i[m, \varepsilon] \tag{5.9}$$

Since the determinants in the numerator of (5.8) are the minors of the elements in the last row of the denominator determinant, (5.9) can be written

$$G[m; n, \varepsilon] = \frac{1}{a_k[n]} \frac{\begin{vmatrix} y_1[n, \varepsilon] \ldots y_k[n, \varepsilon] \\ \ldots\ldots\ldots\ldots \\ \Delta^{k-2} y_1[n, \varepsilon] \ldots \Delta^{k-2} y_k[n, \varepsilon] \\ y_1[m, \varepsilon] \ldots y_k[m, \varepsilon] \end{vmatrix}}{\begin{vmatrix} y_1[m, \varepsilon] \ldots y_k[m, \varepsilon] \\ \Delta y_1[m, \varepsilon] \ldots \Delta y_k[m, \varepsilon] \\ \ldots\ldots\ldots\ldots \\ \Delta^{k-1} y_1[m, \varepsilon] \ldots \Delta^{k-1} y_k[m, \varepsilon] \end{vmatrix}} \tag{5.10}$$

Comparing (5.10) and (3.35) we have

$$G^{*}[n, \varepsilon; m] = G[m; n, \varepsilon] \tag{5.11}$$

which coincides with (5.3).

2.6 ADJOINT SYSTEMS

In this section we shall give the definition of the adjoint sampled data system, basing it on the results obtained in Section 2.1 to 2.5. We shall first establish the connection between the weighting function $W[n, \varepsilon; m]$ and the Green's function $G[n, \varepsilon; m]$ of the corresponding boundary-value problem.

Let a sampled data system be described by the single k-th order difference equation

$$L_n(\Delta, n)\, y[n, \varepsilon] \equiv \sum_{i=0}^{k} a_i[n]\, \Delta^i y[n, \varepsilon] = x[n, 0] \qquad 0 \leqslant \varepsilon \leqslant 1 \tag{6.1}$$

In accordance with the definition the weighting function $W_0[n, \varepsilon; m]$ of this system is determined by the solution of the homogeneous difference equation

$$L_n(\Delta, n)\, W_0[n, \varepsilon; m] = 0, \quad 0 \leqslant \varepsilon \leqslant 1 \tag{6.2}$$

with the initial conditions

$$\begin{aligned} \Delta_n^i W_0[n, \varepsilon; m]\,|_{n=m} &= 0, \quad i = 0, 1, \ldots, k-2, \\ \Delta_n^{k-1} W_0[n, \varepsilon; m]\,|_{n=m} &= \frac{1}{a_k[m]}. \end{aligned} \tag{6.3}$$

We now define the difference boundary-value problem by

$$L_n(\Delta, n)\, y[n, \varepsilon] = 0, \quad 0 \leqslant \varepsilon \leqslant 1 \tag{6.4}$$

and the boundary conditions coinciding with (6.3). Comparing the properties of the weighting function $W_0[n, \varepsilon; m]$ with those of the Green's function $G[n, \varepsilon; m]$ of the boundary-value problem (6.3) and (6.4), we have

$$W_0[n, \varepsilon, m] = \begin{cases} G[n, \varepsilon; m], & n \geqslant m \\ 0 & n < m \end{cases} \tag{6.5}$$

Consequently, by Theorem 2 of Section 2.3 the output variable $y[n, \varepsilon]$ of the sampled data system is determined by

$$y[n, \varepsilon] = \begin{cases} \sum_{m=\nu}^{n} G[n, \varepsilon; m]\, x[m, 0], & n \geqslant m \\ 0 & , n < m \end{cases} \tag{6.6}$$

where ν is the instant at which the input disturbance $x[n, \varepsilon]$ acts, and before which the system was at rest.

If the system is described by an equation of the form

$$\sum_{i=0}^{k} a_i[n]\, \Delta_n^i\, W[n, \varepsilon; m] = \sum_{j=0}^{h} b_j[n, \varepsilon]\, \Delta^j \sigma[n - m, 0] \qquad 0 \leqslant \varepsilon \leqslant 1 \tag{6.7}$$

then its weighting function $W[n, \varepsilon; m]$ is determined by

$$W[n, \varepsilon; m] = \sum_{j=0}^{h} (-1)^j \Delta_m^j \{W_0[n, \varepsilon; m - j]\, b_j[m - j, \varepsilon]\} \qquad 0 \leqslant \varepsilon \leqslant 1 \tag{6.8}$$

or, when (6.5) is used,

$$W[n, \varepsilon; m] = \begin{cases} 0 & , \quad n < m \\ \sum_{j=0}^{h} (-1)^j \Delta_m^j \{G[n, \varepsilon; m - j]\, b_j[m - j, \varepsilon]\} & n \geqslant m \end{cases} \tag{6.9}$$

Denoting

$$\sum_{j=0}^{h} b_j[n, \varepsilon]\, \Delta_n^j = M_n(\Delta, \bar{t}) \tag{6.10}$$

in (6.7), we rewrite (6.8) and (6.9) as follows, using (6.10),

$$W[n, \varepsilon; m] = M_m^*(\Delta, \bar{\tau})\, W_0[n, \varepsilon; m], \quad 0 \leqslant \varepsilon \leqslant 1 \tag{6.11}$$

$$W[n, \varepsilon; m] = \begin{cases} 0 & , n < m \\ M_m^*(\Delta, \bar{\tau})\, G[n\, \varepsilon; m], & n \geqslant m \end{cases} \tag{6.12}$$

where $\bar{\tau} = m + \varepsilon$.

We shall now define the sampled data system which is adjoint to the system described by an equation of the form (6.1).

Definition. The adjoint sampled data system in relation to the system described by the equation

$$L_n(\Delta, n)\, y[n, \varepsilon] = x[n, 0], \quad 0 \leqslant \varepsilon \leqslant 1$$

is a system described by the equation

$$L_n^*(\Delta, n)\, y[n, \varepsilon] = x[n, 0], \quad 0 \leqslant \varepsilon \leqslant 1 \tag{6.13}$$

The weighting function $W_0^*[n, \varepsilon; m]$ of the adjoint sampled data system, on the basis of (5.3) and (6.5), is

$$W_0^*[n, \varepsilon; m] = \begin{cases} G^*[n, \varepsilon; m], & m \geqslant n \\ 0 & m < 0 \end{cases} \tag{6.14}$$

$$W_0^*[n, \varepsilon; m] = W_0[m; n, \varepsilon] \tag{6.15}$$

Thus the weighting function $W_0^*[n, \varepsilon; m]$ of the adjoint sampled data system is a function of m, i.e. the instant at which the σ-function is applied, and is obtained from the weighting function $W_0[n, \varepsilon; m]$ of the original system by changing the roles of the arguments $\overline{t}$ and m.

It was shown in Chapter 1 that the weighting function of a sampled data system can be represented in the form

$$W_0[n, \varepsilon; m] = \sum_{i=1}^{k} y_i^*[m, \varepsilon]\, y_i[n, \varepsilon]$$

where $y[n, \varepsilon]$ is the fundamental set of solutions of (6.4). Consequently, changing the places of the arguments n and m we obtain from the last relation

$$W_0^*[n, \varepsilon; m] = \sum_{i=1}^{k} y_i^*[n, \varepsilon]\, y_i[m, \varepsilon]$$

where $y_i[m, \varepsilon]$ is the fundamental set of solutions of the adjoint equation corresponding to (6.4).

By definition of the weighting function of a sampled data system, the equation for the corresponding adjoint system follows directly from (6.13)

$$L_m^*(\Delta, m)\, W_0[n, \varepsilon; m] = \sigma[n-m, 0], \quad 0 \leqslant \varepsilon \leqslant 1 \qquad (6.16)$$

where, according to the definition of the adjoint operator,

$$L_m^*(\Delta, m)\, W_0[n, \varepsilon; m] = \sum_{i=0}^{k} (-1)^i \Delta_m^i \{W_0[n, \varepsilon; m-i]\, a_i[m-i]\} \qquad (6.17)$$

$$0 \leqslant \varepsilon \leqslant 1$$

The relation (6.14) shows that the weighting function of the adjoint sampled data system determined from (6.16) does not satisfy the condition of physical realisability, since

$$W_0^*[n, \varepsilon; m] = \begin{cases} W_0[m; n, \varepsilon], & m \geqslant n \\ 0, & m < n \end{cases} \qquad (6.18)$$

But it follows from (6.18) that if negative time ($l<0$) is introduced, i.e. the new variable $m=n-l$, then

$$W_0^*[n, \varepsilon; m] = \begin{cases} W_0[n-l; n, \varepsilon], & l \geqslant 0 \\ 0, & l < 0 \end{cases} \qquad (6.19)$$

where l should be taken as the current value of the time.

Thus, for the new variable the weighting function does satisfy the condition of physical realisability. In this case the equation for $W_0[m;\ n,\ \varepsilon]$ assumes the form

$$\sum_{i=0}^{k} (-1)^i \Delta_{n-l}^i \{a_i[n-l-i]\, W_0[n, \varepsilon; n-l-i]\} = \sigma[l, 0] \qquad (6.20)$$

$$0 \leqslant \varepsilon \leqslant 1$$

Taking into account

$$(-1)^i \Delta_{n-l}^i \{a_i[n-l-i]\, W_0[n, \varepsilon; n-l-i]\} = \Delta_l^i \{a_i[n-l]\, W_0[n, \varepsilon; n-l]\},$$

we finally obtain

$$\sum_{i=0}^{k} \Delta_l^i \{a_i[n-l]\, W_0[n, \varepsilon; n-l]\} = \sigma[l, 0], \quad 0 \leqslant \varepsilon \leqslant 1 \qquad (6.21)$$

Equation (6.21) defines the weighting function $W_0[n, \varepsilon; m]$ as a function of m for which $\bar{t}$ is fixed, i.e. $\bar{t}$ is the parameter. (Note that Tartakovskii [22] obtained the equation of the adjoint sampled data system from physical considerations).

Writing (6.19), for $n=N$, in the form

$$W_0^*[N, \varepsilon; N-l] = \begin{cases} W_0[N-l; N, \varepsilon], & l \geqslant 0 \\ 0 & , l < 0 \end{cases} \tag{6.22}$$

we find that the weighting function of the adjoint system is its response to the σ-function applied at the instant $\bar{t}=N+\varepsilon$.

Equation (6.21), defining the weighting function of the adjoint sampled data system, can be obtained by another method. In fact, we assume that the output variable of the system defined by (6.1) is the σ-function, i.e. (for simplicity we consider lattice functions)

$$\sigma[n-m] = \sum_{l=0}^{\infty} W_0[n; l]\, x[l] \tag{6.23}$$

Since in conformity with (6.1), for $y[n-m] = \sigma[n-m]$

$$x[m] = \sum_{i=0}^{k} a_i[n]\, \Delta_n^i \sigma[n-m] \tag{6.24}$$

(6.23) can be written

$$\sigma[n-m] = \sum_{l=0}^{\infty} W_0[n; l] \sum_{i=0}^{k} a_i[l]\, \Delta_l^i \sigma[l-m] \tag{6.25}$$

Applying the property of the σ-function to (6.25)

$$\sum_{l=-\infty}^{\infty} f[l]\, \Delta_l^s \sigma[l-m] = (-1)^s \Delta_m^s f[m-s]$$

we have

$$\sigma[n-m] = \sum_{i=0}^{k} (-1)^i \Delta_m^i \{a_i[m-i]\, W_0[n; m-i]\} \tag{6.26}$$

In (6.26) $\bar{t} = m$ is the independent variable. It follows from the method by which it is obtained that it is satisfied by the same weighting function which satisfies (6.1).

Changing the places of the arguments in (6.26) we have

$$\sigma[m-n]=\sum_{i=0}^{k}(-1)^i\Delta_n^i\{a_i[n-i]W_0^*[n;\ m-i]\} \qquad (6.27)$$

where

$$W_0^*[n;m-i]=W_0[m-i;\ n] \qquad (6.28)$$

It is easy to see that (6.27) is the equation of the adjoint sampled data system, whilst its solution $W_0^*[n;\ m]$ is the weighting function of this system.

Since the weighting function $W_0[n;\ m]$ of the original system satisfies the condition of physical realisability

$$W_0[n;\ m]=\begin{cases} W_0[n;\ m], & n\geqslant m\\ 0, & n<m\end{cases} \qquad (6.29)$$

then, changing the places of m and n in (6.29), we obtain

$$W_0^*[n;\ m]=\begin{cases} W_0[m;\ n], & m\geqslant n\\ 0, & m<n\end{cases} \qquad (6.30)$$

i.e. the weighting function $W_0^*[n,\ m]$ is defined for $n\leqslant m$ and does not satisfy (6.29). Equation (6.30) shows that if the new variable $m-n=l$, which is independent, is introduced in (6.27) we obtain

$$\sigma[l]=\sum_{i=0}^{k}\Delta_l^i\{a_i[m-l]W_0^*[m-l;\ m] \qquad (6.31)$$

i.e. the equation for W_0^* coincides with (6.21) obtained earlier. Note that the weighting function W_0^* satisfies the condition

$$W_0^*[m-l;\ m]=\begin{cases} W_0[m;\ m-l], & l\geqslant 0\\ 0, & l<0\end{cases} \qquad (6.32)$$

Let now the sampled data system be described by a difference equation of the form

$$\sum_{i=0}^{k}a_i[n]\Delta^iy[n]=\sum_{j=0}^{h}b_j[n]\Delta^jx[n] \qquad (6.33)$$

We shall determine the weighting function $W^*[n;\ m]$ of the system whose independent variable is furnished by the instant at which the σ-function is applied.

It has been shown above that the weighting function $W[n; m]$ of the system defined by (6.33) is connected with the weighting function $W_0[n, m]$ of the system defined by (6.1) as follows

$$W[n; m] = \sum_{j=0}^{h} (-1)^j \Delta_m^j \{b_i[m-j] W_0[n; m-j]\} \tag{6.34}$$

Changing the places of n and m in (6.34) we obtain

$$W^*[n; m] = \sum_{i=0}^{h} (-1)^j \Delta_n^j \{b_j[n-j] W_0^*[n; m-j]\} \tag{6.35}$$

where

$$W^*[n; m] = W[m; n]$$

$$W_0^*[n; m] = W_0[m; n]$$

If n is the parameter in (6.34), then the role of the parameter in the relation (6.35) is taken by $m = M$. Introducing the new variable $M - n = l$, we have from (6.35)

$$W^*[M-l; M] = \sum_{j=0}^{h} \Delta_l^j \{b_j[M-l] W_0^*[M-l; M]\} \tag{6.36}$$

Equation on (6.36) determines the weighting function of the adjoint sampled data system in the case where the equation of the original system contains the operator $M_n(\Delta, n)$ on its right-hand side.

We have defined above the adjoint sampled data system, and have obtained the equations which determine its weighting function when the original system is described by a single difference equation. It is easy to show that, in the general case, when the processes in the system are determined by the two equations,

$$\begin{aligned} L_n(\Delta, n) y[n, \varepsilon] &= M_n(\Delta; \bar{t}) x[n, 0], \quad 0 \leqslant \varepsilon \leqslant \gamma \\ S_n(\Delta, n) y[n, \varepsilon] &= G_n(\Delta, \bar{t}) x[n, 0], \quad \gamma \leqslant \varepsilon \leqslant 1 \end{aligned} \tag{6.37}$$

the weighting function of the corresponding adjoint system is determined by

$$L_l^*(\Delta,\ l)W_0[n,\ \varepsilon;\ n-l]=\sigma[l,\ 0]$$
$$W^*[n,\ \varepsilon;\ n-l]=M_l^*(\Delta,\ \bar{\tau})W_0^*[n,\ \varepsilon;\ n-l] \tag{6.38}$$
$$0\leqslant\varepsilon\leqslant\gamma$$

and

$$S_l^*(\Delta,\ l)W_0[n,\ \varepsilon;\ n-l]=\sigma[l,\ 0]$$
$$W^*[n,\ \varepsilon;\ n-l]=G_l^*(\Delta,\ \bar{\tau})W_0^*[n,\ \varepsilon;\ n-l] \tag{6.39}$$
$$\gamma\leqslant\varepsilon\leqslant 1$$

where $\bar{\tau}=l+\varepsilon$.

We have thus defined the adjoint sampled data system and have obtained the corresponding equations which determine its weighting function. It follows that the weighting function of an adjoint system, when considered as a function of m, i.e. of the instant when the input is applied, has all the properties of the weighting function of the original sampled data system, which were proved in Chapter 1.

We shall now consider diagrams of models representing an adjoint sampled data system.

2.7 MODELS OF ADJOINT SYSTEMS

A sampled data system can be specified either by difference equations or by the differential equations of its continuous part supplemented by the form of its sampler. Accordingly the adjoint sampled data system can be represented either by a digital model or by an analogue.

2.7.1 Digital model

If the original sampled data system is described by difference equations of the form (6.1), then the corresponding adjoint system can be determined from

$$\sum_{i=0}^{k}\Delta_l^i\{a_i[n-l]W_0[n,\ \varepsilon;\ n-l]\}=\sigma[l,\ 0] \tag{7.1}$$
$$0\leqslant\varepsilon\leqslant 1$$

A digital model constructed from (6.1) is shown in Fig. 2.1, while a model corresponding to the adjoint sampled data system is shown in Fig. 2.2. It follows from the comparison of the two diagrams that the model of the adjoint sampled data system is obtained from the model of the original system, if we:

1. exchange the roles of the input and output of all the elements of the model, and replace the overall input of the system by its output, and vice versa
2. replace the branch points by the summing points, and vice versa
3. reverse in time the coefficients of the scaling elements.

As follows from (7.1) the reversal in time of the coefficients reduces to the following. If it is required to obtain

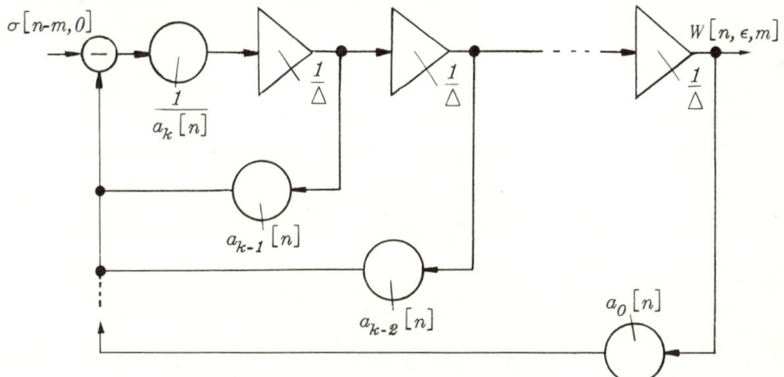

Fig. 2.1 A digital model of the system corresponding to (6.1)

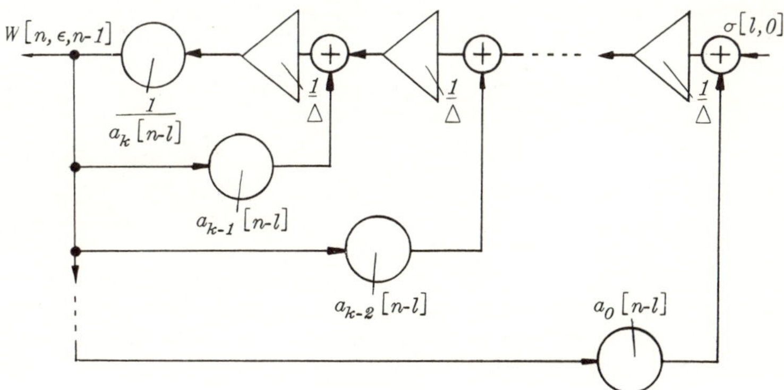

Fig. 2.2 The digital model of a system adjoint to that shown in Fig. 2.1

the weighting function of the adjoint sampled data system for a fixed value $n=N$, then, in order to compute it, the coefficients $a_i[N-l]$ (for l is real time) must be used, i.e. the coefficients must be selected from the table in the direction of decreasing time, commencing with $l=0$.

Following the rules just given we can construct the model of an adjoint sampled data system of arbitrary form.

Analogous rules for constructing diagrams of the digital models of adjoint sampled data systems have been obtained by Tartakovskii [22].

We shall now consider sampled data systems with the weighting functions $W_1[n, \varepsilon; m]$ and $W_2[n, \varepsilon; m]$ connected in series (Fig. 2.3(a)). We shall determine the adjoint sampled data system which corresponds to these two systems connected in series. For the case where the samplers of both elements are synchronised and operate in phase, the relation was obtained in Chapter 1 for the weighting function $W[n, \varepsilon; m]$ for the two systems connected in series

$$W[n, \varepsilon; m] = \sum_{l=-\infty}^{\infty} W_2[n, \varepsilon; l] W_1[l, 0; m] \tag{7.2}$$

Since by definition the weighting function of the adjoint system is

$$W^*[n, \varepsilon; m] = W[m; n, \varepsilon] \tag{7.3}$$

then from (7.2) we obtain

$$\begin{aligned} W^*[n, \varepsilon; m] = W[m; n, \varepsilon] &= \sum_{l=-\infty}^{\infty} W_2[m; l, 0] W_1[l; n, \varepsilon] \\ &= \sum_{l=-\infty}^{\infty} W_1^*[n, \varepsilon; l] W_2^*[l, 0; m] \end{aligned} \tag{7.4}$$

Thus we have

$$W^*[n, \varepsilon; m] = \sum_{l=-\infty}^{\infty} W_1^*[n, \varepsilon; l] W_2^*[l, 0; m] \tag{7.5}$$

whence it follows that the adjoint sampled data system of the two systems connected in series is formed by connecting the corresponding adjoint systems in series, but reversing the order. This is illustrated in Fig. 2.3(b).

In accordance with (7.2), for a series combination of

mutually inverse systems we have

$$\sum_{l=-\infty}^{\infty} W[n, \varepsilon; l] W^{-}[l, 0; m] = \sigma[n - m, 0] \tag{7.6}$$

$$\sum_{l=-\infty}^{\infty} W^{-}[n, \varepsilon; l] W[l, 0; m] = \sigma[n - m, 0] \tag{7.7}$$

It follows from (7.6) and (7.7) that the system of identical transformation is self-adjoint.

By the property of linear difference operators

$$\left(\sum_{i=1}^{m} L_n(\Delta, n)_i\right)^{*} = \sum_{i=1}^{m} L^{*}_{n}(\Delta, n)_i \tag{7.8}$$

we find that a system which is adjoint to a parallel combination of sampled data systems is a parallel combination of the corresponding adjoint systems (Fig. 2.4).

It can also be shown that a system which is adjoint to the

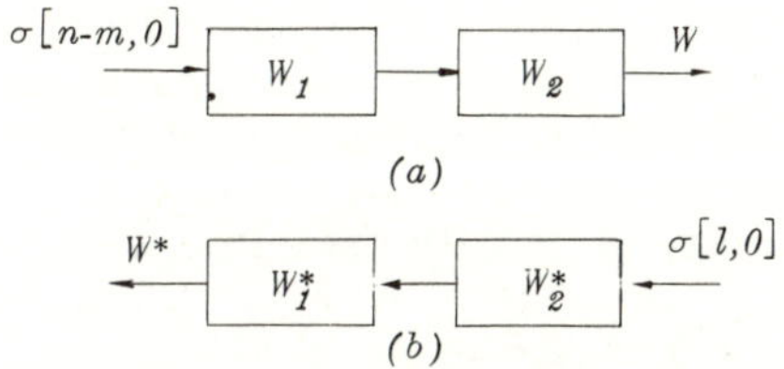

Fig. 2.3 Systems connected in series: (a) - original; (b) - adjoint

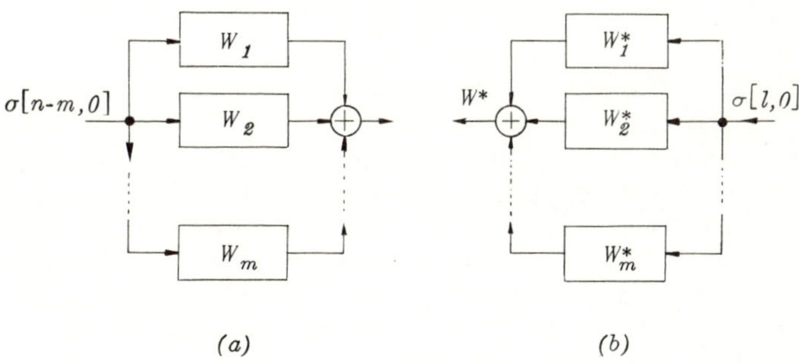

Fig. 2.4 Systems connected in parallel: (a) - original; (b) - adjoint

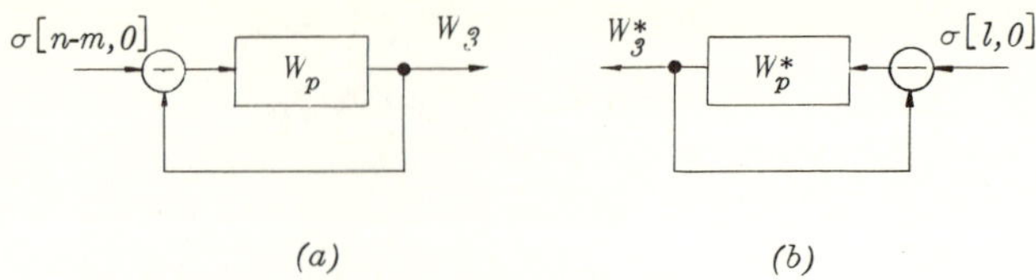

Fig. 2.5 The simplest closed-loop systems: (a) - original; (b) - adjoint

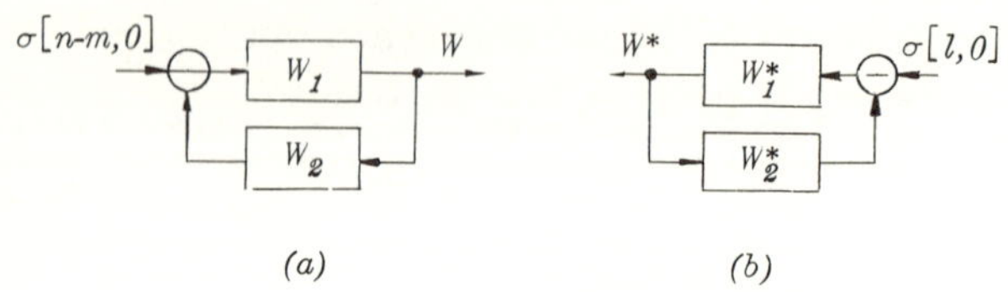

Fig. 2.6 Closed-loop systems: (a) - original; (b) - adjoint

simplest closed-loop system is obtained by closing the adjoint open-loop system and changing the roles of the input and the output. In so doing we replace the branch point by the summing point, and the summing point by the branch point, i.e. the negative feedback loop keeps its role (Fig. 2.5). We can show later, in exactly the same way, that the system whose diagram is shown in Fig. 2.6(a) has the adjoint system shown in Fig. 2.6(b).

Using the rules just discussed we can construct the digital model of an adjoint sampled data system of arbitrary form.

2.7.2 Analogue model

As a rule, a sampled data system is given by differential equations describing its continuous part and by the form of the sampler, so that in the final analysis either the structure of a sampled data system is known, or the transfer function of its continuous part and the type of its sampler are known. Accordingly, it is of practical interest to construct the model of an adjoint sampled data system, using analogue computers. Let us determine the diagram of the analogue of an adjoint sampled data system.

We shall first consider an open-loop system formed by connecting a sampler and a continuous part in series. We

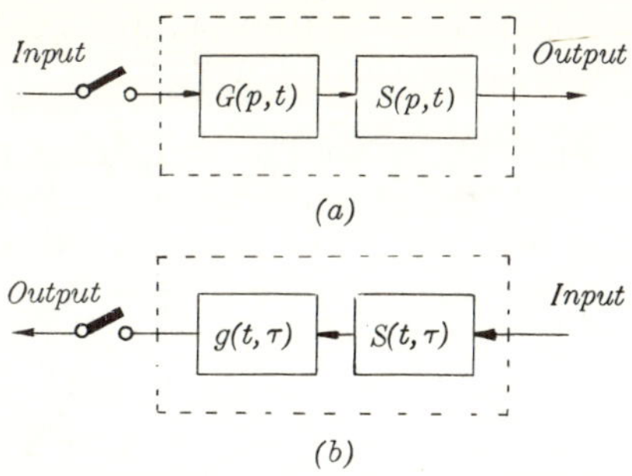

Fig. 2.7 Diagrams of equivalent systems: (a) - original; (b) - adjoint

shall assume that the parameters of the sampler and the continuous part depend on time. The system in question can be transformed into an equivalent system of the simplest form, consisting of a sampler generating the δ-function, and the normalised continuous part. The block diagram of such a system is shown in Fig. 2.7(a). The normalised continuous part of this equivalent system is formed by connecting a shaping filter having the transfer function $G(p, t)$ in series with the original system whose transfer function is denoted by $S(p, t)$. The sampler generating the δ-function is symbolically denoted by the key in Fig. 2.7(a).

The fundamental characteristic of the open-loop system is given by its weighting function $W[n, \varepsilon; m]$. In the previous section the weighting function $W^*[n, \varepsilon; m]$ of the corresponding adjoint system has been obtained from $W[n, \varepsilon; m]$ by changing the roles of the arguments $\bar{t}$ and m, i.e.

$$W^*[n, \varepsilon; m] = W[m; n, \varepsilon] \tag{7.9}$$

The weighting function $W[n, \varepsilon; m]$ of the system coincides with the weighting function of the normalised part [1]. We can therefore write

$$W[n, \varepsilon; m] = W(t; \tau)\Big|_{\substack{t=(n+\varepsilon)T_0 \\ \tau=mT_0}} \tag{7.10}$$

where $W(t; \tau)$ is the response of the normalised continuous part to the δ-function (the weighting function of the normalised continuous part);

T_0 is the sampling period.

Consequently, in order to determine the weighting function

$W[n, \varepsilon; m]$ of the open-loop system we must know the weighting function $W(t; \tau)$ of its normalised part. Let $g(t; \tau)$ be the weighting function of the shaping filter with the transfer function $G(p, t)$, and $s(t; \tau)$ be the weighting function of the continuous part with the transfer function $S(p, t)$. Then the weighting function $W(t; \tau)$ of the normalised continuous part is determined by the superposition formula

$$W(t; \tau) = \int_{-\infty}^{\infty} s(t; \lambda) g(\lambda; \tau) d\lambda \tag{7.11}$$

The weighting function is determined as follows from (7.10):

$$W[n, \varepsilon; m] = \left[\int_{-\infty}^{\infty} s(t; \lambda) g(\lambda; \tau) d\lambda \right]_{\substack{t=(n+\varepsilon) T_0 \\ \tau=mT_0}} \tag{7.12}$$

The weighting function $W^*(t; \tau)$ of the adjoint continuous part is obtained from the weighting function $W(t; \tau)$ by changing the roles of the arguments t and τ [13], i.e.

$$W^*(t; \tau) = W(\tau; t) \tag{7.13}$$

and therefore on the basis of (7.11) we have

$$W^*(t; \tau) = \int_{-\infty}^{\infty} s(\tau; \lambda) g(\lambda; t) d\lambda = \int_{-\infty}^{\infty} g^*(t; \lambda) s^*(\lambda; t) d\lambda, \tag{7.14}$$

whence, when (7.12) is taken into account, we find

$$W^*[n, \varepsilon; m] = \left[\int_{-\infty}^{\infty} g^*(t; \lambda) s^*(\lambda; \tau) d\tau \right]_{\substack{t=(n+\varepsilon) T_0 \\ \tau=mT_0}} \tag{7.15}$$

The relation (7.15) establishes the following rule for constructing the analogue of an open-loop sampled data system: the open-loop adjoint sampled data system consists of the adjoint continuous part and the adjoint shaping filter connected in series, with the order reversed. This rule is illustrated by Fig. 2.7(b).

The presence of the key at the output of the adjoint system emphasises the fact that the weighting function is a lattice function.

The rule obtained for the construction of the analogue of an open-loop adjoint sampled data system is obvious. It follows from the rules for the construction of models of continuous adjoint systems. It should, however, be pointed out that in [22] incorrect results have been obtained in the analysis of problems on the construction of the model of an adjoint sampled data system. It is incorrectly stated there that to form an adjoint sampled data system we must place the sampler at the input of the newly-formed system. But in fact the sampler must be left at the same place in the diagram as in the original system.

We shall consider a particular case of practical importance. Let the sampler generate rectangular pulses whose period equals T_0, the sampling period ($\gamma = 1$). In this case

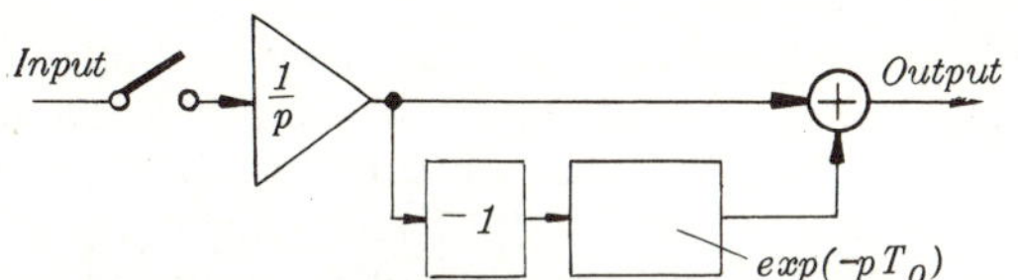

Fig. 2.8 Block diagram of a sampler generating rectangular pulses with the relative duration $\gamma = 1$

the transfer function of the shaping filter $G(p)$ will identically coincide with the Laplace transform of a single rectangular impulse whose length is T_0. If the gain factor of this sampler is unity, then

$$G(p) = \frac{1 - e^{-pT_0}}{p} \tag{7.16}$$

The block diagram of an element whose transfer function is determined by (7.16) is shown in Fig. 2.8, where the element with the symbol e^{-pT_0} denotes the element which purely

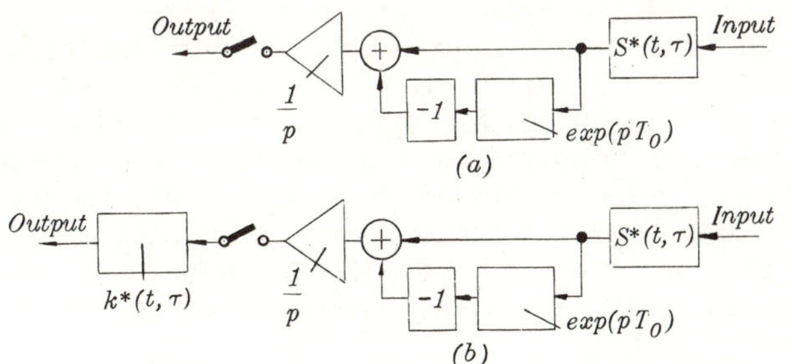

Fig. 2.9 Systems having samplers generating rectangular pulses with $\gamma = 1$

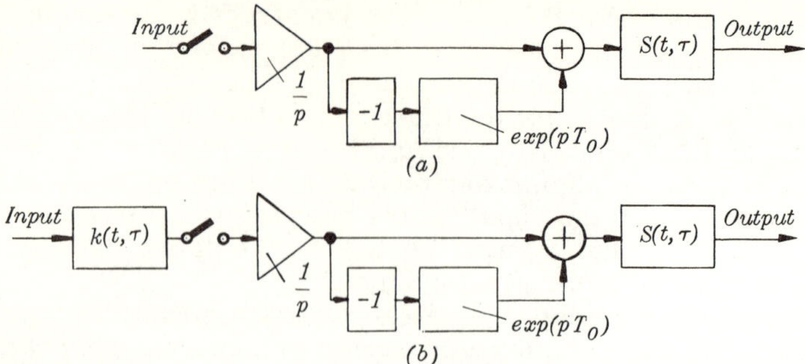

Fig. 2.10 Systems adjoint to that shown in Fig. 2.9

delays the input signal by one sampling period.

It is easy to see that when a signal in the form of the δ-function acts at the input of this sampler, then a single rectangular pulse of unit height and length equal to T_0 appears at its output, i.e. the σ-function appears at the output. If an arbitrary function $x(t)$ acts at the input of such a device, then a step function appears at its output, the height of the steps being equal to the values of the function $x(t)$ at the instants the key is closed. The sampler in question is essentially an ideal element for maintaining signal levels.

It follows from the rules formulated above for the construction of adjoint sampled data systems that the adjoint systems whose diagrams are shown in Fig. 2.10(a) and (b) correspond to the sampled data systems whose diagrams are shown in Fig. 2.9(a) and (b) respectively.

It has been pointed out above that a sampler whose shaping filter has a transfer function of the form (7.16) is an ideal element for maintaining signal levels. In practice, however, it is not possible to realise a structure whose key is closed during infinitely short time. In each particular case it is only possible to reproduce the action of the ideal sampler approximately, assuming, for example, that the closure time for the key is very much less than the duration of the pulses it produces. If the length of the pulses which the sampler must produce is a few seconds, then the closure time must be of the order of a few hundredths (sometimes even tenths) of a second. In such a case it is permissible to assume that the sampler is ideal.

To realise approximately a sampler which generates rectangular pulses with $\gamma = 1$ we can use a single-integrator follow-up system whose forward loop is commutated by the key K (Fig. 2.11). The key K is closed during the time $\tau \ll T_0$ once during the time T_0 which equals the length of the pulses generated, whilst during the rest of the time $T_0 - \tau$ it remains open. While the key is closed we have the usual single-integrator continuous follow-up system whose transfer function equals

$$W(p) = \frac{k}{p+k} \tag{7.17}$$

where k is the gain factor of the integrator.

If we choose k from the condition

$$k \gg \frac{1}{\tau} \quad (\tau \ll T_0) \tag{7.18}$$

then this sampled data follow-up system can be used to realise approximately a sampler generating pulses with

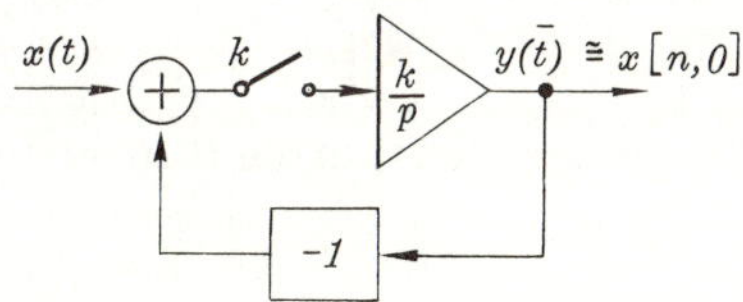

Fig. 2.11 Diagram of a follow-up system having a sampler with $\gamma = 1$

unit relative length. In this case the input action $x(t)$ of the system is approximated by a step function. Thus, we shall assume that during the time τ the input action $x(t)$ varies so little that the following holds:

$$x(t) \approx x(t+\tau)$$

We shall assume further that $y(0) = 0$, that is, prior to the instant $t = 0$, which $x(t)$ acts, the system is at rest. Then from (7.17) we can write

$$y(t) = x(0)(1 - e^{-kt}), \quad 0 \leqslant t < \tau \tag{7.19}$$

During the time interval $T_0 > t \geqslant \tau$ (for the open-loop state of the system) the signal level at the output of the follow-up system remains unchanged:

$$y(t) = x(0)(1 - e^{-k\tau}), \quad \tau \leqslant t < T_0 \tag{7.20}$$

At the next closure of the key the value

$$\mathcal{E}(T_0) \cong x(T_0) - x(0)(1 - e^{-k\tau})$$

appears at the input of the integrator. A signal level of value approximately equal to the value of the input action at the instant $t = T_0$ is maintained at the output;

$$y(t) \cong x(T_0)(1 - e^{-k\tau}) + x(0)\,e^{-k\tau} \approx x(T_0)$$

$$T_0 + \tau \leqslant t \leqslant 2T_0$$

When the key closes periodically a step function approximating the function $x(t)$ appears at the output of the system. It is clear that when the gain factor k of the integrator is increased the time required to process the input action by the sampled data follow-up system is reduced. When $k \to \infty$, instantaneous holding of the input signal is realised.

The choice of the integrator gain factor must in each case be carried out on the basis of the conditions of the problem. In many cases occurring in practice it is sufficient to choose k from the condition

$$k = \frac{3 \text{ to } 4}{\tau} \quad (\tau \ll T_0) \tag{7.21}$$

Thus, for an appropriate choice of the gain factor k and the time of the closed state of the key, an ideal sampler which generates rectangular pulses with $\gamma = 1$ can be realised approximately in the form of a single-integrator sampled data follow-up system.

Applying the rules formulated above for constructing models of an adjoint sampled data system we find that the system whose diagram is shown in Fig. 2.12(a) has the corresponding adjoint system whose diagram is shown in Fig. 2.12(b).

We have considered above the problems of constructing analogues of open-loop adjoint sampled data systems. Using the results obtained in Section 2.7.1, we can determine the following rule for constructing the analogue of a closed-loop adjoint sampled data system: the closed-loop adjoint sampled data system is formed by closing the corresponding open-loop system. The role of the feedback loops of the original system is preserved. This rule is illustrated by Fig. 2.13(a) and (b), as applied to systems whose samplers generate

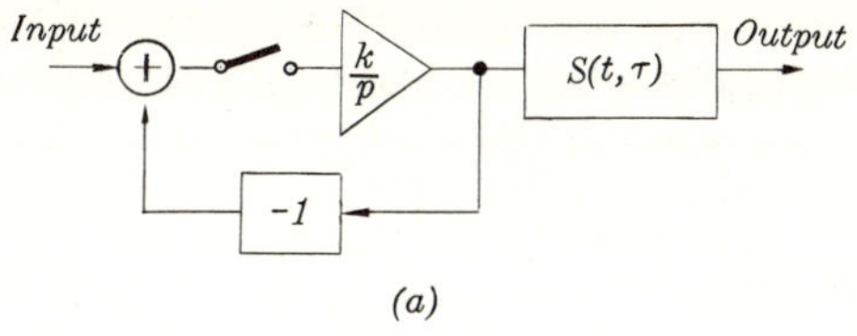

(a)

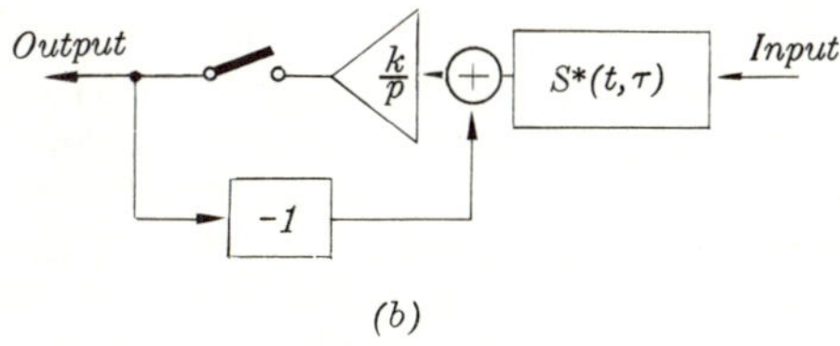

(b)

Fig. 2.12 Open-loop systems: (a) - original; (b) - adjoint

rectangular pulses with the relative duration γ equal to unity. These diagrams are valid in the case where the samplers are realised in the form of a single-integrator follow-up sampled data system.

Applying the rules just given we can construct analogues of any adjoint sampled data system.

We have now considered methods for constructing digital models and analogues of adjoint sampled data systems. The

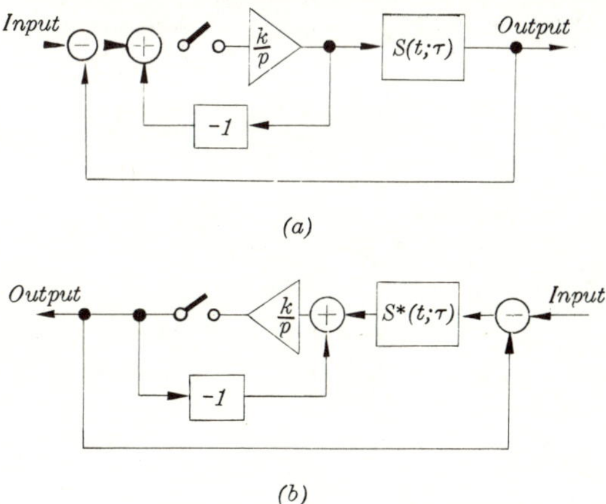

Fig. 2.13 Closed-loop sampled data systems: (a) - original; (b) - adjoint

advisability of using one model or another must be determined in each particular case. Thus, if the sampled data system is given by a difference equation, then clearly we must use a digital model, which can easily be realised in the form of a programme for a digital computer. But in the case where the sampled data system is specified by the differential equations of its continuous part and by the form of the sampler, there is no need to seek the difference equation of this system; an analogue must be used in the investigations. Incidentally, if a non-stationary sampled data system is specified in the form of the differential equations of its continuous part and by the form of its sampler, the determination of the difference equation of such a system is a very difficult problem, which can be solved - and then only approximately - when the coefficients of the differential equations are slowly varying functions. Consequently, the use of analogue computers to simulate sampled data systems is of great practical importance.

The analogues of adjoint sampled data systems can be realised not only by means of analogue computers, but also by digital computers. For this the differential equations describing the continuous part of the system and the equation of the sampler must be represented by an appropriate computation scheme, which is essentially a scheme for solving numerically the differential equations describing the sampled data system.

Chapter 3

STATISTICAL ANALYSIS OF LINEAR SYSTEMS

In this chapter we discuss methods for the statistical analysis of stationary and non-stationary sampled data systems, based on the correlation theory of random processes. The methods given for the determination of statistical characteristics are developed for use on analogue and digital computers. We shall first consider certain general problems concerning the transformation of random functions by means of linear sampled data systems.

3.1 CHARACTERISTICS OF RANDOM PROCESSES

For a wide range of problems, it is often sufficient to characterise the random processes by such statistical characteristics as mathematical expectation and correlation function, and the general definitions of these [13], with the latter regarded as functions of discrete arguments, can be used for the processes in sampled data systems. In what follows we shall consider only real random functions, observed both at the discrete instants $n=0, 1, \ldots$, and at the instants of time $\bar{t}=n+\varepsilon$ $(0 \leqslant \varepsilon \leqslant 1)$.

The mathematical expectation of the random function $X[n]$ of a discrete argument is the term given to such a function $m_x[n]$ whose value for any value of the argument n is equal to the mathematical expectation of the value of the function $X[n]$ for the given n:

$$m_x[n]=M[X[n]] \tag{1.1}$$

The mathematical expectation of X observed at the instant $\bar{t}=n+\varepsilon$, $0\leqslant\varepsilon\leqslant 1$, is, in accordance with the definition (1.1),

$$m_x[n, \varepsilon]=M[X[n, \varepsilon]], \quad 0\leqslant\varepsilon\leqslant 1 \tag{1.2}$$

The correlation function of $X[n]$ is the term given to the function $K_x[n; n_1]$ defined by the relation

$$K_x[n; n_1]=M[\{X[n]-m_x[n]\}\{X[n_1]-m_x[n_1]\}] \tag{1.3}$$

or

$$K_x[n; n_1]=M[\mathring{X}[n]\mathring{X}[n_1]] \tag{1.4}$$

where $\mathring{X}$ is the centralised random function.

When $n=n_1$ (1.3) defines the dispersion of the random function of a discrete argument

$$D_x[n]=K_x[n; n]=M[\mathring{X}^2[n]] \tag{1.5}$$

The correlation function of the random function X observed at the instants of time $\bar{t}=n+\varepsilon$ $(0\leqslant\varepsilon\leqslant 1)$ is determined by

$$\begin{gathered} K_x[n, \varepsilon; n_1, \varepsilon_1]=M[\{X[n, \varepsilon]-m_x[n, \varepsilon]\} \\ \times\{X[n_1, \varepsilon_1]-m_x[n_1, \varepsilon_1]\}] \\ 0\leqslant\varepsilon\leqslant 1, \quad 0\leqslant\varepsilon_1\leqslant 1 \end{gathered} \tag{1.6}$$

Similarly

$$\begin{gathered} D_x[n, \varepsilon]=K_x[n; n, \varepsilon]=M[\mathring{X}^2[n, \varepsilon]] \\ 0\leqslant\varepsilon\leqslant 1 \end{gathered} \tag{1.7}$$

A random function whose mathematical expectation is constant and whose correlation function depends only on the difference of the arguments is called a stationary random function [13]. Thus, for a stationary random function of a discrete argument we have the definition

$$\begin{gathered} m_x[n]=\text{const} \\ K_x[n; n_1]=K_x[n-n_1]=K_x[m], \quad m=n-n_1 \end{gathered} \tag{1.8}$$

It will be shown later that in a sampled data automatic system the stationary random process is furnished by the process observed at the output of the stationary system only at discrete instants of time.

The cross-correlation function of two random functions $X[n]$ and $Y[n]$ of a discrete argument is the function determined by the relation

$$K_{xy}[n;\ n_1] = M[\overset{\circ}{X}[n]\overset{\circ}{Y}[n_1]] \tag{1.9}$$

The cross-correlation function of two random functions X and Y observed at the instants $\bar{t} = n + \varepsilon$ $(0 \leqslant \varepsilon \leqslant 1)$ is determined thus in accordance with (1.9):

$$\begin{gathered} K_{xy}[n,\ \varepsilon;\ n_1,\ \varepsilon_1] = M[\overset{\circ}{X}[n,\ \varepsilon]\overset{\circ}{Y}[n_1,\ \varepsilon_1]] \\ 0 \leqslant \varepsilon \leqslant 1,\quad 0 \leqslant \varepsilon_1 \leqslant 1 \end{gathered} \tag{1.10}$$

Side by side with the correlation function we also have to consider the spectral density defined by the sum of a Fourier series whose coefficients are the values of the correlation function [1]:

$$S_x(j\bar{\omega}) = \sum_{m=-\infty}^{\infty} K_x[m]\,\mathrm{e}^{-j\bar{\omega}m} \tag{1.11}$$

where $\bar{\omega} = \omega T$ is the relative frequency.

For the cross-spectral density we have from (1.11)

$$S_{xy}(j\bar{\omega}) = \sum_{m=-\infty}^{\infty} K_{xy}[m]\,\mathrm{e}^{-j\bar{\omega}m} \tag{1.12}$$

By the definition of a correlation function we have

$$K_x[n,\ \varepsilon;\ n_1,\ \varepsilon_1] = K_x[n_1,\ \varepsilon_1;\ n,\ \varepsilon] \tag{1.13}$$

i.e. it is symmetrical with respect to its arguments. It can be shown for the cross-correlation function that the following relation holds

$$K_{xy}[n,\ \varepsilon;\ n_1,\ \varepsilon_1] = K_{yx}[n_1,\ \varepsilon_1;\ n,\ \varepsilon] \tag{1.14}$$

Taking into account the property of a correlation function (1.13), which is also valid for a stationary random function, we obtain from (1.11)

$$S_x(j\bar{\omega}) = S_x(\bar{\omega}) = 2\sum_{m=0}^{\infty} K_x[m]\cos\bar{\omega}m \tag{1.15}$$

Random functions are called correlated if their cross-correlation function is not identically zero, i.e.

$$K_{xy}[n,\ \varepsilon;\ n_1,\ \varepsilon_1] \not\equiv 0$$

Random functions whose cross-correlation function is identically zero are called non-correlated.

Random functions with non-correlated values, usually termed white noise, play an important part in the theory of random functions. White noise is defined mathematically by the relations [13]

$$m_x(t) \equiv 0,\ \ K_x(t;\ t_1) = G(t)\,\delta(t - t_1) \tag{1.16}$$

where $G(t)$ is the intensity of the white noise;
$\delta(t-\tau)$ is the Dirac δ-function.

In the investigation of sampled data systems a similar part is played by random functions of a discrete argument of the white noise type, which are defined as follows:

$$m_x[n] \equiv 0,\ K_x[n;\ n_1] = G[n]\,\sigma[n - n_1] \tag{1.17}$$

whence it follows that the mathematical expectation of discrete white noise is zero, whilst the correlation function of such a process contains the lattice σ-function as a coefficient for the intensity $G[n]$.

It is clear that discrete white noise can be given also by the relations

$$m_x[n,\ \varepsilon] \equiv 0,\ K_x[n;\ n_1] = G[n]\,\sigma[n - n_1,\ 0] \tag{1.18}$$

i.e. we can write the step σ-function instead of the lattice σ-function.

If the intensity $G[n]$ of discrete white noise is equal to unity, then in accordance with (1.11) the spectral density is

$$S_x(\bar{\omega}) = \sum_{m=-\infty}^{\infty} \sigma[m]\,e^{-j\bar{\omega}m} = 1 \tag{1.19}$$

since by the definition of the lattice σ-function

$$\sigma[m] = \begin{cases} 1 \text{ when } m = 0 \\ 0 \text{ when } m \neq 0 \end{cases}$$

We shall now consider problems connected with the transformation of random functions by means of the simplest operators.

3.2 TRANSFORMATION OF RANDOM FUNCTIONS BY LINEAR DIFFERENCE OPERATORS

Without loss of generality we shall consider random lattice functions.

3.2.1 Determination of the difference of a random function

Let a random lattice function $X[n]$ be given, having the mathematical expectation $m_x[n]$ and the correlation function $K_x[n;\ m]$. We shall determine the mathematical expectation $m_{y1}[n]$ and the correlation function $K_{y1}[n;\ m]$ of a random function $Y_1[n]$ which is connected with $X[n]$ by the relation

$$Y_1[n] = \Delta_n X[n] \tag{2.1}$$

Applying the operation of mathematical expectation to both sides of (2.1) we have

$$\begin{aligned} M[Y_1[n]] &= M[\Delta_n X[n]] = M[X[n+1] - X[n]] \\ &= m_x[n+1] - m_x[n] = \Delta_n m_x[n] \end{aligned} \tag{2.2}$$

Thus

$$m_{y1}[n] = \Delta_n m_x[n] \tag{2.3}$$

i.e. the mathematical expectation of the first difference of the random function X is equal to the first difference of its mathematical expectation.

It follows from (2.2) that $M[\ .\ .\ .]$ (the operator for determining the mathematical expectation) and Δ (operator for determining the difference) are interchangeable, i.e. the following is true:

$$M[\Delta] = \Delta M[\ldots] \tag{2.4}$$

Subtracting (2.3) from (2.1) we obtain

$$\mathring{Y}_1[n] = \Delta_n \mathring{X}[n]$$

Considering the last relation for the two instants n and m we have

$$\mathring{Y}_1[n]\,\mathring{Y}_1[m] = \Delta_n \mathring{X}[n]\,\Delta_m \mathring{X}[m] \tag{2.5}$$

If n and m are regarded as independent variables, then (2.5) can be written

$$\mathring{Y}_1[n]\,\mathring{Y}_1[m] = \Delta_n \Delta_m \mathring{X}[n]\,\mathring{X}[m] \tag{2.6}$$

Applying the operation of mathematical expectation to both sides of (2.6), we find, when (2.4) is taken into account, that

$$K_{y1}[n;\ m] = \Delta_n \Delta_m K_x[n;\ m] \tag{2.7}$$

Thus the correlation function of the first difference of a random function is equal to the second displaced difference of the correlation function of this random function.

Applying (2.3) and (2.7) k times we obtain

$$m_{y_k}[n] = \Delta_n^k m_x[n] \tag{2.8}$$

$$K_{y_k}[n;\ m] = \Delta_n^k \Delta_m^k K_x[n;\ m] \tag{2.9}$$

where y_k denotes that the mathematical expectation and the correlation function are given for the random function

$$Y_k[n] = \Delta_n^k X[n] \tag{2.10}$$

Let us find the cross-correlation function $K_{y1x}[n;\ m]$ of $Y_1[n]$ and $X[n]$. By the definition of the cross-correlation function we have from (2.1)

$$\begin{aligned} K_{y1x}[n;\ m] &= M[\mathring{Y}_1[n]\,\mathring{X}[m]] = M[\Delta_n \mathring{X}[n]\,\mathring{X}[m]] \\ &= \Delta_n M[\mathring{X}[n]\,\mathring{X}[m]] = \Delta_n K_x[n;\ m] \end{aligned} \tag{2.11}$$

Hence we have

$$K_{y_1x}[n;\ m]=\Delta_n K_x[n;\ m] \tag{2.12}$$

In the general case we can show that

$$K_{y_iy_j}[n;\ m]=\Delta_n^i \Delta_m^j K_x[n;\ m] \tag{2.13}$$

The relations derived in this section obviously remain valid for the case where $\varepsilon\neq 0$.

3.2.2 Transformation of a random function by a summation operator

Let a summation operator be given whose kernel is a non-random function $W[n;\ m]$. Suppose also that the mathematical expectation $m_x[n]$ and the correlation function $K_x[n;\ m]$ of a random function $X[n]$ are known. We shall find the mathematical expectation and the correlation function of the random function given by

$$Y[n]=\sum_{m=l}^{n} W[n;\ m]\,X[m] \tag{2.14}$$

Applying the operation of mathematical expectation to both sides of (2.14) we obtain

$$\begin{aligned} M[Y[n]] &= M\left[\sum_{m=l}^{n} W[n;\ m]\,X[m]\right]=M[W[n;l]\,X[l] \\ &\quad +W[n;\ l+1]\cdot X[l+1]+\ldots+W[n;\ n]\,X[n]] \\ &=\sum_{m=l}^{n} W[n;\ m]\,M[X[m]]=\sum_{m=l}^{n} W[n;\ m]\,m_x[m] \end{aligned} \tag{2.15}$$

We have thus

$$m_y[n]=\sum_{m=l}^{n} W[n;\ m]\,m_x[m] \tag{2.16}$$

i.e. the mathematical expectation of $Y[n]$ given by (2.14) is the result of applying the summation operator to the mathematical expectation of $X[n]$.

We point out that according to (2.15) $M[\dots]$ (mathematical expectation operator) and Σ (summation operator) are interchangeable, i.e.

$$M[\Sigma]=\Sigma M[\cdot\cdot\cdot] \tag{2.17}$$

To determine the correlation function of $Y[n]$ we centralise this function by subtracting (2.16) from (2.14) term by term. We then have

$$\overset{\circ}{Y}[n]=\sum_{m=l}^{n} W[n;\ m]\,\overset{\circ}{X}[m] \tag{2.18}$$

Considering the two instants n_1 and n_2, we find

$$\overset{\circ}{Y}[n_1]\,\overset{\circ}{Y}[n_2]=\sum_{m_1=l}^{n_1} W[n_1;\ m_1]\,\overset{\circ}{X}[m_1]\sum_{m_2=l}^{n_2} W[n_2;\ m_2]\,\overset{\circ}{X}[m_2] \tag{2.19}$$

Applying the operation of mathematical expectation to both sides of (2.19), we have, when (2.17) is taken into account,

$$K_y[n_1;\ n_2]=\sum_{m_1=l}^{n_1}\sum_{m_2=l}^{n_2} W[n_1;\ m_1]\,W[n_2;\ m_2]\,K_x[m_1;\ m_2] \tag{2.20}$$

whence it follows that the correlation function $K_y[n_1;\ n_2]$ of $Y[n]$ is obtained by applying the summation operator twice to the correlation function $K_x[m_1;\ m_2]$ of $X[n]$: once with respect to the argument m_1 and the second time with respect to m_2.

We shall now determine the cross-correlation function of the random functions $Y[n]$ and $X[n]$. Multiplying (2.18) by $\overset{\circ}{X}[n_2]$ we have, when $n=n_1$,

$$\overset{\circ}{Y}[n_1]\,\overset{\circ}{X}[n_2]=\sum_{m=l}^{n_1} W[n_1;\ m]\,\overset{\circ}{X}[m]\,\overset{\circ}{X}[n_2] \tag{2.21}$$

Applying now the operation of mathematical expectation to

(2.21) we find

$$K_{yx}[n_1;\ n_2] = \sum_{m=l}^{n_1} W[n_1;\ m]\, K_x[m;\ n_2] \tag{2.22}$$

The cross-correlation function of Y and X is thus the result of once applying the summation operator to the correlation function $K_x[n;\ m]$ of X.

The results obtained in this section remain valid for $\varepsilon \neq 0$.

The general formulae obtained above are analogous to the formulae in the theory of random functions which concern the transformation of the corresponding functions by means of differentiation and integration operators respectively. Thus, for example, (2.8), (2.9) and (2.13) are analogous [13] to:

$$m_{y_k}(t) = \frac{d^k m_x(t)}{dt^k} \tag{2.23}$$

$$K_{y_k}(t;\ t_1) = \frac{\partial^{2k} K_x(t;\ t_1)}{\partial t^k \partial t_1^k} \tag{2.24}$$

$$K_{y_i y_j}(t;\ t_1) = \frac{\partial^{i+j} K_x(t;\ t_1)}{\partial t^i \partial t_1^j} \tag{2.25}$$

where Y_k is defined by

$$Y_k(t) = \frac{d^k}{dt^k} X(t) \tag{2.26}$$

Similarly, (2.20) and (2.22) correspond [13] to:

$$K_y(t_1;\ t_2) = \int\limits_T\int\limits_T W(t_1;\ \tau_1)\, W(t_2;\ \tau_2)\, K_x(\tau_1;\ \tau_2)\, d\tau_1 d\tau_2 \tag{2.27}$$

$$K_{yx}(t_1;\ t_2) = \int\limits_T W(t_1;\ \tau)\, K_x(\tau;\ t_2)\, d\tau \tag{2.28}$$

where the random function $Y(t)$ is given by the integral operator

$$Y(t) = \int\limits_T W(t;\ \tau)\, X(\tau)\, d\tau \tag{2.29}$$

where T is the range of variation of the summation variable.

3.3 TRANSFORMATION OF RANDOM FUNCTIONS BY A LINEAR SYSTEM

3.3.1 General formulae for transformations

It was shown in Chapter 1 that every linear sampled data system transforms input disturbances into functions which characterise the state of the system at each instant of time. Mathematically this property is expressed as follows:

$$Y[m, \varepsilon]=A_n X[n, 0] \tag{3.1}$$

where the symbol A_n denotes the set of mathematical operations carried out on the input action $X[n, \varepsilon]$ of the system. From this transformation we obtain the function $Y[n, \varepsilon]$ which constitutes the output variable of the system.

It was shown in Chapter 2 that the operator A_n of a linear dynamical system is a summation operator whose kernel is the weighting function $W[n, \varepsilon; m]$ or the Green's function $G[n, \varepsilon; m]$ of the corresponding difference boundary-value problem when $n \geqslant m$.

Assuming now that $X[n, \varepsilon]$ is a random function with the given mathematical expectation $m_x[n, \varepsilon]$ and correlation function $K_x[n, \varepsilon; n_1, \varepsilon_1]$, let us find the mathematical expectation $m_y[m, \varepsilon]$ and the correlation function $K_y[m, \varepsilon; m_1, \varepsilon_1]$ of the random function $Y[m,\varepsilon]$ given by (3.1).

Since the operators $M[\ldots]$ and A_n are interchangeable, then, applying the operation of mathematical expectation to both sides of (3.1), we have

$$M[Y[m, \varepsilon]]=M[A_n X[n, 0]]=A_n M[X[n, 0]] \tag{3.2}$$

or

$$m_y[m, \varepsilon]=A_n m_x[n, 0] \tag{3.3}$$

whence it follows that the mathematical expectation of the output variable of a sampled data system is the result of applying the operator of this system to the mathematical expectation of the input action.

Subtracting (3.3) from (3.1) term by term, we obtain the following centralised value of $Y[m, \varepsilon]$

$$\mathring{Y}[m, \varepsilon]=A_n \mathring{X}[n, 0] \tag{3.4}$$

Considering the centralised values of $\mathring{Y}$ at the instants $m+\varepsilon$ and $m_1+\varepsilon_1$ and taking into account the interchangeability property of the operator of mathematical expectation and the operator A_n, we obtain the correlation function of the output variable of the system

$$K_y[m, \varepsilon; m_1, \varepsilon_1] = M[\mathring{Y}[m, \varepsilon]\cdot\mathring{Y}[m_1, \varepsilon_1]]$$

$$= M[A_n\mathring{X}[n, 0]A_{n1}\mathring{X}[n_1, 0]] = M[A_nA_{n1}\mathring{X}[n, 0]\,\mathring{X}[n_1, 0]]$$

$$= A_nA_{n1}M[\mathring{X}[n, 0]\,\mathring{X}[n_1, 0]] = A_nA_{n1}K_x[n, 0; n_1, 0] \tag{3.5}$$

We have

$$K_y[m, \varepsilon; m_1, \varepsilon_1] = A_nA_{n1}K_x[n, 0; n_1, 0] \tag{3.6}$$

whence it follows that the correlation function of a random process at the output of a sampled data system is the result of twice applying the operator of this system to the correlation function of the input disturbance, once with respect to the variable n, and the second time with respect to n_1.

The relations (3.3) and (3.6) are analogous to the relations [13] for transforming random functions by means of continuous systems.

3.3.2 Non-stationary system with a non-stationary random signal at its input

We shall find the expressions for the correlation function and the dispersion of the output variable of a non-stationary sampled data system with the weighting function $W[n, \varepsilon; m]$ when a non-stationary random disturbance with the correlation function $K_x[n, \varepsilon; m, \varepsilon_1]$ acts at the input. Since the operator of a non-stationary system is a summation operator with the kernel $W[n, \varepsilon; m]$, then from (3.6) we have

$$K_y[n_1, \varepsilon_1; n_2, \varepsilon_2] = \sum_{m_1=0}^{n_1}\sum_{m_2=0}^{n_2} W[n_1, \varepsilon_1; m_1]\,W[n_2, \varepsilon_2; m_2] \times K_x[m_1, 0; m_2, 0] \tag{3.7}$$

or

$$K_y[n_1, \varepsilon_1; n_2, \varepsilon_2] = \sum_{m_1=0}^{n_1} W[n_1, \varepsilon_1; m_1] \sum_{m_2=0}^{n_2} W[n_2, \varepsilon_2; m_2] \times K_x[m_1, 0; m_2, 0] \tag{3.8}$$

Equation (3.8) thus determines the correlation function of the output variable $Y[n,\varepsilon]$ of a non-stationary sampled data system.

The dispersion of the output variable can be obtained from (3.8), for $\bar{t}_1 = \bar{t}_2 = n + \varepsilon$: as follows:

$$D_y[n, \varepsilon] = \sum_{m_1=0}^{n} W[n, \varepsilon; m_1] \sum_{m_2=0}^{n} W[n, \varepsilon; m_2] K_x[m_1, 0; m_2\ 0] \tag{3.9}$$

The relations (3.8) and (3.9) are of fundamental importance in the statistical analysis of non-stationary sampled data systems. Sometimes, however, it is more convenient to use a different expression for the dispersion.

The relation (3.9) can be written

$$D_y[n, \varepsilon] = \sum_{m_1=0}^{n} \sum_{m_2=0}^{n} W[n, \varepsilon; m_1] W[n, \varepsilon; m_2] K_x[m_1, 0; m_2, 0] \tag{3.10}$$

whence it follows that the summation here is carried out along the nodes of the square network B whose sides equal n, so that

$$0 \leqslant m_1 \leqslant n \tag{3.11}$$

$$0 \leqslant m_2 \leqslant n \tag{3.12}$$

Further, the diagonal of the network, $m_1 = m_2$ divides it into two regions B_1 and B_2 (Fig. 3.1) defined by the values

$$B_1:\ 0 \leqslant m_1 \leqslant m_2 \leqslant n \tag{3.13}$$

$$B_2:\ 0 \leqslant m_2 \leqslant m_1 \leqslant n \tag{3.14}$$

Thus when (3.13) and (3.14) are used the expression (3.10)

can be written

$$D_y[n_1, \varepsilon] = \sum_{(B)}\sum F(m_1, m_2) = \sum_{(B_1)}\sum F(m_1, m_2) + \sum_{(B_2)}\sum F(m_1, m_2), \tag{3.15}$$

where the double summation of the function

$$F(m_1, m_2) = W[n, \varepsilon; m_1]\, W[n, \varepsilon; m_2]\, K_x[m_1, 0; m_2, 0] \tag{3.16}$$

over the nodes of the regions B_1 and B_2 is expressed in symbolic form. We now point out that the function $F(m_1, m_2)$

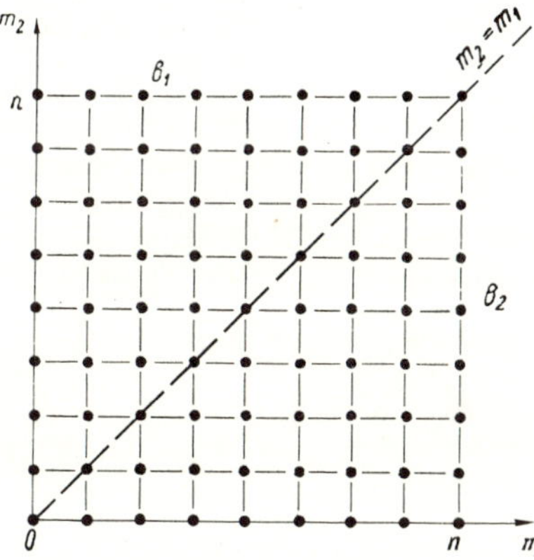

Fig. 3.1 Summation regions in (3.10)

given by (3.16) is symmetrical about the nodes of the diagonal of the square network B, since, by the definition of the correlation function,

$$K_x[m_1, 0; m_2, 0] = K_x[m_2, 0; m_1, 0]$$

and, in addition,

$$W[n, \varepsilon; m_1]\, W[n, \varepsilon; m_2] = W[n, \varepsilon; m_2]\, W[n, \varepsilon; m_1]$$

Consequently, (3.15) can be written as

$$\sum_{(B)}\sum F(m_1, m_2) = 2\sum_{(B_1)}\sum F(m_1, m_2) - \sum_{L} F(m_1, m_2) \tag{3.17}$$

or

$$\sum_{(B)}\sum F(m_1, m_2) = 2\sum_{(B_2)}\sum F(m_1, m_2) - \sum_{L} F(m_1, m_2) \quad (3.18)$$

where summation along the diagonal is denoted by the symbol Σ_L. Thus, using (3.17) and (3.18), we can represent the expression for the dispersion as follows:

$$D_y[n, \varepsilon] = 2\sum_{(B_1)}\sum W[n, \varepsilon; m_2]\, W[n, \varepsilon; m_1]\, K_x[m_1, 0; m_2, 0] - \sum_{m=0}^{n} W^2[n, \varepsilon; m]\, K_x[m, 0; m, 0] \quad (3.19)$$

or

$$D_y[n, \varepsilon] = 2\sum_{(B_2)}\sum W[n, \varepsilon, m_1]\, W[n, \varepsilon; m_2]\, K_x[m_1, 0; m_2, 0] - \sum_{m=0}^{n} W^2[n, \varepsilon; m]\, K_x[m, 0; m, 0] \quad (3.20)$$

Taking into account the definition of the regions B_1 and B_2 and passing onto the repeated summation in the expressions (3.19) and (3.20), we find

$$D_y[n, \varepsilon] = 2\sum_{m_2=0}^{n} W[n, \varepsilon; m_2] \sum_{m_1=0}^{m_2} W[n, \varepsilon; m_1] \times K_x[m_1, 0; m_2, 0] - \sum_{m=0}^{n} W^2[n, \varepsilon; m]\, K_x[m, 0; m, 0] \quad (3.21)$$

and

$$D_y[n, \varepsilon] = 2\sum_{m_1=0}^{n} W[n, \varepsilon; m_1] \sum_{m_2=0}^{m_1} W[n, \varepsilon; m_2] \times K_x[m_1, 0; m_2, 0] - \sum_{m=0}^{n} W^2[n, \varepsilon; m]\, K_x[m, 0; m, 0] \quad (3.22)$$

Introducing the symbols

$$H_1[n, \varepsilon; m_2] = \sum_{m_1=0}^{m_2} W[n, \varepsilon; m_1]\, K_x[m_1, 0; m_2] \quad (3.23)$$

$$H_2[n, \varepsilon; m_1] = \sum_{m_2=0}^{m_1} W[n, \varepsilon; m_2] K_x[m_1, 0; m_2, 0] \tag{3.24}$$

we finally obtain the expressions

$$\begin{aligned} D_y[n, \varepsilon] = 2 \sum_{m_2=0}^{n} W[n, \varepsilon; m_2] H_1[n, \varepsilon; m_2] \\ - \sum_{m=0}^{n} W^2[n, \varepsilon; m] K_x[m, 0; m, 0] \end{aligned} \tag{3.25}$$

$$\begin{aligned} D_y[n, \varepsilon] = 2 \sum_{m_1=0}^{n} W[n, \varepsilon; m_1] H_2[n, \varepsilon; m_1] \\ - \sum_{m=0}^{n} W^2[n, \varepsilon; m] K_x[m, 0; m, 0] \end{aligned} \tag{3.26}$$

which define the dispersion of the non-stationary random process at the output of a sampled data system.

We have obtained above the relations determining the correlation function and the dispersion of the non-stationary random process taking place at the output of a non-stationary sampled data system when a disturbance in the form of a non-stationary random function acts at its input. We shall use these relations to determine the methods of statistical analysis of non-stationary sampled data systems. Let us now consider an important particular case of the non-stationary process which takes place at the output of a stationary sampled data system in a transient state, when a stationary random signal acts at the input of this system.

3.3.3 Stationary system in a transient state with a stationary random signal at its input

Let a stationary random signal $X[n, \varepsilon]$ with the correlation function $K_x[n, \varepsilon]$ act on a stationary sampled data system with the weighting function $W[n, \varepsilon]$. We shall find the expressions for the correlation function $K_y[n_1, \varepsilon_1; n_2, \varepsilon_2]$ and the dispersion $D_y[n, \varepsilon]$ of the output variable $Y[n, \varepsilon]$ of this system in a transient state.

We shall use the basic formula (3.8) which defines the correlation function of a non-stationary random process in a general case where both the system and its random input disturbance are non-stationary. In so doing we shall consider only lattice functions, which, without loss of generality, simplifies the writing of the corresponding formulae. Thus, for lattice functions (3.8) assumes the form

$$K_y[n_1;\ n_2] = \sum_{m_1=0}^{n_1} W[n_1;\ m_1] \sum_{m_2=0}^{n_2} W\{n_2;\ m_2] K_y[m_1;\ m_2] \quad (3.27)$$

where we assume that

$$n_1 < N,\ n_2 < N$$

where N is the relative time of transient processes in the system in question.

Then, from the property of the weighting function of a stationary sampled data system, we have

$$\begin{aligned} W[n_1;\ m_1] &= W[n_1 - m_1] = W[l] \\ W[n_2;\ m_2] &= W[n_2 - m_2] = W[k] \end{aligned} \quad (3.28)$$

where

$$n_1 - m_1 = l,\ n_2 - m_2 = k \quad (3.29)$$

Further, since in the given case the input action of the system is a stationary random signal, then

$$K_x[m_1;\ m_2] = K_x[n_1 - l - n_2 + k] \quad (3.30)$$

Accordingly, when (3.28) and (3.30) are taken into account the expression for the correlation function can be written

$$K_y[n_1;\ n_2] = \sum_{l=0}^{n_1} W[l] \sum_{k=0}^{n_2} W[k] K_x[n_1 - l - n_2 + k] \quad (3.31)$$

Let us transform (3.31). Replacing repeated summation by multiple summation, we have

$$K_y[n_1;\ n_2] = \sum_{l=0}^{n_1} \sum_{k=0}^{n_2} W[l] W[k] K_x[n_1 - l - n_2 + k] \quad (3.32)$$

or

$$K_y[n_1;\ n_2]=\sum_{(B)}\sum F(l,\ k) \tag{3.33}$$

where symbolically the double summation of the function

$$F(l,\ k)=W[l]\,W[k]\,K_x[n_1-l-n_2+k] \tag{3.34}$$

is written for the nodes of the rectangular network B with the sides $l=n_1$ and $k=n_2$ (Fig. 3.2). Expression (3.33) can be written

$$K_y[n_1;\ n_2]=\sum_{(B)}\sum F(l,\ k)=\sum_{(B_1)}\sum F(l,\ k) + \sum_{(B_2)}\sum F(l,\ k)-\sum_{L} F(l,\ k) \tag{3.35}$$

where B_1 and B_2 are the two regions into which the rectangular summation region B is divided by the diagonal $l=k$.

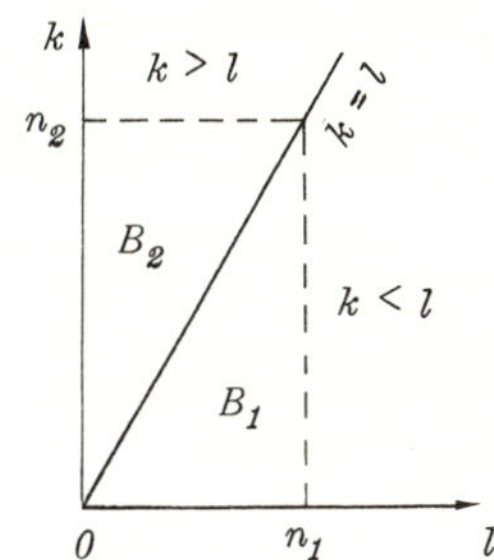

Fig. 3.2 Summation regions in (3.32)

The regions B_1 and B_2 are defined by

$$B_1:\ 0\leqslant k\leqslant l\leqslant n_1 \tag{3.36}$$

$$B_2:\ 0\leqslant l\leqslant k\leqslant n_2 \tag{3.37}$$

We shall first consider summation over B_1. From (3.36) we have

$$\sum_{(B_1)}\sum F(l,\ k)=\sum_{l=0}^{n_1}\sum_{k=0}^{l} F(l,\ k) \tag{3.38}$$

or, with (3.34) taken into account,

$$\sum_{(B_1)}\sum F(l,\ k)=\sum_{l=0}^{n_1}\sum_{k=0}^{l} W[l]\,W[k]\,K_x[n_1-l-n_2+k] \quad (3.39)$$

Expression (3.39) can be written in a different form if we proceed from multiple summation to repeated summation:

$$\sum_{(B_1)}\sum F(l,\ k)=\sum_{l=0}^{n_1} W[l]\sum_{k=0}^{l} W[k]\,K_x[n_1-l-n_2+k] \quad (3.40)$$

We shall now consider summation over B_2. On the basis of (3.37) we have

$$\sum_{(B_2)}\sum F(l,\ k)=\sum_{k=0}^{n_2}\sum_{l=0}^{k} F(l,\ k) \quad (3.41)$$

or, in accordance with (3.34),

$$\sum_{(B_2)}\sum F(l,\ k)=\sum_{k=0}^{n_2} W[k]\sum_{l=0}^{k} W[l]\,K_x[n_1-l-n_2+k] \quad (3.42)$$

We can thus write, for the correlation function of the output variable,

$$\begin{aligned} K_y[n_1;\ n_2]=&\sum_{l=0}^{n_1} W[l]\sum_{k=0}^{l} W[k]\,K_x[n_1-l-n_2+k]\\ &+\sum_{k=0}^{n_2} W[k]\sum_{l=0}^{k} W[l]\,K_x[n_1-l-n_2+k]\\ &-\sum_{h=0}^{|n_1-n_2|} W^2[h]\,K_x[n_1-n_2]. \end{aligned} \quad (3.43)$$

We now assume that $n_1 > n_2$. Then (3.43) can be represented as follows:

$$\begin{aligned} K_y[n_1;\ n_2]=&\sum_{l=0}^{n_2} W[l]\sum_{k=0}^{l} W[k]\,K_x[n_1-l-n_2+k]\\ &+\sum_{l=n_2+1}^{n_1} W[l]\sum_{k=0}^{l} W[k]\,K_x[n_1-l-n_2+k] \quad (3.44)\\ &-\sum_{h=0}^{n} W^2[h]K_x[n_1-n_2]+\sum_{k=0}^{n_2} W[k]\sum_{l=0}^{k} W[l]\,K_x[n_1-l-n_2+k] \end{aligned}$$

Changing the roles of k and l in the last term of the last expression, we obtain

$$K_y[n_1;\ n_2]=\sum_{l=0}^{n_2} W[l]\sum_{k=0}^{l} W[k]K_x[n_1-l-n_2+k]$$
$$+\sum_{l=n_2+1}^{n_1} W[l]\sum_{k=0}^{l} W[k]K_x[n_1-l-n_2+k] \tag{3.45}$$
$$+\sum_{l=0}^{n_2} W[l]\sum_{k=0}^{l} W[k]K_x[n_1-k-n_2+l]-\sum_{h=0}^{n_2} W^2[h]K_x[n_1-n_2]$$

Because

$$K_x[n_1-l-n_2+k]=K_x[n_1-k-n_2+l]$$

the expression (3.45) can be written

$$K_y[n_1;\ n_2]=2\sum_{l=0}^{n_2} W[l]\sum_{k=0}^{l} W[k]K_x[n_1-l-n_2+k] \tag{3.46}$$
$$+\sum_{l=n_2+1}^{n_1} W[l]\sum_{k=0}^{l} W[k]K_x[n_1-l-n_2+k]-\sum_{h=0}^{n_2} W^2[h]K_x[n_1-n_2]$$

whence it follows that, to determine the correlation function, we must carry out summation along the nodes of a square network with sides $l=n_2$ and $k=n_2$, and along the nodes of a rectangular network with sides $l=n_1$ and $k=n_1-n_2-1$.

Expression (3.46) is in fact the required expression for the correlation function of the output variable of a stationary sampled data system in a transient state. This has been obtained for lattice functions. It is easy to show that, in general, (3.46) has the form

$$K_y[n_1,\ \varepsilon_1;\ n_2,\ \varepsilon_2]=2\sum_{l=0}^{n_2} W[l,\ \varepsilon_1]\sum_{k=0}^{l} W[k,\ \varepsilon_2]$$
$$\times K_x[n_1-l-n_2+k,\ 0]$$
$$+\sum_{l=n_2+1}^{n_1} W[l,\ \varepsilon_1]\sum_{k=0}^{l} W[k,\ \varepsilon_2]K_x[n_1-l-n_2+k_1,\ 0] \tag{3.47}$$
$$-\sum_{h=0}^{n_2} W^2[h,\ \varepsilon]K_x[n_1-n_2,\ 0]$$

We now determine the expression for the dispersion of the output variable of a stationary system in a transient state. Putting $n_1 = n_2 = n$, in (3.46) we obtain

$$D_y[n] = 2\sum_{l=0}^{n} W[l] \sum_{k=0}^{l} W[k] K_x[k-l] - \sum_{h=0}^{n} W^2[h] K_x[0] \tag{3.48}$$

whence it follows that, to determine the dispersion, we must sum over the region B_1 of the square network B with

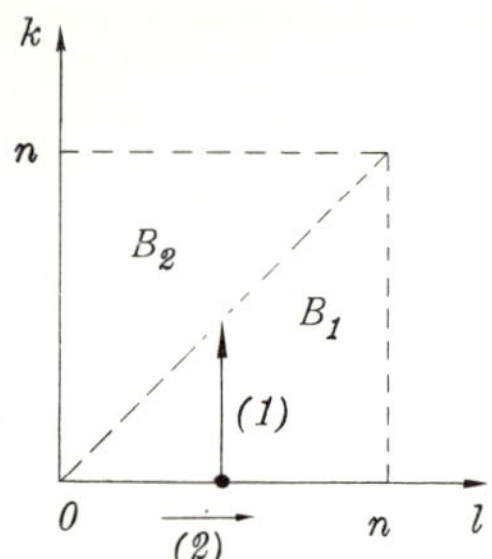

Fig. 3.3 Summation regions in (3.48)

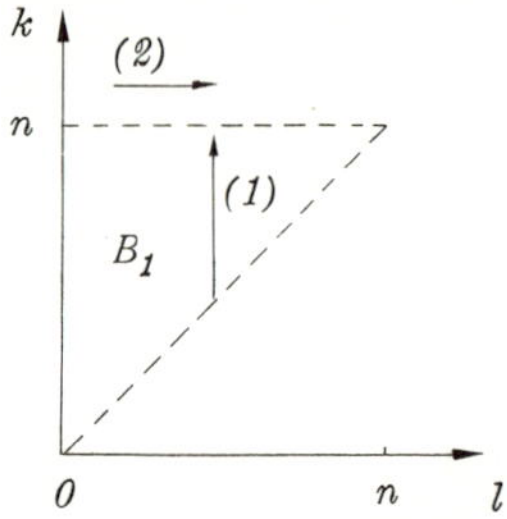

Fig. 3.4 The order of summation in (3.48)

the sides $l = k = n$ (Fig. 3.3). If the summation is carried out over the region B_2, where the order of summation is chosen as shown in Fig. 3.4, then the dispersion is determined by

$$D_y[n] = 2\sum_{l=0}^{n} W[l] \sum_{k=0}^{n} W[k] K_x[k-l] - \sum_{h=0}^{n} W^2[h] K_x[0] \tag{3.49}$$

We now introduce in (3.49) the summation variable

$$\xi = k - l$$

Then,

$$k = \xi + l$$

$$\xi = \begin{cases} 0 & \text{when } k = l \\ n - l & \text{when } k = n \end{cases} \tag{3.50}$$

Thus, when (3.50) is taken into account, (3.49) can be written

$$D_y[n] = 2\sum_{l=0}^{n}\sum_{\xi=0}^{n-l} W[l]\,W[l+\xi]\,K_x[\xi] - \sum_{h=0}^{n} W^2[h]\,K_x[0] \quad (3.51)$$

Since the variable l assumes values from 0 to n, it follows from (3.50) that the summation in (3.51) is carried out over the region S_2 of the square network S with the sides $\xi = n$ and $l = n$ (Fig. 3.5). Replacing now summation over S_2 in (3.51) by summation over S_1, we obtain

$$D_y[n] = 2\sum_{\xi=0}^{n}\sum_{l=0}^{n-\xi} W[l]\,W[l+\xi]\,K_x[\xi] - \sum_{h=0}^{n} W^2[h]\,K_x[0] \quad (3.52)$$

We now introduce the function L given by

$$L[n;\,\xi] = \sum_{l=0}^{n} W[l]\,W[l+\xi] \quad (3.53)$$

It follows from the definition (3.53) that L depends on the

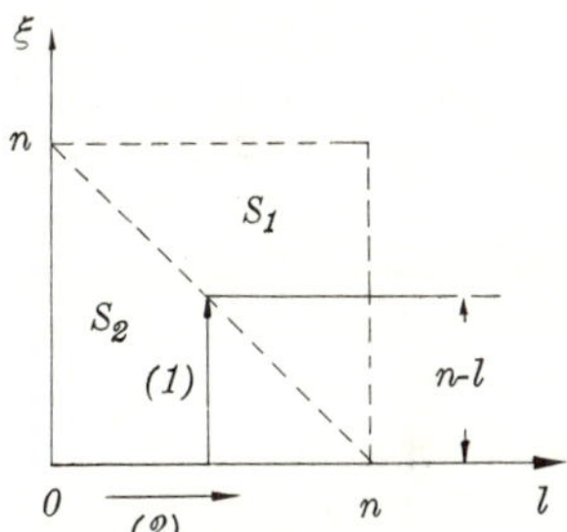

Fig. 3.5 Summation regions in (3.51)

parameter ξ and has the upper limit n of summation for its argument, so that

$$\sum_{l=0}^{n\pm m} W[l]\,W[l+\xi] = L[n\pm m;\,\xi] \qquad m = 0,1,\ldots \quad (3.54)$$

In accordance with (3.54), denoting

$$\sum_{l=0}^{n-\xi} W[l]\,W[l+\xi] = L[n-\xi;\,\xi] \quad (3.55)$$

in (3.52), we have the final expression

$$D_y[n] = 2\sum_{\xi=0}^{n} L[n-\xi;\ \xi] K_x[\xi] - \sum_{h=0}^{n} W^2[h] K_x[0] \quad (3.56)$$

which gives the dispersion of the output variable of the stationary system in a transient state.

Expression (3.56) has been obtained for lattice functions. In the general case it is easy to show that the relation

$$D_y[n,\ \varepsilon] = 2\sum_{\xi=0}^{n} L[n-\xi;\ \xi,\ \varepsilon] K_x[\xi,\ 0] - \sum_{h=0}^{n} W^2[h,\ \varepsilon] K_x[0,\ 0] \quad (3.57)$$

is valid, where

$$L[n-\xi;\ \xi,\ \varepsilon] = \sum_{l=0}^{n-\xi} W[l,\ \varepsilon] W[l+\xi,\ \varepsilon] \quad (3.58)$$

3.3.4 Stationary system in a steady state with a stationary random signal at its input

Suppose we are given a stationary system with the weighting function $W[n,\ \varepsilon]$. We shall find the expressions for the correlation function and the dispersion of the output variable of this system in a steady state, if a stationary random function $X[n,\ \varepsilon]$ acts at the input with the correlation function $K_x[n,\ \varepsilon]$.

It was shown in Chapter 1 that the output variable of a sampled data system with constant parameters is determined by

$$y[n,\ \varepsilon] = \sum_{m=0}^{\infty} W[m,\ \varepsilon] x[n-m,\ 0] \quad (3.59)$$

Assuming now that $x[n,\ \varepsilon]$ is a stationary random signal, and applying the operation of mathematical expectation to both sides of (3.59), we find

$$\begin{aligned} m_y[n,\ \varepsilon] &= \sum_{m=0}^{\infty} W[m,\ \varepsilon] m_x[n-m,\ 0] \\ &= m_x \sum_{m=0}^{\infty} W[m,\ \varepsilon] = m_y(\varepsilon) \end{aligned} \quad (3.60)$$

since by the condition $X[n, \varepsilon]$ is a stationary random function.

Centralising $Y[n, \varepsilon]$ and considering its values at the instants $\bar{t}_1 = n_1 + \varepsilon_1$ and $\bar{t}_2 = n_2 + \varepsilon_2$, we find, in accordance with the definition (3.6), the correlation function of the random process $Y[n, \varepsilon]$

$$K_y[l, \varepsilon] = \sum_{m_1=0}^{\infty} W[m_1, \varepsilon_1] \sum_{m_2=0}^{\infty} W[m_2, \varepsilon_2] K_x[l + m_1 - m_2, 0] \tag{3.61}$$

When the summation along the inner sum is carried out, ε_2 varies between the limits of the closed interval [0; 1]. This is true also by analogy for summation with respect to m_1. Therefore (3.61) can be written

$$K_y[l, \varepsilon] = \sum_{m_1=0}^{\infty} W[m_1, \varepsilon] \sum_{m_2=0}^{\infty} W[m_2, \varepsilon] K_x[l + m_1 - m_2, 0] \tag{3.62}$$

The expressions (3.60) for the mathematical expectation and (3.62) for the correlation function show that the output variable of a system in a steady state, when a stationary random process acts at its input, is a non-stationary random function between the limits $0 \leqslant \varepsilon \leqslant 1$, i.e. between the limits of the sampling period. Hence it follows from (3.60) and (3.62) that the output variable, considered only at the discrete instants $\bar{t} = n + \varepsilon_0, \varepsilon_0 = \text{const}$, is a stationary random function. In particular, putting $\varepsilon = 0$, we have

$$m_y(\varepsilon)|_{\varepsilon=0} = m_y(0) = \text{const} \tag{3.63}$$

$$K_y[l, \varepsilon]|_{\varepsilon=0} = K_y[l, 0] = \sum_{m_1=0}^{\infty} W[m_1, 0] \times \sum_{m_2=0}^{\infty} W[m_2, 0] K_x[l + m_1 - m_2, 0] \tag{3.64}$$

whence we see that the mathematical expectation of the random function $Y[n, \varepsilon]|_{\varepsilon=0}$ is constant, whilst its correlation function depends only on the difference of the arguments $l = n_1 - n_2$.

The dispersion of the output variable can be obtained

from (3.62) if we put $l=0$:

$$K_y[0, \varepsilon] = D_y(\varepsilon) = \sum_{m_1=0}^{\infty} W[m_1, \varepsilon] \sum_{m_2=0}^{\infty} W[m_2, \varepsilon] K_x[m_1 - m_2, 0] \tag{3.65}$$

Consequently, the dispersion of the output variable of a stationary system depends on the parameter ε, which is a direct consequence of the random process $Y[n, \varepsilon]$ being non-stationary between the limits of the sampling interval. Putting $\varepsilon=0$ in (3.65) we have

$$D_y(0) = \sum_{m_1=0}^{\infty} W[m_1, 0] \sum_{m_2=0}^{\infty} W[m_2, 0] K_x[m_1 - m_2, 0] \tag{3.66}$$

whence it follows that the dispersion of the output variable, observed at the discrete instants of time, is constant.

The formulae obtained in this section, which define the correlation function and the dispersion of the output variable of a stationary system, are of fundamental importance in the analysis of systems with constant parameters when random disturbances act at the input of these systems.

3.3.5 Multi-dimensional system

Let a multi-dimensional system be given having l inputs and k outputs. We shall find the correlation functions and the dispersion of the random functions $Y_i[n, \varepsilon], i=1, \ldots, k$ at the outputs of this system, if the random disturbances X_j, $j=1, 2, \ldots, l$ act at its l inputs.

It was shown in Chapter 1 that the connection between the output variable Y_i and the input disturbances X_j of a multi-dimensional system is established by

$$Y_i[n, \varepsilon] = \sum_{j=1}^{l} \sum_{m=-\infty}^{n} W_{ij}[n, \varepsilon; m] X_j[m, 0] \tag{3.67}$$

$$i = 1, 2, \ldots, k$$

where W_{ij} are the weighting functions of the multi-dimensional system, forming the rectangular matrix

$$\|W_{ij}\| \quad \begin{matrix} i=1, 2, \ldots, k \\ j=1, 2, \ldots, l \end{matrix}$$

Considering for simplicity only lattice functions, we obtain from (3.67), by applying the operation of mathematical expectation to both sides of this equation,

$$m_{yi}[n]=\sum_{j=1}^{l}\sum_{m=-\infty}^{n} W_{ij}[n;\ m]\, m_{xj}[m] \qquad i=1,\ 2,\ \ldots,\ k \tag{3.68}$$

Centralising now the output variables $Y_i[n]$, and considering the values of these variables for the two instants $\bar{t}_1=n_1$ and $\bar{t}_2=n_2$, we have

$$\overset{\circ}{Y}_i[n_1]=\sum_{j=1}^{l}\sum_{m_1=-\infty}^{n_1} W_{ij}[n_1;\ m_1]\, \overset{\circ}{X}_j[m_1] \qquad i=1,\ 2,\ \ldots,\ k \tag{3.69}$$

$$\overset{\circ}{Y}_h[n_2]=\sum_{\alpha=1}^{l}\sum_{m_2=-\infty}^{n_2} W_{h\alpha}[n_2,\ m_2]\, \overset{\circ}{X}_\alpha[m_2] \qquad h=1,\ 2,\ \ldots,\ k \tag{3.70}$$

Applying the operation of mathematical expectation to both sides of the equation determined as the product of (3.69) and (3.70), we obtain

$$K_{yih}[n_1;\ n_2]=\sum_{j=1}^{l}\sum_{\alpha=1}^{l}\sum_{m_1=-\infty}^{n_1} W_{ij}[n_1;\ m_1]\sum_{m_2=-\infty}^{n_2} W_{h\alpha}[n_2;\ m_2] \times K_{xj\alpha}[m_1;\ m_2],\ i,\ h=1,\ 2,\ \ldots,\ k, \tag{3.71}$$

which give the second-order moments of the functions $\overset{\circ}{Y}_i$ and $\overset{\circ}{Y}_h$ in terms of the second-order moments of the random functions $\overset{\circ}{X}_j$ and $\overset{\circ}{X}_\alpha$.

In the particular case where

$$K_{xj\alpha}[m_1,\ m_2]=\begin{cases} 0 & \text{when } j\neq\alpha, \\ K_{xj}[m_1;\ m_2] & \text{when } j=\alpha \end{cases} \tag{3.72}$$

the following equality holds:

$$K_{yih}[n_1;\ n_2]=\begin{cases} 0 & \text{when } i\neq h, \\ K_{yi}[n_1;\ n_2] & \text{when } i=h, \end{cases} \tag{3.73}$$

Accordingly, we have from (3.71)

$$K_{yi}[n_1;\ n_2]=\sum_{i=1}^{l}\sum_{m_1=-\infty}^{n_1} W_{ij}[n_1;\ m_1]\sum_{m_2=-\infty}^{n_2} W_{ij}[n_2;\ m_2] \times K_{x_j}[m_1;\ m_2],\quad i=1,\ 2,\ \ldots,\ k \tag{3.74}$$

Equation (3.74) defines the correlation functions of the output variables of a multi-dimensional sampled data system with the condition that (3.72) and (3.73) apply.

To determine the dispersion of Y_i we put $n_1=n_2=n$ in (3.74). We then find that

$$D_{yi}[n]=\sum_{j=1}^{l}\sum_{m_1=-\infty}^{n} W_{ij}[n;\ m_1]\sum_{m_2=-\infty}^{n} W_{ij}[n;\ m_2]\,K_{x_j}[m_1;\ m_2] \qquad i=1,\ 2,\ \ldots,\ k \tag{3.75}$$

whence it follows that the dispersion of the output variable is determined by the sum

$$D_{yi}[n]=\sum_{j=1}^{l}\bar{D}_{yj}[n],\quad i=1,\ 2,\ \ldots,\ k \tag{3.76}$$

where

$$\bar{D}_{yj}[n]=\sum_{m_1=-\infty}^{n} W_{ij}[n;\ m_1]\sum_{m_2=-\infty}^{n} W_{ij}[n;\ m_2]\,K_{x_j}[m_1;\ m_2] \qquad j=1,\ 2,\ \ldots,\ l \tag{3.77}$$

is the dispersion of the response of the system at the i-th output to the disturbance X_j applied to the j-th input, with the condition that $X_\beta=0,\ \beta\neq j$. This is a consequence of the linearity of the system in question.

The expressions for the correlation function and the dispersion have been obtained for lattice functions. They remain of course valid also in the general case.

3.3.6 Particular cases of random processes

In this section we shall consider a case often met in practice, when the sampled data system is subjected to a random disturbance in the form of discrete white noise. Let a random function X with zero mathematical expectation and a correlation function of the form

$$K_x[n;\ m]=c^2\sigma[n-m,\ 0] \tag{3.78}$$

act at the input of a stationary sampled data system with the weighting function $W[n,\ \varepsilon]$. We shall find expressions for the correlation function and the dispersion of the output variable of this system in a steady state.

To determine the correlation function we use (3.62)

$$K_y[l,\ \varepsilon]=\sum_{m_1=0}^{\infty}W[m_1,\ \varepsilon]\sum_{m_2=0}^{\infty}W[m_2,\ \varepsilon]K_x[l+m_1-m_2,\ 0] \tag{3.79}$$

Putting

$$K_x[l+m_1-m_2,\ 0]=c^2\sigma[l+m_1-m_2,\ 0] \tag{3.80}$$

in (3.79), we obtain

$$\begin{aligned}K_y[l,\ \varepsilon]&=\sum_{m_1=0}^{\infty}W[m_1,\ \varepsilon]\sum_{m_2=0}^{\infty}W[m_2,\ \varepsilon]c^2\sigma[l+m_1-m_2,\ 0]\\&=c^2\sum_{m_1=0}^{\infty}W[m_1,\ \varepsilon]W[l+m_1,\ \varepsilon]\end{aligned} \tag{3.81}$$

which gives the correlation function of the random process at the output of the stationary system in the steady state when discrete white noise with the intensity c^2 acts at its input.

When $l=0$, (3.81) gives the dispersion of the random process in question:

$$D_y[0,\ \varepsilon]=c^2\sum_{m_1=0}^{\infty}W^2[m_1,\ \varepsilon] \tag{3.82}$$

which, as follows from (3.82), depends on the parameter ε $(0\leqslant\varepsilon\leqslant 1)$. When $\varepsilon=0$ we obtain from (3.82) the expression for the dispersion at the instant the sampler key closes

$$D_y(0)=c^2\sum_{m_1=0}^{\infty}W^2[m_1,\ 0] \tag{3.83}$$

Let the random function X with the correlation function

(3.78) act at the input of a non-stationary system with the weighting function $W[n, \varepsilon; m]$. We shall find the correlation function and the dispersion of the output variable of this system. We use (3.8), writing it for lattice functions as

$$K_y[n_1; n_2] = \sum_{m_1=0}^{n_1} W[n_1; m_1] \sum_{m_2=0}^{n_2} W[n_2; m_2] K_x[m_1; m_2] \tag{3.84}$$

Since the random function in this case is stationary, then

$$K_x[m_1; m_2] = K_x[m_1 - m_2]$$

and (3.84) assumes the form

$$K_y[n_1; n_2] = \sum_{m_1=0}^{n_1} W[n_1; m_1] \sum_{m_2=0}^{n_2} W[n_2; m_2] K_x[m_1 - m_2] \tag{3.85}$$

Putting now

$$K_x[m_1 - m_2] = c^2 \sigma[m_1 - m_2]$$

in (3.85) we have

$$\begin{aligned} K_y[n_1; n_2] &= \sum_{m_1=0}^{n_1} W[n_1; m_1] \sum_{m_2=0}^{n_2} W[n_2; m_2] c^2 \sigma[m_1 - m_2] \\ &= c^2 \sum_{m_1=0}^{n_1} W[n_1; m_1] W[n_2; m_1] \end{aligned} \tag{3.86}$$

Equation (3.86) determines the correlation function of the random process at the output of a non-stationary system when discrete white noise acts at its input. This formula has been obtained for lattice functions. In the general case we have for the correlation function

$$K_y[n_1, \varepsilon_1; n_2, \varepsilon_2] = c^2 \sum_{m_1=0}^{n_1} W[n_1, \varepsilon_1; m_1] W[n_2, \varepsilon_2; m_2] \tag{3.87}$$

Putting $n_1 = n_2 = n$ in (3.86) we find

$$D_y[n] = c^2 \sum_{m=0}^{n} W^2[n; m] \tag{3.88}$$

which gives the dispersion of the random process at the

output of a non-stationary system when white noise with constant intensity acts at the input. It can also be shown that in the general case ($\varepsilon \neq 0$) the dispersion can be found from

$$D_y[n, \varepsilon] = c^2 \sum_{m=0}^{n} W^2[n, \varepsilon; m] \tag{3.89}$$

The relations obtained in this section confirm that the expressions for calculating the correlation function and the dispersion of the random process at the output of a sampled data system are particularly simple when discrete white noise acts at the input of the system. We shall show later that the statistical analysis of sampled data systems, when arbitrary (in some sense or other) random disturbances act at their inputs, can be reduced to the analysis of certain equivalent systems subjected to the action of discrete white noise. From this viewpoint the formulae just obtained are of fundamental importance in the statistical analysis of stationary and non-stationary sampled data systems.

3.4 CORRELATION FUNCTION OF THE OUTPUT RANDOM PROCESS OF A NON-STATIONARY SYSTEM WITH DISCRETE WHITE NOISE AT ITS INPUT

Let a non-stationary sampled data system be described by a single k-th order difference equation of the form

$$L_n(\Delta, n) y[n, \varepsilon] = M_n(\Delta, \bar{t}) x[n, 0], \; 0 \leqslant \varepsilon \leqslant 1 \tag{4.1}$$

We assume that a random function X with zero mathematical expectation and the correlation function

$$K_x[n; m] = c^2 \sigma[n-m, 0], \; c^2 = 1 \tag{4.2}$$

acts at the input of this system, i.e. we assume that X is white noise with constant intensity equal to unity.* We shall determine the properties of the correlation function of the random process at the output for the given disturbance X. If we take (3.8), the correlation function of the output process of a non-stationary system with the weighting function

* The restriction $c^2 = 1$ is not fundamental. It is introduced only for convenience in calculation.

$W[n, \varepsilon; m]$ is

$$K_y[n, \varepsilon; m, \varepsilon_1] = \sum_{l=0}^{n} W[n, \varepsilon; l] \sum_{\lambda=0}^{m} W[m, \varepsilon_1; \lambda] K_x[l, 0; \lambda, 0] \tag{4.3}$$

Substituting the value of the correlation function of the input disturbance determined by (4.2) into (4.3), we have

$$K_y[n, \varepsilon; m, \varepsilon_1] = \sum_{l=0}^{n} W[n, \varepsilon; l] \sum_{\lambda=0}^{m} W[m, \varepsilon_1; \lambda] \sigma[l-\lambda, 0] \tag{4.4}$$

Carrying out the summation in (4.4) we have to deal with two cases, $n>m$ and $n<m$. We shall consider each case separately.

Suppose first that the case $n<m$ applies. Then according to (4.4) the summation variables vary between the limits

$$l=0 \text{ to } n, \ \lambda=0 \text{ to } m$$

as shown in Fig. 3.6(a). Since by the definition of the σ-function

$$\sigma[l-\lambda, 0] = \begin{cases} 0 & \text{when } l \neq \lambda \\ 1 & \text{when } l = \lambda \end{cases}$$

the summation in (4.4) should first be carried out with respect to λ. We have thus

$$K_y[n, \varepsilon; m, \varepsilon_1] = \sum_{l=0}^{n} W[n, \varepsilon; l] W[m, \varepsilon_1; l], \ n<m \tag{4.5}$$

Suppose now that the case $n>m$ applies. Then according to (4.4) the summation variables vary between the limits shown in Fig. 3.6(b). Consequently, in this case we must

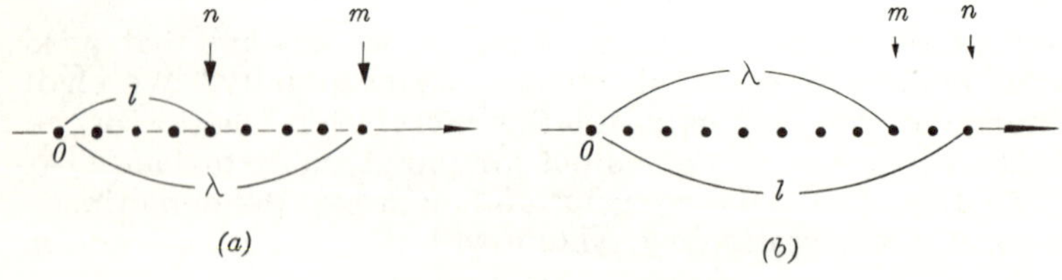

Fig. 3.6 Variation ranges of the summation variables in (4.4)

first carry out summation with respect to l, and then with respect to the variable λ. Accordingly in this case we have

$$K_y[n, \varepsilon; m, \varepsilon_1] = \sum_{\lambda=0}^{m} W[n, \varepsilon; \lambda] W[m, \varepsilon_1; \lambda], \ n > m \quad (4.6)$$

Using (4.5) and (4.6) we can finally write the expression (4.4) for the correlation function as

$$K_y[n, \varepsilon; m, \varepsilon_1] = \begin{cases} \sum_{l=0}^{n} W[n, \varepsilon; l] W[m, \varepsilon_1; l], & n < m \\ \sum_{l=0}^{m} W[n, \varepsilon; l] W[m, \varepsilon_1; l], & n > m \end{cases} \quad (4.7)$$

We shall now apply the difference operator $L_n(\Delta, n)$ with respect to the variable n to the second equation of (4.7). We now have

$$\begin{aligned} L_n(\Delta, n) K_y[n, \varepsilon; m, \varepsilon_1] \\ = L_n(\Delta, n) \sum_{l=0}^{m} W[n, \varepsilon; l] W[m, \varepsilon_1; l] \\ = \sum_{l=0}^{m} L_n(\Delta, n) W[n, \varepsilon; l] W[m, \varepsilon_1; l], \ n > m \end{aligned} \quad (4.8)$$

It was shown in Chapter 1 that the weighting function $W[n, \varepsilon; m]$ of a sampled data system satisfies the difference equation describing the system, provided the disturbing function is replaced by the σ-function. Thus, on the basis of (4.1), we have

$$L_n(\Delta, n) W[n, \varepsilon; m] = M_n(\Delta, \bar{t}) \sigma[n-m, 0), \ 0 \leqslant \varepsilon \leqslant 1 \quad (4.9)$$

Consequently, when (4.9) is used, (4.8) gives

$$L_n(\Delta, n) K_y[n, \varepsilon; m, \varepsilon_1] = \sum_{l=0}^{m} W[m, \varepsilon_1; l] M_n(\Delta, \bar{t}) \sigma[n-l, 0] \quad (4.10)$$

$$n > m$$

Since

$$\sigma[n-l, 0] = \begin{cases} 0 & \text{when } n \neq l \\ 1 & \text{when } n = l \end{cases}$$

we obtain from (4.10)

$$L_n(\Delta, n) K_y[n, \varepsilon; m, \varepsilon_1] = 0, \quad n > m \tag{4.11}$$

since (4.10) is defined for $n>m$ (the σ-function is zero for $n>m$)

Consequently, we obtain the important result that the correlation function $K_y[n, \varepsilon; m, \varepsilon_1]$ of the random process at the output of a sampled data system, when discrete white noise acts at the input, satisfies, for $n>m$, the homogeneous difference equation (4.11) which corresponds to the equation of the system.

Let us now consider the first equation of (4.7)

$$K_y[n, \varepsilon; m, \varepsilon_1] = \sum_{l=0}^{n} W[n, \varepsilon; l] W[m, \varepsilon_1; l], \; n < m \tag{4.12}$$

It is easy to see that $K_y[n, \varepsilon; m, \varepsilon_1]$ for $n<m$ is the response of a sampled data system with the weighting function $W[n, \varepsilon; l]$ to a disturbance which coincides with the weighting function of the corresponding adjoint sampled data system, i.e. from (4.12) we have

$$\begin{aligned} K_y[n, \varepsilon; m, \varepsilon_1] &= \sum_{l=0}^{n} W[n, \varepsilon; l] W[m, \varepsilon_1; l] \\ &= \sum_{l=0}^{n} W[n, \varepsilon; l] W^*[l; m, \varepsilon_1], \; n < m \end{aligned} \tag{4.13}$$

As a consequence of this we can write

$$L_n(\Delta, n) K_y[n, \varepsilon; m, \varepsilon_1] = M_n(\Delta, \overline{t}) W[m, \varepsilon_1; n], \; n < m \tag{4.14}$$

We thus obtain the other important result, that the correlation function $K_y[n, \varepsilon; m, \varepsilon_1]$ of the random process at the output of a sampled data system, when discrete white noise acts at its input, satisfies the non-homogeneous difference equation of this system for $n<m$, with the condition that the disturbing function is replaced by the weighting function of the corresponding adjoint system.

Expression (4.14) can also be written

$$L_n(\Delta, n) K_y[n, \varepsilon; m, \varepsilon_1] = M_n(\Delta, \overline{t}) M^*_m(\Delta, \overline{\tau}) W_0[n, \varepsilon_1; m] \quad n < m \tag{4.15}$$

where $\bar{\tau}=m+\varepsilon_1$ and $W_0[n, \varepsilon; m]$ is the weighting function of the system defined by

$$L_n(\Delta, n) W_0[n, \varepsilon; m]=\sigma[n-m, 0], \quad 0\leqslant\varepsilon\leqslant 1 \tag{4.16}$$

Considering the correlation function $K_y[n, \varepsilon; m, \varepsilon_1]$ as a function of m, we can show that it satisfies the following difference equations:

$$\overset{*}{L}_m(\Delta, m) K_y[n, \varepsilon; m, \varepsilon_1]=0, \quad m>n \tag{4.17}$$

$$\overset{*}{L}_m(\Delta, m) K_y[n, \varepsilon; m, \varepsilon_1]=\overset{*}{M}_m(\Delta, \bar{\tau}) W[n, \varepsilon_1; m], \quad m<n \tag{4.18}$$

The last relation can also be written

$$\overset{*}{L}_m(\Delta, m) K_y[n, \varepsilon; m, \varepsilon_1]=\overset{*}{M}_m(\Delta, \bar{\tau}) M_n(\Delta, \bar{t}) W_0[n, \varepsilon_1; m] \quad m<n \tag{4.19}$$

It should be pointed out that, in order to determine the correlation function, the difference equation (4.14) must be solved for zero initial conditions, if prior to the instant at which the action is applied the system was in the undisturbed state. On the other hand, (4.11) should be solved for the given values of the dispersion and the $(n-1)$ differences of the correlation function when $n=m$.

Let us now formulate the properties of the correlation function of the random process at the output of a sampled data system when a random signal in the form of white noise acts at its input. These properties follow directly from (4.11) to (4.18) and from the properties of the weighting functions of the original and adjoint systems.

1. The number of the piecewise-continuous differences of the correlation function $K_y[n, \varepsilon; m, \varepsilon_1]$ with respect to n over the region $n>m$, and with respect to m over the region $n<m$, is equal to the number of piecewise-continuous differences of the weighting function of the corresponding sampled data system.
2. The correlation function K_y can be represented in the form

$$K_y[n, \varepsilon; m, \varepsilon_1]=\sum_{i=1}^{k} \overset{*}{\varphi}_i[m, \varepsilon_1] \varphi_i[n, \varepsilon], \; n>m \tag{4.20}$$

$$K_y[n, \varepsilon; m, \varepsilon_1] = \sum_{i=1}^{k} \varphi_i^*[n, \varepsilon]\, \varphi_i[m, \varepsilon_1], \quad n < m \tag{4.21}$$

where k is the order of the difference equation of the system;
$\varphi_i[n, \varepsilon]$ is the fundamental set of solutions of the difference equation (4.11);
$\varphi_i^*[m, \varepsilon_1]$ are functions determined by the initial conditions.

3. For a sampled data system which satisfies the condition of stability, i.e.

$$\sum_{m=-\infty}^{\infty} |W[n, \varepsilon; m]| < \infty, \quad n > m \tag{4.22}$$

we have the relation

$$\lim_{n-m\to\infty} W[n, \varepsilon; m] = 0 \tag{4.23}$$

Hence from (4.4) we obtain

$$\lim_{n-m\to\infty} \Delta_n^r K_y[n, \varepsilon; m, \varepsilon_1] = \lim_{n-m\to-\infty} \Delta_n^r K_y[n, \varepsilon; m, \varepsilon_1] = 0 \tag{4.24}$$
$$r = 0, 1, \ldots$$

The first condition in (4.24) follows directly from (4.23), whilst the second is the consequence of the correlation function being symmetrical:

$$K_y[n, \varepsilon; m, \varepsilon_1] = K_y[m, \varepsilon_1; n, \varepsilon] \tag{4.25}$$

4. It follows from (4.14) and (4.19) that the correlation function $K_y[n, \varepsilon; m, \varepsilon_1]$ is connected with the Green's function $G[n, \varepsilon; m]$ of the self-adjoint boundary value problem

$$L_n^*(\Delta, n)\, L_n(\Delta, n)\, G[n, \varepsilon; m] = 0$$

$$\lim_{|n-m|\to\infty} \Delta_n^r G[n, \varepsilon; m] = 0, \quad r = 0, 1, \ldots, k-1$$

by the following relation:

$$K_y[n, \varepsilon; n, \varepsilon_1] = M_n(\Delta, \overline{t})\, M_n^*(\Delta, \overline{t})\, G[n, \varepsilon; m] \tag{4.26}$$

The properties of the correlation function have been proved for the case where the sampled data system is described by a single difference equation. It is easy to show that all this remains valid in the general case where processes in the sampled data system are defined by two difference equations for the instants $0 \leqslant \varepsilon \leqslant \gamma$ and $\gamma \leqslant \varepsilon \leqslant 1$.

We point out also that cross-correlation functions possess analogous properties, which can easily be verified if we use the methods just discussed.

The results obtained in this section have great theoretical importance. In particular, using these results we shall later give a method of constructing discrete shaping filters which are used to form random processes with given characteristics from discrete white noise. In addition, these results enable us to solve a number of important problems connected with the synthesis of sampled data systems.

3.5 SHAPING FILTERS FOR DISCRETE RANDOM PROCESSES

3.5.1 Formulation and solution of the problem

In many problems dealing with the statistical investigation of sampled data automatic systems it is necessary to obtain random discrete processes with given correlation functions. A random discrete process with the required statistical characteristics, e.g. with the given correlation function, can be obtained by means of a linear sampled data system. Thus, let the operator of the sampled data system be A_n. Then the output variable Y, when a disturbance X acts at the input of the system, is the result of applying the operator of the system to the function X:

$$Y = A_n X \tag{5.1}$$

If X is a random function (non-stationary in the general case) with the correlation function $K_x[n;\ m]$, then, as was shown in Section 3.3, the correlation function $K_y[n;\ m]$ of the random function Y is the result of twice applying the operator A to the correlation function K_x, once with respect to n, and the second time with respect to m:

$$K_y\,[n;\ m] = A_n A_m K_x\,[n;\ m] \tag{5.2}$$

It follows from (5.2) that a random process Y with a given

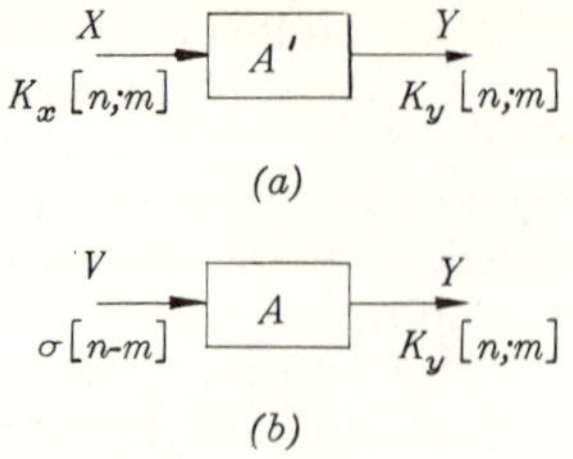

(a)

(b)

Fig. 3.7 Formation of a random process with a given correlation function

correlation function $K_y[n;\ m]$ can be obtained, generally speaking, from any random process X (Fig. 3.7(a)) by an appropriate choice of the system operator A. But the operator A has the simplest form in the case where the random process Y is obtained from discrete white noise (Fig. 3.7(b)), which by definition is a discrete random function with non-correlated values.

A sampled data system which transforms discrete white noise into a discrete random process with a given correlation function is called a discrete shaping filter.

It should be borne in mind that we are considering a class of random processes which can be obtained by solving the respective difference equations.

We shall now proceed directly to the formulation and solution of the problem of determination of a discrete shaping filter. We shall assume that the filter is determined if its difference equations are found.

Assuming that in the general case the shaping filter can be non-stationary, we write its equation in the form

$$\Phi_n(\Delta,\ n)\,Y[n] = \Psi_n(\Delta,\ n)\,V[n] \tag{5.3}$$

where V is discrete white noise with unit intensity;
Y is the random process to be formed having the correlation function $K_y[n;\ m]$;
$\Phi_n(\Delta, n)$ and $\Psi_n(\Delta, n)$ are difference operators defined by the relations

$$\Phi_n(\Delta,\ n) = \sum_{i=0}^{k} a_i[n]\,\Delta_n^i \tag{5.4}$$

$$\Psi_n(\Delta,\ n] = \sum_{i=0}^{h} b_i[n]\,\Delta_n^i \tag{5.5}$$

In a particular case, when the random function Y is stationary, the difference operators $\Phi_n(\Delta, n)$ and $\Psi_n(\Delta, n)$ have constant coefficients.

Thus if the correlation function $K_y[n; m]$ of the random process Y to be shaped is given, the problem of determining the shaping filter then reduces to the determination of the difference operators $\Phi_n(\Delta, n)$ and $\Psi_n(\Delta, n)$.

It was shown in the preceding section that the correlation function of the random process at the output of a sampled data system, when discrete white noise acts at its input, satisfies the following difference equations:

$$\Phi_n(\Delta, n) K_y[n; m] = 0, \quad n > m \tag{5.6}$$

$$\Phi_m(\Delta, m) K_y[n; m] = \Psi_m(\Delta, m) \Psi^*_m(\Delta, m) W_0[n; m] \qquad n > m \tag{5.7}$$

where $\Psi^*_m(\Delta, m)$ is a difference operator, adjoint to $\Psi_n(\Delta, n)$, which is defined by the relation

$$\Psi^*_m(\Delta, m) W_0[n; m] = \sum_{i=0}^{h} (-1)^i \Delta^i_m \{b_i[m-i] W_0[n; m-i]\} \tag{5.8}$$

whilst $W_0[n; m]$ is the solution of the homogeneous difference equation corresponding to (5.3)

$$\Phi_n(\Delta, n) W_0[n; m] = 0, \quad n > m \tag{5.9}$$

with the initial conditions

$$\Delta^r_n W_0[n; m]|_{n=m} = 0, \quad r = 0, 1, \ldots, k-2$$
$$\Delta^{k-1}_n W_0[n; m]|_{n=m} = \frac{1}{a_k[m]} \tag{5.10}$$

The difference equations (5.6) and (5.7) which are satisfied by the correlation function $K_y[n; m]$ of the required random process are the starting points for the determination of the operators $\Phi_n(\Delta, n)$ and $\Psi_n(\Delta, n)$.

As was shown in Section 3.4, the correlation function $K_y[n; m]$ as a solution of the difference equation (5.6) can be represented in the form

$$K_y[n; m] = \sum_{i=1}^{k} \lambda_i[n]\,\eta_i[m], \quad n > m \tag{5.11}$$

where $\lambda_i[n]$ is the fundamental set of solutions of equation (5.6);

$\eta_i[m]$ are functions determined by the initial conditions;

k is the order of the difference operator $\Phi_n(\Delta, n)$.

We can show that for linear difference equations the following theorem is true.

Theorem. The fundamental set of solutions $\lambda_i[n]$, $i=1, 2, \ldots, k$ completely defines the linear k-th order difference equation with the coefficient of the highest difference equal to unity. This equation is given by

$$\Phi_n(\Delta, n)\,y[n] = \begin{vmatrix} \lambda_1[n] & \lambda_2[n] & \ldots & \lambda_k[n] & y[n] \\ \Delta\lambda_1[n] & \Delta\lambda_2[n] & \ldots & \Delta\lambda_k[n] & \Delta y[n] \\ \cdots & \cdots & \cdots & \cdots & \cdots \\ \Delta^k\lambda_1[n] & \Delta^k\lambda_2[n] & \ldots & \Delta^k\lambda_k[n] & \Delta^k y[n] \end{vmatrix} = 0 \tag{5.12}$$

From (5.11) and the theorem just formulated it follows that the difference operator $\Phi_n(\Delta, n)$ can be determined, if we represent the given correlation function $K_y[n; m]$ in the form (5.11), then set up a determinant of the type (5.12) and expand it by the elements of its last column. In so doing the order of the difference operator $\Phi_n(\Delta, n)$ is determined by the number of the functions λ_i into which we decompose the given correlation function.

We shall now proceed to determine the operator $\Psi_n(\Delta, n)$. It was shown in Chapter 1 that the weighting function $W_0[n; m]$ can be given in terms of the fundamental set of solutions $\lambda_i[n]$ of the difference equation (5.9) as follows:

$$W_0[n; m] = (-1)^{k-1} \frac{1}{a_k[m]\,D[m]} \times \begin{vmatrix} \lambda_1[m] & \lambda_2[m] & \ldots & \lambda_k[m] \\ \Delta\lambda_1[m] & \Delta\lambda_2[m] & \ldots & \Delta\lambda_k[m] \\ \cdots & \cdots & \cdots & \cdots \\ \Delta^{k-2}\lambda_1[m] & \Delta^{k-2}\lambda_2[m] & \ldots & \Delta^{k-2}\lambda_k[m] \\ \lambda_1[n] & \lambda_2[n] & \ldots & \lambda_k[n] \end{vmatrix} \tag{5.13}$$

where $a_k[m]$ is the coefficient of the highest operator $\Phi_n(\Delta, n)$ for $n = m$;

$$D[m] = \begin{vmatrix} \lambda_1[m+1] & \lambda_2[m+1] & \dots & \lambda_k[m+1] \\ \Delta\lambda_1[m+1] & \Delta\lambda_2[m+1] & \dots & \Delta\lambda_k[m+1] \\ \dots & \dots & \dots & \dots \\ \Delta^{k-1}\lambda_1[m+1] & \Delta^{k-1}\lambda_2[m+1] & \dots & \Delta^{k-1}\lambda_k[m+1] \end{vmatrix} \quad (5.14)$$

Knowing the operator $\Phi_n(\Delta, n)$ we can find the result of applying the 'product' of the adjoint operators $\Psi_n(\Delta, n)$ and $\Psi^*_n(\Delta, n)$ to the function $W_0[n; m]$ as shown by (5.7), i.e. we have

$$\Psi_m(\Delta, m)\Psi^*_m(\Delta, m)W_0[n; m] = \Phi_m(\Delta, m)K_y[n; m] \qquad n > m \quad (5.15)$$

Determining now the form of the 'product' of the adjoint operation Ψ and Ψ^* from (5.15) and decomposing this product into adjoint factors, we obtain the required operator $\Psi_n(\Delta, n)$.

From the foregoing we can thus formulate the following rules for determining the difference equation of a discrete filter which forms a random process Y with the given correlation function $K_y[n; m]$ from discrete white noise.

1. Represent the correlation function $K_y[n; m]$ of the random process Y in the form of a sum of the products of the λ_i and η_i,

$$K_y[n; m] = \sum_{i=1}^{k} \eta_i[m]\lambda_i[n]$$

2. Determine the difference operator $\Phi_n(\Delta, n)$, set up and expand a determinant of the type (5.12).
3. Find the function $W_0[n; m]$ from (5.13).
4. Find the result of applying the operator $\Psi_m(\Delta, m)\times$ $\Psi^*_m(\Delta, m)$ to the function $W_0[n; m]$ in accordance with the formula

$$\Phi_m(\Delta, m)K_y[n; m] = \Psi_m(\Delta, m)\Psi^*_m(\Delta, m)W_0[n; m] \qquad n > m$$

5. Determine the form of the 'product' of the operators $\Psi_n(\Delta, n)$ and $\Psi^*_n(\Delta, n)$, and find the operator $\Psi_n(\Delta, n)$ by decomposing this 'product' into adjoint factors.

The theory given above enables us to determine the difference equations of filters which form both stationary and non-stationary random processes out of white noise. It is important to point out that in the general case a stationary random process can be formed out of white noise by means of a stationary filter with an infinite memory, i.e. if the weighting function of the shaping filter is $W_f[n-m]$, then the required random process Y is connected with the white noise V by the relation

$$Y[n] = \sum_{m=-\infty}^{n} W_f[n-m]V[m] \tag{5.16}$$

or, after changing the summation variable,

$$Y[n] = \sum_{\lambda=0}^{\infty} W_f[\lambda]V[n-\lambda] \tag{5.17}$$

It follows from (5.17) that only when $\bar{t} \to \infty$ will the output process of the shaping filter approach the random process desired. But in practice, since shaping filters are stable and their weighting functions decay sufficiently rapidly when n increases, the output processes of the filters are very close to those desired when $n = 3$ to 5. Hence it is clearly not advisable to obtain stationary random processes by means of non-stationary filters with a finite memory.

We make a brief remark concerning a filter which is inverse to the shaping filter. Let the difference equation of a shaping filter be

$$\Phi_n(\Delta, n)Y[n] = \Psi_n(\Delta, n)X[n] \tag{5.18}$$

Then the weighting function $W_f[n; m]$ of this filter satisfies the difference equation

$$\Phi_n(\Delta, n)W_f[n; m] = \Psi_n(\Delta, n)\sigma[n-m] \tag{5.19}$$

with zero initial conditions. If we now determine a sampled data system which is inverse to the shaping filter, then by the definition of an inverse system, the equation for its weighting function has the form

$$\Phi_n(\Delta, n)\sigma[n-m) = \Psi_n(\Delta, n)W_f[n; m] \tag{5.20}$$

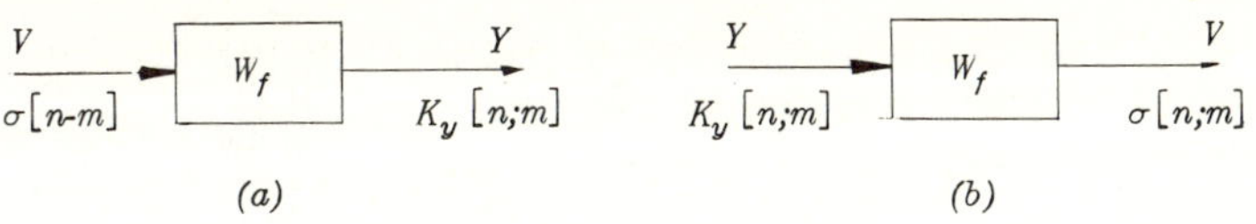

Fig. 3.8 Definition of a system inverse to the shaping filter

Hence we obtain the following important result, that if a sampled data system (discrete shaping filter) transforms a random input signal of the white noise type into a random process Y with the correlation function $K_y[n;\ m]$, then a system which is inverse to it transforms the random process Y with the correlation function $K_y[n;\ m]$ into a random process of the white noise type. This is illustrated by Fig. 3.8.

We shall now consider examples of the determination of a shaping filter.

Example 1. Find the difference equation of a discrete shaping filter for a random process with the correlation function

$$K_y[n-m] = De^{-\alpha|n-m|}, \quad \alpha > 0 \tag{5.21}$$

When $n > m$ we have from (5.21)

$$K_y[n-m] = De^{-\alpha(n-m)}, \quad n \geqslant m \tag{5.22}$$

where

$$\begin{aligned} \lambda_1[n] &= e^{-\alpha n} \\ \eta_1[m] &= De^{\alpha m} \end{aligned} \tag{5.23}$$

In the given case ($k=1$) the determinant (5.12) has the form

$$\Phi_n(\Delta,\ n)y[n] = \begin{vmatrix} \lambda_1[n] & y[n] \\ \Delta\lambda_1[n] & \Delta y_1[n] \end{vmatrix} = \begin{vmatrix} e^{-\alpha n} & y[n] \\ e^{-\alpha n}(e^{-\alpha}-1) & \Delta y[n] \end{vmatrix} \tag{5.24}$$

Expanding (5.24) we find

$$\Phi_n(\Delta,\ n)y[n] = e^{-\alpha n}\Delta y[n] + (1-e^{-\alpha})e^{-\alpha n}y[n]$$

and, with the common factor omitted, we finally obtain

$$\Phi_n(\Delta,\ n)\,y[n]=\Delta y[n]+(1-e^{-\alpha})\,y[n] \qquad (5.25)$$

We shall now find the weighting function $W_0[n;\ m]$ defined by (5.13). Since

$$a_1[m]=1$$

$$D[m]=e^{-\alpha(m+1)}$$

then $W_0[n;\ m]$ is determined by

$$W_0[n;\ m]=e^{-\alpha(n-m-1)} \qquad (5.26)$$

We now determine the result of applying the operator $\Psi_m(\Delta,\ m)\Psi^*_m(\Delta,\ m)$ to $W_0[n;\ m]$. On the basis of (5.7) we have

$$\Phi_m(\Delta,\ m)\,K_y[n;\ m]=D(e^{\alpha}-e^{-\alpha})\,e^{-\alpha(n-m)},\quad n>m \qquad (5.27)$$

Consequently,

$$\Psi_m(\Delta,\ m)\Psi^*_m(\Delta,\ m)\,e^{-\alpha(n-m-1)}=D(e^{\alpha}-e^{-\alpha})\,e^{-\alpha(n-m)}$$

whence we obtain

$$\Psi_m(\Delta,\ m)\Psi^*_m(\Delta,\ m)=D(e^{\alpha}-e^{-\alpha})\,e^{-\alpha}$$

and

$$\Psi_m(\Delta,\ m)=\sqrt{D(e^{\alpha}-e^{-\alpha})\,e^{-\alpha}} \qquad (5.28)$$

The difference equation of the shaping filter, when (5.25), (5.26) and (5.28) are used, can thus be written in the form

$$\Delta y[n]+(1-e^{-\alpha})\,y[n]=\sqrt{D(e^{\alpha}-e^{-\alpha})\,e^{-\alpha}}\;x[n+1] \qquad (5.29)$$

To check the result just obtained we shall find the correlation function of the random process at the output of the filter described by (5.29), when white noise with unit intensity acts at its input. We use (3.81), which in our case assumes the form

$$K_y[l]=\sum_{m=0}^{\infty}W_f[m]\,W_f[l+m] \qquad (5.30)$$

where $W_f[n]$ is the weighting function of the filter in question. Putting $x[n]=\sigma[n]$ in (5.29) we obtain the equation for the weighting function W_f

$$\Delta W_f[n]+(1-e^{-\alpha})W_f[n]=\sqrt{D(e^{\alpha}-e^{-\alpha})e^{-\alpha}}\,\sigma[n+1]$$

whence we find the expression for the weighting function of the shaping filter

$$W_f[n]=\sqrt{D(1-e^{-2\alpha})}\,e^{-\alpha n},\quad n\geqslant 0$$

Substituting this expression into (5.30) we obtain

$$K_y[l]=D(1-e^{-2\alpha})e^{-\alpha l}\sum_{m=0}^{\infty}e^{-2\alpha m}$$

Since

$$\sum_{m=0}^{\infty}e^{-2\alpha m}=\frac{1}{1-e^{-2\alpha}}$$

then finally we obtain the following expression for the correlation function $K_y[l]$

$$K_y[l]=De^{-\alpha l}$$

which coincides with the given expression (5.21), for $l=n-m$ and $n\geqslant m$.

Example 2. Determine the difference equation of a shaping filter for forming a non-stationary random process with a correlation function of the form

$$K_y[n;\ m]=\varphi[n]\varphi[m]e^{-\alpha|n-m|} \tag{5.31}$$

For $n\geqslant m$ we have from (5.31)

$$K_y[n;\ m]=\varphi[n]\varphi[m]e^{-\alpha(n-m)},\quad n\geqslant m$$

where

$$\lambda_1[n]=\varphi[n]e^{-\alpha n}$$

$$\eta_1[m]=\varphi[m]e^{\alpha m}$$

In this case the determinant (5.12) has the form

$$\Phi_n(\Delta,\ n)\, y[n] = \begin{vmatrix} \lambda_1[n] & y[n] \\ \Delta\lambda_1[n] & \Delta y[n] \end{vmatrix} \tag{5.32}$$

Since

$$\begin{aligned} \Delta\lambda_1[n] &= \Delta\{\varphi[n]\, e^{-\alpha n}\} = \Delta\varphi[n]\cdot e^{-\alpha n} \\ &+ \varphi[n+1]\,(e^{-\alpha(n+1)} - e^{-\alpha n}) \\ &= \varphi[n+1]\, e^{-\alpha} e^{-\alpha n} - \varphi[n]\, e^{-\alpha n} \end{aligned}$$

the determinant (5.32) equals

$$\Phi_n(\Delta,\ n) = \begin{vmatrix} \varphi[n]\, e^{-\alpha n} & y[n] \\ \varphi[n+1]\, e^{-\alpha(1+n)} - \varphi[n]\, e^{-\alpha n} & \Delta y[n] \end{vmatrix}$$

whence we obtain the difference operator $\Phi_n(\Delta,\ n)$

$$\Phi_n(\Delta,\ n)\, y[n] \equiv \varphi[n]\, \Delta y[n] - (\varphi[n+1]\, e^{-\alpha} - \varphi[n])\, y[n] \tag{5.33}$$

To determine the difference operator $\Psi_n(\Delta,\ n)$, we have

$$a_k[m] = \varphi[m]$$

$$D[m] = \lambda_1[m+1] = \varphi[m+1]\, e^{-\alpha(m+1)}$$

Consequently, the function $W_0[n;\ m]$ is

$$W_0[n;\ m] = \frac{\varphi[n]}{\varphi[m]\,\varphi[m+1]}\, e^{-\alpha(n-m-1)}$$

Applying (5.15) we find the difference operator

$$\Psi_n(\Delta,\ n)\, v[n] \equiv \varphi[n]\, \varphi[m]\, (\Delta v[n] + v[n])$$

The difference equation of the required shaping filter thus has the form

$$\begin{aligned} \varphi[n]\, \Delta y[n] - (\varphi[n+1]\, e^{-\alpha} - \varphi[n])\, y[n] \\ = \varphi[n]\, \varphi[m]\, (\Delta v[n] + v[n]) \end{aligned} \tag{5.34}$$

This equation can be written in a different form if we replace the differences in (5.34) by the corresponding displaced functions. Thus, putting

$$\Delta y[n] = y[n+1] - y[n]$$
$$\Delta v[n] = v[n+1] - v[n]$$

in (5.34) we find

$$\varphi[n]\, y[n+1] - \varphi[n+1]\, y[n]\, e^{-\alpha} = \varphi[n]\, \varphi[m]\, v[n+1]$$

This equation is essentially an algorithm by which we can obtain a random process with a correlation function of the form (5.31) from the non-correlated random sequence V.

The examples just considered show that the procedure given for determining the difference equation of a discrete shaping filter enable us to find the operators $\Phi_n(\Delta, n)$ and $\Psi_n(\Delta, n)$ quite simply in cases where $\Psi_n(\Delta, n)$ is of low order. But if it is greater than or equal to two, considerable difficulties arise in determining the form of the 'product' of $\Psi_n(\Delta, n)$ and $\Psi^*_n(\Delta, n)$. Difficulties are also met with in decomposing this 'product' into adjoint factors. This can easily be confirmed by a simple example.

Suppose that it is required to determine a discrete filter to form a random process Y with a correlation function of the form

$$K_y[n-m] = c_1 e^{-\alpha_1 |n-m|} + c_2 e^{-\alpha_2 |n-m|} \tag{5.35}$$

When $n \geqslant m$,

$$\begin{aligned} \lambda_1[n] &= e^{-\alpha_1 n}, & \lambda_2[n] &= e^{-\alpha_2 n} \\ \eta_1[m] &= c_1 e^{\alpha_1 m}, & \eta_2[m] &= c_2 e^{\alpha_2 m} \end{aligned} \tag{5.36}$$

and the determinant (5.12) has the form

$$\Phi_n(\Delta, n)\, y[n]$$
$$= \begin{vmatrix} e^{-\alpha_1 n}(e^{-\alpha_1} - 1) & e^{-\alpha_2 n}(e^{-\alpha_2} - 1) & \Delta y[n] \\ e^{-\alpha_1 n} & e^{-\alpha_2 n} & y[n] \\ e^{-\alpha_1 n}(e^{-\alpha_1} - 1)^2 & e^{-\alpha_2 n}(e^{-\alpha_2} - 1)^2 & \Delta^2 y[n] \end{vmatrix} \tag{5.37}$$

Expanding the determinant (5.37) by the elements of the last column and eliminating the common factor we find the operator

$$\Phi_n(\Delta, n)\,y[n] \equiv \Delta^2 y[n] - \Delta y[n](e^{-\alpha_1} + e^{-\alpha_2} - 2) + y[n](e^{-\alpha_1} - 1)(e^{-\alpha_2} - 1) \tag{5.38}$$

The weighting function $W_0[n; m]$ in the given case, in accordance with (5.13), is

$$W_0[n; m] = \frac{1}{e^{-(\alpha_1+\alpha_2)}(e^{-\alpha_2} - e^{-\alpha_1})}\left(e^{-\alpha_1(n-m)} - e^{-\alpha_2(n-m)}\right) \quad n > m \tag{5.39}$$

Subsequently, in order to determine $\Psi_n(\Delta, n)$, we must first apply the operator (5.38) to the correlation function (5.35), determine then the form of the product $\Psi_n(\Delta, n)$ $\Psi^*_n(\Delta, n)$, and by decomposing this product into adjoint factors find the right-hand side of the difference equation of the shaping filter. As follows from (5.38) and (5.39), the process of determining $\Psi_n(\Delta, n)$ will be laborious.

In the case where the shaping filter must be determined for a non-stationary random process the method given above turns out to be even more cumbersome. Because of this it is often advisable to determine the difference equations of shaping filters by and another method, which will now be discussed.

3.5.2 Determination of discrete from continuous shaping filters

It was shown in Section 3.4 that the correlation function $K_y[n; m]$ of the random process Y at the output of a sampled data system, when white noise acts at its input, satisfies the corresponding difference equation, and is connected in a definite manner with the Green's function of the difference boundary-value problem

$$L^*_n(\Delta, n)\,L_n(\Delta, n)\,K_y[n; m] = 0$$

$$\lim_{n-m\to\infty} \Delta^r_n K_y[n; m] = \lim_{n-m\to-\infty} \Delta^r_n K_y[n; m] = 0$$

$$r = 0, 1, \ldots, k-1$$

Consequently, the correlation function $K_y[n; m]$ can be regarded as the weighting function of a certain self-adjoint sampled data system. But since the correlation function $K_y[n;\ m]$ is the result of twice applying the summation operator A to the correlation function of discrete white noise (once with respect to the argument n, and the second time with respect to m), then a self-adjoint system whose weighting function is the correlation function $K_y[n;\ m]$ is formed by two systems (original and adjoint) connected in series. The latter have weighting functions which form the kernel of the summation operator A (with respect to the arguments n and m respectively). Hence the problem of determining a discrete shaping filter can be reduced to either determining the kernel of the summation operator, or, in the final analysis, determining the structure of a system whose weighting function coincides identically with the kernel of the summation operator.

We shall now determine the structure of a sampled data system whose difference equation coincides with the difference equation of the discrete filter which forms a random process with a given correlation function out of white noise.

It has been shown in [1] that the weighting function of an open-loop sampled data system coincides with the weighting function of the normalised continuous part of this system. The continuous part consists of an element determined by the type of the sampler and a continuous element connected in series. Denoting the weighting function of the normalised continuous part by $W_N\ (t;\ \tau)$, we express this in the following form

$$W[n,\ \varepsilon;\ m] = W_N(t;\ \tau)_{\substack{t=(n+\varepsilon)T_0,\\ \tau=mT_0}} \tag{5.40}$$

where $W[n,\ \varepsilon;\ m]$ is the weighting function of the open-loop system.

If the system is represented by a δ-pulse element and a continuous part connected in series, then the weighting function $W_N(t;\ \tau)$ of the normalised continuous part differs only in scale from the weighting function of the corresponding continuous part. Thus, let $W_S\ (t;\ \tau)$ be the weighting function of the element which is determined by the form of the sampler, and let $W_C\ (t;\ \tau)$ be the weighting function of the continuous part. Then from Fig. 3.9 we have the following equations for the normalised continuous part:

$$z(t) = \int_{-\infty}^{t} W_S\ (t;\ \lambda)\, x(\lambda)\, d\lambda, \tag{5.41}$$

$$y(t)=\int_{-\infty}^{t} W_C(t;\ \alpha)\, z(\alpha)\, d\alpha \tag{5.42}$$

Here $x(t)$ and $z(t)$ are respectively the input and output of the element with the weighting function $W_S(t;\ \tau)$, while $y(t)$ is the output variable of the normalised continuous part. Since

$$z(\alpha)=\int_{-\infty}^{\alpha} W_S(\alpha;\ \lambda)\, x(\lambda)\, d\lambda \tag{5.43}$$

we have from (5.42)

$$y(t)=\int_{-\infty}^{t} W_C(t;\ \alpha)\left\{\int_{-\infty}^{\alpha} W_S(\alpha;\ \lambda)\, x(\lambda)\, d\lambda\right\} d\alpha \tag{5.44}$$

To determine the weighting function $W_N(t;\ \tau)$ of the normalised continuous part, we put

$$x(t)=\delta(t),$$

in (5.44). We then obtain

$$\begin{aligned} W_N(t;\ \tau) &= \int_{-\infty}^{t} W_C(t;\ \alpha)\left\{\int_{-\infty}^{\alpha} W_S(\alpha;\ \lambda)\, \delta(\lambda-\tau)\, d\lambda\right\} d\alpha \\ &= \int_{-\infty}^{t} W_C(t;\ \alpha)\, W_S(\alpha;\ \tau)\, d\alpha \end{aligned} \tag{5.45}$$

If the sampler generates the δ -function with the weights k_S, i.e.

$$W_S(t;\ \tau)=k_S\,\delta(t-\tau) \tag{5.46}$$

then from (5.45) $W_N(t;\ \tau)$ is

$$W_N(t;\ \tau)=\int_{-\infty}^{t} W_C(t;\ \alpha)\, k_S\,\delta(\alpha-\tau)\, d\alpha = k_S W_C(t;\ \tau) \tag{5.47}$$

whence we can conclude that the weighting function of the normalised continuous part differs only in scale from the weighting function of the continuous part of the system.

On the basis of (5.40) and (5.47) we have

$$W[n,\ \varepsilon;\ m]=k_S W_C(t;\ \tau)\Big|_{\substack{t=(n+\varepsilon)T_0 \\ \tau=mT_0}} \tag{5.48}$$

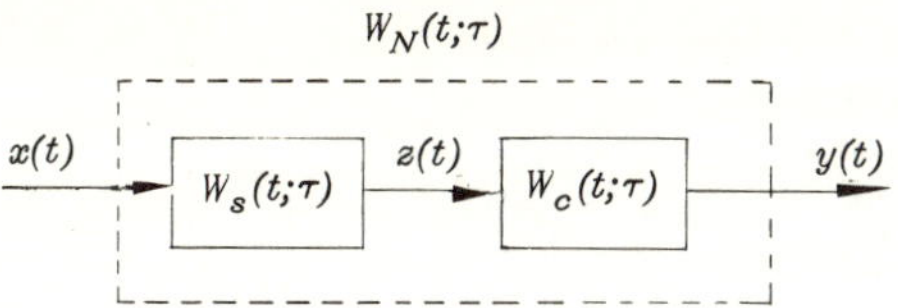

Fig. 3.9 An open-loop system

i.e. the weighting function of a sampled data system with a δ-pulse element differs only in scale from the weighting function of the continuous part.

Suppose now that the continuous part of a system is represented by a filter forming a random process with the correlation function $K_y(t;\ \tau)$ out of white noise. This function is obtained from the correlation function $K_y[n,\ \varepsilon;\ m]$ by substituting

$$nT_0 = t \quad \text{and} \quad mT_0 = \tau$$

where T_0 is the sampling period. Accordingly, from (5.48) the correlation function of the discrete random process at the output of the sampled data system, when discrete white noise acts at its input is generally

$$K_y[n;\ m] = RK_y(t;\ \tau)\big|_{\substack{t=nT_0\\ \tau=mT_0}} \tag{5.49}$$

where R is a coefficient of proportionality. Hence we obtain the following important result, that the weighting function of an open-loop sampled data system formed by connecting a δ-pulse element in series with a continuous filter, forming from white noise a continuous random process with the correlation $K_y(t;\ \tau)\big|_{\substack{t=nT_0\\ \tau=mT_0}} = K_y[n;\ m]$, is the kernel of the summation operator which when applied twice to the correlation function of discrete white noise, gives the correlation function $K_y[n;\ m]$ of the discrete random process, accurate up to the constant factor R.

The difference equation of a discrete filter forming from white noise a random process Y with the correlation function $K_y[n;\ m]$ can thus be obtained from the transfer function of a sampled data system formed by connecting a δ-pulse element and a continuous shaping filter in series. We emphasise once more that the correlation function of the process at the output of such a sampled data system does not coincide with the given one, but differs from it, as follows from (5.49), by a constant factor.

The coefficient of proportionality R can be unity only in a particular case. This is due to the difference between continuous and discrete random processes. We shall demonstrate this by a simple example.

Let the continuous part of a sampled data system be a filter with the transfer function

$$W(p)=\frac{\sqrt{2\alpha}}{p+\alpha} \tag{5.50}$$

This continuous filter with the transfer function (5.50) forms from white noise V of unit intensity and the correlation function

$$K_v(t-\tau)=\delta(t-\tau)$$

a random process Y with a correlation function of the form

$$K_y(t-\tau)=e^{-\alpha|t-\tau|} \tag{5.51}$$

In fact, the correlation function of the random process at the output of a stationary system, when white noise acts at the input, is

$$K_y(t-\tau)=\int_{-\infty}^{t} W(t-\lambda)W(\tau-\lambda)\,d\lambda \tag{5.52}$$

where $W(t-\tau)$ is the weighting function of the system. It is easy to see that the weighting function of the filter with the transfer function (5.50) is determined as follows:

$$W_f(t-\tau)=\sqrt{2\alpha}e^{-\alpha(t-\tau)}, \quad t\geqslant\tau \tag{5.53}$$

Substituting the expression (5.50) of the weighting function into (5.52) we find

$$K_y(t-\tau)=\int_{-\infty}^{t} 2\alpha e^{-\alpha(t-\lambda)}e^{-\alpha(\tau-\lambda)}d\lambda$$

$$=2\alpha e^{-\alpha(t+\tau)}\int_{-\infty}^{t} e^{2\alpha\lambda}d\lambda=e^{-\alpha|t-\tau|}$$

which coincides with Equation (5.51).

We now connect a δ-pulse element with the gain factor $k_s=1$ in series with the filter (5.50). Then from (5.48) and (5.53), the weighting function of the discrete filter, $W_f[n-m]$, equals ($\varepsilon=0$)

$$W_f[n-m]=\sqrt{2\alpha}\,e^{-\alpha(n-m)}, \quad n\geqslant m \tag{5.54}$$

We shall next find the correlation function of the random process at the output of this system, when white noise with unit intensity and the correlation function

$$K_v[n-m]=\sigma[n-m]$$

acts at its input. When $\varepsilon=0$,we have from (3.87)

$$\begin{aligned} K_y[l]&=\sum_{m=0}^{\infty} W_f[m]\,W_f[l+m]=2\alpha e^{-\alpha l}\sum_{m=0}^{\infty} e^{-2\alpha m} \\ &=\frac{2\alpha}{1-e^{-2\alpha}}\,e^{-\alpha l}=\frac{2\alpha}{1-e^{-2\alpha}}\,e^{-\alpha t}, \quad t=0,\ 1,\dots \end{aligned} \tag{5.55}$$

Comparing (5.51) and (5.55) we find that in this case the gain factor is given by

$$R=\frac{2\alpha}{1-e^{-2\alpha}}$$

Later we shall assume that a discrete shaping filter is a sampled data system (for $\varepsilon=0$) with a δ-pulse element whose gain factor k_s is not known and has to be determined. The method of determining the gain factor will be given below.

Thus, having included the coefficient of proportionality in the gain factor k_s of the sampler, we write (5.49) in the form

$$K_y[n;\ m]=k_s K_y(t;\ \tau)\Big|_{\substack{t=nT_0\\ \tau=mT_0}} \tag{5.56}$$

It will subsequently be seen that (5.49) represented in the form (5.56) is very convenient.

From what has been said we can formulate the rules for determining the difference equation of a discrete filter which forms a discrete random process with the given correlation function $K_y[n;\ m]$ out of white noise.

1. Write the given correlation function in the natural time scale, using the relations

$$nT_0=t, \ mT_0=\tau$$

2. Using any known methods [17, 23], determine the differential equation (transfer function $W(p, t)$) of a continuous filter forming from white noise a random process $Y(t)$ with the correlation function

$$K_y(t; \tau)=K_y[n; m]\Big|_{\substack{nT_0=t \\ mT_0=\tau}}$$

3. Find the transfer function

$$W^*(q, n, 0)=\frac{Y^*(q, 0)}{X^*(q, 0)} \tag{5.57}$$

of the sampled data system formed by connecting a δ-pulse element and the continuous shaping filter in series. (Methods of determining the transfer functions are discussed by Tsypkin [1]).

4. Proceeding to the originals in (5.57), find the difference equation of the discrete shaping filter

$$\Phi_n(\Delta, n)Y[n]=\Psi_n(\Delta, n)X[n] \tag{5.58}$$

Here the operator $\Psi_n(\Delta, n)$ contains the unknown gain factor k_s.

5. Determine k_s. To determine the gain factor k_s of the sampler we can use a method which, in spite of being somewhat cumbersome, has the advantage of enabling the correctness of the results thus obtained to be verified within limits. The main points of this method consist of the following.

The weighting function $W_f[n; m]$ of the shaping filter is determined by the solution (numerical or analytical) of equation (5.58). The expression for the weighting function has the form

$$W_f[n; m]=k_s \cdot G[n; m] \tag{5.59}$$

Then, using the expression for the correlation function of the random process at the output when discrete white noise acts at the input, we find

$$\overline{K}_y[n; m]=A_nA_m\sigma[n-m]=k_s^2 S(n, m) \tag{5.60}$$

where the summation operators have the weighting function $W_f[n;\ m]$ given by (5.59) as their kernel. Since by the condition of the problem the dispersion of the given random process,

$$D_y[n]=K_y[n;\ m]$$

must equal the dispersion at the output of the sampled data system, $\bar{D}_y[n]$, determined by (5.60) when $m=n$, we obtain the following expression for the gain factor k_s

$$k_s=\sqrt{\frac{K_y[n;\ n]}{S(n,\ n)}} \tag{5.61}$$

We shall consider an example of determining a discrete shaping filter to illustrate the use of the method just given.

Example. Find the difference equation of a discrete filter which forms a random process Y with a correlation function of the form

$$K_y[n-m]=De^{-\alpha\,|n-m|} \tag{5.62}$$

from white noise of unit intensity. Using the relations

$$nT_0=t \text{ and } mT_0=\tau$$

we obtain from (5.62), when $n\geqslant m$,

$$K_y(t-\tau)=De^{-\frac{\alpha}{T_0}(t-\tau)},\ t\geqslant\tau \tag{5.63}$$

It can be shown that the continuous filter which forms a random process with the correlation function (5.63) from white noise has the following transfer function:

$$W(p)=\sqrt{D\frac{2\alpha}{T_0}}\ \frac{1}{p+\frac{\alpha}{T_0}} \tag{5.64}$$

Let us now find the transfer function of a sampled data system having a δ-pulse element and a continuous part with the transfer function (5.64). For this we use the formulae of

Tsypkin [1]

$$W^*(q, \varepsilon) = \sum_{\nu=1}^{l} c'_{\nu 0} \frac{e^q}{e^q - e^{q_\nu}} e^{q_\nu \varepsilon} \tag{5.65}$$

where $W^*(q, \varepsilon)$ is the transfer function of the sampled data system;

q_ν are the roots of the characteristic equation of the transfer function $W(p)$ when $p = \frac{q}{T_0}$:

$$W\left(\frac{q}{T_0}\right) = \frac{P\left(\frac{q}{T_0}\right)}{Q\left(\frac{q}{T_0}\right)} \tag{5.66}$$

$$c'_{\nu 0} = \frac{k_S}{T_0} \frac{P\left(\frac{q_\nu}{T_0}\right)}{Q'\left(\frac{q_\nu}{T_0}\right)}, \quad \nu = 1, 2, \ldots, l \tag{5.67}$$

We point out that (5.65) is valid in the case of the simple roots q_ν, $\nu = 1, 2, \ldots, l$.

Thus, using the substitution $p = \frac{q}{T_0}$, we obtain from (5.64)

$$W(q) = \frac{T_0 \sqrt{D \frac{2\alpha}{T_0}}}{q + \alpha} \tag{5.68}$$

whence we find the root

$$q_1 = -\alpha$$

of the characteristic equation.

In accordance with (5.67), coefficient c'_{10} is

$$c'_{10} = \frac{k_S}{T_0} \frac{T_0 \sqrt{D \frac{2\alpha}{T_0}}}{1} = k_S \sqrt{D \frac{2\alpha}{T_0}} \tag{5.69}$$

Using (5.65) we determine the transfer function of the sampled data system for $\varepsilon = 0$

$$W^*(q) = k_S \sqrt{D \frac{2\alpha}{T_0}} \frac{e^q}{e^q - e^{-\alpha}} = \frac{Y^*(q)}{X^*(q)} \tag{5.70}$$

Proceeding to the originals in Equation (5.70), we find the

difference equation

$$y[n+1]-e^{-\alpha}y[n]=k_S\sqrt{D\frac{2\alpha}{T_0}}x[n+1] \qquad (5.71)$$

which includes the unknown gain factor k_S of the sampler. To determine k_S we must, following the method given above, find the weighting function of the filter $W_f[n-m]$ given by (5.71).

Using the new independent variable

$$n+1=k$$

in (5.71), we obtain the equation

$$y[k]=e^{-\alpha}y[k-1]+k_S\sqrt{D\frac{2\alpha}{T_0}}x[k] \qquad (5.72)$$

whose solution, when $x[k]=\sigma[k]$, is the weighting function. We have thus

$$W_f[k]=e^{-\alpha}W_f[k-1]+k_S\sqrt{D\frac{2\alpha}{T_0}}\sigma[k] \qquad (5.73)$$

whence we obtain

$$W_f[0]=\begin{cases} k_S\sqrt{D\frac{2\alpha}{T_0}}, & n=0 \\ e^{-\alpha n}W_f[n-1], & n\geqslant 1 \end{cases} \qquad (5.74)$$

or, in the final form,

$$W_f[n-m]=k_S\sqrt{D\frac{2\alpha}{T_0}}e^{-\alpha(n-m)}, \quad n>m \qquad (5.75)$$

To determine the coefficient k_S we find the correlation function of the process at the output of a filter with the weighting function (5.75). From (3.81) we have for lattice functions

$$\bar{K}_y[l]=\sum_{m=0}^{\infty}W_f[m]W_f[l+m]$$
$$=k_S^2D\frac{2\alpha}{T_0}e^{-\alpha l}\sum_{m=0}^{\infty}e^{-2\alpha m}=k_S^2\frac{D\frac{2\alpha}{T_0}}{1-e^{-2\alpha}}e^{-\alpha l}, \quad l>0 \qquad (5.76)$$

Using (5.61) we find, with (5.62) and (5.76) taken into account,

$$k_s = \sqrt{\frac{K_y[0]}{S(0)}} = \sqrt{\frac{\dfrac{D}{2D\dfrac{\alpha}{T_0}}}{1-e^{-2\alpha}}} = \sqrt{\frac{1-e^{-2\alpha}}{\dfrac{2\alpha}{T_0}}} \qquad (5.77)$$

where in accordance with (5.60)

$$\bar{K}_y[0] = k_s^2\, S(0)$$

In this way, substituting the value of the gain factor k_s from (5.77) into (5.71), we obtain

$$y[n+1] - e^{-\alpha} y[n] = \sqrt{D(1-e^{-2\alpha})}\, x[n+1] \qquad (5.78)$$

The difference equation (5.78) is in fact the final form of the equation of the required discrete shaping filter. This equation coincides with (5.29), obtained in Section 3.5.1 by another method.

Thus, using two methods, we have solved the problem of determining discrete filters which form stationary and non-stationary random processes from discrete white noise. The more effective method for determining discrete shaping filters is given by the second one, which is based on the use of appropriate shaping filters. When the second method is used, however, we must overcome the difficulties connected with the determination of the difference equation from the corresponding differential equation with variable coefficients. Certain difficulties will also arise in determining the gain factor k_s of a sampled data system which constitutes the discrete shaping filter. The fact is that a discrete filter which forms non-stationary processes can be found in closed form only in a very few comparatively simple cases. It should, however, be pointed out that in practice it is not always necessary to determine the difference equations of shaping filters. Often it turns out to be sufficient to know only the differential equations of the continuous part of the filter. In this case the problem of forming a given discrete random process reduces to the computation of the response of a sampled data system with a δ-pulse element to discrete white noise. The solution of this problem can be carried out with the aid of a digital computer.

It therefore appears that to form a given discrete random

process using a digital computer we have no need to find the difference equations of the filter from the transfer function of the corresponding sampled data system. It is also obvious that simulation of a sampled data system (shaping filter) on a digital computer enables the gain factor k_S of this system to be found easily.

3.6 STATISTICAL ANALYSIS OF STATIONARY SYSTEMS IN STEADY STATE

In this section we shall discuss methods of obtaining the correlation function and dispersion of the random process at the output of a stationary sampled data system when a stationary random function acts at its input. Since the analytical methods for studying sampled data systems, based on the connection between the spectral densities of the random input and output processes, are cumbersome and often cannot be applied because of the complexity of the systems, we shall devote our main attention to those methods of statistical analysis which are based on the use of modern computational techniques.

In determining the methods for obtaining the statistical characteristics of a random process at the output of a sampled data system, we shall proceed from the fundamental relation (3.62) obtained in this chapter. In so doing we shall assume that a random function X with zero mathematical expectation acts at the input of a system with the weighting function $W[n, \varepsilon]$. The problem of determining the mathematical expectation of the output process of the system in a steady state reduces to the usual problem of determining the response of this system to a non-random function constituting the mathematical expectation of the random input function.

The basic expression which we shall use, when determining the methods for obtaining the correlation function and dispersion of the random process Y at the output of the stationary system, is thus given by the relation which establishes a connection between the correlation functions of the input and output processes:

$$K_y[l, \varepsilon] = \sum_{m_1=0}^{\infty} W[m_1, \varepsilon] \sum_{m_2=0}^{\infty} W[m_2, \varepsilon] K_x[l + m_1 - m_2, 0] \tag{6.1}$$

Depending on the physical interpretation of (6.1), we can use two methods for the determination of the correlation function $K_y[l, \varepsilon]$.

First method. Let the correlation function $K_x[n]$ of the random input process be given either analytically or in a tabular form. We shall determine the structure of an experimental system which enables us to find the correlation function $K_y[l, \varepsilon]$ of the output process of the system characterised by the weighting function $W[n, \varepsilon]$.

Let us change the variables in the formula (6.1), denoting

$$l+m_1=-k$$

Then we have (considering, for simplicity, only lattice functions)

$$K_y[l]=\sum_{k=-\infty}^{-l} W[-k-l] \sum_{m_2=0}^{\infty} W[m_2] K_x[-k-m_2] \tag{6.2}$$

Since by the definition of the correlation function

$$K_y[l]=K_y[-l]$$

and since summing a variable with the opposite sign is equivalent to changing the sign of the sum, then we write (6.2) in the following form:

$$K_y[l]=\sum_{k=-\infty}^{l} W[l-k] \sum_{m_2=0}^{\infty} W[m_2] K_x[-k-m_2] \tag{6.3}$$

If we denote the inner sum in (6.3) by

$$\sum_{m_2=0}^{\infty} W[m_2] K_x[k-m_2]=z[k] \tag{6.4}$$

then (6.3) for the correlation function assumes the form

$$K_y[l]=\sum_{k=-\infty}^{l} W[l-k] z[-k] \tag{6.5}$$

Let us analyse the results just obtained. It follows from (6.4) that the signal $z[k]$ represents the response of the system in question to an action which coincides with the correlation function $K_x[n]$ of the random function X. Furthermore,

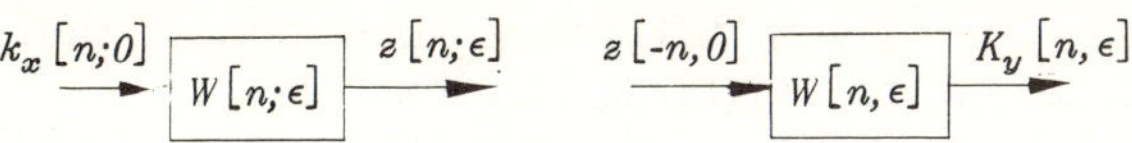

Fig. 3.10 Determination of the correlation function by the series method

as follows from (6.5), the correlation function $K_y[l]$ of the output process of this system represents its response to an action of the form $z[k]$ when $k<0$. Thus, in order to obtain the correlation function of the random process at the output of a stationary system, we must first obtain the signal $z[k]$, and then from (6.5), after reversing the time in the signal $z[k]$, determine the desired correlation function $K_y[l]$ for a fixed value of l. The sequence for determining the correlation function is illustrated by Fig, 3.10.

Above we have given the general procedure for determining the correlation function $K_y[l]$. In practice a sampled data system can either be given by the differential equations of its continuous part and the equation of the sampler, or it can be given by difference equations, though the latter is encountered less often. Because of this, different procedures for determining the correlation function, based on the use of either analogue or digital computers, must be used.

If a sampled data system is specified by the differential equations of its continuous part and by the form of the sampler, then the procedure for determining the correlation by means of an analogue computer is reduced to the following.

1. The sampled data system in question is simulated by means of analogue computers.
2. An action coinciding with the correlation function $K_x[n]$ is fed to the input of the model. The response of the model, representing the signal z given by (6.4), is recorded.
3. After reversing the time the signal $z[-n]$ is again fed to the input of the model of the system under consideration. The output signal represents in this case the value of the correlation function $K_y[l, \varepsilon]$ at the instant $\bar{t}=l+\varepsilon$. When $l=0$ the output signal gives the dispersion of the random process at the output of the system in a steady state.

In the case where the sampled data system is given in the form of difference equations, it is advisable to use a digital computer to determine the correlation function. The procedure for determining $K_y[l, \varepsilon]$ reduces to the following.

1. Using the given difference equations of the system, determine its weighting function $W[n, \varepsilon]$.
2. Find the response of the system to a signal of the form $K_x[n]$ in accordance with

$$z[n]=\sum_{m=0}^{N} W[m] K_x[n-m] \tag{6.6}$$

where the upper limit of summation, N, is chosen from the condition

$$W[\lambda]\approx 0 \text{ when } \lambda \geqslant N$$

3. Determine the correlation function $K_y[l, \varepsilon]$, using the formula

$$K_y[l, \varepsilon]=\sum_{k=-N}^{l} W[l-k, \varepsilon] z[-k, 0] \tag{6.7}$$

where $z[-k, 0]$ is the response of the system to the signal $K_x[n]$, taken in the direction of decreasing time. Then in accordance with (6.7) the correlation function $K_y[l, \varepsilon]$ can be computed only for discrete values of ε: $\varepsilon_1, \varepsilon_2, \ldots (0 \leqslant \varepsilon \leqslant 1)$.

The use of digital computers to determine the correlation function by the procedure just given may turn out to be suitable not only when the sampled data system is given by difference equations, but also when the system is given by the differential equations of its continuous part and by the form of the sampler. We have two ways of using a digital computer. The first consists of replacing the analogue of the sampled data system by the corresponding schemes for numerically solving the equations of the continuous part and sampler. In so doing we use the procedure for determining the correlation function by means of analogue computers.

The second way of using a digital computer is based on the circumstance that the weighting function of a system can be determined as its response to the σ-function directly from the given differential equations of the continuous part, and from the equation of the sampler. After the weighting function of the system has been determined, by numerically solving the corresponding equations, the correlation function $K_y[l, \varepsilon]$ is determined by the computer.

The method just given for the determination of the correlation function at the output of a stationary system has a serious shortcoming which is particularly noticeable when analogue computers are used. This is that, to obtain the correlation function, we must first obtain the intermediate signal z which has to be recorded, then 'reversed' in time and fed to the input of the model of the system being investigated. This is undesirable, as it involves preparation of the components of variable coefficients. It is a consequence of the fact that $K_y[l, \varepsilon]$ is obtained because the signal $K_x[n]$ passes twice in succession through one and the same system. This method of determining the correlation function can therefore be called a series method.

We shall now give a method which is free of the above shortcoming.

Second method. We shall consider another physical interpretation of the expression for the correlation function. For this purpose we change the variables in (6.2)

$$l+m_1=k$$

We then obtain

$$K_y[l]=\sum_{k=l}^{\infty} W[k-l] \sum_{m_2=0}^{\infty} W[m_2] K_x[k-m_2] \tag{6.8}$$

Denoting

$$z[k]=\sum_{m_2=0}^{\infty} W[m_2] K_x[k-m_2] \tag{6.9}$$

in (6.8) we have the following expression for the correlation function

$$K_y[l]=\sum_{k=l}^{\infty} W[k-l]\, z[k] \tag{6.10}$$

Let us analyse these results. It follows from (6.9) that the signal $z[k]$ is the response of the system in question at the instant $\bar{t}=k$ to an action which coincides with the correlation function $K_x[k-m_2]$ of the random function X. Further, it follows from (6.10) that the correlation function $K_y[l]$ of the random process at the output of the system is the result of summing with respect to k the function

$$\psi(k, l)=W[k-l]\, z[k]$$

where $W[k-l]$ is the weighting function of this system, i.e. the response of the system to the σ-function applied to it at the instant $\bar{t}=l$.

In the case $\varepsilon \neq 0$ we find from (6.9) and (6.10) that

$$z[k, \varepsilon] = \sum_{m_2=0}^{\infty} W[m_2, \varepsilon] K_x[k-m_2, 0] \tag{6.11}$$

and

$$K_y[l, \varepsilon] = \sum_{k=l}^{\infty} W[k-l, \varepsilon] z[k, \varepsilon] \tag{6.12}$$

The relations (6.11) and (6.12) are used as the starting points to determine the correlation function of the random process at the output of a stationary system when a random function X with the correlation function $K_x[n]$ acts at its input. If the system being investigated is given by the differential equations of its continuous part and by the form of the sampler, then the procedure of determining the correlation function by means of analogue computers reduces to the following.

1. Using the analogues we simulate two identical sampled data systems under consideration (Fig. 3.11).
2. An action of the form $K_x[n]$ is fed to the input of one of the models at the instant $\bar{t}=0$. In accordance with (6.11) the output variable of this model represents the signal $z[n, \varepsilon]$. The σ-function is fed to the input of the second model at time $\bar{t}=l$. The output variable of this model is given by the weighting function $W[n-l, \varepsilon]$ of the system.
3. The output signals of both models are fed to the holding circuits (F) whose keys operate synchronously and in phase with the sampler keys of the models. At the output of the holding circuits we have step functions whose

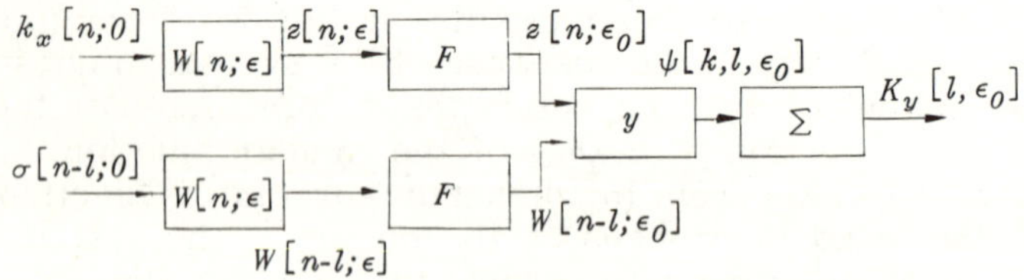

Fig. 3.11 Determination of the correlation function by the parallel method

values over the intervals of time $n \leqslant \bar{t} < n+1$ are equal to the values of the signals $z[n, \varepsilon]$ and $W[n-l, \varepsilon]$ at the discrete instants ($\varepsilon = 0$).

4. From the outputs of the holding circuits the signals are fed to the multiplying device ($\mathcal{Y}$), at the output of which we have the function

$$\psi(k, l) = W[k-l, 0]\, z[k, 0]$$

where

$$\psi(k, l) = 0 \quad \text{for} \quad k < l$$

5. The function $\psi(k, l)$, entering the summing device (Σ), is summed with respect to the variable k; as the result of this we have the value of the correlation function

$$K_y[l, 0] = \sum_{k=l}^{\infty} W[k-l, 0]\, z[k, 0] \qquad (6.13)$$

at the output of the summator for a fixed value of l when $\varepsilon = 0$.

It is obvious that when $l = 0$ we obtain from (6.13) the dispersion of the output process

$$D_y(0) = K_y[0, 0] = \sum_{k=l}^{\infty} W[k, 0]\, z[k, 0] \qquad (6.14)$$

In this way, to obtain the dispersion of the output variable of a stationary sampled data system we must, in accordance with the procedure just outlined, feed actions of the form $K_x[n]$ and $\sigma[n]$ simultaneously to the inputs of the models.

The diagram of the model just described (Fig. 3.11) can also be used to obtain the correlation function $K_y[l, \varepsilon]$ when $\varepsilon \neq 0$. For this the keys of the holding circuits should close the corresponding circuits with delay ε_0 ($0 < \varepsilon_0 \leqslant 1$) relative to the instants at which the sampler keys of the models of the system are closed. In this case at the output of the holding circuits we have step functions which over the intervals of time $n \leqslant t < n+1$ are equal to the signals $z[n, \varepsilon]$ and $W[n-l, \varepsilon]$ when $\varepsilon = \varepsilon_0$.

Varying the quantity ε_0 between the limits 0 and 1, we obtain the correlation function K_y between the instants when the sampler key is closed from

$$K_y[l,\ \varepsilon_0] = \sum_{k=l}^{\infty} \psi(k,\ l,\ \varepsilon_0) \qquad (6.15)$$

where

$$\psi(k,\ l,\ \varepsilon_0) = W[k-l,\ \varepsilon_0]\, z[k,\ \varepsilon_0]$$

When $l=0$ (the σ-function and a signal of the form $K_x[n]$ are simultaneously fed to the corresponding inputs of the models) we obtain the value of the dispersion of the output process between the instants when the sampler key is closed

$$D_y(\varepsilon_0) = K_y[0,\ \varepsilon_0] = \sum_{k=0}^{\infty} \psi(k,\ 0,\ \varepsilon_0) \qquad (6.16)$$

An experimental set-up which is constructed from Fig. 3.11 enables the process to be fully automated to obtain the correlation function (dispersion) of the output variable of the sampled data system being investigated. If the test set-up does not contain the required summing device then the signal from the output of the multiplier must be recorded and the corresponding ordinates of the function $\psi(k,\ l,\ \varepsilon)$ summed manually. This process clearly does not present any difficulties.

Above we have given a procedure for determining the correlation function and dispersion of the random process at a stationary system in the case where the system is given by the differential equations of its continuous part and where analogue computers are used to determine the correlation function.

If a sampled data system is given by the difference equations, then, to determine the correlation function $K_y[l,\ \varepsilon]$, it is advisable to use digital computers, and the procedure for determining $K_y[l,\ \varepsilon]$ amounts to the following.

1. From the given difference equations of the system we calculate the weighting function

   $$W[n,\ \varepsilon],\ n = 0,\ 1, \ldots,\ N$$

   where N is chosen from the condition that $W[\lambda,\ \varepsilon] \approx 0$ for $\lambda \geqslant N$.

2. In accordance with

$$z[k,\ \varepsilon] = \sum_{m=0}^{N} W[m,\ \varepsilon]\, K_x[k-m,\ 0],\ \varepsilon = \varepsilon_1,\ \varepsilon_2, \ldots \qquad (6.17)$$

$$(0 \leqslant \varepsilon \leqslant 1)$$

we determine the response of the system to the signal $K_x[n]$, acting at its input at the instant

$$\bar{t}=n+\varepsilon=0$$

3. We calculate the product of the functions

$$\psi(k,\ l,\ \varepsilon)=W[k-l,\ \varepsilon]\,z[k,\ \varepsilon],\ k=0,\ 1,\ldots,\ N \quad (6.18)$$

for a fixed value of l.

4. From the formula

$$K_y[l,\ \varepsilon]=\sum_{k=l}^{N}\psi(k,\ l,\ \varepsilon),\ \varepsilon=\varepsilon_1,\ \varepsilon_2,\ldots \quad (6.19)$$

we determine the correlation function for the discrete values of the parameter ε.

Varying the values $l=0,\ 1,\ \ldots,$ we can obtain by a number of switchings of the model the correlation function

$$K_y[l,\ \varepsilon],\ 1=0,\ 1,\ \ldots$$

When $l=0$ we obtain the dispersion

$$D_y(\varepsilon)=\sum_{k=0}^{N}\psi(k,\ 0,\ \varepsilon),\ \varepsilon=\varepsilon_1,\ \varepsilon_2,\ldots \quad (6.20)$$

Digital computers can also be used to determine the correlation function in the case where the system is given not by difference equations, but by the differential equations of the continuous part and by the sampler. In so doing we can use either the analogue computer procedure or that using digital computers.

From what has been said it is clear that the second method of determining the correlation function, which we may call a parallel method, gains the advantage over the series method by 'automating' the process. This, however, is achieved by nearly doubling the equipment needed to construct the test installation, since it now includes two models of the system being studied. Consequently, neither the first nor the second method is really successful for determining the correlation function and dispersion of the output process of a complex system. A more effective and convenient method of analysis is furnished by using the method which is based on the use of discrete shaping filters.

3.7 USE OF DISCRETE SHAPING FILTERS TO DETERMINE THE CORRELATION FUNCTION IN STEADY STATE

As has been mentioned in Section 3.6 the problem of analysis of stationary sampled data systems is formulated in the following manner. A stationary random function X with zero mathematical expectation and correlation function $K_x[n, \varepsilon]$ acts at the input of a system whose weighting function is $W[n, \varepsilon]$. It is required to determine the correlation function of the output process Y in steady state. This is shown diagrammatically in Fig. 3.12(a).

Methods of solving this problem by analogue and digital computers were given in the preceding section. Here we

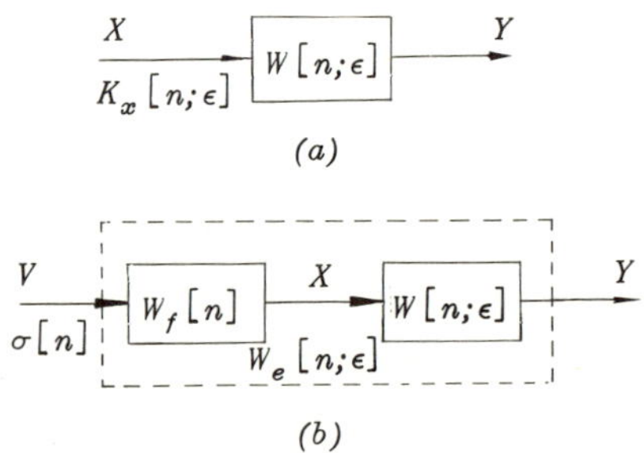

Fig. 3.12 Determination of the correlation function by the method of shaping filters

shall give another method for determining the correlation function, based on the use of discrete shaping filters. For this we form a random function X with the correlation function $K_x[n, \varepsilon]$ from the white noise V, using a discrete shaping filter with weighting function $W_f[n]$. The problem formulated above is then reduced to the determination of the correlation function of the output process of a sampled data filter with the weighting function $W_f[n]$ and the system being investigated connected in series, when discrete white noise with the correlation function $\sigma[n]$ acts at the input. The problem of analysis thus formulated is shown diagrammatically in Fig. 3.12(b).

We denote by $W_e[n, \varepsilon]$ the weighting function of the system formed when the discrete shaping filter and the system being investigated in series is connected, i.e.

$$W_e[n, \varepsilon] = \sum_{l=-\infty}^{n} W[n-l, \varepsilon] W_f[l, 0] = \sum_{m=0}^{\infty} W[m, \varepsilon] W_f[n-m, 0] \tag{7.1}$$

We now use (6.1) which in the given case assumes the form (for lattice functions)

$$K_y[l] = \sum_{m_1=0}^{\infty} W_e[m_1] \sum_{m_2=0}^{\infty} W_e[m_2] K_x[l-m_2+m_1] \tag{7.2}$$

Since the disturbance acting on the equivalent system is white noise with the correlation function

$$K_v[n-m] = \sigma[n-m]$$

then from (7.2) we find

$$K_y[l] = \sum_{m_1=0}^{\infty} W_e[m_1] \sum_{m_2=0}^{\infty} W_e[m_2] \sigma[l-m_2+m_1] = \sum_{m_1=0}^{\infty} W_e[m_1] W_e[l+m_1] \tag{7.3}$$

Putting

$$m_1 + l = k$$

in (7.3), we obtain the final expression for the correlation function of the output process of the system under consideration

$$K_y[l] = \sum_{k=l}^{\infty} W_e[k-l] W_e[k] \tag{7.4}$$

Expression (7.4) gives the rule for obtaining the correlation function $K_y[l]$ for a fixed value of l. When $l=0$, we have from (7.4) the dispersion of the output process

$$D_y(0) = K_y[0] = \sum_{k=0}^{\infty} W_e^2[k] \tag{7.5}$$

The values of the correlation function over the interval

between the instants when the sampler key is closed can be obtained by

$$K_y[l, \varepsilon] = \sum_{k=l}^{\infty} W_e[k-l, \varepsilon] W_e[k, \varepsilon] \tag{7.6}$$

Putting $l=0$ in (7.6) we find the dispersion

$$D_y(\varepsilon) = K_y[0, \varepsilon] = \sum_{k=0}^{\infty} W_e^2[k, \varepsilon] \tag{7.7}$$

Figure 3.13 shows a model for determining the correlation function in the case where the system is given by differential equations, and analogue computers are used to

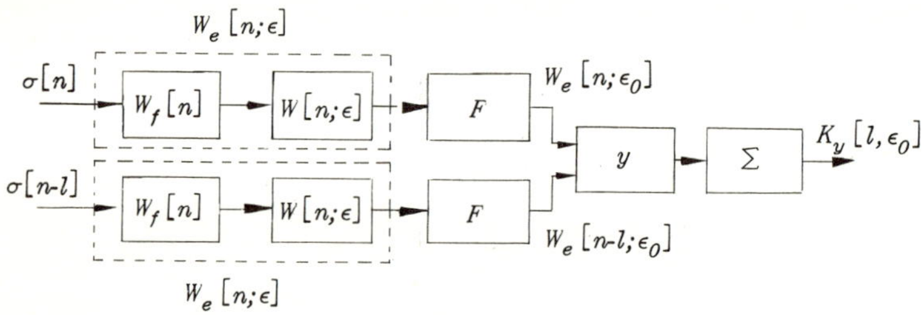

Fig. 3.13 Determination of the correlation function by the method of shaping filters

determine the correlation function. It follows from the diagram that the procedure for determining the correlation function of the output process reduces to the following:

1. Two identical equivalent systems are simulated by means of analogue computers.
2. The σ-function is fed to the input of one of the models at the instant $\bar{t}=0$, at $\bar{t}=l$ the σ-function is fed to the input of the other model.
3. The output signals of both systems (their weighting functions displaced relative to each other by the time $\bar{t}=l$) are held until $\varepsilon=\varepsilon_0$ $(0 \leqslant \varepsilon_0 \leqslant 1)$ and then multiplied together, yielding the function

$$S(n, l, \varepsilon_0) = W_e[n-l, \varepsilon_0] W_e[n, \varepsilon_0] \\ n = 0, 1, \ldots \tag{7.8}$$

4. Summing the function $S(n, l, \varepsilon_0)$ with respect to the

variable n, according to (7.6), gives the value of the correlation function for a fixed value of l

$$K_y[l, \varepsilon_0] = \sum_{n=l}^{\infty} S(n, l, \varepsilon_0) \tag{7.9}$$

When $l=0$ (the σ-function is fed simultaneously to the inputs of both models) we obtain from (7.9) the dispersion of the output process

$$D_y(\varepsilon_0) = \sum_{n=0}^{\infty} S(n, 0, \varepsilon_0) \tag{7.10}$$

The procedure just described is a general one. Where digital computers are used to determine the correlation function, the corresponding elements of the diagram (Fig. 3.13) are sections of the programme for the numerical solution of the differential (or difference) equations of the system.

It follows from (7.8) that the function $S(n, l, \varepsilon_0)$ is obtained by multiplying together the weighting functions of the equivalent systems, which are displaced relative to each other by l sampling periods. Consequently, there is no need to use a simulation scheme with two models of the system as shown in Fig. 3.13. The correlation function $K_y[l, \varepsilon]$ can be obtained by the scheme shown in Fig. 3.14, where the element with the symbol lT_0 denotes the element for ideal delay of the input signal by l sampling periods.

Let us compare now two methods of determining the correlation function: the second method and the method based on the use of shaping filters. It follows from the comparison that the use of discrete shaping filters allows an experimental set-up constructed for the determination of the correlation function with approximately half the equipment

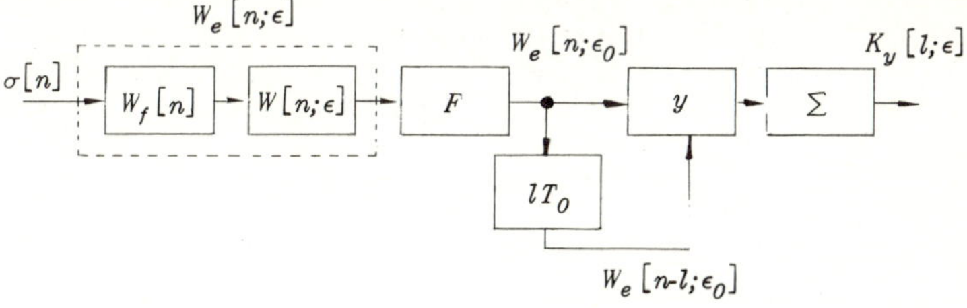

Fig. 3.14 Determination of the correlation function by the method of shaping filters

required by the second method. This is particularly important in cases where the system being investigated is fairly complex.

We shall make a few comments concerning the simulation of a shaping filter. If we use analogue computers to determine the correlation function of the output process, then it is advisable to replace the discrete shaping filter with the weighting function $W_f[n]$ by a continuous filter. Simulation of the latter by means of an analogue computer does not present any difficulties.

In accordance with (7.6), to determine the correlation function $K_y[l, \varepsilon]$, we must know the weighting function $W_e[n, \varepsilon]$ of the equivalent system formed according to the formula of sampled data systems connected in series:

$$W_e[n, \varepsilon] = \sum_{m=0}^{\infty} W[m, \varepsilon] W_f[n-m, 0] \tag{7.11}$$

whence it follows that we need know $W_f[n, 0]$ only at the discrete points $n=0, 1, \ldots$ to determine the correlation function.

If we now construct a continuous filter whose response to the δ-function will coincide with the weighting function W_f (even if this is true only at the discrete instants when the sampler key is closed), then the problem of simulating the discrete filter is reduced to the simulation of the corresponding continuous filter. We shall now give a method for determining a continuous filter which is equivalent (in the sense just stated) to the discrete shaping filter.

Let us formulate this problem in more detail: find the differential operator $W_f(p), p \equiv \frac{d}{dt}$, of a continuous filter whose response to the δ-function $W_f(t-\tau)$, coincides with the given weighting function $W_f[n-m]$ of the discrete shaping filter at the discrete instants of time. Suppose, then, that the weighting function $W_f[n-m]$ of a discrete shaping filter is given. The condition of the problem requires that

$$W_f(t-\tau) = W_f[n-m]\big|_{\substack{nT_0=t \\ mT_0=\tau}} \tag{7.12}$$

The differential operator $W_f(p)$ is required in the form

$$W_f(p) = \frac{M(p)}{L(p)} \tag{7.13}$$

where $M(p)$ and $L(p)$ are polynomials in p.

It is known from automatic control theory [13] that the weighting function $W_f(t-\tau)$, in accordance with (7.13), satisfies the differential equation

$$L(p)W_f(t-\tau)=M(p)\delta(t-\tau),\ t\geqslant\tau \tag{7.14}$$

with zero initial conditions. In addition

$$W_f(t-\tau)=M^*(p)W_{f_0}(t-\tau),\ t\geqslant\tau \tag{7.15}$$

where $M^*(p)$ is a differential operator adjoint to the operator $M(p)$;

$W_{f_0}(t-\tau)$ is the fundamental set of solutions of the equation

$$L(p)W_{f_0}(t-\tau)=0,\ t\geqslant\tau \tag{7.16}$$

with the corresponding initial conditions. Thus, from the given function $W_f(t-\tau)$, which satisfies (7.12) we must determine the differential operators $L(p)$ and $M(p)$. We shall now proceed to solve this problem.

It was shown in the first chapter that the weighting function $W[n-m]$ of a sampled data system can be represented in the form

$$W[n-m]=\sum_{i=1}^{k}\eta_i[m]\lambda_i[n],\ n\geqslant m \tag{7.17}$$

where λ_i form the fundamental set of solutions of the corresponding difference equation;

η_i are functions determined by the initial conditions.

Using (7.17) we can write

$$W_f[n-m]=\sum_{i=1}^{k}\eta_i[m]\lambda_i[n],\ n\geqslant m \tag{7.18}$$

Using the natural time scale in (7.18), i.e. assuming that

$$nT_0=t \quad \text{and} \quad mT_0=\tau$$

we find

$$W_f(t-\tau)=\sum_{i=1}^{k}\eta_i(\tau)\lambda_i(t),\ t\geqslant\tau \tag{7.19}$$

where $\lambda_i(t)$ form the fundamental set of solutions of the homogeneous differential equation

$$L(p)\lambda(t)=0,\ t\geqslant 0 \qquad (7.20)$$

with the corresponding initial conditions.

To determine the differential operator $L(p)$ we use the familiar theorem in the theory of differential equations, which states that the fundamental set of solutions $\lambda_i(t)$, $i=1, 2, \ldots, k$ completely defines a k-th order homogeneous differential equation with the coefficient of the highest derivative equal to unity. This differential equation is given as follows:

$$L(p)\lambda(t)=\begin{vmatrix} \lambda_1(t) & \lambda_2(t) & \ldots & \lambda_k(t) & \lambda(t) \\ \lambda_1'(t) & \lambda_2'(t) & \ldots & \lambda_k'(t) & \lambda'(t) \\ \ldots & \ldots & \ldots & \ldots & \ldots \\ \lambda_1^{(k)}(t) & \lambda_2^{(k)}(t) & \ldots & \lambda_k^{(k)}(t) & \lambda^{(k)}(t) \end{vmatrix}=0 \qquad (7.21)$$

Consequently, setting up a determinant of the form (7.21) and expanding it by the elements of its last column, we find the operator $L(p)$. Further, as follows from (7.15), we must find the function $W_{f_0}(t-\tau)$ in order to determine the operator $M(p)$. It is known [13] that the function $W_{f_0}(t-\tau)$ as the solution of (7.20) is determined in the following manner:

$$W_{f_0}(t-\tau)=\frac{(-1)^{k-1}}{a_k D(\tau)}\begin{vmatrix} \lambda_1(t) & \lambda_2(t) & \ldots & \lambda_k(t) \\ \lambda_1(\tau) & \lambda_2(\tau) & \ldots & \lambda_k(\tau) \\ \lambda_1'(\tau) & \lambda_2'(\tau) & \ldots & \lambda_k'(\tau) \\ \ldots & \ldots & \ldots & \ldots \\ \lambda_1^{(k-2)}(\tau) & \lambda_2^{(k-2)}(\tau) & \ldots & \lambda_k^{(k-2)}(\tau) \end{vmatrix} \qquad (7.22)$$

where a_k is the coefficient of the highest derivative of the operator $L(p)$

$D(\tau)$ is the Wronskian determinant corresponding to (7.20).

Thus, determining $W_{f_0}(t-\tau)$ in accordance with (7.22) and (7.19), and then substituting it into (7.15), we find the operator $M^*(p)$ from the condition that the result of applying $M^*(p)$ to $W_{f_0}(t-\tau)$ coincides with the weighting function $W_f(t-\tau)$. Then, making use of certain properties of the adjoint differential operators with constant coefficients

$$\begin{aligned} M^*(p) &= M(-p) \\ M^*(-p) &= M(p) \end{aligned} \tag{7.23}$$

we find the required operator $M(p)$.

In accordance with the foregoing the procedure for determining a continuous filter whose response to the δ-function coincides with the weighting function of the discrete shaping filter reduces to the following:

1. The given weighting function $W_f[n-m]$ is written in the natural time scale and represented in the form

$$W_f\,[n-m]\big|_{\substack{nT_0=t \\ mT_0=\tau}} = \sum_{i=1}^{k} \eta_i(\tau)\lambda_i(t) \tag{7.24}$$

2. The differential operator $L(p)$ is determined from

$$L(p)\lambda(t) = \begin{vmatrix} \lambda_1(t) & \lambda_2(t) & \dots & \lambda_k(t) & \lambda(t) \\ \lambda_1'(t) & \lambda_2'(t) & \dots & \lambda_k'(t) & \lambda'(t) \\ \dots & \dots & \dots & \dots & \dots \\ \lambda_1^{(k)}(t) & \lambda_2^{(k)}(t) & \dots & \lambda_k^{(k)}(t) & \lambda^{(k)}(t) \end{vmatrix} \tag{7.25}$$

3. The function $W_{f0}(t-\tau)$ is determined from (7.22).
4. From the relation

$$W_f(t-\tau) = M^*(p)\,W_{f0}(t-\tau)\,,\ M(p) = M^*(-p) \tag{7.26}$$

the operator $M(p)$ is finally determined.

Returning to the problem of determining the correlation function of the random process at the output of a stationary system by means of shaping filters, we note that, when the discrete shaping filter is replaced by an equivalent continuous filter, Fig. 3.14 can be transformed into that shown in Fig. 3.15. In so doing we must apply an action in the form of the δ-function to the input of the model.

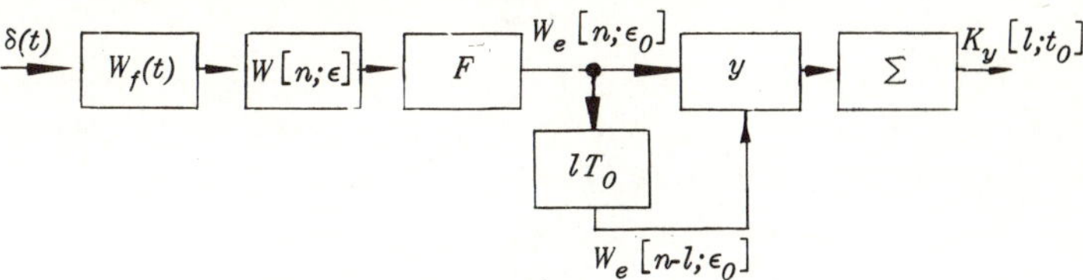

Fig. 3.15 Determination of the correlation function by an equivalent continuous shaping filter

3.8 STATISTICAL ANALYSIS OF THE TRANSIENT MOTION OF A STATIONARY SYSTEM

In this section we shall give a method for determining the dispersion of the output process of a system in transient motion when a stationary random disturbance acts at the input. We shall formulate this problem as follows. Suppose we have a stationary system with the weighting function $W[n, \varepsilon]$, processes in which terminate after the time $t=NT_0$, where T_0 is the sampling period. It is required to determine the dispersion $D_y[n, \varepsilon]$, $n=0, 1, \ldots, N$, of the output process Y of the system when a random disturbance X with the correlation function $K_x[n; m]$ acts at its input.

The weighting function of a stationary system is a function only of the instant $\bar{t}=n+\varepsilon$ at which the signal is observed at the output. But over the time interval $0<\bar{t}<N$, i.e. during the time of the transient process, the system should be regarded as non-stationary. In general any stationary system being in a state of rest before the instant $\bar{t}=0$, and being disturbed at the instant $\bar{t}=0_+$, can be considered as a particular case of a non-stationary system whose parameters vary with a jump when the input disturbance is applied. Thus, considering the stationary system over the time interval $0<\bar{t}<N$ as non-stationary, we write the expression for the dispersion of its output process

$$D_y[n]=\sum_{m_1=-\infty}^{n} W[n; m_1] \sum_{m_2=-\infty}^{n} W[n; m_2] K_x[m_1; m_2] \tag{8.1}$$

(For simplicity we shall consider lattice functions. The results can easily be extended to the general case).

If X is a stationary random function, then from (8.1) we obtain

$$D_y[n]=\sum_{m_1=-\infty}^{n} W[n; m_1] \sum_{m_2=-\infty}^{n} W[n; m_2] K_x[m_2-m_1] \tag{8.2}$$

In a particular case, when the input disturbance is discrete white noise with constant intensity c^2, the expression for the dispersion can be reduced to the form

$$D_y[n]=c^2 \sum_{m=-\infty}^{n} W^2[n; m] \tag{8.3}$$

Thus, if the input disturbance of the system is a random process of the white noise type, the dispersion of the output process can be determined very simply.

Using discrete shaping filters, we can replace the original system being investigated, whose input disturbance is a random process X with the correlation function $K_x[n;\ m]$, by an equivalent system whose input disturbance is discrete white noise V. The equivalent system is a shaping filter with the weighting function $W_f[n;\ m]$ connected in series with the original system under consideration.

Denoting the weighting function of the equivalent system by $W_e[n;\ m]$ we find the expression for the dispersion of the output process in the form

$$D_y[n] = \sum_{m=-\infty}^{n} W_e^2[n;\ m] \tag{8.4}$$

It follows that to determine the dispersion of the output process at the instant $\bar{t}=n$, $0 \leqslant n \leqslant N$, the weighting function of the equivalent system, $W_e[n;\ m]$, considered as a function of the second argument m, must be squared and the resulting products summed over all the instants at which the input disturbance acts. In so doing the problem of determining the dispersion can be solved by either analogue or digital computers.

The methods for determining the dispersion based on (8.4) will be considered in the statistical analysis of non-stationary sampled data systems. Here we shall give another method which is applicable only to systems with constant parameters, in the case where the input action of the system is stationary and has a correlation function of the form

$$K_x[n-m] = e^{-\alpha|n-m|} \tag{8.5}$$

An analogous method for determining the dispersion for continuous systems has been derived by Laning [23].

Accordingly, we shall solve the problem of determining the dispersion of the output process of a stationary system in transient motion when a stationary random disturbance X with a correlation function of the form (8.5) acts at its input. It has been shown in Section 3.5 that a filter, which forms a random process X with the correlation function (8.5) out of white noise V, is described by the difference equation

$$\Delta x[n]+(1-e^{-\alpha})x[n]=\sqrt{1-e^{-2\alpha}}\,v[n+1] \tag{8.6}$$

The weighting function of this filter is given by

$$W_f[n]=\sqrt{1-e^{-2\alpha}}\,e^{-\alpha n},\ n\geqslant 0 \tag{8.7}$$

Consequently, the original system being investigated can be replaced by an equivalent one which is formed by connecting in series a filter with the weighting function (8.7) and the original system with the weighting function $W[n, \varepsilon]$ (Fig. 3.16).

It should be borne in mind that the random process X is obtained from the discrete white noise V in accordance with the formula

$$X[n]=\sum_{m=-\infty}^{n} W_f[n-m]v[m] \tag{8.8}$$

Accordingly we must consider the case where the action on the system occurs before the instant $n=0$, i.e. before the

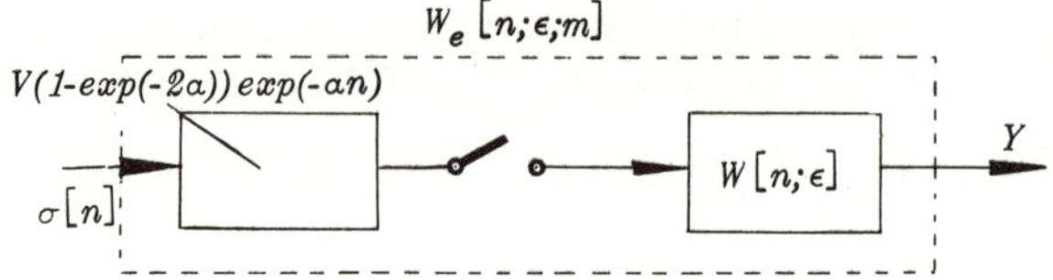

Fig. 3.16 Determination of the dispersion in a transient state

system is switched on ($m<0$), and after the system has been switched on ($m>0$). This is represented by the presence of a switch in Fig. 3.16, which is closed at $n=0$.

When the signal $\sigma[n-m]$ acts at the input of the filter its output variable $z[n]$ is given by

$$z[n-m]=\sqrt{1-e^{-2\alpha}}\,e^{-\alpha(n-m)},\ n>m \tag{8.9}$$

The weighting function $W_e[n;\ m]$ of the equivalent system is determined in this case as follows

$$\begin{aligned} W_e[n;\ m] &= \sum_{m_1=0}^{n} W[n-m_1]z[m_1-m] \\ &= \sqrt{1-e^{-2\alpha}}\,e^{\alpha m}\sum_{m_1=0}^{n} W[n-m_1]e^{-\alpha m_1},\ m<0 \end{aligned} \tag{8.10}$$

since the action at the input of the system having the weighting function $W[n-m]$ commences at the instant the switch is closed ($n=0$).

When $m>0$ the action at the input of the system commences not at the instant the switch is closed, but at the instant the σ-function acts. Consequently, for $m>0$ we have

$$\begin{aligned} W_e[n; m] &= \sum_{m_1=m}^{n} W[n-m_1]\sqrt{1-e^{-2\alpha}}\, e^{-\alpha(m_1-m)} \\ &= \sqrt{1-e^{-2\alpha}} \sum_{m_1=m}^{n} W[n-m_1]\, e^{-\alpha(m_1-m)}, \quad m>0 \end{aligned} \tag{8.11}$$

It follows from (8.10) that the function

$$F[n] = \sum_{m_1=0}^{n} W[n-m_1]\, e^{-\alpha m_1} \tag{8.12}$$

is the response of the system to the signal $e^{-\alpha n}$ when $n>0$, i.e. when the system is switched on, and is the response to zero signal when $n<0$. Thus the expression for the weighting function of the equivalent system can be written in the following form for $m<0$, with (8.12) taken into account,

$$W_e[n; m] = \sqrt{1-e^{-2\alpha}}\, e^{\alpha m} F[n] \tag{8.13}$$

When $m>0$ we obtain from (8.11)

$$\begin{aligned} W_e[n; m] &= \sqrt{1-e^{-2\alpha}} \sum_{m_1=0}^{n-m} W[n-m-m_1]\, e^{-\alpha m_1} \\ &= \sqrt{1-e^{-2\alpha}}\, F[n-m] \end{aligned} \tag{8.14}$$

Consequently, we have

$$W_e[n; m] = \begin{cases} e^{\alpha m}\sqrt{1-e^{-2\alpha}}\, F[n], & -\infty < m \leqslant -1 \\ \sqrt{1-e^{-2\alpha}}\, F[n-m], & 0 \leqslant m \leqslant n \end{cases} \tag{8.15}$$

Substituting (8.15) into the expression (8.4) for the dispersion, we find

$$D_y[n] = (1-e^{-2\alpha}) \sum_{m=-\infty}^{-1} e^{2\alpha m} F^2[n] + (1-e^{-2\alpha}) \sum_{m=0}^{n} F^2[n-m] \tag{8.16}$$

Carrying out summation in (8.16) we have

$$(1-e^{-2\alpha})\sum_{m=-\infty}^{-1} e^{2\alpha m} = e^{-2\alpha} \tag{8.17}$$

$$\sum_{m=0}^{n} F^2[n-m] = \sum_{m=0}^{n} F^2[m] \tag{8.18}$$

Hence, when (8.17) and (8.18) are used, the expression for the dispersion can be written in the following form:

$$D_y[n] = e^{-2\alpha}F^2[n] + (1-e^{-2\alpha})\sum_{m=0}^{n} F^2[m] \tag{8.19}$$

Since $F[n]$ is the response of the system to a signal of the form $e^{-\alpha n}$ when $n > 0$, the dispersion $D_y[n]$ can be obtained as a function of the time n by means of a model whose diagram is shown in Fig. 3.17. The procedure for determining the dispersion then reduces to the following.

An action of the form $e^{-\alpha n}$ $(n \geqslant 0)$ is fed to the input of the model. The response, constituting the function $F[n, \varepsilon]$, enters the holding circuit whose key works synchronously and in phase with the sampler key of the system. The output of the holding circuit is a step function whose values over the interval $n \leqslant \bar{t} < n+1$ equal the values of the function $F[n, \varepsilon]$ at the discrete instants $(\varepsilon = 0)$.

The step function $F[n, 0]$ enters the squaring device (K), from the output of which $F^2[n, 0]$ is fed to the scale element with the coefficient $e^{-2\alpha}$ and hence to the summing device.

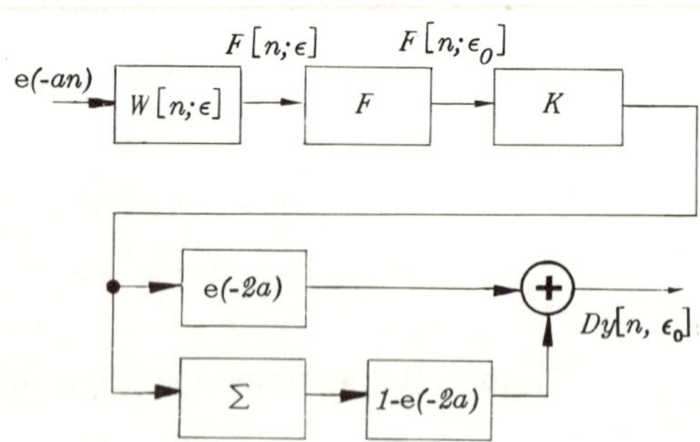

Fig. 3.17 Determination of the dispersion at the output of a stationary system in a transient state

The signal $\sum_{m=0}^{n} F^2[m, 0]$ is multiplied by the constant coefficient $(1-e^{-2\alpha})$ and is then added to the signal $e^{-2\alpha}F^2[n, 0]$. As a result, the value of the dispersion is obtained at the output of the model in accordance with (8.19).

It is clear that the dispersion is a function of the discrete argument $n=0, 1, \ldots, N$. When $n \geqslant N$ a steady state process takes place in the system, and $D_y[n]$ is constant.

The dispersion of the output process, $D_y[n, \varepsilon]$, when $\varepsilon \neq 0$, can be found from

$$D_y[n, \varepsilon] = e^{-2\alpha}F^2[n, \varepsilon] + (1-e^{-2\alpha}) \sum_{m=0}^{n} F^2[m, \varepsilon] \qquad (8.20)$$

$$0 \leqslant \varepsilon \leqslant 1$$

To determine $D_y[n, \varepsilon]$ when $\varepsilon = \varepsilon_0$ $(0 \leqslant \varepsilon_0 \leqslant 1)$ we can use the model just described, with the only difference that the key of the holding circuit must not operate in phase with the sampler key of the system, but must close the circuit with a delay ε_0.

The model shown in Fig. 3.17 can be represented in a different form, noting that the input action of the model, $e^{-\alpha n}$, is the weighting function of a filter given by the difference equation

$$\Delta y[n] + (1-e^{-\alpha})\, y[n] = x[n+1] \qquad (8.21)$$

Thus, replacing $e^{-\alpha n}$ in the filter response (8.21) by the σ-function and changing the order in which the filter and the system are connected up (this is possible since their parameters are constant) we obtain the diagram of the model shown in Fig. 3.18.

Above we have given a procedure for determining the

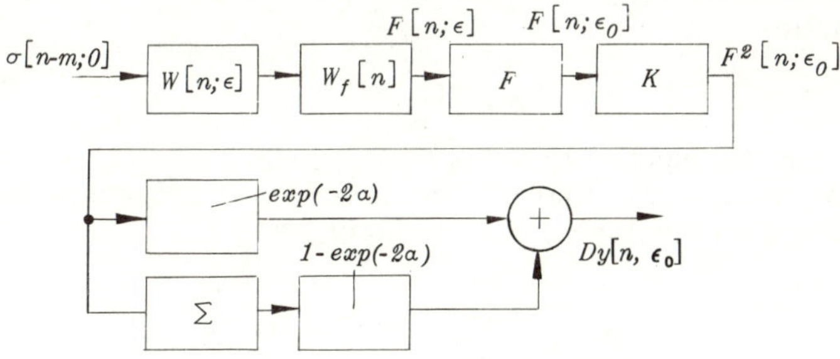

Fig. 3.18 Determination of the dispersion at the output of a stationary system in a transient state

dispersion of the output process of a stationary system in transient motion by means of analogue computers. But, as before, digital computers may be used. In this case the corresponding blocks of the diagrams shown in Figs. 3.17 and 3.18 constitute programmes for the numerical solution of the corresponding equations which describe the operation of these blocks.

In concluding we point out that the method just given can easily be extended to the general case where the correlation function of the random function X has the form

$$K_x[n] = \sum_{i=1}^{k} \eta_i e^{-\alpha_i |n|} \tag{8.22}$$

where η_i are constants. (It can be shown that a correlation function of any stationary random process can be represented in the form (8.22)).

Let us consider a simple example illustrating the use of the above method.

Example. Determine the dispersion $D_y[n, \varepsilon]$ of the output process of a sampled data system in transient motion, if the weighting function of this system is given by

$$W[n, \varepsilon] = \begin{cases} -1, & n=0 \\ 3, & n=1 \\ -3, & n=2 \\ 1, & n=3 \\ 0, & n \geqslant 4 \end{cases} \tag{8.23}$$

and a stationary random function X with the correlation function

$$K_x[n] = e^{-\alpha |n|} \tag{8.24}$$

acts at the input of this system.

We must first find the function $F[n, \varepsilon]$, defined by (8.12). For a step function it has the form

$$F[n, \varepsilon] = \sum_{m=0}^{n} W[n-m, \varepsilon] e^{-\alpha m} \tag{8.25}$$

Putting $n=0, 1, \ldots$ in (8.25), and using (8.23) and (8.24), we find

$$
\begin{aligned}
F[0,\ \varepsilon] &= -1 \\
F[1,\ \varepsilon] &= 3 - e^{-\alpha} \\
F[2,\ \varepsilon] &= -3 + 3e^{-\alpha} - e^{-2\alpha} \\
F[3,\ \varepsilon] &= 1 - 3e^{-\alpha} + 3e^{-2\alpha} - e^{-3\alpha}
\end{aligned}
\tag{8.26}
$$

The subsequent values of $F[n,\ \varepsilon]$, when $n>3$, are of no interest to us, since the processes in the system considered terminate after a time equal to three sampling periods.

Using (8.20) we now find the values of the dispersion $D_y[n,\ \varepsilon]$ for $n=0,\ 1,\ 2$ and 3.

3.9 STATISTICAL ANALYSIS OF NON-STATIONARY SYSTEMS

3.9.1 Stationary random disturbance acting at the input

Suppose that a non-stationary sampled data system with the weighting function $W[n,\ \varepsilon;\ m]$ is given, and a stationary random function X with zero mathematical expectation and the correlation function $K_x[n-m]$ acts at the input of this system. Considering for simplicity the case of lattice functions, we have in accordance with (3.8) the expression for the correlation function of the output process

$$
K_y[n_1;\ n_2] = \sum_{m_1=0}^{n_1} W[n_1;\ m_1] \sum_{m_2=0}^{n_2} W[n_2;\ m_2] K_x[m_1-m_2] \tag{9.1}
$$

Let us now assume that the random function X acting at the input is white noise with unit intensity. Then from (9.1) we find the following expression for the correlation function

$$
K_y[n_1;\ n_2] = \sum_{m_1=0}^{n_1} W[n_1;\ m_1] W[n_2;\ m_1] \tag{9.2}
$$

It follows from (9.2) that, in order to determine the correlation function, we must have the weighting function of the system as a function of m_1 (the instant at which the action is applied) for fixed values of n_1 and n_2.

When (9.2) is used, the procedure for determining the correlation function by means of analogue computers reduces to the following.

1. The system in question is simulated by means of analogue computers.
2. The σ-function is fed to the input of the model at the successive instants $m_1 = 0, 1, \ldots, n$ and the output signals, representing the weighting functions $W[n; m_{1j}]$, are recorded.
3. The relations $W[n_1; m_1]$ and $W[n_2; m_1]$ are constructed from the set of these weighting functions $W[n, m_{1j}]$ as functions of m_1 when the values of n_1 and n_2 are fixed.
4. The values of $W[n_1; m_1]$ and $W[n_2; m_1]$ are multiplied together and summed with respect to m_1; as the result the correlation function $K_y[n_1; n_2]$ is obtained in accordance with (9.2).

Repeating the above procedure for different values of n_1 and n_2, we obtain a table of values of the correlation function depending on both arguments.

Putting $\bar{t}_1 = n_1 + \varepsilon_1$ and $\bar{t}_2 = n_2 + \varepsilon_2$ in (9.2), we find

$$K_y[n_1, \varepsilon_1; n_2, \varepsilon_2] = \sum_{m_1=0}^{n_1} W[n_1, \varepsilon_1; m_1] W[n_2, \varepsilon_2; m_1] \tag{9.3}$$

whence it follows that, to determine the correlation function when $\varepsilon_1 \neq 0$ and $\varepsilon_2 \neq 0$, we must know the weighting functions for the fixed values of $\bar{t}_1 = n_1 + \varepsilon_1$ and $\bar{t}_2 = n_2 + \varepsilon_2$.

The dispersion of the output process can be obtained from (9.3), when $\bar{t}_1 = \bar{t}_2 = \bar{t}$, by

$$D_y[n, \varepsilon] = K_y[\bar{t}; \bar{t}] = \sum_{m=0}^{n} W^2[n, \varepsilon; m] \tag{9.4}$$

Expression (9.4) shows that, to determine the dispersion of the output process of a non-stationary system by analogue computer, we must carry out $n+1$ switchings-on of the model, where n is the instant when the dispersion is determined. The situation is similar when digital computers are used: when the dispersion is determined at the instant $\bar{t} = n + \varepsilon$ on a digital computer, the corresponding difference equations or differential equations of the system must be solved $n+1$ times.

Suppose now that a stationary random disturbance X with the correlation function $K_x[n-m]$ acts at the input of a non-stationary system. When discrete shaping filters are used the problem how to determine the correlation function and dispersion of the output process can be reduced to the problem of determining these functions for the equivalent system with discrete white noise acting at its input. We shall merely point out that if necessary the discrete shaping filter can be replaced by an equivalent continuous filter. The procedure for determining the corresponding equivalent continuous filter has been given in Section 3.7.

3.9.2 Non-stationary random disturbance action at the input

We assume, as before, that a function X with zero mathematical expectation acts at the input of a non-stationary system. We shall give a method for determining the dispersion of the output process, when X is a non-stationary random function with the correlation function $K_x[n;\ m]$. From (3.9) we have the expression for the dispersion (in the case of lattice functions)

$$D_y[n]=\sum_{m_1=0}^{n} W[n;\ m_1]\sum_{m_2=0}^{n} W[n;\ m_2]K_x[m_1;\ m_2] \tag{9.5}$$

We shall give a physical interpretation of (9.5). Its inner sum can be regarded as the response of a system with weighting function $W[n;\ m_2]$ to an action of the form $K_x[m_1;\ m_2]$, applied to the input of the system at the instant m_1, with the value of m_2 fixed, i.e. we have

$$S(n;\ m_1)=\sum_{m_2=0}^{n} W[n;\ m_2]K_x[m_1;\ m_2] \qquad m_1=\text{const} \tag{9.6}$$

Further, it follows from (9.5), when (9.6) is taken into account, that the dispersion of the output process at the instant n is the response of the system at this instant to a disturbance of the form $S(n;\ m_1)$, considered as a function of the argument m_1, for a fixed value of n. We have thus

$$D_y[n]=\sum_{m_1=0}^{n} W[n;\ m_1]S(n;\ m_1) \tag{9.7}$$

In the general case the dispersion for the instant $n+\varepsilon$

is given by

$$D_y[n, \varepsilon] = \sum_{m_1=0}^{n} W[n, \varepsilon; m_1] S(n, \varepsilon; m_1) \tag{9.8}$$

where

$$S(n, \varepsilon; m_1) = \sum_{m_2=0}^{n} W[n, \varepsilon; m_2] K_x[m_1, 0; m_2, 0] \tag{9.9}$$

Expressions (9.8) and (9.9) determine the procedure for obtaining the dispersion of the output process of the system under consideration. When analogue computers are used this procedure reduces to the following.

1. From the given correlation function $K_x[m_1; m_2]$ construct the set of functions

$$K_x[m_{10}; m_2], K_x[m_{11}; m_2], \ldots, K_x[m_{1n}; m_2]$$

depending on the argument m_2 for fixed values of $m_1 = 0, 1, \ldots, n$.

2. Feeding a signal of the form $K_x[m_{1j}; m_2]$ to the input of the system, obtain by means of the $n+1$ switchings-on of the model the following set of functions

$$\{S(n, \varepsilon; m_{1j})\}, \; j = 0, 1, \ldots, n$$

3. From the set $\{S(n, \varepsilon; m_{1j})\}$ construct the following relation $S(n, \varepsilon; m_1)$ as a function of m_1.
4. Feed an action of the form $S(n, \varepsilon; m_1)$ to the input of the model; the output variable of the model at the instant $n+\varepsilon$ gives the dispersion $D_y[n, \varepsilon]$.

To obtain the dispersion of the output process of a non-stationary system at the instant $n+\varepsilon$, when a non-stationary random disturbance acts at its input, we must thus carry out $n+2$ switchings-on of the model.

If digital computers are used to determine the dispersion, then, to obtain the final result, we must solve the differential equations or the difference equations of the system $n+2$ times for the corresponding disturbing functions.

From the method described for obtaining the dispersion it is clear how the correlation function is obtained; we need not therefore dwell on this in detail.

It is clear that the method given for determining the statistical characteristics of the output process is cumbersome.

The use of discrete shaping filters enables the problem just considered to be reduced to a simpler problem of determining the statistical characteristics of the output process of an equivalent system whose input disturbance is discrete white noise. Using the procedure given in Section 3.5 to determine a filter that forms a random process X with a given correlation function $K_x[n-m]$, out of white noise with unit intensity, and replacing the system under consideration by an equivalent one, we can determine the required dispersion from

$$D_y[n, \varepsilon] = \sum_{m=0}^{n} W_e^2[n, \varepsilon; m] \tag{9.10}$$

where $W_e[n, \varepsilon; m]$ is the weighting function of the equivalent system, formed by a shaping filter with the weighting function $W_f[n; m]$ and the original system connected in series (Fig. 3.19(a)).

It is apparent that the use of shaping filters considerably simplifies the problem of determining the dispersion (the correlation function). It should be pointed out that if necessary the discrete filter can be replaced by a corresponding continuous filter. The procedure for determining a continuous filter which is equivalent to the discrete one, given in Section 3.7, can easily be extended to the case of non-stationary shaping filters.

Although the use of shaping filters reduces the labour considerably, this problem nevertheless remains fairly laborious. In fact, in sampled data control systems the length of the discreteness period of commands can be of the order of 5 to 10 seconds, whilst the control process may continue for scores of minutes. If it is required to determine the dispersion of a certain coordinate at the conclusion of the control

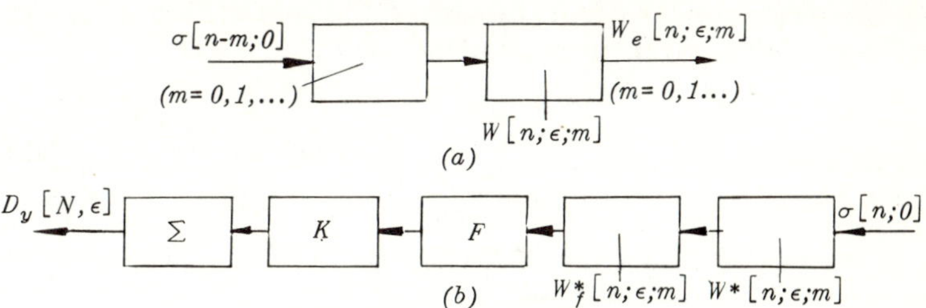

Fig. 3.19 Determination of the dispersion at the output of a non-stationary system

process when $\bar{t}=N+\varepsilon$, then it is obvious that in this case N has the order of several tens or even hundreds of units. This means that, in order to determine the dispersion $D[n, \varepsilon]$ at the point $n=N$, we must either repeatedly ($N+1$ times) switch on the model, or solve the same number of times the differential equations of the system by numerical methods.

The complexity of the problem is due to the fact that the weighting function $W[n, \varepsilon; m]$ of the system in question (or the equivalent system), depending on the second argument m, must be known at the instant the action is applied, with $n=N$ kept at a fixed value.

It has been shown in Chapter 2 that if the original sampled data system with the weighting function $W[n, \varepsilon; m]$ is replaced by its adjoint, then the response of the adjoint system to the σ-function is given by the weighting function $W^*[n, \varepsilon; m]$ as a function of the second argument m for a fixed value of the first argument n. Consequently, the problem can be considerably simplified if adjoint sampled data systems are used. Indeed, when adjoint systems are used the required characteristic (the dispersion, for example) can be obtained by a single switching-on of the model, or by a single solution of the adjoint difference or differential equations on a digital computer.

We shall illustrate the use of adjoint systems by determining the dispersion of the output variable of a system when a random process X with the correlation function $K_x[n; m]$ acts at its input. Determining the discrete filter which forms the random process out of the white noise V, and connecting it (or an equivalent continuous filter) in series with the system under consideration, we reduce the problem to that of determining the dispersion of the process at the output of the equivalent system when discrete white noise acts at its input (Fig. 3.19(a)). In this case the dispersion at the instant $n=N$ is given in accordance with (9.10) by

$$D_y[N, \varepsilon]=\sum_{m=0}^{N} W_e^2[N, \varepsilon; m] \tag{9.11}$$

If we now replace the shaping filter by the corresponding adjoint filter, and replace the system by the corresponding adjoint system, we then obtain the equivalent adjoint system. The dispersion of the output process of this system can be written in the form

$$D_y[N, \varepsilon] = \sum_{m=0}^{N} W_e^{*2}[N, \varepsilon; m] = \sum_{m=0}^{N} W_e^2[N, \varepsilon; N - m] \quad (9.12)$$

where $W_e^*[N, \varepsilon; m]$ is the weighting function of the equivalent adjoint system.

It follows from (9.12) that the dispersion at the instant $N+\varepsilon$ represents the response of the model shown in Fig. 3.19(b) to the σ-function. Consequently, it can be obtained by a single switching-on. Similarly, if digital computers are used to determine the dispersion, the final result can be obtained by solving only once the corresponding adjoint difference or differential equation. In this way, the use of discrete shaping filters and adjoint systems enables the amount of work to be reduced several times in determining the statistical characteristics of non-stationary sampled data systems, by comparison with the direct methods of statistical analysis.

The methods of analysis given above have been developed for one-dimensional sampled data systems and for systems with a single sampler, but the results can easily be extended to multi-dimensional systems and to systems with several samplers. For this we must use suitable relations which establish a connection between the correlation function of the input and output processes of a multi-dimensional system; we must also use relations establishing a connection between the weighting functions of sampled data systems with several samplers.

3.10 MATHEMATICAL EXPECTATION AND DISPERSION AT THE OUTPUT OF A NON-STATIONARY SYSTEM WITH GIVEN INITIAL CONDITIONS

The method given in the preceding section enables us to obtain the dispersion of the random output process of a non-stationary system at the required instant $n = N$ by means of a single switching-on of the model (or by a single numerical solution of the corresponding equations on a digital computer). It has been assumed that at the instant $n = n_0$ when the system is switched on it is at rest, i.e. it is not disturbed.

In practice there arises a necessity to investigate the

accuracy of a system whose state at the beginning of the control process (at the instant it is switched on) is characterised by certain values of the state parameters. In other words, at the instant when the control system is switched on the output coordinates of the controlled system are non-zero. In the general case the output coordinates of the system can have random values at the beginning of the control process. Furthermore, these values can depend on the instant of switching-on, i.e. in the general case they are functions of time.

Mathematically the problem of investigating the accuracy of a disturbed sampled data system reduces to the problem of determining the probability characteristics of random functions, given by difference equations with random initial conditions which are functions of time.

Subsequently we shall assume that at the instant n_0, the instant at which the system is switched on, its state is characterised by a set of random functions $c_i[n_0]$, which are the values of certain random functions $c_i[n]$ for $n=n_0$. In the general case we should assume that the random functions $c_i[n]$ are non-stationary, so that

$$m_{c_i}=M[c_i[n]]=m_{c_i}[n]$$

and

$$D_{c_i}=M[c_i^2[n]]-m_{c_i}^2[n]=D_{c_i}[n]$$

Since we assume that the mathematical expectation m_{c_i} and the dispersion D_{c_i} of the random functions $c_i[n]$ depend on time, then the mathematical expectation and dispersion of the output coordinate of the system at a certain instant, $n=N$ depend on both the instant n_0 when the system was switched and the chosen instant $n=N$. Denoting the output coordinate of the system by Y we have

$$m_y[N]=M[Y[N]]=m_y[N,\ n_0]$$

$$D_y[N]=M[Y^2[N]]-m_y^2[N]=D_y[N,\ n_0]$$

i.e. the mathematical expectation and the dispersion are functions of the two variables N and n_0. We shall now proceed to formulate and solve this problem. The method given presupposes that the sampled data system is given by a difference equation. This restriction is not fundamental and the results obtained are easily extended to the case where the system is given by differential-difference equations.

Suppose the system in question is given by a single k-th-order difference equation of the form (for simplicity we consider lattice functions)

$$L_n(\Delta, n)\, y[n] = M_n(\Delta, n)\, x[n] \tag{10.1}$$

and that its state at the instant $n = n_0$ is characterised by the set of values

$$\Delta_n^i y[n]\,|_{n=n_0} = c_i[n]\,|_{n=n_0}, \quad i = 0, 1, \ldots, k-1 \tag{10.2}$$

We assume that $\{x[n]\} = X$ is a non-stationary random function with the mathematical expectation $m_x[n]$ and the dispersion $D_x[n]$.

As for c_i, $i=0, 1, \ldots, k-1$, we assume that they are random functions which are independent in pairs, i.e.

$$\begin{gathered} M[c_i c_j] = 0, \; i \neq j \\ i, j = 0, 1, \ldots, k-1 \end{gathered} \tag{10.3}$$

and independent of the random function X, i.e.

$$M[c_i X] = 0, \; i = 0, 1, \ldots, k-1 \tag{10.4}$$

The mathematical expectation and the dispersion of the random functions $c_i[n]$ will be denoted by $m_{c_i}[n]$ and $D_{c_i}[n]$ respectively.

Using the above conditions, let us determine the mathematical expectation $m_y[N, n_0]$ and the dispersion $D_y[N, n_0]$ of the output coordinate of the system given by (10.1), assuming them to be functions of n_0 for the fixed instant $n = N$.

When the disturbance X acts at the input of the system with the initial conditions (10.2), then a forced response due to the action of the disturbance X occurs at its output together with a free response (a transient process) due to the non-zero initial conditions (10.2). If the disturbance begins at the input of the system at the instant $n = n_0$, then the output coordinate Y at the instant $n = N$ is

$$Y[N] = Y_{f_o}[N] + Y_{f_r}[N] \tag{10.5}$$

where Y_{f_o} is the forced response of the system

$$Y_{f_0}[N] = \sum_{m=n_0}^{N} W[N;\, m]\, X[m] \tag{10.6}$$

($W[n;\ m]$ is the weighting function of the system given by (10.1)), and Y_{f_r} is the free response (transient process) of the system

$$Y_{f_r}[N] = \sum_{i=0}^{k-1} c_i[n_0]\, w_i[N] \tag{10.7}$$

where $w_i[n]$ are the output functions of the system in the particular case where the disturbance X is identically zero. whilst $c_i[n_0] = \sigma[n - n_0]$ when $c_r[n_0] \equiv 0$ $(r = 0, 1, \ldots, i-1, i+1, \ldots, k-1)$.

The functions $w_r[n]$ are determined by solving the homogeneous difference equation which corresponds to (10.1),

$$\sum_{r=0}^{k} a_r[n]\, \Delta^r w_i[n] = 0, \quad i = 0, 1, \ldots, k-1 \tag{10.8}$$

with the initial conditions

$$\Delta^r w_i[n] = \begin{cases} 0 & \text{when } i \neq r \\ \sigma[n - n_0] & \text{when } i = r \end{cases} \tag{10.8'}$$

In this way, the total value of the output variable of the system in question, when (10.5), (10.6) and (10.7) are taken into account, can be written in the following form:

$$Y[N] = \sum_{i=0}^{k-1} c_i[n_0]\, w_i[N] + \sum_{m=n_0}^{N} W[N; m]\, X[m] \tag{10.9}$$

Applying the operation of mathematical expectation to both sides of (10.9) we obtain

$$m_y[N, n_0] = \sum_{i=0}^{k-1} m_{c_i}[n_0]\, w_i[N] + \sum_{m=n_0}^{N} W[N; m]\, m_x[m] \tag{10.10}$$

Relation (10.10) enables us to determine the mathematical expectation $m_y[N, n_0]$ of the output coordinate of the system for a fixed value $n = N$ and the given value of the instant n_0 at which the system is switched on. As follows from (10.10), the mathematical expectation $m_y[N, n_0]$ is given by two components. The first component

$$\sum_{i=0}^{k-1} m_{c_i}[n_0]\, w_i[N]$$

is due to the values of $m_{c_i}[n_0]$ at the instant n_0 of switching on and constitutes the transient process of the system. The second component

$$\sum_{m=n_0}^{N} W[N; m]\, m_x[m]$$

is the forced response of the system to the mathematical expectation m_x of the random excitation that acts on the system, beginning from the instant $n=n_0$.

If the system being investigated is fairly complex, then analogue or digital computers may be used to determine the mathematical expectation $m_y[N, n_0]$. In so doing, if it is required to determine the dependence of the mathematical expectation on the instant n_0 of switching on for a fixed value of N, then (10.10) must be solved repeatedly for different values of n_0. The appropriate block diagram of the model and the algorithm for determining $m_y[N, n_0]$ by means of a digital computer are easily obtained from (10.10).

In a particular case where

$$M[c_i[n]] \equiv 0, \quad i=0, 1, \ldots, k-1, \quad 0 \leqslant n \leqslant N$$

the problem of determining the mathematical expectation reduces to the usual problem of determining the response of a non-stationary sampled data system with zero initial conditions to a non-random function. From (10.10) we have in this case

$$m_y[N, n_0] = \sum_{m=n_0}^{N} W[N; m]\, m_x[m] \qquad (10.11)$$

If at the instant of switching on the state of the system is specified by a set of non-random values of c_i, i.e.

$$M[c_i[n]] \equiv c_i = \text{const}, \quad i=0, 1, \ldots, k-1$$
$$n_0 \leqslant n \leqslant N$$

the mathematical expectation is determined by

$$m_y[N, n_0] = \sum_{i=0}^{k-1} c_i w_i[N] + \sum_{m=n_0}^{N} W[N; m] m_x[m] \qquad (10.12)$$

It follows from the foregoing discussion that, for a given instant of observation, (10.10), (10.11) and (10.12) enable us to determine the mathematical expectation only at a fixed value n_0. The relationship between the mathematical expectation and n_0 can be obtained by repeatedly solving these equations.

The method of adjoint systems enables us to find the mathematical expectation $m_y[N, n_0]$ as a function of n_0 by a single solution of the corresponding equations. We shall consider the use of this method to solve the problem formulated above. We shall proceed from (10.10).

We first point out that the functions w_i in (10.10) depend not only on the instant of observation, $n=N$, but also on the instant at which the system is switched on, n_0, as the system in question is non-stationary. Consequently, we must put in (10.10)

$$w_i = w_i[N, n_0], \quad i = 0, 1, \ldots, k-1 \qquad (10.13)$$

Thus, taking into account (10.13), we have

$$m_y[N, n_0] = \sum_{i=0}^{k-1} m_{c_i}[n_0] w_i[N, n_0] + \sum_{m=n_0}^{N} W[N; m] m_x[m] \qquad (10.14)$$

Let us introduce the new independent variable $l = N - m$ in (10.14) and put $l = N - n_0$. We then obtain

$$\begin{aligned} m_y[N, N-l] &= \sum_{i=0}^{k-1} m_{c_i}[N-l] w_i[N, N-l] \\ &+ \sum_{l=0}^{N-n_0} W[N; N-l] m_x[N-l] \end{aligned} \qquad (10.15)$$

Relation (10.15) can also be written in the form

$$m_y[l, N] = \sum_{i=0}^{k-1} m_{c_i}[N-l] w_i^*[l, N] + \sum_{l=0}^{N-n_0} W^*[l; N] m_x[N-l] \qquad (10.16)$$

where $W^*[l;N]$ is the weighting of a system which is adjoint to the original one;
$w_i^*[l,N]$ are functions of the independent variable l determined by solving the homogeneous difference equation

$$\sum_{r=0}^{k}(-1)^r\Delta_n^r\{a_i[n-r]w_i^*[n-r]\}=0 \qquad i=0,1,\ldots,k-1 \tag{10.17}$$

which is adjoint to (10.8), with the initial conditions

$$\Delta^r w_i^*[n]=\begin{cases}0 & \text{when } i\neq r\\ \sigma[n-n_0] & \text{when } i=r\end{cases} \tag{10.18}$$

Equation (10.16) gives the mathematical expectation m_y as a function of $n_0 (n_0 \leqslant N)$. Both analogue and digital computers can be used to determine the mathematical expectation. The block diagram of the model and the scheme of the algorithm can easily be obtained from (10.16).

In a particular case, when the initial conditions of the system are such that

$$M[c_i[n]]\equiv 0, \quad i=0,1,\ldots,k-1$$
$$n_0\leqslant n\leqslant N$$

then (10.16) assumes the form

$$m_y[l,N]=\sum_{l=0}^{N-n_0}W^*[l;N]\,m_x[N-l] \tag{10.19}$$

This expression defines the mathematical expectation of the output coordinate of the system as a function of the instants n_0 at which the system is switched on, the instant of observation, $n=N$, being fixed.

If at the instant of switching on the system is characterised by a set of non-random functions c_i, i.e.

$$M[c_i[n]]=c_i=\text{const}, \quad i=0,1,\ldots,k-1$$

$$n_0\leqslant n\leqslant N$$

then the mathematical expectation as a function of the instants of switching on can be found by

$$m_y[N, n_0] = \sum_{i=0}^{k-1} c_i w_i^*[l, N] + \sum_{l=0}^{N-n_0} W^*[l; N]\, m_x[N-l] \quad (10.20)$$

We shall now solve the problem of determining the dispersion at the output of the system described by (10.1). To do this we shall find the expression for the dispersion, assuming that at the instant of switching on the state of the system is characterised by a certain set of random initial conditions.

For centralised random functions, using (10.9) and taking into account (10.13), we can write

$$\overset{\circ}{Y}[N] = \sum_{i=0}^{k-1} c_i[n_0]\, w_i[N, n_0] + \sum_{m=n_0}^{N} W[N; m]\, \overset{\circ}{X}[m]$$

Then the dispersion $D_y[N, n_0]$ is determined as follows:

$$D_y[N, n_0] = M\left[\left\{\sum_{i=0}^{k-1} c_i[n_0]\, w_i[N, n_0] + \sum_{m=n_0}^{N} W[N; m]\, \overset{\circ}{X}[m]\right\}^2\right] \quad (10.21)$$

Carrying out certain transformations in (10.21) we find that

$$D_y[N, n_0] = \sum_{i=0}^{k-1} D_{c_i}[n_0] w_i^2[N, n_0] + \sum_{m_1=n_0}^{N} W[N; m_1] \sum_{m_2=n_0}^{N} W[N; m_2]\, K_x[m_2; m_1] \quad (10.22)$$

Expression (10.22) enables us to determine the dispersion $D_y[N, n_0]$ of the output coordinate of the system for a fixed value of the instant of observation, $n = N$, and for a given value of the instant n_0 of switching on. As follows from (10.22), the dispersion is given by two components. The first component

$$\sum_{i=0}^{k-1} D_{c_i}[n_0]\, w_i^2[N, n_0]$$

is due to the random initial conditions $c_i[n_0]$ with the dispersions $D_{c_i}[n_0]$ at the instant of switching on. The second component is the dispersion of the forced motion of the system, due to the action of the random disturbance X with the correlation function $K_x[n;\ m]$.

Analogue and digital computers can be used to determine the dispersion in accordance with (10.22). In so doing, if it is necessary to obtain the relationship between the dispersion and its instant n_0 of switching on for a fixed value of N, then (10.22) must be solved repeatedly for different values of n_0.

In a particular case where at the instant of switching on the state of the system is characterised by zero initial conditions

$$c_i[n] \equiv 0,\ i = 0, 1, \ldots, k-1$$

$$n_0 \leqslant n \leqslant N$$

the problem of determining the dispersion reduces to the problem considered in the preceding section, concerning the dispersion of a non-stationary system with zero initial conditions. Using (10.22), we obtain in this case

$$D_y[N; n_0] = \sum_{m_1=n_0}^{N} W[N; m_1] \sum_{m_2=n_0}^{N} W[N; m_2] K_x[m_2; m_1] \quad (10.23)$$

If at instant of switching on the state of the system is characterised by a set of non-random values of c_i, then the dispersion at the output of the system can also be found from (10.23).

It follows from the foregoing that, for a given instant of observation $n = N$, (10.22) and (10.23) enable the dispersion to be determined only for a fixed value of n_0. The relationship between the dispersion and n_0 can be obtained by repeatedly solving these equations.

The method of adjoint systems enables the dispersion $D_y[N, n_0]$ to be found as a function of n_0 by a single solution of the corresponding equation. We shall now give this method.

Let us consider (10.22), which is written in the form

$$D_y[N, n_0] = D_1[N, n_0] + D_2[N, n_0] \quad (10.24)$$

where

$$D_1[N, n_0] = \sum_{i=0}^{k-1} D_{c_i}[n_0]\, \varpi_i^2[N, n_0] \quad (10.25)$$

and

$$D_2[N, n_0] = \sum_{m_1=n_0}^{N} W[N; m_1] \sum_{m_2=n_0}^{N} W[N; m_2] K_x[m_2; m_1] \quad (10.26)$$

It follows from (10.26) that the component $D_2[N, n_0]$ of the dispersion is obtained as the result of twice summing the function

$$F(N, m_1, m_2) = W[N; m_1] W[N; m_2] K_x[m_2; m_1]$$

along the nodes of the square network with the sides $N_2 - n_0$. It has been shown in Section 3.3 that summation along the nodes of a square network can be replaced by summation along the nodes of a triangular network. Carrying out the necessary transformations in (10.26) we obtain the expression for the component $D_2[N, n_0]$ of the dispersion

$$D_2[N, n_0] = 2 \sum_{m_1=n_0}^{N} W[N; m_1] \sum_{m_2=m_1}^{N} W[N; m_2] \times K_x[m_2; m_1] - \sum_{m=n_0}^{N} W^2[N; m] K_x[m; m] \quad (10.27)$$

The inner summation in (10.27) is carried out with respect to the variable $m_2 \geqslant m_1$. Therefore we can assume that

$$K_x[m_1; m_2] = \begin{cases} 0 & \text{when } m_2 < m_1 \\ K_x[m_2; m_1] & \text{when } m_2 \geqslant m_1 \end{cases} \quad (10.28)$$

It follows from (10.28) that the correlation function $K_x[m_1; m_2]$, when $m_2 \geqslant m_1$ can be regarded as the weighting function of a sampled data system. Denoting this weighting function by $W_f[m_1; m_2]$ we have from (10.28)

$$W_f[m_2; m_1] = \begin{cases} 0 & \text{when } m_2 < m_1 \\ K_x[m_2; m_1] & \text{when } m_2 \geqslant m_1 \end{cases} \quad (10.29)$$

Using (10.29), we can write (10.27) as follows:

$$D_2[N, n_0] = 2 \sum_{m=n_0}^{N} W[N; m] H[N; m] - \sum_{m=n_0}^{N} W^2[N; m] K_x[m; m] \quad (10.30)$$

where

$$H[N;m[=\sum_{m_2=m_1}^{N} W[N;m_2]\,W_f\,[m_2;\,m_1] \qquad (10.31)$$

is the weighting function of a system formed by a sampled data system (with the weighting function $W_f[n;\ m]$) and the system under consideration (with the weighting function $W[n;m]$) connected in series.

Thus the final expression for the dispersion at the output of the system can be written

$$D_y[N,n_0]=\sum_{i=0}^{k-1} D_{c_i}[n_0]\,w_i^2[N,n_0]$$
$$+2\sum_{m=n_0}^{N} W[N;m]\,H[N;m]-\sum_{m=n_0}^{N} W^2[N;m]\,K_x[m;m] \qquad (10.32)$$

Since $K_x[m;m]$ in (10.32) gives the dispersion $D_x[m]$, we can also write

$$D_y[N,n_0]=\sum_{i=0}^{k-1} D_{c_i}[n_0]\,w_i^2[N,n_0]$$
$$+2\sum_{m=n_0}^{N} W[N;m]\,H[N;m] \qquad (10.33)$$
$$-\sum^{N} W^2[N;m]\,D_x[m]$$

We now introduce the new variable $m=N-l$ in (10.33) and put $l=N-n_0$. We then have

$$D_y[N,\ N-l]=\sum_{i=0}^{k-1} D_{c_i}[N-l]\,w_i^2[N,N-l]$$
$$+2\sum_{l=0}^{N-n_0} W[N;N-l]\,H[N;N-l] \qquad (10.34)$$
$$-\sum_{l=0}^{N-n_0} W^2[N;N-l]\,D_x[N-l]$$

This relation can also be written in the form

$$D_y[l,N]=\sum_{i=0}^{k-1} D_{c_i}[N-l]\,w_i^{*2}[l,N]$$
$$+2\sum_{l=0}^{N-n_0} W^*[l,N]\,H^*[l;N] \qquad (10.35)$$
$$-\sum_{l=0}^{N-n_0} W^{*2}[l;N]\,D_x[N-l]$$

where $W^*[l;N]$ is the weighting function of a system which is adjoint to the one being considered;

$H^*[l;N]$ is the weighting function of a function which is adjoint to the system formed by a filter with weighting function $W_f[n;m]$ and the system being investigated (with weighting function $W[n;m]$) connected in series;

$w_i^*[l;N]$ are functions determined by solving (10.17) with the initial conditions (10.18).

Expression (10.35) gives the dispersion $D_y[l,N]$ as a function of the instants $n_0 (n_0 \leqslant N)$ at which the system is switched on, when the value of the instant of observation, $n=N$, is fixed. Analogue and digital computers can be used to determine the dispersion in accordance with (10.35).

If at the instant of switching on the state of the system is characterised by zero initial conditions, then the dispersion at the output of the system can be obtained from

$$D_y[l,N]=2\sum_{l=0}^{N-n_0} W^*[l;N]\,H^*[l;N]$$
$$-\sum_{l=0}^{N-n_0} W^{*2}[l;N]\,D_x[M-l] \qquad (10.36)$$

Expression (10.36) gives the dispersion also in the case where the state of the system at the instant of switching on is characterised by non-random initial conditions.

Thus the problem has been solved above for determining the mathematical expectation and the dispersion at the output of a non-stationary sampled data system with random initial conditions. The method given enables us to determine, with the required accuracy, the characteristics of the system as functions of the instants it is switched on.

A similar method for the analysis of continuous systems has been developed by Krut'ko [27].

3.11 DISPERSION OF THE OUTPUT COORDINATE AS A FUNCTION OF TIME

When assessing the quality of operation of sampled data systems we must sometimes determine the dispersion of the system coordinate in question as it varies continuously with time for the entire duration of the control process. If we use the method given in Section 3.9, based on the use of discrete shaping filters and adjoint systems, the dispersion $D_y[n, \varepsilon]$ of the output coordinate Y of the system as a function of the time $\bar{t}=n+\varepsilon$ can be obtained by repeatedly solving (by an analogue or digital computer) the corresponding adjoint equations of the system for the given instants of time $n=0,1,\ldots$ and $0\leqslant\varepsilon\leqslant1$. It is clear that this method is laborious, and it is of interest to develop a method enabling the dispersion to be found directly as a function of the time of observation $\bar{t}=n+\varepsilon$, by means of a single solution of the corresponding equations on an analogue or digital computer. We shall give this method for the case where a sampled data system is given by the corresponding difference equations.

Let a sampled data system be given by a single k-th order difference equation of the form

$$L_n(\Delta, n)\,y[n, \varepsilon]=M_n(\Delta, \bar{t})\,x[n, 0], \quad 0\leqslant\varepsilon\leqslant1 \tag{11.1}$$

Proceeding from (11.1), we have to find a difference equation whose solution is given by the dispersion $D_y[n, \varepsilon]$ of the output process Y of the system, when a random disturbance X, which in the general case is non-stationary, acts at its input.

Before solving this problem in the general form, let us consider the simplest case, where the system is described by a first-order equation, whilst the input distrubance is white noise.

Accordingly, let us find the difference equation whose solution is given by the dispersion D_y of the output process Y of the system described by the equation

$$\Delta y[n]+a_1[n]\,y[n]=v[n] \tag{11.2}$$

where v is discrete white noise with a correlation function of the form (considering, for simplicity, lattice functions)

$$K_v[n-m]=D_v\sigma[n-m], \quad D_v=\text{const} \tag{11.3}$$

Determining the first difference of the square of the output variable of the system in question, we find

$$\Delta\{y^2[n]\}=\Delta y[n]\{2y[n]+\Delta y[n]\} \qquad (11.4)$$

Taking into account that by (11.2)

$$\Delta y[n]=v[n]-a_1[n]y[n] \qquad (11.5)$$

we transform equation (11.4) into the form

$$\begin{aligned}\Delta\{y^2[n]\}=v^2[n]+2(1-a_1[n])v[n]y[n]\\ -a_1[n](2-a_1[n])y^2[n]\end{aligned} \qquad (11.6)$$

or

$$\begin{aligned}\Delta\{y^2[n]\}+a_1[n](2-a_1[n])y^2[n]\\ =2(1-a_1[n])v[n]y[n]+v^2[n]\end{aligned} \qquad (11.7)$$

Applying the operation of mathematical expectation to both sides of (11.7) we find

$$\begin{aligned}\Delta D_y[n]+a_1[n](2-a_1[n])D_y[n]=M[v^2[n]]\\ +2(1-a_1[n])M[v[n]y[n]]\end{aligned} \qquad (11.8)$$

where

$$D_y[n]=M[y^2[n]]$$

When (11.3) is taken into account we have

$$M[v^2[n]]=D_v=\text{const}$$

where D_v is the intensity of discrete white noise.

To calculate $M[v[n]y[n]]$ we use the relation

$$y[n]=\sum_{m=-\infty}^{n} W[n;m]v[m] \qquad (11.9)$$

where $W[n;\ m]$ is the weighting function of the sampled data system in question.

Multiplying both sides of (11.9) by $v[n]$ and applying then the operation of mathematical expectation to both sides of the equation thus obtained, we find

$$M[v[n]y[n]]=\sum_{m=-\infty}^{n} W[n;m]\,M[v[n]v[m]]$$

Since for discrete white noise

$$M[v[n]v[m]]=D_v\sigma[n-m]$$

the last relation assumes the form

$$\begin{aligned}M[v[n]y[n]]&=\sum_{m=-\infty}^{n} W[n;m]\,D_v\sigma[n-m]\\&=D_vW[n;m]\big|_{m=n}=0\end{aligned} \tag{11.10}$$

due to the fact that in the case considered $W[m;m]=0$, which follows directly from (11.2).

Thus the difference equation (11.8) for the dispersion can finally be written in the following form:

$$\Delta D_y[n]+a_1[n](2-a_1[n])D_y[n]=D_v \tag{11.11}$$

The solution of this equation can be written in the form of a recursive relation

$$D_y[n]=D_v+\{1-a_1[n-1](2-a_1[n-1])\}D_y[n-1] \tag{11.12}$$

which, for a given initial value of $D_y[0]$, enables the values of the dispersion to be succesively calculated for all discrete instants of time over the interval $[0;n]$.

Let us now generalise this procedure for the case where the system is described by a k-th order difference equation of the form (11.1), and the action X is a random process with the correlation function $K_x[n;\ m]$. As before we shall consider lattice functions. The results are easily extended to the case $\varepsilon\neq 0$.

We assume that the random process X with the correlation function $K_x[n;m]$ can be obtained from the discrete white noise V with unit intensity by means of a discrete filter, so that

$$\Phi_n(\Delta,n)\,x[n]=\Psi_n(\Delta,n)\,v[n] \tag{11.13}$$

where the difference operator $\Phi_n(\Delta, n)$ has the l-th order.

Using the rules given in Chapter 1 for operating on non-stationary difference operators, we find the $(k+l)$-th-order difference equation:

$$S_n(\Delta, n)\, y[n] = G_n(\Delta, n)\, v[n] \tag{11.14}$$

which describes the processes in the equivalent sampled data system when discrete white noise acts at its input. Thus, for a sampled data system given by the $(h=k+l)$-th-order difference equation (11.14) we must find a difference equation whose solution is given by the dispersion $D_y[n]$ of the output variable Y, when the white noise V with unit intensity acts at the input of this system.

It has been shown in Chapter 1 that an h-th order equation of the form (11.14) can be reduced to a system of h first-order difference equations:

$$\begin{aligned} y[n] &= y_1[n] + F_0[n]\, v[n] \\ \Delta y_1[n] &= y_2[n] + F_1[n]\, v[n] \\ &\cdots\cdots\cdots\cdots \\ \Delta y_{h-1}[n] &= y_h[n] + F_{h-1}[n]\, v[n] \\ \Delta y_h[n] &= -a_{h-1}[n]\, y_h[n] - \ldots - a_0[n]\, y_1[n] \\ &\quad + F_h[n]\, v[n] \end{aligned} \tag{11.15}$$

where a_i are the coefficients of the operator $S_n(\Delta, n)$, whilst the functions $F_i[n]$ are determined by the corresponding system of equations.

A sampled data system described by the difference equations (11.15) may be regarded as a multi-dimensional system consisting of h holding circuits connected with each other (sampled data follow-up systems with a single integrator, with $\gamma=1$).

The connection between the output variable $y_i[n]$ of the i-th circuit and the input actions x_j ($j=1, \ldots, h$) is established, as was shown in Chapter 1, by the following relation:

$$y_i[n] = \sum_{j=1}^{h} \sum_{m=-\infty}^{n} W_{ij}[n; m]\, x_j[m] \tag{11.16}$$

Here $W_{ij}[n; m]$ is the weighting function representing the response at the instant $\bar{t}=n$ at the output of the i-th circuit when the σ-function is fed to the input of the j-th circuit at the instant $\bar{t}=m$, with the condition that the input actions of all the remaining circuits are zero, i.e. with the condition that

$$x_\lambda \equiv 0,\ \lambda = 1, 2, \ldots, j-1,\ j+1, \ldots, h$$

It was established in Section 1.9 that the weighting functions $W_{ij}[n;\ m]$, when $n=m+1$, form the square unit matrix

$$W = \begin{Vmatrix} 1 & 0 & 0 & \ldots & 0 \\ 0 & 1 & 0 & \ldots & 0 \\ \cdot & \cdot & \cdot & \cdot & \cdot \\ 0 & 0 & 0 & \ldots & 1 \end{Vmatrix}$$

i.e. they satisfy the relations

$$\begin{gathered} W_{ij}[m+1,\ m] = \begin{cases} 0, & i \neq j \\ 1, & i = j \end{cases} \\ W_{ij}[m;\ m] = 0,\ i,\ j = 1, 2, \ldots, h \end{gathered} \tag{11.17}$$

We shall now solve the problem of forming the difference equation for the dispersion of the output variable of the system given by (11.14). For this we set up the system of equations

$$\begin{gathered} \left.\begin{gathered} \Delta[y_i y_j] = y_i \Delta y_j + \Delta y_i (y_j + \Delta y_j) \\ i, j = 1, 2, \ldots, h-1 \end{gathered}\right\} \\ \Delta[y_i y_h] = y_i \Delta y_h + \Delta y_i (y_h + \Delta y_h) \\ i = 1, 2, \ldots, h-1 \\ \Delta[y_h^2] = y_h \Delta y_h + \Delta y_h (y_h + \Delta y_h) \end{gathered} \tag{11.18}$$

Substituting the values Δy_i from (11.15) into (11.18), we find

$$\begin{gathered} \Delta[y_i y_j] = y_i (y_{j+1} + F_j v) + y_j (y_{i+1} + F_i v) \\ + (y_{i+1} + F_i v)(y_{j+1} + F_j v);\ i, j = 1, 2, \ldots, h-1; \end{gathered} \tag{11.19}$$

$$\Delta[y_i y_h] = y_i\left(F_h v - \sum_{l=1}^{h} a_{h-l} y_{h+1-l}\right) + (y_{i+1} + F_i v)\, y_h + (y_{i+1} + F_i v) \times\left(F_h v - \sum_{l=1}^{h} a_{h-l} y_{h+1-l}\right); \quad i = 1, 2, \ldots, h-1 \tag{11.19}$$

$$\Delta[y_h^2] = 2y_h\left(F_h v - \sum_{l=1}^{h} a_{h-l} y_{h+1-l}\right) + \left(F_h v - \sum_{l=1}^{h} a_{h-l} y_{h+1-l}\right)^2$$

We shall apply the operation of mathematical expectation to both sides of (11.9). We obtain

$$M[y_i y_j] = D_{ij} \qquad M[F_i F_j v^2[n]] = F_i F_j \tag{11.20}$$

where $D_{i,j}$ is the correlation moment of the product of the random functions Y_i and Y_j.

We now calculate $M[v[n] y_i[n]]$. For this we use (11.16). It is obvious that in this case we obtain

$$y_i[n] = \sum_{j=1}^{h} \sum_{m=-\infty}^{n} W_{ij}[n; m] F_j[m] v[m] \qquad i = 1, 2, \ldots, h \tag{11.21}$$

Multiplying both sides of (11.21) by $v[n]$ and applying the operation of mathematical expectation to both sides of the equation thus obtained, we find

$$M[v[n] y_i[n]] = \sum_{j=1}^{h} \sum_{m=-\infty}^{n} W_{ij}[n; m] F_j[m] M[v[n] v[m]] \tag{11.22}$$

Since V is white noise with unit intensity, then

$$M[v[n] v[m]] = \sigma[n - m]$$

and (11.22) assumes the form

$$M[v[n]\,y_i[n]] = \sum_{j=1}^{h} \sum_{m=-\infty}^{n} W_{ij}[n;m]\,F_j[m]\,\sigma[n-m]$$
$$= \sum_{j=1}^{h} W_{ij}[n;n]\,F_j[n] \tag{11.23}$$

Using (11.17), we obtain from (11.23)

$$M[v[n]\,y_i[n]] = 0 \tag{11.24}$$

If (11.20) and (11.24) are used, (11.19) can be written as follows:

$$\Delta D_{ij} = F_i F_j + D_{i,j+1} + D_{j,i+1} + D_{i+1,j+1}$$
$$i, j = 1, 2, \ldots, h-1$$
$$\Delta D_{i,h} = D_{i+1,h} - \sum_{l=1}^{h} a_{h-l} D_{i,h+1-l}$$
$$- \sum_{l=1}^{h} a_{h-l} D_{i+1,h+1-l} \tag{11.25}$$
$$i = 1, 2, \ldots, h-1$$
$$\Delta D_{h,h} = \sum_{l=1}^{h} \sum_{m=1}^{h} a_{h-l} a_{h-m} D_{h+1-l,h+1-m}$$
$$- 2 \sum_{l=1}^{h} a_{h-l} D_{h,h+l-1}$$

Since $D_{i,j} = D_{j,i}$, then (11.25) yields $\frac{h(h+1)}{2}$ first-order equations; solving these with the initial conditions $D_{i,j}[0]$ we determine the correlation moments $D_{i,j}[n]$.

The dispersion of the process Y is obtained, if we use the first equation of (11.15)

$$D_y[n] = M[(y_1 + F_0 v)^2] = D_{1,1}[n]$$
$$+ 2F_0[n]\,M[y_1 v] + F_0^2\,\sigma[n]$$

But since according to (11.24)

$$M[y_1[n]\,v[n]]=0$$

we finally have

$$D_y[n]=D_{1,1}[n]+F_0^2[n]\,\sigma[n] \tag{11.26}$$

where by definition

$$\sigma[n]=\begin{cases}1 \text{ when } n=0\\ 0 \text{ when } n\neq 0\end{cases}$$

Thus, if the above procedure is followed, determination of the dispersion as a function of time of the output process of a sampled data system reduces to the following.

1. Determine the discrete shaping filter for the given random disturbance of the system.
2. Find the difference equation of the equivalent system.
3. Replace the difference equation thus obtained by a system of first-order equations of the form (11.15), and then set up a system of equations of the form (11.25); the correlation moments $D_{i,j}[n]$ of the output coordinates are determined by solving this system.

The dispersion of the output coordinate is then determined from (11.26).

Since (11.16) was used to determine the correlation moments $M[v[n[y_i[n]]$, the results obtained are also true for the case where the system in question is at rest up to the instant at which the disturbance X is applied.

Using similar methods we can write the corresponding equations for the dispersion in the case where the system is in the disturbed state, characterised by non-random and random values of the state parameters.

The above method is convenient in those cases where the system in question is described by low-order difference equations.

A similar method, applicable to continuous systems, has been developed by Pugachev [9]. Using this method we can easily obtain the corresponding equations for the dispersion in the case where the sampled data system is given by the differential equations of its continuous part together with the form of its sampler, i.e. when it is given by differential-difference equations.

Chapter 4

STATISTICAL ANALYSIS OF NON-LINEAR SYSTEMS

The analysis of non-linear sampled data systems, like the analysis of continuous non-linear systems, is exceptionally complex, since the principle of superposition is not applicable to sampled data systems described by non-linear difference equations. At the present time there is a number of methods for the statistical analysis of non-linear continuous systems, enabling the probability characteristics to be determined for the output coordinates of a system subjected to random disturbances. Included among these are a group of approximate methods based on the use of correlation theory and an exact method based on the theory of Markov process.

This chapter discusses the most effective methods of analysis which are applicable to non-linear sampled data systems. Of the approximate methods we present the method of direct linearisation of non-linear relations and the method of statistical linearisation worked out by Kazakov [36] for continuous systems. The possibility of using the method of statistical linearisation for the analysis of sampled data systems containing non-linear elements was shown by Pyatnitskii [39]. The method of statistical linearisation was first used by Tsypkin [40] to evaluate the accuracy of quantisation in digital systems.

Besides the direct development of the method of statistical linearisation, this chapter deals with the generalisation of this method, from which the method of statistical linearisation arises as a first approximation. This more general method is based on the correlation function of the random process at the output of a non-linear element. All the methods given can be used in the statistical investigation of non-linear sampled data systems. The choice of a particular method must depend on the nature of the problem and computational facilities available.

In the analysis of non-linear sampled data systems there is no need to consider their physical nature; it is sufficient to consider the equations describing the processes in these systems, regardless of whether they are digital computers with 'non-linear' programmes, or automatic control systems consisting of linear and non-linear elements. But since in many practical problems the non-linear sampled data systems are ordinary sampled data control systems whose loops include non-linear elements, we shall, to make the problem more concrete, consider the block diagrams of a non-linear sampled data system. We assume that a non-linear system is formed by including a sampler of the first type in the loop of a continuous non-linear system. In this case, as was shown by Tsypkin [41], the order in which the sampler is included (before or after the non-linear element) has no effect on the processes in the system.

4.1 EFFECT OF DIRECT LINEARISATION OF NON-LINEAR RELATIONSHIPS

Direct linearisation of non-linear relationships is one of the simplest and crudest methods of analysis of non-linear systems. It does not admit any refinements, and has comparatively narrow limits of application, but does have the advantage of simplicity, and it can be used to obtain preliminary approximate assessments of the quality of operation of non-linear sampled data systems (see Pugachev's book on the direct linearisation method for the approximate study of the accuracy of continuous non-linear systems [49]).

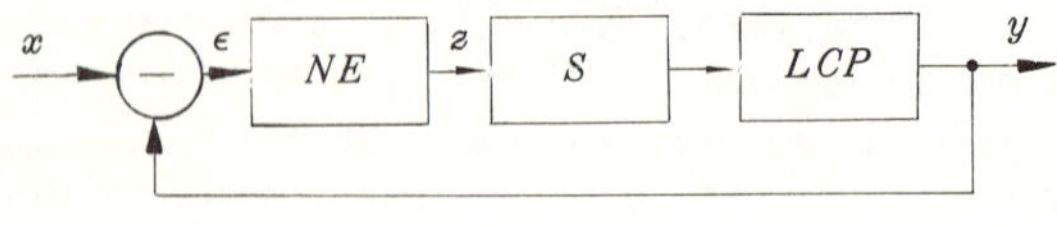

Fig. 4.1 A non-linear system

4.1.1 Physical basis of the method

Let us consider the non-linear sampled data follow-up system shown in Fig. 4.1. Suppose the non-linear element included in the loop has the characteristic $g(\mathcal{E})$ (Fig. 4.2) which admits the existence of the derivative $\frac{dg(\mathcal{E})}{d\mathcal{E}}$ over a certain range of values of $\mathcal{E}$. Then, if this system is fairly

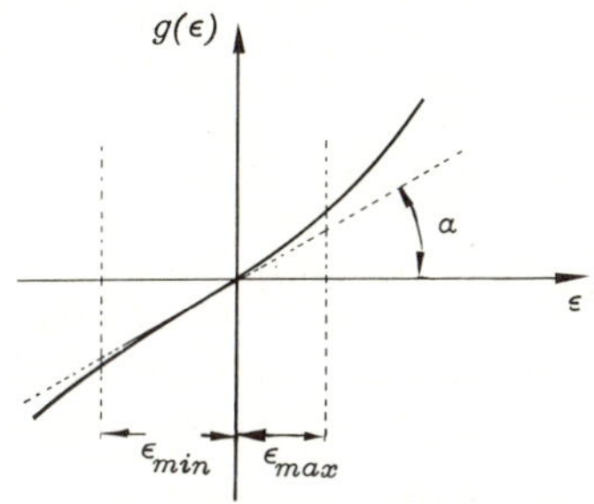

Fig. 4.2 Determination of the gain factor of a linear element

quick-acting, the error $\mathcal{E}$ of the system is small, so that between the limits $\mathcal{E}_{min} \leqslant \mathcal{E} \leqslant \mathcal{E}_{max}$ the non-linear element with characteristic $g(\mathcal{E})$ can be replaced by a linear inertia-free amplifying element with gain factor

$$k = \left.\frac{dg(\mathcal{E})}{d\mathcal{E}}\right|_{\mathcal{E}=0} = \tan\alpha \tag{1.1}$$

Consequently, replacing the non-linear element by a linear element with transfer coefficient (1.1) reduces the non-linear system to a linear one to the first approximation.

It is clear from the foregoing that direct linearisation demands that the differentiability condition of the non-linear relation $g(\mathcal{E})$ is satisfied. The validity of the results is determined not only by the form of the function $g(\mathcal{E})$, but by the system itself, by its quick-acting property.

We shall now discuss the application of direct linearisation to the determination of the mathematical expectation and the correlation function of the output coordinates.

4.1.2 Linearised equations of a non-linear system

We shall consider a non-linear sampled data system with l inputs, as shown in Fig. 4.3. Let the non-linear element (NE) be given by its characteristic $z=g(\mathcal{E})$.

Using the method given in Chapter 1 we can reduce the

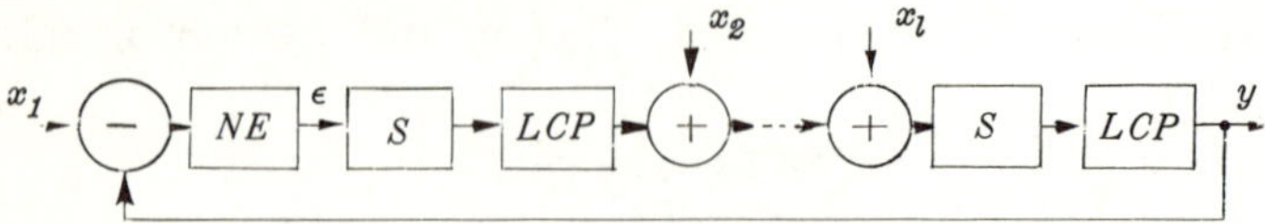

Fig. 4.3 A multi-dimensional linear system

k-th order equation of the sampled data system in question to a system of k first-order equations of normal form. Considering for the sake of simplicity lattice functions, we write this system of equations in the form

$$\Delta Y_i[n] = F_i(n, Y_1, Y_2, \ldots, Y_k, X_1, X_2, \ldots, X_l) \qquad (1.2)$$
$$i = 1, 2, \ldots, k$$

where Y_i are the output variables (coordinates) of the system under consideration;

X_j, $j = 1, 2, \ldots, l$, are random input disturbances, constituting real random functions of time;

F_i are non-linear functions determined by the form of the function $g(\mathcal{E})$.

The functions F_i are assumed to possess the derivatives

$$\frac{\partial F_i}{\partial Y_j}, \ \frac{\partial F_i}{\partial X_\alpha}, \ i, j = 1, 2, \ldots, k, \ \alpha = 1, 2, \ldots, l$$

We assume that the random disturbances X_j have the mathematical expectations m_{x_j}, so that

$$X_j = m_{x_j} + \overset{\circ}{X}_j, \ j = 1, 2, \ldots, l \qquad (1.3)$$

where $\overset{\circ}{X}_i$ are centralised random functions.

To determine the linearised equations of the system we expand the non-linear functions F_i in a Taylor series near the mean values of the arguments and limit ourselves in this expansion to the linear terms.

Representing the output coordinates Y_i of the system in

the form

$$Y_i = m_{y_i} + \overset{\circ}{Y}_i, \ i = 1, 2, \ldots, k \tag{1.4}$$

we can choose the mathematical expectations m_{x_j} and m_{y_i} for the mean values of X_j and Y_i, and the expansion of the functions F_i can be carried out for the centralised functions $\overset{\circ}{X}_j$ and $\overset{\circ}{Y}_i$. In this case we have

$$\begin{aligned} &F_i(n, Y_1, Y_2, \ldots, Y_k; \ X_1, X_2, \ldots, X_l) \\ &\approx F_i(n, m_{y_1}, m_{y_2}, \ldots, m_{y_k}, m_{x_1}, m_{x_2}, \ldots, m_{x_l}) \\ &+\sum_{h=1}^{k} \left[\frac{\partial F_i}{\partial m_{y_h}}\right]_0 \overset{\circ}{Y}_h + \sum_{\lambda=1}^{l} \left[\frac{\partial F_i}{\partial m_{x\lambda}}\right]_0 \overset{\circ}{X}_\lambda, \ i = 1, 2, \ldots, k \end{aligned} \tag{1.5}$$

where

$$\left[\frac{\partial F_i}{\partial m_{x\lambda}}\right]_0 = \frac{\partial F_i(n, m_{y_1}, m_{y_2}, \ldots m_{y_k}, m_{x_1}, m_{x_2}, \ldots, m_{xl})}{\partial m_{x\lambda}} \Bigg|_{\substack{\overset{\circ}{Y}_i = 0 \\ \overset{\circ}{X}_j = 0}}$$

$$i = 1, 2, \ldots, k; \ j = 1, 2, \ldots, l$$

Substituting (1.5) into (1.2) we obtain

$$\begin{aligned} &\Delta m_{y_i}[n] + \Delta \overset{\circ}{Y}_i[n] \\ &= F_i(n, m_{y_1}, m_{y_2}, \ldots, m_{y_k}, m_{x_1}, m_{x_2}, \ldots, m_{x_l}) \\ &+\sum_{h=1}^{k} \left[\frac{\partial F_i}{\partial m_{y_h}}\right]_0 \overset{\circ}{Y}_h[n] + \sum_{\lambda=1}^{l} \left[\frac{\partial F_i}{\partial m_{x_\lambda}}\right]_0 \overset{\circ}{X}_\lambda[n], \ i = 1, 2, \ldots, k \end{aligned} \tag{1.6}$$

Let us apply the operation of mathematical expectation to both sides of (1.6). As the result we obtain

$$\begin{gathered} \Delta m_{y_i}[n] = F_i(n, m_{y_1}, m_{y_2}, \ldots, m_{y_k}, m_{x_1}, m_{x_2}, \ldots, m_{x_l}) \\ i = 1, 2, \ldots, k \end{gathered} \tag{1.7}$$

The system (1.7) so obtained gives the mathematical expectations m_{y_i} of the output coordinates of the non-linear system in terms of the mathematical expectations m_{x_j} of the input disturbances. This system is obtained from the original system if we replace the random functions Y_i and X_j by the mathematical expectations m_{y_i} and m_{x_j} respectively.

It follows from (1.7) that the required mathematical expectations represent the response of the system to the mathematical expectations of the input disturbances.

The difference equations for the centralised random functions $\overset{\circ}{Y}_i$ can be obtained if the equations of (1.7) are subtracted from the corresponding equations of (1.6). Carrying out this subtraction term by term, we find

$$\Delta \overset{\circ}{Y}_i[n] = \sum_{\lambda=1}^{l} \left[\frac{\partial F_i}{\partial m_{x\lambda}}\right]_0 \overset{\circ}{X}_\lambda[n] + \sum_{h=1}^{k} \left[\frac{\partial F_i}{\partial m_{y_h}}\right]_0 \overset{\circ}{Y}_h[n], \; i = 1, 2, \ldots, k \tag{1.8}$$

The expression (1.8) is a system of linear first-order difference equations with variable coefficients, which enables us to determine the correlation functions of the output coordinates of the system by the methods of the linear theory.

For a non-linear system with a single input, (1.7) and (1.8) assume the form

$$\Delta m_{y_i}[n] = F_i(n, m_{y_1}, m_{y_2}, \ldots, m_{y_k}, m_x)$$

$$\Delta \overset{\circ}{Y}_i[n] = \sum_{h=1}^{k} a_{ih}[n] \overset{\circ}{Y}_h[n] + b_i[n] \overset{\circ}{X}[n], \; i = 1, 2, \ldots, k \tag{1.9}$$

where in accordance with (1.8)

$$a_{ih}[n] = \frac{\partial F_i(n, m_{y_1}, m_{y_2}, \ldots, m_{y_k}, m_x)}{\partial m_{y_h}} \tag{1.10}$$

$$b_i[n] = \frac{\partial F_i(n, m_{y_1}, m_{y_2}, \ldots, m_{y_k}, m_x)}{\partial m_x} \tag{1.11}$$

$$i, h = 1, 2, \ldots, k$$

Thus the determination of the accuracy of a non-linear sampled data system by the method of direct linearisation reduces to the determination of the accuracy of a linear system (to a first approximation) with respect to the centralised random functions, and that of a non-linear system with respect to the mathematical expectation.

At the same time it should be borne in mind that if the initial values of the state parameters of the system are non-random, then the equations for the mathematical expectation should be solved with non-zero initial conditions, while the equations connecting the centralised random functions should be solved with zero initial conditions.

If the initial state of the system is characterised by a set of random values of the state parameters, however, then all equations must be solved with non-zero conditions.

To solve (1.7) and (1.8) (or (1.9) in the case where the system has a single input) we can use both analogue and digital computers.

It should be pointed out that the method of direct linearisation allows the mathematical expectation and the correlation functions of the output coordinates of the system to be determined with acceptable accuracy only in those cases where the terms discarded, after the functions F_i have been expanded in a Taylor series, are small. Consequently, the results can be approximately evaluated by determining the squared term of the expansion. In doing so we can take the distribution law of the quantities Y_i as normal.

As already pointed out, the method of direct linearisation is applicable only in those cases where the characteristics of the non-linear elements are differentiable with respect to the output variable. In practical problems, however, the loop of the non-linear sampled data system usually contains non-linear elements whose characteristics represent either piecewise-continuous or even discontinuous functions. In these cases it is not possible to use the method of direct linearisation. This seriously limits the use of the method.

The direct substitution of a linear element by an inertia-free amplifier with the corresponding gain factor in many cases does not enable us to investigate the non-linear system with the accuracy which is required in practice. In these cases the validity of the results can be increased considerably if the non-linear element is replaced by a linear element which is equivalent to the non-linear element in a probabilistic sense. Such linearisation of non-linear elements, called statistical linearisation, has been worked out by Kazakov [36, 37]. Less general results have been obtained by Booton [43].

4.2 STATISTICAL LINEARISATION OF NON-LINEAR RELATIONSHIPS

4.2.1 Foundations of the method

The method of statistical linearisation is based on replacing the essentially non-linear element, which transforms the random function, by a linear element which is equivalent to it in a certain sense. Let the non-linear element be given by its characteristic

$$Y = g(X) \tag{2.1}$$

where Y is the output and X the input of the non-linear element. Let us assume that X is a scalar random function of a discrete argument

$$X[n] = m_x[n] + \overset{\circ}{X}[n] \tag{2.2}$$

In this case the output quantity of the non-linear element will also be a random function. We represent this function also as the sum of its mathematical expectation $m_y[n]$ and the centralised random function $\overset{\circ}{Y}[n]$:

$$Y[n] = m_y[n] + \overset{\circ}{Y}[n] \tag{2.3}$$

The main feature of statistical linearisation is that the non-linear element with the characteristic (2.1) is replaced by an inertia-free linear element, so that the random function $Y^*[n]$, being an approximation of the true random function $Y[n]$, appears at its output. We write $Y^*[n]$ in the following form:

$$Y^*[n] = \varphi_0[n] + k_1[n]\,\overset{\circ}{X}[n] \tag{2.4}$$

where φ_0 and k_1 are called the coefficients of statistical linearisation. These coefficients characterise the inertia-free element replacing the non-linear element, and determine the degree of approximation of $Y[n]$ by $Y^*[n]$.

To determine the coefficients of statistical linearisation φ_0 and k_1 we must have two equations connecting the characteristics of $Y[n]$ and $Y^*[n]$. Usually it is stipulated that the mathematical expectation and the dispersion of $Y^*[n]$ be equal respectively to the mathematical expectation and the dispersion of $Y[n]$, i.e.

$$M[Y^*[n]] = M[Y[n]] \tag{2.5}$$

$$M[\overset{\circ}{Y}{}^{*2}[n]] = M[\overset{\circ}{Y}{}^2[n]] \tag{2.6}$$

The conditions (2.5) and (2.6) imposed on $Y^*[n]$ are the main feature of the first method of approximation of the true random function $Y[n]$.

We shall determine the expression for $\varphi_0[n]$ and $k_1[n]$, proceeding from the conditions (2.5) and (2.6). Substituting (2.3) and (2.4) into (2.5), we have

$$\varphi_0[n] = m_y[n] \tag{2.7}$$

whence it follows that φ_0 is the transfer coefficient of the linear element in terms of the mathematical expectation. Thus if the linear element which replaces the non-linear element is to produce the mathematical expectation of the true random function $Y[n]$ exactly, its gain factor must be chosen to equal the mathematical expectation $m_y[n]$.

To determine the coefficient k_1 we substitute the values of the centralised random functions $\overset{\circ}{Y}$ and $\overset{\circ}{Y}{}^*$ into (2.6). We then have

$$M[\{k_1[n]\overset{\circ}{X}[n]\}^2] = M[\overset{\circ}{Y}{}^2[n]]$$

or

$$k_1^2[n]\,D_x[n] = D_y[n]$$

where D_x and D_y are the dispersions of the functions X and Y respectively. From the last equation we obtain

$$k_1[n] = \pm\sqrt{\frac{D_y[n]}{D_x[n]}} \tag{2.8}$$

It follows from (2.8) that the coefficient of statistical linearisation k_1 characterises the transfer of the random term of the random function X by means of the linear element.

We point out that the sign of the coefficient k_1 is chosen such that the signs of X and Y coincide. In practice this condition is satisfied if in (2.8) we choose the sign which coincides with the sign of the derivative of the function $g(x)$, i.e.

$$\operatorname{sign} k_1 = \operatorname{sign}\frac{dg(x)}{dx}$$

For simplicity we shall subsequently write the plus sign in (2.8).

Thus if the linear element is to produce the dispersion of the true random function Y exactly, we must choose its transfer coefficient for the random term in accordance with (2.8).

Thus the choice of the coefficients φ_0 and k_1 to accord

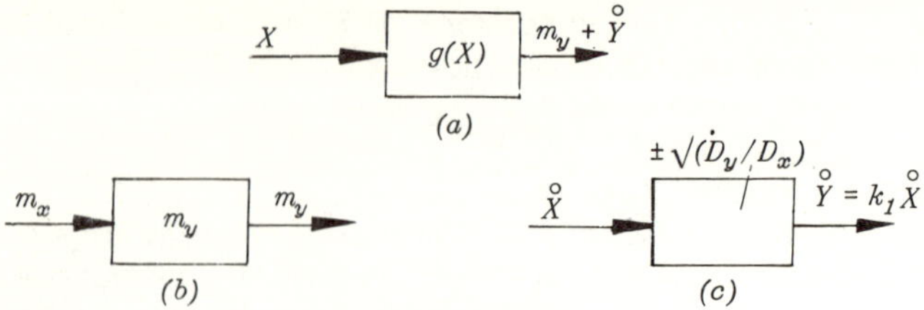

Fig. 4.4 Determination of the gain factor of equivalent linear elements

with Equations (2.7) and (2.8) means the replacement of the non-linear element with the characteristic $g(x)$ by two linear inertia-free elements for the mathematical expectation and the random term (Fig. 4.4). Note that the transfer coefficients of the linear elements are functions of time if X is non-stationary. In the case where X is a stationary random function m_x = const, D_x = const) the coefficients φ_0 and k_1 are constant.

Since the probability characteristics of the random function Y (its mathematical expectation m_y and dispersion D_y) at the output of the non-linear element are determined by the form of the non-linear transformation of X and its probability characteristics, the coefficients of statistical linearisation φ_0 and k_1 are in fact determined by the form of $g(x)$ and the probability characteristics of X. As follows from (2.7) and (2.8), φ_0 and k_1 are determined by the mathematical expectation and the dispersion of the true random function Y at the output of the non-linear element, as well as by the mathematical expectation and dispersion of the random input function X.

The mathematical expectation m_y of the random function Y at the output of the non-linear element can be calculated if, for each instant $n=0, 1, \ldots,$ we know the law of differential distribution of the random function acting at the input of the non-linear element. If the transformation of X

$$Y=g(X)$$

is single-valued, then in accordance with the theory of transformation of random functions [49], the mathematical expectation of the random function Y can be calculated by

$$m_y=\int_{-\infty}^{\infty} g(x) f(x)\, dx \tag{2.9}$$

where $f(x)$ is the law of differential distribution of X, given for each value of the discrete argument n from the set of all the values for which X is defined.

Thus, on the basis (2.7) and (2.9) the transfer coefficient of the equivalent linear element for the mathematical expectation equals

$$\varphi_0 = \int_{-\infty}^{\infty} g(x) f(x)\, dx \qquad (2.10)$$

Subsequently, when writing the coefficients φ_0 and k_1, we shall drop the symbol denoting their dependence on the argument n.

To determine k_1 we must first find the expression for the dispersion D_y of the random function Y at the output of the non-linear element. Following the theory of transformation of random functions we have

$$D_y = \int_{-\infty}^{\infty} (y - m_y)^2 f(x)\, dx$$

but since Y is the result of transforming the X, and the mathematical expectation m_y is given by (2.9), then for the dispersion (in the case of a single-valued characteristic of $g(x)$) we obtain the expression

$$D_y = \int_{-\infty}^{\infty} [g(x) - \int_{-\infty}^{\infty} g(x) f(x)\, dx]^2 f(x)\, dx \qquad (2.11)$$

Substituting (2.10) into (2.11), we have

$$D_y = \int_{-\infty}^{\infty} [g(x) - \varphi_0]^2 f(x)\, dx = \int_{-\infty}^{\infty} g^2(x) f(x)\, dx$$

$$- 2\varphi_0 \int_{-\infty}^{\infty} g(x) f(x)\, dx + \varphi_0^2 \int_{-\infty}^{\infty} f(x)\, dx$$

Since

$$\int_{-\infty}^{\infty} f(x)\, dx = 1$$

$$\int_{-\infty}^{\infty} g(x) f(x)\, dx = \varphi_0$$

we obtain from the last equation

$$D_y = \int_{-\infty}^{\infty} g^2(x) f(x)\,dx - \varphi_0^2 \tag{2.12}$$

Thus, using (2.8) and (2.12) we write the expression for the transfer coefficient for the random term of the equivalent linear element as follows:

$$k_1 = \left\{ \frac{\int_{-\infty}^{\infty} g^2(x) f(x)\,dx - \varphi_0^2}{\int_{-\infty}^{\infty} (x - m_x)^2 f(x)\,dx} \right\}^{\frac{1}{2}} \tag{2.13}$$

In the case where the transformation $g(x)$ of X is not single-valued, to determine the coefficients of statistical linearisation of the non-linear element with the characteristic $g(x)$ we must know the two-dimensional differential distribution law of the set of two random quantities X and $\dot{X} = \frac{dX}{dt}$, given for each instant n of time. Denoting this distribution law by $f(x, \dot{x})$ we have, in accordance with the theory of transformation of random functions,

$$\varphi_0 = \int_{-\infty}^{\infty}\int_{-\infty}^{\infty} g(x) f(x, \dot{x})\,dx\,d\dot{x} \tag{2.14}$$

Analogously for the transfer coefficient k_1 we have, using (2.8),

$$k_1 = \left\{ \frac{\int_{-\infty}^{\infty}\int_{-\infty}^{\infty} g(x) f(x, \dot{x})\,dx\,d\dot{x} - \varphi_0^2}{\int_{-\infty}^{\infty}\int_{-\infty}^{\infty} (x - m_x)^2 f(x, \dot{x})\,dx\,d\dot{x}} \right\}^{\frac{1}{2}} \tag{2.15}$$

The expressions (2.10), (2.13), (2.14) and (2.15), establishing the relation between the coefficients of statistical linearisation and the probability characteristics of X by means of the $g(x)$, were obtained by Kazakov [36], from the condition that the approximation of the true random function Y by the random function Y^* is carried out by equating their mathematical expectations and dispersions, i.e. proceeding from (2.5) and (2.6). But in fact other conditions can be imposed on Y^*. For example, we can stipulate the mathematical expectation of the square of the difference between Y and Y^* to be minimal, i.e.

$$M[(Y - Y^*)^2] = \min \tag{2.16}$$

The condition (2.16) is the main feature of the second method of approximation used in statistical linearisation [36].

Let us find the expression for the coefficient of statistical linearisation, proceeding from (2.16). Substituting the expression of the random quantity into (2.16), we have

$$\begin{aligned} M[(Y[n] - \varphi_0[n] - \mathrm{k}_1[n]\overset{\circ}{X}[n])^2] = -2\varphi_0 m_y + M[Y^2[n]] \\ + \varphi_0^2[n] + \mathrm{k}_1^2[n]D_x - 2\mathrm{k}_1[n]M[Y[n]\overset{\circ}{X}[n]] \end{aligned} \tag{2.17}$$

To determine the coefficients φ_0 and k_1 we must have two equations, which can be obtained if we equate to zero the partial derivatives of (2.17) with respect to φ_0 and k_1. The differentiation of (2.17) gives

$$\frac{\partial M[(Y - Y^*)^2]}{\partial \varphi_0} = 2\varphi_0 - 2m_y = 0 \tag{2.18}$$

$$\frac{\partial M[(Y - Y^*)^2]}{\partial \mathrm{k}_1} = 2\mathrm{k}_1 D_x - 2K_{y\overset{\circ}{x}}[n;n] = 0 \tag{2.19}$$

where

$$K_{y\overset{\circ}{x}}[n;n] = M[Y[n]\overset{\circ}{X}[n]]$$

We have from the first equation

$$\varphi_0 = m_y \tag{2.20}$$

whence it follows that the transfer coefficient of the equivalent element for the mathematical expectation, when determined by the second method of approximation, coincides with the coefficient φ_0 determined according to the first method of approximation.

The transfer coefficient of the equivalent linear element

for the random term is determined from (2.19). We then have

$$\mathrm{k}_1 = \frac{K_{y\overset{\circ}{x}}[n;\, n]}{K_x[n;\, n]} \tag{2.21}$$

where $K_x[n;\, m]$ is the correlation function of the random function X.

It can be shown that φ_0 and k_1, determined by (2.20) and (2.21) transform $M[(Y-Y^*)^2]$ into a minimum, and not into a maximum.

Since for the single-valued transformation $Y=g(X)$ the mathematical expectation $M[Y\overset{\circ}{X}]$ is determined by

$$M[Y\overset{\circ}{X}] = K_{y\overset{\circ}{x}}[n;\, n] = \int_{-\infty}^{\infty} g(x)(x-m_x)f(x)\,dx$$

then in accordance with (2.21), the transfer coefficient k_1 for the random component is determined by

$$\mathrm{k}_1 = \frac{\int_{-\infty}^{\infty} g(x)(x-m_x)f(x)\,dx}{\int_{-\infty}^{\infty} (x-m_x)^2 f(x)\,dx} \tag{2.22}$$

If the transformation $Y=g(X)$ is not single-valued, the expression for the transfer coefficient k_1 has the form

$$\mathrm{k}_1 = \frac{\int_{-\infty}^{\infty}\int_{-\infty}^{\infty} g(x)(x-m_x)f(x,\dot{x})\,dx\,d\dot{x}}{\int_{-\infty}^{\infty}\int_{-\infty}^{\infty} (x-m_x)f(x,x)\,dx\,d\dot{x}} \tag{2.23}$$

The formulae given above thus enable us to find the expressions for the coefficients of statistical linearisation for non-linear elements given by the characteristic $g(X)$. For this, of course, we must know the distribution laws (uni-dimensional distribution law in the case of the single-valued characteristic $g(X)$, and two-dimensional distribution law in the case of the non-single-valued characteristics $g(X)$) of the random functions X which represent the input actions of the non-linear elements. For non-linear elements having skew symmetric characteristics, it is advisable to choose the approximating function in the form

$$Y^*[n] = k_0[n]\, m_x[n] + k_1[n]\, \overset{\circ}{X}[n] \tag{2.24}$$

In this case it is easy to show that for both methods of approximation the transfer coefficient for the mathematical expectation must be calculated by

$$k_0[n] = \frac{m_y[n]}{m_x[n]} \tag{2.25}$$

whence it follows that $\varphi_0[n]$ is connected with $k_0[n]$ by

$$k_0[n] = \frac{\varphi_0[n]}{m_x[n]} \tag{2.26}$$

4.3 GENERALISED NON-LINEAR CHARACTERISTICS: STATISTICAL LINEARISATION OF NON-LINEAR ELEMENTS

Certain formulae have been obtained for the coefficients of statistical linearisation for certain typical non-linear elements [36]. These formulae have been obtained for the case where the input action X of the non-linear element has a normal probability density distribution for each instant of time, i.e.

$$f(x) = \frac{1}{\eta_x \sqrt{2\pi}}\, e^{-\frac{(x-m_x)^2}{2\eta_x^2}} \tag{3.1}$$

where η_x is the mean-square deviation of the random quantity X from its mathematical expectation m_x.

Since the non-linear element is usually included in the closed-loop of a control system side by side with inertial linear elements, the random process in the closed-loop system is transformed by the linear inertial elements in such a way that the distribution law of its probability density approaches the normal. This is in fact the physical basis on which we can assume that the probability density of the input action X of the non-linear element, which is included in the closed-loop, follows the normal distribution law.

The formulae for the coefficients of statistical linearisation have been obtained at the present time for a limited number of non-linear elements. Here we shall introduce the generalised non-linear element whose characteristic will be called the generalised non-linear characteristic. A graphical representation is shown in Fig. 4.5.

It can be seen that, for definite values of the parameters of the generalised non-linear characteristic, we can obtain the characteristics of different typical non-linear elements which are encountered when solving practical problems. This enables us to determine easily the expressions for the coefficients of statistical linearisation of these typical non-linear elements, proceeding from the expressions for the coefficients of statistical linearisation of the non-linear elements with the generalised characteristic.

The characteristics of typical non-linear elements are shown in Figs 4.6, 4.7 and 4.8. These characteristics are obtained from the general non-linear characteristic for definite values of its parameters. The values of these parameters are shown in the figures.

The above characteristics of typical non-linear elements do not exhaust the variety of characteristics which can be thus obtained.

Let us now determine the coefficients of statistical linearisation for the non-linear element with the generalised non-linear characteristic shown in Fig. 4.5. We shall assume that the probability density of the random function X, being

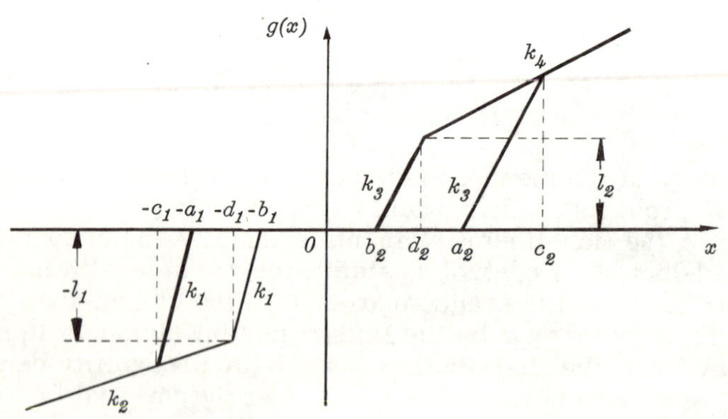

Fig. 4.5 A generalised non-linear characteristic

the input action of the non-linear element, and the probability density of the set of the random functions X and $\dot{X}$ are distributed normally at each instant of time.

The transfer coefficient φ_0 of the non-linear element for the mathematical expectation, in the case of a non-single-valued characteristic, is determined from

$$\varphi_0 = \int_{-\infty}^{\infty}\int_{-\infty}^{\infty} g(x) f(x, \dot{x})\, dx d\dot{x} \qquad (3.2)$$

Since the distribution law $f(x, \dot{x})$, which is assumed to be normal, is symmetric, (3.2) is considerably simplified.

Let us consider the non-linear element whose characteristic is shown in Fig. 4.9. In this case we can show that

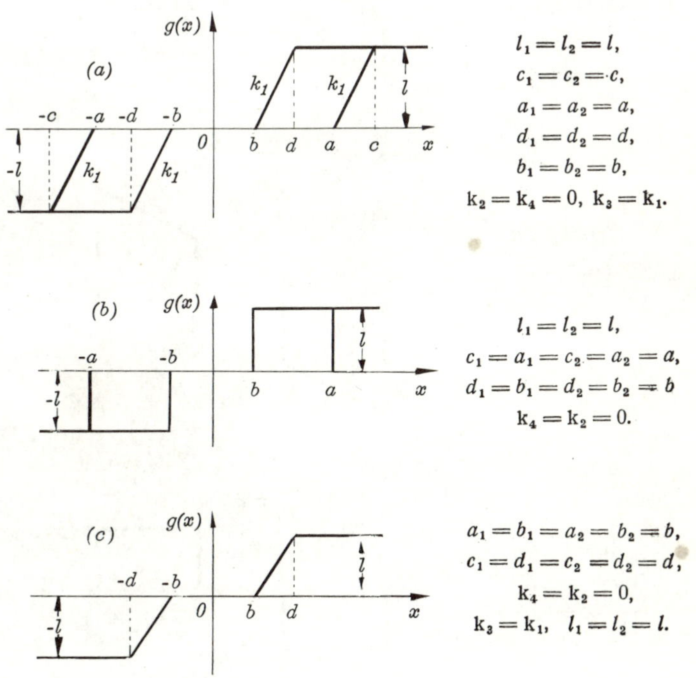

Fig. 4.6 Non-linear characteristics obtained from the generalised non-linear characteristic

$$l_1 = l_2 = l,$$
$$a_1 = b_1 = a_2 = b_2 = b,$$
$$c_1 = d_1 = c_2 = d_2 = d,$$
$$k_3 = k_1, \quad k_4 = k_2 > 0.$$

$$l_1 = l_2 = l,$$
$$a_1 = b_1 = a_2 = b_2 = b,$$
$$c_1 = d_1 = c_2 = d_2 = d,$$
$$k_3 = k_1, \quad k_4 = k_2 < 0.$$

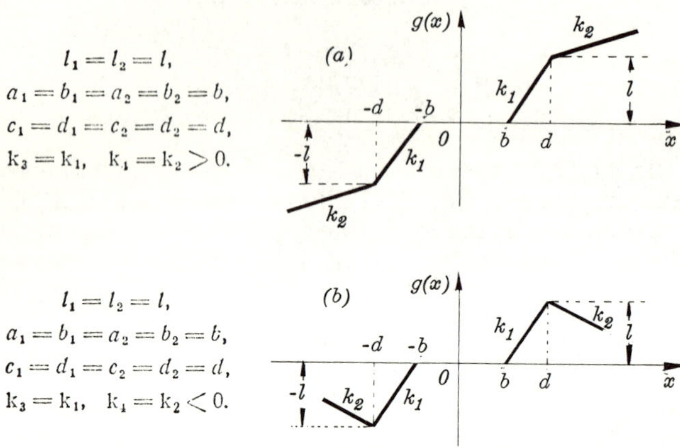

$$l_1 = l_2 = l,$$
$$c_1 = a_1 = d_1 = b,$$
$$c_2 = a_2 = d_2 = b,$$
$$k_4 = k_2, \; k_3 = k_1 = \infty.$$

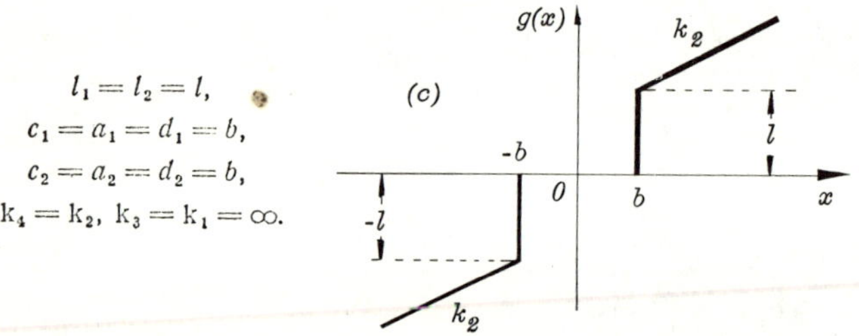

$$l_1 = l_2 = l,$$
$$c_1 = a_1 = d_1 = b,$$
$$c_2 = a_2 = d_2 = b,$$
$$k_4 = k_2 = 0, \; k_3 = k_1 = \infty.$$

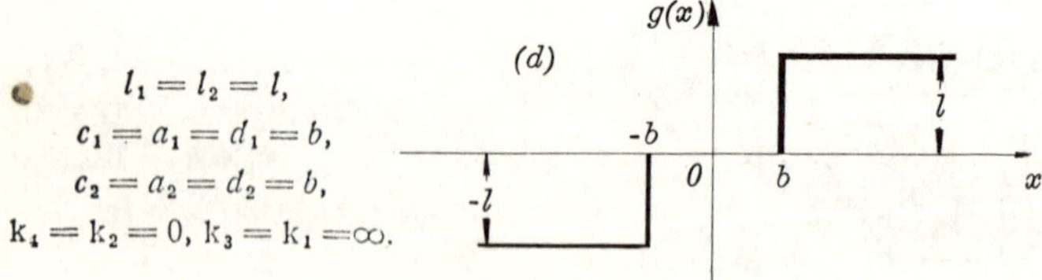

Fig. 4.7 Non-linear characteristics obtained from the generalised non-linear characteristic

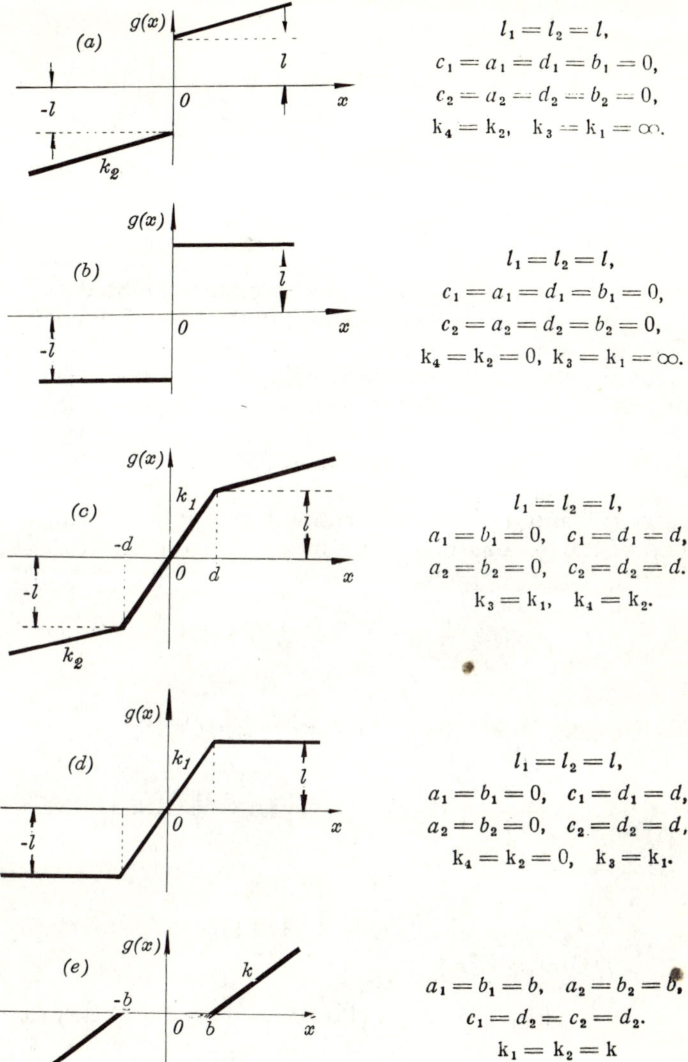

Fig. 4.8 Non-linear characteristics obtained from the generalised non-linear characteristic

$$I=\int_{-\infty}^{\infty}\int_{-\infty}^{\infty} g(x)f(x,\dot{x})\,dx\,d\dot{x}=\int_{-\infty}^{-a} g_4(x)f(x)\,dx$$
$$+\int_{-a}^{b}\frac{1}{2}[g_1(x)+g_2(x)]f(x)\,dx+\int_{b}^{\infty} g_3(x)f(x)\,dx \qquad (3.3)$$

where

$$f(x)=\int_{-\infty}^{\infty} f(x,\dot{x})\,d\dot{x}$$

is the distribution law for X, having the form (3.1).

We next write the expression for the transfer coefficient φ_0 of the linear element with a generalised non-linear characteristic. In accordance with Fig. 4.5 and the rule for calculating integrals of the type (3.3) we have

$$\varphi_0=I_0^{(-)}+I_0^{(+)} \qquad (3.4)$$

where $I_0^{(-)}$ and $I_0^{(+)}$ are determined by applying (3.3) to both sides of the generalised non-linear characteristic

$$I_0^{(-)}=\int_{-\infty}^{-c_1}(k_2x-l_1)f(x)\,dx$$
$$+\frac{1}{2}\int_{-c_1}^{-a_1}[(k_1+k_2)x-l_1]f(x)\,dx \qquad (3.5)$$
$$+\frac{1}{2}\int_{-a_1}^{-d_1}(k_2x-l_1)f(x)\,dx+\frac{1}{2}\int_{-d_1}^{-b_1}k_1xf(x)\,dx$$

$$I_0^{(+)}=\frac{1}{2}\int_{b_2}^{d_2}k_3xf(x)\,dx+\frac{1}{2}\int_{d_2}^{a_2}(k_4x+l_2)f(x)\,dx$$
$$+\frac{1}{2}\int_{a_2}^{c_2}[(k_3+k_4)x+l_2]f(x)\,dx+\int_{c_2}^{\infty}(k_4x+l_2)f(x)\,dx \qquad (3.6)$$

Thus, in order to find the expression for the transfer coefficient, for the mathematical expectation, of the non-linear element with a generalised characteristic, we must calculate the integrals (3.5) and (3.6). The integrals $I_0^{(-)}$ and

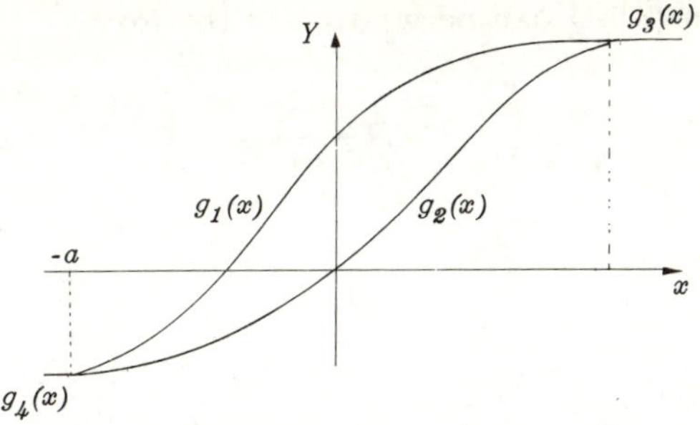

Fig. 4.9 Two-valued non-linear characteristic

$I_0^{(+)}$ can be calculated easily, provided the distribution law $f(x)$ is normal, that is, of the form (3.1).

As an example, let us find the integral

$$i = \int_{-\infty}^{-c_1} (k_2 x - l_1) f(x)\, dx. \tag{3.7}$$

Substituting the expression of $f(x)$, given by (3.1), into (3.7) we have

$$i = \frac{k_2}{\eta_x \sqrt{2\pi}} \int_{-\infty}^{-c_1} x e^{-\frac{(x-m_x)^2}{2\eta_x^2}} dx - \frac{l_1}{\eta_x \sqrt{2\pi}} \int_{-\infty}^{-c_1} e^{-\frac{(x-m_x)^2}{2\eta_x^2}} dx \tag{3.8}$$

To calculate the integral

$$i_1 = \frac{1}{\eta_x \sqrt{2\pi}} \int_{-\infty}^{-c_1} e^{-\frac{(x-m_x)^2}{2\eta_x^2}} dx \tag{3.9}$$

we introduce the new variable

$$\lambda = \frac{x - m_x}{\eta_x} \tag{3.10}$$

Then we have from (3.9)

$$i_1 = \frac{1}{\sqrt{2\pi}} \int_{-\infty}^{-\frac{c_1 + m_x}{\eta_x}} e^{-\frac{\lambda^2}{2}} d\lambda \tag{3.11}$$

Expression (3.11) can be written in the form

$$i_1 = \frac{1}{\sqrt{2\pi}} \int_{-\infty}^{0} e^{-\frac{\lambda^2}{2}} d\lambda - \frac{1}{\sqrt{2\pi}} \int_{-\frac{c_1 + m_x}{\eta_x}}^{0} e^{-\frac{\lambda^2}{2}} d\lambda$$

or

$$i_1 = \frac{1}{\sqrt{2\pi}} \int_{0}^{\infty} e^{-\frac{\lambda^2}{2}} d\lambda - \frac{1}{\sqrt{2\pi}} \int_{0}^{\frac{c_1 + m_x}{\eta_x}} e^{-\frac{\lambda^2}{2}} d\lambda$$

whence

$$i_1 = \Phi(\infty) - \Phi\left(\frac{c_1 + m_x}{\eta_x}\right) \tag{3.12}$$

where

$$\Phi(u) = \frac{1}{\sqrt{2\pi}} \int_{0}^{u} e^{-\frac{\lambda^2}{2}} d\lambda$$

is the probability integral. Tables of values of the probability integral can be found, for example in Pugachev's book [49].

Since $\Phi(\infty) = \frac{1}{2}$, (3.12) can be written in the final form

$$i_1 = \frac{1}{2} - \Phi\left(\frac{c_1 + m_x}{\eta_x}\right) \tag{3.13}$$

To calculate the integral

$$i_2 = \frac{1}{\eta_x\sqrt{2\pi}} \int_{-\infty}^{-c_1} x e^{-\frac{(x-m_x)^2}{2\eta_x^2}} dx$$

we represent it in the form of the sum of two integrals

$$i_2 = \frac{1}{\eta_x\sqrt{2\pi}} \int_{-\infty}^{-c_1} (x - m_x) e^{-\frac{(x-m_x)^2}{2\eta_x^2}} dx + \frac{m_x}{\eta_x\sqrt{2\pi}} \int_{-\infty}^{-c_1} e^{-\frac{(x-m_x)^2}{2\eta_x^2}} dx$$

the second of which is an integral of the type (3.9). Introducing the new variable (3.10) for the first integral we can reduce it to the form

$$\frac{1}{\eta_x\sqrt{2\pi}}\int_{-\infty}^{-c_1}(x-m_x)\,e^{-\frac{(x-m_x)^2}{2\eta_x^2}}dx=\frac{1}{\sqrt{2\pi}}\int_{-\infty}^{-\frac{c_1+m_x}{\eta_x}}\lambda e^{-\frac{\lambda^2}{2}}d\lambda$$

Calculation of this integral is not difficult.

Using the above methods of integration we can find the expression for the transfer coefficient φ_0 for the mathematical expectation:

$$\begin{aligned}\varphi_0=&\frac{k_2m_x}{2}+\frac{m_x(k_1-k_2)+l_1}{2}\left[\Phi\left(\frac{c_1+m_x}{\eta_x}\right)+\Phi\left(\frac{d_1+m_x}{\eta_x}\right)\right]\\&-\frac{k_1m_x}{2}\left[\Phi\left(\frac{a_1+m_x}{\eta_x}\right)+\Phi\left(\frac{b_1+m_x}{\eta_x}\right)\right]\\&+\eta_x\frac{k_1-k_2}{2\sqrt{2\pi}}\left[e^{-\frac{(c_1+m_x)^2}{2\eta_x^2}}+e^{-\frac{(d_1+m_x)^2}{2\eta_x^2}}\right]\\&-\eta_x\frac{k_1}{2\sqrt{2\pi}}\left[e^{-\frac{(d_1+m_x)^2}{2\eta_x^2}}+e^{-\frac{(b_1+m_x)^2}{2\eta_x^2}}\right]-\frac{l_1-l_2}{2}+\frac{k_4m_x}{2}\\&+\frac{m_x(k_3-k_4)-l_2}{2}\left[\Phi\left(\frac{c_2-m_x}{\eta_x}\right)+\Phi\left(\frac{d_2-m_x}{\eta_x}\right)\right]\\&-\frac{k_3m_x}{2}\left[\Phi\left(\frac{a_2-m_x}{\eta_x}\right)+\Phi\left(\frac{b_2-m_x}{\eta_x}\right)\right]\\&-\eta_x\frac{k_3-k_4}{2\sqrt{2\pi}}\left[e^{-\frac{(c_2-m_x)^2}{2\eta_x^2}}+e^{-\frac{(d_2-m_x)^2}{2\eta_x^2}}\right]\\&+\eta_x\frac{k_3}{2\sqrt{2\pi}}\left[e^{-\frac{(a_2-m_x)^2}{2\eta_x^2}}+e^{-\frac{(b_2-m_x)^2}{2\eta_x^2}}\right]\end{aligned}\qquad(3.14)$$

Using (3.14) we can easily find the formula for φ_0 for any non-linear element whose characteristic is obtained from the generalised non-linear characteristic.

For a non-linear element whose characteristic is obtained

from the generalised characteristic when $l_1 = l_2 = l$, $k_3 = k_1$ and $k_4 = k_2$, is determined from

$$\varphi_0 = k_2 m_x + \frac{m_x(k_1 - k_2) + l}{2}\left[\Phi\left(\frac{c_1 + m_x}{\eta_x}\right) + \Phi\left(\frac{d_1 + m_x}{\eta_x}\right)\right]$$
$$+ \frac{m_x(k_1 - k_2) - l}{2}\left[\Phi\left(\frac{c_2 - m_x}{\eta_x}\right) + \Phi\left(\frac{d_2 - m_x}{\eta_x}\right)\right]$$
$$- \frac{k_1 m_x}{2}\left[\Phi\left(\frac{a_1 + m_x}{\eta_x}\right) + \Phi\left(\frac{b_1 + m_x}{\eta_x}\right)\right.$$
$$\left. + \Phi\left(\frac{a_2 - m_x}{\eta_x}\right) + \Phi\left(\frac{b_2 - m_x}{\eta_x}\right)\right]$$
$$+ \eta_x \frac{k_1 - k_2}{2\sqrt{2\pi}}\left[e^{-\frac{(c_1+m_x)^2}{2\eta_x^2}} + e^{-\frac{(d_1+m_x)^2}{2\eta_x^2}}\right. \tag{3.15}$$
$$\left. - e^{-\frac{(c_2-m_x)^2}{2\eta_x^2}} - e^{-\frac{(d_2-m_x)^2}{2\eta_x^2}}\right] + \eta_x \frac{k_1}{2\sqrt{2\pi}}\left[e^{-\frac{(a_2-m_x)^2}{2\eta_x^2}}\right.$$
$$\left. + e^{-\frac{(b_2-m_x)^2}{2\eta_x^2}} - e^{-\frac{(a_1+m_x)^2}{2\eta_x^2}} - e^{-\frac{(b_1+m_x)^2}{2\eta_x^2}}\right]$$

Setting the parameters of the characteristic equal to $c_1 = a_1 = c_2 = a_2 = a$, $d_1 = b_1 = d_2 = b_2 = b$ and $k_2 = 0$, we obtain the characteristic shown in Fig. 4.6b. For these values of the parameters we find from (3.15)

$$\varphi_0 = \frac{l}{2}\left[\Phi\left(\frac{a + m_x}{\eta_x}\right) - \Phi\left(\frac{a - m_x}{\eta_x}\right)\right.$$
$$\left. + \Phi\left(\frac{b + m_x}{\eta_x}\right) - \Phi\left(\frac{b - m_x}{\eta_x}\right)\right] \tag{3.16}$$

Expression (3.16) coincides with the corresponding formula obtained by Kazakov [36].

Let us now put $a = b = 0$ in (3.16). For these values of the parameters of the generalised characteristic we obtain the characteristic of the inertia-free relay shown in Fig. 4.8b. In this case we have

$$\varphi_0 = \frac{l}{2}\left[\Phi\left(\frac{m_x}{\eta_x}\right) - \Phi\left(-\frac{m_x}{\eta_x}\right) + \Phi\left(\frac{m_x}{\eta_x}\right) - \Phi\left(-\frac{m_x}{\eta_x}\right)\right]$$

Since the probability integral is an odd function of its upper limit, i.e.

$$\int_0^x e^{-\frac{\lambda^2}{2}} d\lambda = -\int_0^{-x} e^{-\frac{\lambda^2}{2}} d\lambda$$

we find from the last relation that

$$\varphi_0 = 2l\Phi\left(\frac{m_x}{\eta_x}\right) \tag{3.17}$$

This expression for φ_0 also coincides with that obtained in [36].

Proceeding in a similar manner we can find the expression for φ_0 of any non-linear element whose characteristic can be derived from the generalised non-linear characteristic.

Let us now determine the transfer coefficient k_1, for the random term, of the non-linear element with the generalised characteristic.

Using the first method of approximation we determine the coefficient $k_1^{(1)}$ (the symbol (1) denotes the first method of approximation) from the expression

$$k_1^{(1)} = \frac{1}{\eta_x} \sqrt{\int_{-\infty}^{\infty}\int_{-\infty}^{\infty} g^2(x) f(x, \dot{x})\, dx d\dot{x} - \varphi_0^2} \tag{3.18}$$

Writing the double integral in the form

$$\int_{-\infty}^{\infty}\int_{-\infty}^{\infty} g^2(x) f(x, \dot{x})\, dx d\dot{x} = I_1 = I_1^{(-)} + I_1^{(+)} \tag{3.19}$$

we have the generalised non-linear characteristic

$$\begin{aligned} I_1^{(-)} = & \int_{-\infty}^{-c_1} (k_2 x - l_1)^2 f(x)\, dx \\ & + \frac{1}{2} \int_{-c_1}^{-a_1} [k_1^2 x^2 + (k_2 x - l_1)^2] f(x)\, dx \\ & + \frac{1}{2} \int_{-a_1}^{-d_1} (k_2 x - l_1)^2 f(x)\, dx + \frac{1}{2} \int_{-d_1}^{-b_1} k_1^2 x^2 f(x)\, dx \end{aligned} \tag{3.20}$$

$$I_1^{(+)} = \frac{1}{2}\int_{b_2}^{d_2} k_3^2 x^2 f(x)\,dx + \frac{1}{2}\int_{d_2}^{a_2} (k_4 x + l_2)^2 f(x)\,dx$$
$$+\frac{1}{2}\int_{a_2}^{c_2} [k_3^2 x^2 + (k_4 x + l_2)^2] f(x)\,dx + \int_{c_2}^{\infty} (k_4 x + l_2)^2 f(x)\,dx \qquad (3.21)$$

Substituting the expression of the distribution law $f(x)$ given by (3.1) into (3.20) and (3.21), and carrying out the integration, we find

$$I_1^{(-)} = \frac{k_2^2(\eta_x^2 + m_x^2) - 2k_2 m_x l_1 + l_1^2}{2}$$
$$+\frac{(k_1^2 + k_2^2)(\eta_x^2 + m_x^2) + 2k_2 m_x l_1 - l_1^2}{2}\Phi\left(\frac{c_1 + m_x}{\eta_x}\right)$$
$$+\frac{(k_1^2 - k_2^2)(\eta_x^2 + m_x^2) + 2k_2 m_x l_1 - l_1^2}{2}\Phi\left(\frac{d_1 + m_x}{\eta_x}\right)$$
$$-\frac{k_1^2(\eta_x^2 + m_x^2)}{2}\left[\Phi\left(\frac{a_1 + m_x}{\eta_x}\right) + \Phi\left(\frac{b_1 + m_x}{\eta_x}\right)\right] \qquad (3.22)$$
$$+\eta_x \frac{k_1^2(a_1 - m_x)}{2\sqrt{2\pi}} e^{-\frac{(a_1+m_x)^2}{2\eta_x^2}} + \eta_x \frac{k_1^2(b_1 - m_x)}{2\sqrt{2\pi}} e^{-\frac{(b_1+m_x)^2}{2\eta_x^2}}$$
$$+\eta_x \frac{2k_2 l_1 - (k_1^2 - k_2^2)(c_1 - m_x)}{2\sqrt{2\pi}} e^{-\frac{(c_1+m_x)^2}{2\eta_x^2}}$$
$$+\eta_x \frac{2k_2 l_1 - (k_1^2 - k_2^2)(d_1 - m_x)}{2\sqrt{2\pi}} e^{-\frac{(d_1+m_x)^2}{2\eta_x^2}}$$

and

$$I_1^{(+)} = \frac{k_4^2(\eta_x^2 + m_x^2) - 2k_4 m_x l_2 + l_2^2}{2}$$
$$+\frac{(k_3^2 + k_4^2)(\eta_x^2 + m_x^2) + 2k_4 m_x l_2 - l_2^2}{2}\Phi\left(\frac{c_2 - m_x}{\eta_x}\right) \qquad (3.23)$$
$$-\frac{(k_3^2 - k_4^2)(\eta_x^2 + m_x^2) + 2k_4 m_x l_2 - l_2^2}{2}\Phi\left(\frac{d_2 - m_x}{\eta_x}\right)$$

$$-\frac{k_3^2(\eta_x^2+m_x^2)}{2}\left[\Phi\left(\frac{a_2-m_x}{\eta_x}\right)+\Phi\left(\frac{b_2-m_x}{\eta_x}\right)\right]$$

$$+\eta_x\frac{k_3^2(a_2+m_x)}{2\sqrt{2\pi}}e^{-\frac{(a_2+m_x)^2}{2\eta_x^2}}+\eta_x\frac{k_3^2(b_2+m_x)}{2\sqrt{2\pi}}e^{-\frac{(b_2+m_x)^2}{2\eta_x^2}}$$

$$+\eta_x\frac{2k_4l_2-(k_3^2-k_4^2)(c_2+m_x)}{2\sqrt{2\pi}}e^{-\frac{(c_2-m_x)^2}{2\eta_x^2}} \tag{3.23}$$

$$+\eta_x\frac{2k_4l_2-(k_3^2-k_4^2)(d_2+m_x)}{2\sqrt{2\pi}}e^{-\frac{(d_2-m_x)^2}{2\eta_x^2}}$$

Thus, the coefficient $k_1^{(1)}$ of the non-linear element with the generalised characteristic can be found from

$$k_1^{(1)}=\frac{1}{\eta_x}\sqrt{I_1^{(-)}+I_1^{(+)}-\varphi_0^2} \tag{3.24}$$

where $I_1^{(-)}$ and $I_1^{(+)}$ are given by (3.22) and (3.23) respectively, whilst the coefficient φ_0 is given by (3.14).

The results obtained above enable us to find easily the expression for $k_1^{(1)}$ of any non-linear element whose characteristic is derived from the generalised characteristic.

Thus, putting $l_1=l_2=l$, $k_3=k_1$, and $k_4=k_2$ for the generalised characteristic, we obtain from (3.22) and (3.23)

$$\begin{aligned}I=I_1^{(-)}+I_1^{(+)}&=k_2^2(\eta_x^2+m_x^2)-2k_2m_xl+l^2\\&+\frac{(k_1^2+k_2^2)(\eta_x^2+m_x^2)+2k_2m_xl-l^2}{2}\\&\times\left[\Phi\left(\frac{c_1+m_x}{\eta_x}\right)+\Phi\left(\frac{c_2-m_x}{\eta_x}\right)\right]\\&+\frac{(k_1^2-k_2^2)(\eta_x^2+m_x^2)+2k_2m_xl-l^2}{2}\\&\times\left[\Phi\left(\frac{d_1+m_x}{\eta_x}\right)+\Phi\left(\frac{d_2-m_x}{\eta_x}\right)\right]\\&-\frac{k_1^2(\eta_x^2+m_x^2)}{2}\left[\Phi\left(\frac{a_1+m_x}{\eta_x}\right)+\Phi\left(\frac{b_1+m_x}{\eta_x}\right)\right.\end{aligned} \tag{3.25}$$

$$+\Phi\left(\frac{a_2-m_x}{\eta_x}\right)+\Phi\left(\frac{b_2-m_x}{\eta_x}\right)\Big]$$

$$+\eta_x\frac{2k_2l-(k_1^2-k_2^2)(c_1-m_x)}{2\sqrt{2\pi}}e^{-\frac{(c_1+m_x)^2}{2\eta_x^2}}$$

$$+\eta_x\frac{2k_2l-(k_1^2-k_2^2)(c_2+m_x)}{2\sqrt{2\pi}}e^{-\frac{(c_2-m_x)^2}{2\eta_x^2}}$$

$$+\eta_x\frac{2k_2l-(k_1^2-k_2^2)(d_1-m_x)}{2\sqrt{2\pi}}e^{-\frac{(d_1+m_x)^2}{2\eta_x^2}} \qquad (3.25)$$

$$+\eta_x\frac{2k_2l-(k_1^2-k_2^2)(d_2+m_x)}{2\sqrt{2\pi}}e^{-\frac{(d_2-m_x)^2}{2\eta_x^2}}$$

$$+\eta_x\frac{k_1^2(a_1-m_x)}{2\sqrt{2\pi}}e^{-\frac{(a_1+m_x)^2}{2\eta_x^2}}+\eta_x^2\frac{k_1^2(a_2+m_x)}{2\sqrt{2\pi}}e^{-\frac{(a_2-m_x)^2}{2\eta_x^2}}$$

$$+\eta_x\frac{k_1^2(b_1-m_x)}{2\sqrt{2\pi}}e^{-\frac{(b_1+m_x)^2}{2\eta_x^2}}+\eta_x\frac{k_1^2(b_2+m_x)}{2\sqrt{2\pi}}e^{-\frac{(b_2-m_x)^2}{2\eta_x^2}}$$

Let us now put $c_1=a_1=c_2=a_2=a$, $d_1=b_1=d_2=b_2=b$ and $k_2=0$ in the expression (3.25). In this case, making use of (3.24), we obtain the expression for $k_1^{(1)}$ of the non-linear element whose characteristic is shown in Fig. 4.6b:

$$k_1^{(1)}=\frac{1}{\eta_x}\left\{l^2-\varphi_0^2-\frac{l^2}{2}\left[\Phi\left(\frac{a+m_x}{\eta_x}\right)+\Phi\left(\frac{a-m_x}{\eta_x}\right)\right.\right.$$
$$\left.\left.+\Phi\left(\frac{b+m_x}{\eta_x}\right)+\Phi\left(\frac{b-m_x}{\eta_x}\right)\right]\right\}^{\frac{1}{2}} \qquad (3.26)$$

This expression coincides with the corresponding expression obtained in [36].

Subsequently, assuming that $a=b=0$ in (3.26), we find

$$k_1^{(1)}=\frac{1}{\eta_x}(l^2-\varphi_0^2)^{\frac{1}{2}} \qquad (3.27)$$

Expression (3.27) gives the transfer coefficient of a non-linear element with the relay characteristic (Fig. 4.8b).

Since the coefficient φ_0 for a non-linear element with the relay characteristic is given by (3.17), then the expression for $k_1^{(1)}$ can finally be written in the form

$$k_1^{(1)}=\frac{l}{\eta_x}\left\{1-4\Phi\left(\frac{m_x}{\eta_x}\right)\right\}^{\frac{1}{2}} \tag{3.28}$$

Proceeding in a similar manner and using (3.22) and (3.23) we can find the transfer coefficient, for the random term, of any non-linear element whose characteristic is derived from the non-linear characteristic.

Let us next determine the expression for the transfer coefficient $k_1^{(2)}$, for the random term, of the non-linear element with the generalised characteristic, using the second method of approximation.

The coefficient $k_1^{(2)}$, as has been shown in Section 4.2 is determined as follows:

$$k_1^{(2)}=\frac{1}{\eta_x^2}\int_{-\infty}^{\infty}\int_{-\infty}^{\infty} g(x)(x-m_x)f(x,\dot{x})\,dx\,d\dot{x} \tag{3.29}$$

Introducing the symbol

$$I_2=\int_{-\infty}^{\infty}\int_{-\infty}^{\infty} g(x)(x-m_x)f(x,\dot{x})\,dx\,d\dot{x} \tag{3.30}$$

and using the rule for calculating integrals of the form (3.30) when the distribution law $f(x,\dot{x})$ is symmetric, we have for the generalised non-linear characteristic

$$I_2=I_2^{(-)}+I_2^{(+)}$$

where

$$\begin{aligned} I_2^{(-)}=&\int_{-\infty}^{-c_1}(k_2x-l_1)(x-m_x)f(x)\,dx\\ &+\frac{1}{2}\int_{-c_1}^{-a_1}[(k_1+k_2)x-l_1](x-m_x)f(x)\,dx\\ &+\frac{1}{2}\int_{-a_1}^{-d_1}(k_2x-l_1)(x-m_x)f(x)\,dx\\ &+\frac{1}{2}\int_{-d_1}^{-b_1}k_1x(x-m_x)f(x)\,dx \end{aligned} \tag{3.31}$$

$$I_2^{(+)} = \frac{1}{2}\int_{b_2}^{d_2} k_3 (x - m_x) f(x)\, dx$$
$$+ \frac{1}{2}\int_{d_2}^{a_2} (k_4 x + l_2)(x - m_x) f(x)\, dx$$
$$+ \frac{1}{2}\int_{a_2}^{c_2} [(k_3 + k_4) x + l_2](x - m_x) f(x)\, dx \tag{3.32}$$
$$+ \int^{\infty} (k_4 x + l_2)(x - m_x) f(x)\, dx$$

Substituting the expression of the distribution law given by (3.1) into (3.31) and (3.32), and carrying out the differentiation, we find

$$I_2^{(-)} = \eta_x^2 \frac{k_2}{2} + \eta_x^2 \frac{k_1 - k_2}{2}\left[\Phi\left(\frac{c_1 + m_x}{\eta_x}\right) + \Phi\left(\frac{d_1 + m_x}{\eta_x}\right)\right]$$
$$- \eta_x^2 \frac{k_1}{2}\left[\Phi\left(\frac{a_1 + m_x}{\eta_x}\right) + \Phi\left(\frac{b_1 + m_x}{\eta_x}\right)\right]$$
$$+ \eta_x \frac{l_1 - (k_1 - k_2) c_1}{2\sqrt{2\pi}} e^{-\frac{(c_1 + m_x)^2}{2\eta_x^2}} + \eta_x \frac{l_1 - (k_1 - k_2) d_1}{2\sqrt{2\pi}} e^{-\frac{(d_1 + m_x)^2}{2\eta_x^2}}$$
$$+ \eta_x \frac{k_1 a_1}{2\sqrt{2\pi}} e^{-\frac{(a_1 + m_x)^2}{2\eta_x^2}} + \eta_x \frac{k_1 b_1}{2\sqrt{2\pi}} e^{-\frac{(b_1 + m_x)^2}{2\eta_x^2}} \tag{3.33}$$

and

$$I_2^{(+)} = \eta_x^2 \frac{k_4}{2} + \eta_x^2 \frac{k_3 - k_4}{2}\left[\Phi\left(\frac{c_2 - m_x}{\eta_x}\right) + \Phi\left(\frac{d_2 - m_x}{\eta_x}\right)\right]$$
$$- \eta_x^2 \frac{k_3}{2}\left[\Phi\left(\frac{a_2 - m_x}{\eta_x}\right) + \Phi\left(\frac{b_2 - m_x}{\eta_x}\right)\right]$$
$$+ \eta_x \frac{l_2 - (k_3 - k_4) c_2}{2\sqrt{2\pi}} e^{-\frac{(c_2 - m_x)^2}{2\eta_x^2}} + \eta_x \frac{l_2 - (k_3 - k_4) d_2}{2\sqrt{2\pi}} e^{-\frac{(d_2 - m_x)^2}{2\eta_x^2}}$$
$$+ \eta_x \frac{k_3 a_2}{2\sqrt{2\pi}} e^{-\frac{(a_2 - m_x)^2}{2\eta_x^2}} + \eta_x \frac{k_3 b_2}{2\sqrt{2\pi}} e^{-\frac{(b_2 - m_x)^2}{2\eta_x^2}} \tag{3.34}$$

Thus the transfer coefficient $k_1^{(2)}$, for the random term, of the non-linear element with the generalised characteristic can be found from

$$k_1^{(2)} = \frac{2}{\eta_x^2}\left(I_2^{(-)} + I_2^{(+)}\right) \tag{3.35}$$

where $I_2^{(-)}$ and $I_2^{(+)}$ are determined by the expressions (3.33) and (3.34) respectively.

Using (3.33) to (3.35) we can easily find the formulae for the coefficient $k_1^{(2)}$ of any non-linear element whose characteristic is obtained from the generalised non-linear characteristic.

Thus, for example, for a non-linear element whose characteristic is obtained from the generalised characteristic when $l_1 = l_2 = l$, $k_3 = k_1$ and $k_1 = k_2$, the coefficient $k_1^{(2)}$ is determined by

$$\begin{aligned}
k_1^{(2)} = {} & k_2 + \frac{k_1 - k_2}{2}\left[\Phi\left(\frac{c_1 + m_x}{\eta_x}\right) + \Phi\left(\frac{d_1 + m_x}{\eta_x}\right)\right. \\
& \left. + \Phi\left(\frac{c_2 - m_x}{\eta_x}\right) + \Phi\left(\frac{d_2 - m_x}{\eta_x}\right)\right] - \frac{k_1}{2}\left[\Phi\left(\frac{a_1 + m_x}{\eta_x}\right)\right. \\
& \left. + \Phi\left(\frac{b_1 + m_x}{\eta_x}\right) + \Phi\left(\frac{a_2 - m_x}{\eta_x}\right) + \Phi\left(\frac{b_2 - m_x}{\eta_x}\right)\right] \\
& + \frac{1}{\eta_x}\,\frac{l - (k_1 - k_2)\,c_1}{2\sqrt{2\pi}}\, e^{-\frac{(c_1 + m_x)^2}{2\eta_x^2}} \\
& + \frac{1}{\eta_x}\,\frac{l - (k_1 - k_2)\,d_1}{2\sqrt{2\pi}}\, e^{-\frac{(d_1 + m_x)^2}{2\eta_x^2}} \\
& + \frac{1}{\eta_x}\,\frac{l - (k_1 - k_2)\,c_2}{2\sqrt{2\pi}}\, e^{-\frac{(c_2 - m_x)^2}{2\eta_x^2}} \\
& + \frac{1}{\eta_x}\,\frac{l - (k_1 - k_2)\,d_2}{2\sqrt{2\pi}}\, e^{-\frac{(d_2 - m_x)^2}{2\eta_x^2}} \\
& + \frac{1}{\eta_x}\,\frac{k_1}{2\sqrt{2\pi}}\left[a_1 e^{-\frac{(a_1 + m_x)^2}{2\eta_x^2}} + b_1 e^{-\frac{(b_1 + m_x)^2}{2\eta_x^2}}\right. \\
& \left. + a_2 e^{-\frac{(a_2 - m_x)^2}{2\eta_x^2}} + b_2 e^{-\frac{(b_2 - m_x)^2}{2\eta_x^2}}\right]
\end{aligned} \tag{3.36}$$

Now, putting $c_1 = a_1 = c_2 = a_2 = a$, $d_1 = b_1 = d_2 = b_2 = b$ and $k_2 = 0$ in (3.36), we obtain the expression for the transfer coefficient $k_1^{(2)}$ of the non-linear element with the characteristic shown in Fig. 4.6(b):

$$k_1^{(2)} = \frac{1}{\eta_x} \frac{l}{2\sqrt{2\pi}} \left[e^{-\frac{(a+m_x)^2}{2\eta_x^2}} + e^{-\frac{(b+m_x)^2}{2\eta_x^2}} + e^{-\frac{(a-m_x)^2}{2\eta_x^2}} + e^{-\frac{(b-m_x)^2}{2\eta_x^2}} \right] \tag{3.37}$$

This coincides with that obtained by Kazakov [36].

If we assume that $a = b = 0$ we obtain from (3.37)

$$k_1^{(2)} = \frac{1}{\eta_x} \frac{2l}{\sqrt{2\pi}} e^{-\frac{m_x^2}{2\eta_x^2}} \tag{3.38}$$

which determines the transfer coefficient, for the random type, of a non-linear element with the relay characteristic (Fig. 4.8(b)).

In the same way we could find the expression for the coefficient $k_1^{(2)}$ of other non-linear elements whose characteristics are derived from the generalised non-linear characteristic.

4.4 STATISTICAL LINEARISATION IN ERROR ANALYSIS OF STATIONARY NON-LINEAR SYSTEMS IN STEADY STATE

4.4.1 Linear systems equivalent to the original non-linear system

We shall consider the non-linear system with constant parameters shown in Fig. 4.10. Let the linear continuous part of the system have the transfer function

$$W(p) = \frac{Q(p)}{P(p)} \tag{4.1}$$

For simplicity we assume that processes in the system are given by a single difference equation, i.e. the sampler generates rectangular pulses with the relative duration

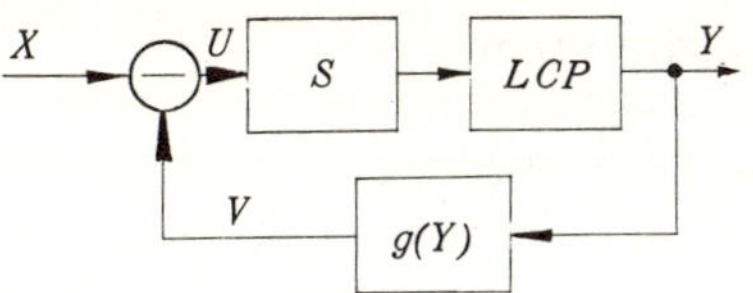

Fig. 4.10 A system with a single non-linear element

$\gamma=1$ or δ-pulses, so that the transfer function of the continuous part of the system has the form

$$W^*(q, \varepsilon)=\frac{M^*(q, \varepsilon)}{L^*(q)}, \quad 0\leqslant\varepsilon\leqslant 1 \tag{4.2}$$

Let the non-linear element be given by its characteristic

$$v = g(y) \tag{4.3}$$

which for simplicity is assumed to be skew symmetric.

We assume that a stationary random function X acts at the input of the system. It is required to find the mathematical expectation of the output coordinate.

Using (4.2) and (4.3) we can write the equations describing processes in the closed-loop system under consideration. These equations have the form (considering lattice functions for simplicity):

$$\left.\begin{aligned} L(\Delta)Y[n] &= M(\Delta)U[n] \\ U[n] &= X[n]-V[n] \\ V[n] &= g(Y[n]) \end{aligned}\right\} \tag{4.4}$$

The difference operators $L(\Delta)$ and $M(\Delta)$ in (4.4) are obtained from (4.2) in accordance with the rules given in Chapter 1, so that

$$L^*(q)=L(\Delta)|_{\Delta=e^q-1}, \quad M^*(q)=M(\Delta)|_{\Delta=e^q-1}$$

Representing the random functions X, Y, U and V by the sum of their mathematical expectations and the corresponding centralised random functions, i.e.

$$\left.\begin{aligned} X &= m_x+\mathring{X} \\ Y &= m_y+\mathring{Y} \\ U &= m_u+\mathring{U} \\ V &= m_v+\mathring{V} \end{aligned}\right\} \tag{4.5}$$

we write (4.4) in the form

$$\left.\begin{aligned} L(\Delta)(m_y[n]+\overset{\circ}{Y}[n]) &= M(\Delta)(m_u[n]+\overset{\circ}{U}\cdot[n]) \\ m_u[n]+\overset{\circ}{U}[n] &= m_x[n]+\overset{\circ}{X}[n]-m_v[n]-\overset{\circ}{V}[n] \\ m_v[n]+\overset{\circ}{V}[n] &= g(m_y+\overset{\circ}{Y}) \end{aligned}\right\} \quad (4.6)$$

Since the non-linear equation $V=g(Y)$ is contained by (4.6), then, to determine the mathematical expectation m_y and the dispersion D_y of the random function Y which is the output coordinate of the system, we cannot use the theory of linear sampled data systems given in Chapter 3. But to calculate approximately m_y and D_y we can use statistical linearisation of $g(Y)$, replacing the non-linear element by two linear elements with appropriate transfer coefficients. At the same time we can assume that the probability density of Y for each instant approximately follows the normal law. The distribution law of the random process is closer to the normal the more inertial is the linear part of the system.

Accordingly, we put

$$V = \mathrm{k}_0 m_y + \mathrm{k}_1 \overset{\circ}{Y} \quad (4.7)$$

where k_1 is the transfer coefficient of the non-linear element for the random term, determined by either the first or the second method of approximation, and k_0 is the transfer coefficient for the mathematical expectation, given in terms of φ_0 according to (2.26):

$$\mathrm{k}_0 = \frac{\varphi_0}{m_y}$$

Using (4.7), we can write (4.6) in the following form:

$$\left.\begin{aligned} L(\Delta)(m_y[n]+\overset{\circ}{Y}[n]) &= M(\Delta)(m_u[n]+\overset{\circ}{U}[n]) \\ m_u[n]+\overset{\circ}{U}[n] &= m_x[n]+\overset{\circ}{X}[n]-m_v[n]-\overset{\circ}{V}[n] \\ m_v[n]+\overset{\circ}{V}[n] &= \mathrm{k}_0 m_y[n]+\mathrm{k}_1\overset{\circ}{Y}[n] \end{aligned}\right\} \quad (4.8)$$

This system contains only linear equations.

We now apply mathematical expectation to both sides of all the equations in (4.8). Taking into account the linearity of the operators $L(\Delta)$ and $M(\Delta)$, we have:

$$\left.\begin{aligned} L(\Delta)\, m_y[n] &= M(\Delta)\, m_u[n] \\ m_u[n] &= m_x[n]-m_v[n] \\ m_v[n] &= \mathrm{k}_0 m_y[n] \end{aligned}\right\} \quad (4.9)$$

Equations (4.9) give the mathematical expectation m_y of the output coordinate Y of the original system.

Subtracting (4.9) from (4.8) we can find the equations for the centralised random functions:

$$\left.\begin{aligned} L(\Delta)\mathring{Y}[n] &= M(\Delta)\mathring{U}[n] \\ \mathring{U}[n] &= \mathring{X}[n] - \mathring{V}[n] \\ \mathring{V}[n] &= \mathrm{k}_1\mathring{Y}[n] \end{aligned}\right\} \quad (4.10)$$

Thus, as the result of the statistical linearisation of $g(Y)$ we have obtained two systems of linear equations, (4.9) and (4.10), whence the mathematical expectation m_y and the dispersion D_y of Y for the original non-linear system can be determined approximately. Consequently, the determination of the accuracy of a non-linear system by statistical linearisation reduces to the study of two linear systems, which are equivalent in a definite probabilistic sense to the original non-linear system. The block diagrams of these equivalent linear systems are given in Fig. 4.11(a) and (b).

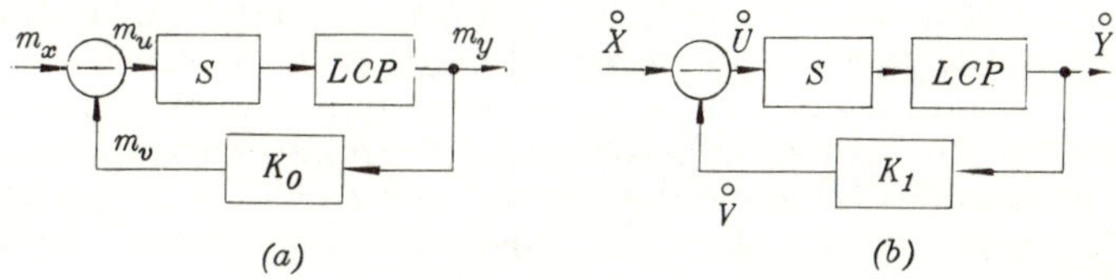

Fig. 4.11 Linear systems equivalent in the sense of statistical linearisation to the non-linear system shown in Fig. 4.10

The input action of the first system, i.e. the system described by (4.9), is a non-random function which constitutes the mathematical expectation of the input disturbance of the original system. The input action of the second system, i.e. the system described by (4.10) is a random function with zero mathematical expectation. Owing to the fact that the coefficients of statistical linearisation, k_0 and k_1, are functions of m_y and the mean-square value $\eta_y = \sqrt{D_y}$, that is,

$$\left.\begin{aligned} \mathrm{k}_0 &= \mathrm{k}_0(m_y, \eta_y) \\ \mathrm{k}_1 &= \mathrm{k}_1(m_y, \eta_y) \end{aligned}\right\} \quad (4.11)$$

the systems (4.9), (4.10) are found to be mutually connected.

In fact, in order to determine m_y from (4.9) the value of the coefficient $k_0(m_y, \eta_y)$ must be known; to calculate the latter we must know the dispersion D_y which can be determined from (4.10). Further, to determine D_y from (4.10), we must know the value of $k_1(m_y, \eta_y)$, for the calculation of which m_y must be known.

Thus, the problem of determining m_y and D_y of the output coordinate of a non-linear system using statistical linearisation leads to the solution of two systems of equations, (4.9) and (4.10), and two equations, (4.11), providing the connection.

Equations (4.9) to (4.11) are most easily solved for steady-state motion when there are no self-excited oscillations in the system. In such a case the difference equations (4.9) transform into algebraic equations.

4.4.2 Non-linear system in a steady state in the absence of self-excited oscillations

Let us consider methods for solving (4.9) to (4.11) for a steady-state motion of a non-linear system when self-excited oscillations are absent. We shall assume that the system is subjected to a stationary random disturbance X with mathematical expectation m_x and correlation function $K_x[n]$.

Since $m_x =$ const and there are no self-excited oscillations in the system, then for steady-state motion we have

$$\Delta^r m_y[n] = \Delta^r m_u[n] = 0, \quad r = 1, 2, \ldots, k$$

and (4.9) is transformed into a system of algebraic equations of the following form:

$$\left.\begin{aligned} L(0)\, m_y &= M(0)\, m_u \\ m_u &= m_x - m_v \\ m_v &= k_0 m_y \end{aligned}\right\} \qquad (4.12)$$

where

$$L(0) = L(\Delta)|_{\Delta=0}; \quad M(0) = M(\Delta)|_{\Delta=0}$$

Consequently, the problem of determining the mathematical expectation and the dispersion reduces to the solution of

(4.12) and (4.10). Equations (4.12) and (4.10) are connected with each other, as before, by (4.11).

Since (4.10) is a system of linear equations, then, to determine the dispersion, we can use linear theory and apply the procedure used when solving the corresponding equations which give the dispersion at the output of a continuous non-linear system.

Let us denote the spectral density of X by $S_x(\bar{\omega})$. Then the spectral density $S_y(\bar{\omega})$ of Y at the output of the linear system given by (4.10) equals

$$S_y(\bar{\omega}) = |W^*(j\bar{\omega})|^2 S_x(\bar{\omega}) \tag{4.13}$$

where $W^*(j\bar{\omega})$ is the frequency characteristic of this system, which from (4.10) is

$$W^*(j\bar{\omega}) = \frac{M^*(j\bar{\omega})}{L^*(j\bar{\omega}) + \mathrm{k}_1(m_y, \eta_y)\, M^*(j\bar{\omega})} \tag{4.14}$$

From the known spectral density $S_y(\bar{\omega})$ we can find the dispersion using the formula (see for example Tsypkin [41])

$$D_y = \frac{1}{\pi}\int_0^{\pi} S_y(\bar{\omega})\, d\bar{\omega} \tag{4.15}$$

Thus, to determine approximately m_y and D_y at the output of the system we have the following equations:

$$m_y = \frac{M(0)}{L(0) + \mathrm{k}_0(m_y, \eta_y)\, M(0)}\, m_x \tag{4.16}$$

$$D_y = \frac{1}{\pi}\int_0^{\pi} \left| \frac{M^*(j\bar{\omega})}{L^*(j\bar{\omega}) + \mathrm{k}_1(m_y, \eta_y)\, M^*(j\bar{\omega})} \right|^2 S_x(\bar{\omega})\, d\bar{\omega} \tag{4.17}$$

$$\left.\begin{aligned} \mathrm{k}_0 &= \mathrm{k}_0(m_y, \eta_y), \\ \mathrm{k}_1 &= \mathrm{k}_1(m_y, \eta_y). \end{aligned}\right\} \tag{4.18}$$

Equations (4.16) to (4.18) can be solved either by successive approximations or graphically, as in the case of continuous systems [36, 38]. The graphical method is extremely laborious and can be used only with simple systems.

A practical method for solving (4.16) to (4.18) is given by the method of successive approximations. We choose as the first approximation the initial values of the coefficients of statistical linearisation $k_0 = k_{01}$ and $k_1 = k_{11}$. Using these values we calculate the mathematical expectation m_{y_1} and the dispersion $\eta_{y_1}^2$ of the first approximation from (4.16) and (4.17). The values of m_{y_1} and $\eta_{y_1}^2$ thus calculated are the starting points for determining the second approximation of the coefficients k_0 and k_1 by (4.18)

$$k_{02} = k_{02}(m_{y_1}, \eta_{y_1}), \qquad k_{12} = k_{12}(m_{y_1}, \eta_{y_1})$$

The second approximation is completed by calculating m_{y2} and η_{y2} from (4.16) and (4.17). The cycle is repeated until the following conditions are satisfied.

$$|m_{y\,(n-1)} - m_{yn}| \leqslant \delta_m \qquad |\eta^2_{y\,(n-1)} - \eta^2_{yn}| \leqslant \delta_\eta$$

or

$$|k_{0\,(n-1)} - k_{0n}| \leqslant \delta_0 \qquad |k_{1\,(n-1)} - k_{1n}| \leqslant \delta_1$$

where δ_m, δ_η, δ_0 and δ_1 are small positive numbers.

The given procedure of determining m_y and η_y^2 easily lends itself to setting up the corresponding algorithm. Accordingly the whole calculation process can be carried out by a digital computer.

The calculation of the mathematical expectation and the dispersion from (4.16) and (4.17) requires laborious preliminary work to determine the square of the modulus of the frequency characteristic $\overline{W^*(j\omega)}$ for the equivalent linear sampled data system. This procedure is particularly laborious in those cases where the system is given by the differential equations of its continuous part.

Because of this another method is recommended for the determination of m_y and D_y, which eliminates the need to find the frequency characteristic and its modulus for the corresponding system. This method is based on calculating the weighting function and on the methods of analysis given in Chapter 3.

4.5 MATHEMATICAL EXPECTATION AND DISPERSION AT THE OUTPUT OF A NON-LINEAR SYSTEM IN A TRANSIENT STATE

In the preceding section we considered methods for determining the mathematical expectation and dispersion at the output of a stationary system in a steady-state. In practice it is often required to determine the accuracy of a non-linear system in a transient state.

During transient processes a system with constant parameters must be regarded as a non-stationary system whose parameters vary with a jump at the instant of switching on, and the problem therefore becomes considerably more complex. In this section we shall give one of the possible methods of solution.

As before, we assume that a stationary random function X with mathematical expectation $m_x =$ const and correlation function $K_x[n]$ acts on the system shown in Fig. 4.10.

In steady-state motion, without any self-excited oscillations in the system, the mathematical expectation m_y is constant and equal to the steady-state response of the system to a step input coinciding with the mathematical expectation m_x as given by equation (4.9). In a transient state m_y is a function of time.

$$m_y = m_y[n], \ 0 \leqslant n \leqslant N$$

where N is the time of transient processes in the system.

Analogously the dispersion D_y in a transient state also depends on time. Hence the following two problems concerning the accuracy of a non-linear system in a transient state may arise in practice:

1. the definition of the mathematical expectation and dispersion at the output of a sampled data system at a particular instant $\bar{t} = n \, (0 < n < N)$;
2. the more general problem of determining the mathematical expectation and the dispersion as functions of the time $\bar{t}$.

We shall give here the solution of the first problem, where m_y and D_y are to be determined at a particular instant. In so doing we assume that the correlation function $K_x[n]$ of the random function X, which acts at the input of the system, has the form

$$K_x[n] = A e^{-\alpha |n|}, \ \alpha > 0 \tag{5.1}$$

Let us write (4.9) and (4.10) of the linear systems, which are equivalent to the original non-linear system, in the following form:

$$(L(\Delta) + k_0 M(\Delta)) \, m_y[n] = M(\Delta) \, m_x[n] \tag{5.2}$$

$$(L(\Delta)+k_1 M(\Delta))\overset{\circ}{Y}[n]=M(\Delta)\overset{\circ}{X}[n] \tag{5.3}$$

The mathematical expectation m_y at the instant $\bar{t}=n$ is the response of the system, given by (5.2), to a disturbance which coincides with m_x and can be calculated from

$$m_y[n]=m_x\sum_{k=0}^{n} W_m[k] \tag{5.4}$$

where $W_m[n]$ is the weighting function of the system. In addition, the mathematical expectation can be found by solving numerically the difference equations or the corresponding differential-difference equations, with the assumption that the disturbing function identically coincides with m_x.

To determine the dispersion D_y we can use the method of shaping filters. It was shown in Chapter 3 that in a transient state the dispersion $D_y[n]$ of the output coordinate of a linear system, when a random disturbance with a correlation function of the form (5.1) acts at its input, is determined as follows:

$$D_y[n]=Ae^{-2\alpha}F^2[n]+(1-e^{-2\alpha})A\sum_{m=0}^{n}F^2[m] \tag{5.5}$$

where $F[n]$ is the response of the system to a disturbance of the form $e^{-\alpha n}$ when $n\geqslant 0$. Consequently, in order to determine the dispersion at the instant $\bar{t}=n$ we must first find the value of the response $F[m]$ of the system to an exponential action for the instants $m=0, 1, \ldots, n$, and then apply (5.5).

The response $F[m]$, $m=0,1,\ldots, n$ can be found by a numerical solution of the corresponding difference or differential-difference equations which describe the processes in the system, with the assumption that the input action of this system is $e^{-\alpha n}$.

Thus, the method given above enables us to determine m_y and D_y at a fixed instant $\bar{t}=n$ for prescribed values of the coefficients of statistical linearisation $k_0(m_y, \eta_y)$ and $k_1(m_y, \eta_y)$.

Supplementing (5.2) and (5.3) by the connection equation (4.18) and using the method of successive approximations, we can find the mathematical expectation and dispersion at the instant we are interested in.

The mathematical expectation and dispersion as functions

of time can be found by determining m_y and D_y for the various instants $0<n<N$ by means of the above method.

Accordingly, making use of this method we can solve the second, more general problem concerning the accuracy of a non-linear sampled data system in a transient state: the problem of determining the mathematical expectation and dispersion as functions of time.

It was assumed above that the correlation function $K_x[n]$ of the random function X which acts at the input of the system has the form (5.1). But since the correlation function of any stationary random process can be represented in the form

$$K_x[n] = \sum_{i=1}^{k} A_i e^{-\alpha_i |n|}, \quad \alpha_i > 0 \tag{5.6}$$

the method given above is easily extended to the case of an arbitrary random stationary disturbance.

In fact, if the correlation function $K_x[n]$ has the form (5.6), the dispersion can be determined from

$$D_y[n] = \sum_{i=1}^{k} A_i e^{-2\alpha_i} F_i^2[n] + \sum_{i=1}^{k} A_i (1 - e^{-2\alpha_i}) \sum_{m=0}^{n} F_i^2[m] \tag{5.7}$$

where $F_i[l]$ is the response of the system at the instant $\bar{t} = l$ to a disturbance of the form $e^{-\alpha_i n}$, $n > 0$.

As before, the mathematical expectation m_y is given by formula (5.4).

The process of determining the mathematical expectation and the dispersion by means of the above method can be carried out by a digital computer.

With slight modifications the methods just given can be extended to the case where X has a mathematical expectation in the form of a slowly varying function of time

These methods are also valid for the case where the block diagram of the system has the form shown in Fig. 4.10, i.e. where a non-linear element is in the feedback loop. Note that the methods discussed can easily be extended to other cases, e.g. where a non-linear element is included in the forward loop of the system.

Note also that the methods for determining mathematical expectation and dispersion have been obtained with the assumption that there is only a single non-linear element in the loop of the non-linear system. But these methods are in fact easily generalised for the case where the system contains several non-linear elements. The amount of computational work necessary to obtain the accuracy characteristics of the system then increases considerably.

If a non-linear system contains several non-linear elements, it may turn out that the method of statistical linearisation, being the simplest and most convenient method for engineering calculations, requires more computational work than the direct method of statistical experimentation, the Monte Carlo method. A preliminary analysis of the problems which have to be solved usually enables us to choose one method or another. It is difficult to give general rules.

4.6 METHOD OF STATISTICAL LINEARISATION IN STUDIES OF THE ACCURACY OF NON-STATIONARY SYSTEMS

The method of statistical linearisation is at present the simplest engineering method for an approximate study of the accuracy of non-linear stationary systems. But the use of this method to assess the accuracy of non-stationary systems, when the latter are subjected to the action of non-stationary random disturbances, encounters serious difficulties.

The procedures given in the preceding sections for the determination of the mathematical expectation and dispersion at the output of the system can of course be applied to non-stationary systems, but the convergence of the process of successive approximations is very slow.

To improve the convergence we can choose the coefficients k_0 and k_1 of the first approximation from a preliminary analysis of the system in question. For this we can determine a number of values $k_0[n_i]$ and $k_1[n_i]$ for fixed instants n_i, assuming that the system is stationary and its parameters correspond to those of the original non-stationary system at the instants $\bar{t}=n_i$. We can also assume that the disturbances acting on the system are stationary.

Considering then the coefficients of statistical linearisation as known functions of time, we can determine the mathematical expectation and dispersion at the instants under consideration, for given characteristics of the random disturbances, using the theory of linear non-stationary sampled data systems. Using the values of m_y and D_y thus found we can find more accurate values for $k_0[n]$ and $k_1[n]$ and again determine the mathematical expectation and dispersion.

Continuing this process, we can determine the mathematical expectation and dispersion with any desired accuracy within the limits of the method.

It follows from the foregoing that, the use of statistical linearisation for the approximate analysis of non-stationary systems involves a great deal of preparatory work and a large number of computations. Hence the use of statistical linearisation for the analysis of non-stationary systems (in particular, systems with several non-linear elements) is hardly suitable. In these cases direct analysis, i.e. the method of statistical experimentation (the Monte Carlo method), requires less machine time, as well as less preliminary work.

4.7 CORRELATION FUNCTION OF A RANDOM PROCESS AT THE OUTPUT OF A NON-LINEAR ELEMENT

Statistical linearisation of the non-linear relationships is based on replacing the non-linear element by two inertia-free linear elements whose transfer coefficients are chosen to satisfy the condition of keeping the mathematical expectation and dispersion the same. At the same time no account is taken of the transformation of the frequency spectrum of the random signal by means of the non-linear element. This circumstance accounts for the comparatively low accuracy in determining the mathematical expectation and dispersion of a non-linear system when the method of statistical linearisation is used.

Increased accuracy can be achieved if we take into account the transformation of the frequency spectrum by the non-linear element. We shall now consider an approximate method for analysing the accuracy of non-linear systems, from which the method of statistical linearisation arises as a first approximation. This method is based on the knowledge of the correlation function of the random process at the output of the non-linear element.

Let us determine the correlation function of the random function at the output of the non-linear element. Suppose that a stationary random function X with mathematical expectation m_x and correlation function $K_x[n]$ acts at the input of the non-linear element. Denoting the characteristic of the non-linear element by $g(x)$ and using the theory of transforming random functions, we can write the following expression which defines the correlation function $K_y[x]$ of the random process at the output of the element in question:

$$K_y[m]=\int_{-\infty}^{\infty}\int_{-\infty}^{\infty} g(x_1)\,g(x_2)\,f(x_1,\ x_2)\,dx_1dx_2-m_y^2 \qquad (7.1)$$

where m_y is the mathematical expectation of the random process Y;

$f(x_1,\ x_2)$ is the distribution law of the set of random quantities $x_1=x[n]$ and $x_2=x[n+m]$.

As before, we assume that the distribution law $f(x_1,\ x_2)$ is normal and can be written in the following form:

$$f(x_1,\ x_2)=\frac{1}{2\pi\eta_x^2\sqrt{1-\rho_x^2[m]}}\exp\left\{-\frac{(x_1-m_x)^2+(x_2-m_x)^2-2\rho_x(x_1-m_x)(x_2-m_x)}{2\eta_x^2(1-\rho_x^2[m])}\right\} \qquad (7.2)$$

where η_x^2 and m_x are the dispersion and the mathematical expectation of X, whilst $\rho_x[m]=\frac{K_x[m]}{\eta_x^2}$ is the normalised correlation function of X.

To determine the correlation function we use ideas of Amiantov and Tikhonov [44] and Bunimovich [45]. These are based on the representation proved by Cramer [46], of the normal two-dimensional distribution law in the form of an infinite series

$$\frac{1}{2\pi\eta_x^2\sqrt{1-\rho_x^2}}\exp\left\{-\frac{x_1^2+x_2^2-2\rho_x x_1x_2}{2\eta_x^2(1-\rho_x^2)}\right\}=\frac{1}{\eta_x^2}\sum_{l=0}^{\infty}\Phi^{(l+1)}\left(\frac{x_1}{\eta_x}\right)\Phi^{(l+1)}\left(\frac{x_2}{\eta_x}\right)\frac{\rho_x^2}{l!} \qquad (7.3)$$

where

$$\Phi^{(l+1)}(z)=(-1)^l H_l(z)\frac{1}{\sqrt{2\pi}}e^{-\frac{z^2}{2}} \qquad (7.4)$$

and $H_l(z)$ are Chebychev-Hermite polynomials satisfying the equality

$$\frac{d^l}{dz^l} e^{-\frac{z^2}{2}} = (-1)^l H_l(z) e^{-\frac{z^2}{2}} \tag{7.5}$$

In what follows we shall use a form of expansion of the two-dimensional distribution law which differs slightly from (7.3). Using (7.4) and (7.5), we can write

$$\frac{1}{\sqrt{1-\rho_x^2}} \exp\left\{-\frac{\xi^2+\vartheta^2-2\rho_x\xi\vartheta}{2(1-\rho_x^2)}\right\} = \sum_{l=0}^{\infty} \frac{\rho_x^l}{l!} \frac{d^l}{d\xi^l} e^{-\frac{\xi^2}{2}} \frac{d^l}{d\vartheta^l} e^{-\frac{\vartheta^2}{2}} \tag{7.6}$$

Introducing the symbols

$$\frac{x_1-m_x}{\eta_x} = \xi \text{ and } \frac{x_2-m_x}{\eta_x} = \vartheta \tag{7.7}$$

the distribution law (7.2) can be written in the form analogous to (7.6)

$$f(\eta_x\xi+m_x, \eta_x\vartheta+m_x) = \frac{1}{2\pi\eta_x^2} \sum_{l=0}^{\infty} \frac{\rho_x^l}{l!} \frac{d^l}{d\xi^l} e^{-\frac{\xi^2}{2}} \frac{d^l}{d\vartheta^l} e^{-\frac{\vartheta^2}{2}} \tag{7.8}$$

Substituting now the expression for the two-dimensional distribution law (7.8) into the (7.6) we have

$$K_y[m] = \sum_{l=0}^{\infty} \frac{1}{2\pi} \frac{\rho_x^l}{l!} \int_{-\infty}^{\infty}\int_{-\infty}^{\infty} g(\eta_x\xi+m_x) g(\eta_x\vartheta+m_x) \times \frac{d^l}{d\xi^l} e^{-\frac{\xi^2}{2}} \frac{d^l}{d\vartheta^l} e^{-\frac{\vartheta^2}{2}} d\xi\, d\vartheta - m_y^2 \tag{7.9}$$

It follows from (7.9) that the double integral can be represented as a product of two single integrals. Accordingly we have

$$K_y[m] = \sum_{l=0}^{\infty} \frac{1}{2\pi} \frac{\rho_x^l}{l!} \int_{-\infty}^{\infty} g(\eta_x\xi+m_x) \frac{d^l}{d\xi^l} e^{-\frac{\xi^2}{2}} d\xi \times \int_{-\infty}^{\infty} g(\eta_x+m_x) \frac{d^l}{d\vartheta^l} e^{-\frac{\vartheta^2}{2}} d\vartheta - m_y^2$$

or, in the final form,

$$K_y[m] = \sum_{l=0}^{\infty} a_l^2 \rho_x^l[m] - m_y^2 \tag{7.10}$$

where

$$a_l = \frac{1}{\sqrt{2\pi l!}} \int_{-\infty}^{\infty} g(\eta_x \xi + m_x) \frac{d^l}{d\xi^l} e^{-\frac{\xi^2}{2}} d\xi \tag{7.11}$$

When $l = 0$ we obtain from (7.11)

$$a_0 = \frac{1}{\sqrt{2\pi}} \int_{-\infty}^{\infty} g(\eta_x \xi + m_x) e^{-\frac{\xi^2}{2}} d\xi = m_y$$

Consequently, the correlation function of the random process at the output of the non-linear element is determined by

$$K_y[m] = \sum_{l=1}^{\infty} a_l^2 \rho_x^l[m] \tag{7.12}$$

Expression (7.12) has been obtained also by Amiantov and Tikhonov [44].

The terms of (7.12) take into account the distortion of the spectrum of the input signal by the non-linear element when $l > 1$.

When $m = 0$ we obtain from (7.12) the following formula giving the dispersion of the random process at the output of the non-linear element

$$\eta_y^2 = K_y[0] = \sum_{l=1}^{\infty} a_l^2 \tag{7.13}$$

since $\rho_x^l[0] = 1$.

If a linear system with weighting function $W[n; m]$ is connected in series with the non-linear element, then the correlation function $K_z[n; m]$ of the random process Z at the output is determined by

$$K_z[n; m] = \sum_{m_1=0}^{n_1} W[n_1; m_1] \sum_{m_2=0}^{n} W[n; m_2] \sum_{l=1}^{\infty} a_l^2 \rho_x^l[m_1 - m_2] \tag{7.14}$$

Similarly we can find the expression for the cross-correlation function $K_{zx}[n; m]$ of Z (at the output of a system formed by connecting in series the non-linear element and the sampled data system with the weighting function $W[n;\ m]$) and the random function X, which constitutes the input action of the system thus formed. In the case of a stationary system this expression has the form

$$K_{zx}[n] = -\eta_x a_1 \sum_{m=0}^{n} W[n-m]\,\rho_x[m] \tag{7.15}$$

Putting $W[n-m] = \sigma[n-m]$ in (7.15) we find

$$K_{zx}[n] = -\eta_x a_1 \sum_{m=0}^{n} \sigma[n-m]\,\rho_x[m] = -\eta_x a_1 \rho_x[n] \tag{7.16}$$

whence it follows that the cross-correlation function of the random function Z at the output of an inertia-free non-linear element and the random input function X is directly proportional to the correlation function of X. It also should be borne in mind that the results obtained above are valid for the case where the distribution law $f(x_1,\ x_2)$ is normal.

As follows from (7.12), in order to determine the correlation function of the random process at the output of the non-linear element, we must know the coefficients a_l, which are determined by the characteristic of the non-linear element. For typical non-linear elements the corresponding expressions for a_l can be obtained beforehand. For a small number of non-linear elements with characteristics of the relay, limiter, limiter with a dead zone, dead zone, cubic parabola and skew-symmetric quadratic parabola type, the corresponding expressions for determining a_l have been derived by Amiantov and Tikhonov [44] and Pupkov [47]. These do not exhaust the large variety of non-linear elements encountered in practice.

In the next section expressions will be determined for the coefficients a_l of the generalised non-linear characteristic. These expressions enable us to find a_l for a large number of non-linear elements whose characteristics are derived from the generalised non-linear characteristics.

4.8 COEFFICIENTS FOR A NON-LINEAR ELEMENT WITH A GENERALISED CHARACTERISTIC

It was shown in the preceding section that the correlation function of the random process at the output of a non-linear element can be represented in the form of a series

$$K_y[m]=\sum_{l=1}^{\infty} a_l^2 \rho_x^l [m] \tag{8.1}$$

where the coefficients a_l are determined from

$$a_l=\frac{1}{\sqrt{2\pi l!}} \int_{-\infty}^{\infty} g(\eta_x \xi + m_x) \frac{d^l}{d\xi^l} e^{-\frac{\xi^2}{2}} d\xi \tag{8.2}$$

Expression (8.2) is valid for the case where the characteristic $g(x)$ of the non-linear element is single-valued and the distribution $f(x_1, x_2)$ law is normal.

If the characteristic of the non-linear element is not single-valued, then, in order to determine the correlation function K_y, we must know the distribution law $f(x_1, \dot{x}_1, x_2, \dot{x}_2)$ of the set of random quantities $x_1=x[n]$, $\dot{x}_1 = \dot{x}[n]$, $x_2=x[n+m]$ and $\dot{x}_2=\dot{x}[n+m]$. It can be shown that for a multiple-valued characteristic (see, for example, Fig. 4.9), when the distribution law is symmetric, the correlation function is calculated from

$$K_y[m]-\int_{-\infty}^{-a}\int_{-\infty}^{-a} g_4(x_1) g_4(x_2) f(x_1, x_2)\, dx_1 dx_2$$

$$+\int_{-a}^{b}\int_{-a}^{b} \frac{1}{2}[g_1(x_1) g_1(x_2)+g_2(x_1) g_2(x_2)] f(x_1, x_2)\, dx_1 dx_2$$

$$+\int_{b}^{\infty}\int_{b}^{\infty} g_3(x_1) g_3(x_2) f(x_1, \dot{x}_2)\, dx_1 dx_2 - m_y^2 \tag{8.3}$$

If the distribution law $f(x_1, x_2)$ is normal, then, using the method given in the preceding section, and proceeding from (8.3), we can show that a_l over the multiple-valued sections are determined from

$$a_l=\frac{1}{\sqrt{2\pi l!}} \int_{-\frac{a+m_x}{\eta_x}}^{\frac{b-m_x}{\eta_x}} \frac{1}{2}[g_1(\eta_x \xi + m_x)+g_2(\eta_x \xi + m_x)] \frac{d^l}{d\xi^l} e^{-\frac{\xi^2}{2}} d\xi \tag{8.4}$$

Expressions (8.2) and (8.4) demonstrate that for a given distribution law (normal in the case considered) the coefficients are determined solely by the characteristic of the non-linear element. For typical non-linear elements the expressions for a_l can thus be determined beforehand.

In this section we shall find the expressions for the coefficients a_l $(l=0,1,2, ..., 5)$ of a non-linear element with the generalised characteristic. Using these expressions we can then determine a_l for any other non-linear element whose characteristic is derived from the generalised characteristic. We shall confine ourselves to the first six coefficients. For this the accuracy of the correlation function obtained from (8.1) is sufficiently high. This is explained by a fairly rapid convergence of the series (8.1), since its terms contain the factor $1/l!$.

Putting $l=0,1, ..., 5$ in (8.2) we have

$$a_0=\frac{1}{\sqrt{2\pi}}\int_{-\infty}^{\infty} g(\eta_x\xi+m_x)\,e^{-\frac{\xi^2}{2}}\,d\xi \tag{8.5}$$

$$a_1=-\frac{1}{\sqrt{2\pi}}\int_{-\infty}^{\infty} \xi g(\eta_x\xi+m_x)\,e^{-\frac{\xi^2}{2}}\,d\xi \tag{8.6}$$

$$a_2=\frac{1}{\sqrt{2\pi\cdot 2!}}\int_{-\infty}^{\infty} (\xi^2-1)\,g(\eta_x\xi+m_x)\,e^{-\frac{\xi^2}{2}}\,d\xi \tag{8.7}$$

$$a_3=\frac{1}{\sqrt{2\pi\cdot 3!}}\int_{-\infty}^{\infty} (3\xi-\xi^3)\,g(\eta_x\xi+m_x)\,e^{-\frac{\xi^2}{2}}\,d\xi \tag{8.8}$$

$$a_4=\frac{1}{\sqrt{2\pi\cdot 4!}}\int_{-\infty}^{\infty} (3-6\xi^2+\xi^4)\,g(\eta_x\xi+m_x)\,e^{-\frac{\xi^2}{2}}\,d\xi \tag{8.9}$$

$$a_5=\frac{1}{\sqrt{2\pi\cdot 5!}}\int_{-\infty}^{\infty} (-15\xi+10\xi^3-\xi^5)\,g(\eta_x\xi+m_x)\,e^{-\frac{\xi^2}{2}}\,d\xi \tag{8.10}$$

For convenience in practical calculations it is advisable to transform the expressions for a_l in such a way that the coefficients with higher numbers are expressed in terms of coefficients with lower numbers. Carrying out certain transformations, we obtain the following expressions for the coefficients:

$$a_2=\frac{1}{\sqrt{2!}}[I(\xi^2)-a_0] \tag{8.11}$$

$$a_3 = -\frac{1}{\sqrt{3!}}[I(\xi^3) + 3a_1] \tag{8.12}$$

$$a_4 = \frac{1}{\sqrt{4!}}[I(\xi^4) - 6\sqrt{2!}a_2 - 3a_0] \tag{8.13}$$

$$a_5 = -\frac{1}{\sqrt{5!}}[I(\xi^5) + 10\sqrt{3!}a_3 + 15a_1] \tag{8.14}$$

where

$$I(\xi^l) = \frac{1}{\sqrt{2\pi}} \int_{-\infty}^{\infty} \xi^l g(\eta_x \xi + m_x)\, e^{-\frac{\xi^2}{2}}\, d\xi \tag{8.15}$$

Equations (8.11) to (8.14) are also convenient for calculating a_l for non-linear elements with multiple-valued characteristics. In this case only $I(\xi^l)$ must be determined in another way. If on the section $[c;\ d]$ the characteristic $g(x)$ is such that when $\dot{x} > 0$ $g(x) = g_1(x)$, and when $\dot{x} < 0$ $g(x) = g_2(x)$, then (8.15) assumes the form

$$I(\xi^l) = \frac{1}{\sqrt{2\pi}} \int_{\frac{c-m_x}{\eta_x}}^{\frac{d-m_x}{\eta_x}} \frac{1}{2}\, \xi^l [g_1(\eta_x\xi + m_x) + g_2(\eta_x\xi + m_x)]\, e^{-\frac{\xi^2}{2}}\, d\xi \tag{8.16}$$

It should also be borne in mind that on multiple-valued sections the coefficients a_0 and a_1 in (8.11) to (8.14) must be calculated from

$$a_0 = \frac{1}{\sqrt{2\pi}} \int_{\frac{c-m_x}{\eta_x}}^{\frac{d-m_x}{\eta_x}} \frac{1}{2}\, [g_1(\eta_x\xi + m_x) + g_2(\eta_x\xi + m_x)]\, e^{-\frac{\xi^2}{2}}\, d\xi \tag{8.17}$$

$$a_1 = -\frac{1}{\sqrt{2\pi}} \int_{\frac{c-m_x}{\eta_x}}^{\frac{d-m_x}{\eta_x}} \frac{1}{2}\, \xi [g_1(\eta_x\xi + m_x) + g_2(\eta_x\xi + m_x)]\, e^{-\frac{\xi^2}{2}}\, d\xi \tag{8.18}$$

The calculation of the integrals is lengthy. Because of this we shall give only the final expressions for a_l.

In accordance with (8.5) and (8.15) the coefficient a_0 equals

$$a_0 = I(\xi^0) = I^{(-)}(\xi^0) + I^{(+)}(\xi^0) \tag{8.19}$$

where $I^{(-)}(\xi^0)$ and $I^{(+)}(\xi^0)$ are the values of the integral (8.5) for the left-hand and right-hand parts of the generalised characteristic.

In fact

$$\begin{aligned} I^{(-)}(\xi^0) = & \frac{(k_1 - k_2)\eta_x}{2\sqrt{2\pi}}\left[e^{\frac{(c_1+m_x)^2}{2\eta_x^2}} + e^{-\frac{(d_1+m_x)^2}{2\eta_x^2}}\right] \\ & - \frac{k_1\eta_x}{2\sqrt{2\pi}}\left[e^{-\frac{(a_1+m_x)^2}{2\eta_x^2}} + e^{-\frac{(b_1+m_x)^2}{2\eta_x^2}}\right] \\ & + \frac{(k_1 - k_2)m_x + l_1}{2}\left[\Phi\left(\frac{c_1+m_x}{\eta_x}\right) + \Phi\left(\frac{d_1+m_x}{\eta_x}\right)\right] \\ & - \frac{k_1 m_x}{2}\left[\Phi\left(\frac{a_1+m_x}{\eta_x}\right) + \Phi\left(\frac{b_1+m_x}{\eta_x}\right)\right] + k_2 m_x - l_1 \end{aligned} \tag{8.20}$$

and

$$\begin{aligned} I^{(+)}(\xi^0) = & -\frac{(k_3 - k_4)\eta_x}{2\sqrt{2\pi}}\left[e^{-\frac{(c_2-m_x)^2}{2\eta_x^2}} + e^{-\frac{(d_2-m_x)^2}{2\eta_x^2}}\right] \\ & + \frac{k_3\eta_x}{2\sqrt{2\pi}}\left[e^{-\frac{(a_2-m_x)^2}{2\eta_x^2}} + e^{-\frac{(b_2-m_x)^2}{2\eta_x^2}}\right] \\ & + \frac{(k_3 - k_4)m_x - l_2}{2}\left[\Phi\left(\frac{c_2-m_x}{\eta_x}\right) + \Phi\left(\frac{d_2-m_x}{\eta_x}\right)\right] \\ & - \frac{k_3 m_x}{2}\left[\Phi\left(\frac{a_2-m_x}{\eta_x}\right) + \Phi\left(\frac{b_2-m_x}{\eta_x}\right)\right] + k_4 m_x + l_2 \end{aligned} \tag{8.21}$$

The coefficient a_1 given by (8.6) is determined as follows:

$$a_1 = I^{(-)}(\xi^1) + I^{(+)}(\xi^1) \tag{8.22}$$

where

$$I^{(-)}(\xi^1) = \frac{(k_1 - k_2)c_1 - l_1}{2\sqrt{2\pi}} e^{-\frac{(c_1+m_x)^2}{2\eta_x^2}} - \frac{k_1 a_1}{2\sqrt{2\pi}} e^{-\frac{(a_1+m_x)^2}{2\eta_x^2}}$$

$$+ \frac{(k_1 - k_2)d_1 - l_1}{2\sqrt{2\pi}} e^{-\frac{(d_1+m_x)^2}{2\eta_x^2}} - \frac{k_1 b_1}{2\sqrt{2\pi}} e^{-\frac{(b_1+m_x)^2}{2\eta_x^2}} \tag{8.23}$$

$$- \frac{(k_1 - k_2)\eta_x}{2}\left[\Phi\left(\frac{c_1 + m_x}{\eta_x}\right) + \Phi\left(\frac{d_1 + m_x}{\eta_x}\right)\right]$$

$$+ \frac{k_1\eta_x}{2}\left[\Phi\left(\frac{a_1 + m_x}{\eta_x}\right) + \Phi\left(\frac{b_1 + m_x}{\eta_x}\right)\right] - k_2\eta_x$$

and

$$I^{(+)}(\xi^1) = \frac{(k_3 - k_4)c_2 - l_2}{2\sqrt{2\pi}} e^{-\frac{(c_2-m_x)^2}{2\eta_x^2}} - \frac{k_3 a_2}{2\sqrt{2\pi}} e^{-\frac{(a_2-m_x)^2}{2\eta_x^2}}$$

$$+ \frac{(k_3 - k_4)d_2 - l_2}{2\sqrt{2\pi}} e^{-\frac{(d_2-m_x)^2}{2\eta_x^2}} - \frac{k_3 b_2}{2\sqrt{2\pi}} e^{-\frac{(b_2-m_x)^2}{2\eta_x^2}} \tag{8.24}$$

$$- \frac{(k_3 - k_4)\eta_x}{2}\left[\Phi\left(\frac{c_2 - m_x}{\eta_x}\right) + \Phi\left(\frac{d_2 - m_x}{\eta_x}\right)\right]$$

$$+ \frac{k_3\eta_x}{2}\left[\Phi\left(\frac{a_2 - m_x}{\eta_x}\right) + \Phi\left(\frac{b_2 - m_x}{\eta_x}\right)\right] - k_4\eta_x$$

The coefficient a_2 given by (8.11) is determined by

$$a_2 = \frac{1}{\sqrt{2!}}\left[I^{(-)}(\xi^2) + I^{(+)}(\xi^2) - a_0\right] \tag{8.25}$$

where

$$I^{(-)}(\xi^2) = U_2^{(-)}(c_1) e^{-\frac{(c_1+m_x)^2}{2\eta_x^2}} - V_2^{(-)}(a_1) e^{-\frac{(a_1+m_x)^2}{2\eta_x^2}}$$

$$+ U_2^{(-)}(d_1) e^{-\frac{(d_1+m_x)^2}{2\eta_x^2}} - V_2^{(-)}(b_1) e^{-\frac{(b_1+m_x)^2}{2\eta_x^2}} \tag{8.26}$$

$$+ \frac{(k_1 - k_2)m_x + l_1}{2}\left[\Phi\left(\frac{c_1 + m_x}{\eta_x}\right) + \Phi\left(\frac{d_1 + m_x}{\eta_x}\right)\right]$$

$$- \frac{k_1 m_x}{2}\left[\Phi\left(\frac{a_1 + m_x}{\eta_x}\right) + \Phi\left(\frac{b_1 + m_x}{\eta_x}\right)\right] + k_2 m_x - l_1$$

Here

$$U_2^{(-)}(z)=\frac{(k_1-k_2)(z+m_x)^2-[(k_1-k_2)m_x+l_1](z+m_x)+2\eta_x^2(k_1-k_2)}{2\eta_x\sqrt{2\pi}} \tag{8.27}$$

$$V_2^{(-)}(z)=\frac{(z+m_x)^2-(z+m_x)m_x+2\eta_x^2}{2\eta_x\sqrt{2\pi}}k_1 \tag{8.28}$$

$$U_2^{(-)}(c_1)=U_2^{(-)}(z)\,|_{z=c_1};\ U_2^{(-)}(d_1)=U_2^{(-)}(z)\,|_{z=d_1} \tag{8.29}$$

$$V_2^{(-)}(b_1)=V_2^{(-)}(z)\,|_{z=b_1};\ V_2^{(-)}(a_1)=V_2^{(-)}(z)\,|_{z=a_1} \tag{8.30}$$

$$\begin{aligned}I^{(+)}(\xi^2)=&-U_2^{(+)}(c_2)\,e^{-\frac{(c_2-m_x)^2}{2\eta_x^2}}+V_2^{(+)}(a_2)\,e^{-\frac{(a_2-m_x)^2}{2\eta_x^2}}\\&-U_2^{(+)}(d_2)\,e^{-\frac{(d_2-m_x)^2}{2\eta_x^2}}+V_2^{(+)}(b_2)\,e^{-\frac{(b_2-m_x)^2}{2\eta_x^2}}\\&+\frac{(k_3-k_4)m_x-l_2}{2}\left[\Phi\left(\frac{c_2-m_x}{\eta_x}\right)+\Phi\left(\frac{d_2-m_x}{\eta_x}\right)\right]\\&-\frac{k_3m_x}{2}\left[\Phi\left(\frac{a_2-m_x}{\eta_x}\right)+\Phi\left(\frac{b_2-m_x}{\eta_x}\right)\right]+k_4m_x+l_2\end{aligned} \tag{8.31}$$

where $U_2^{(+)}(z)$ and $V_2^{(+)}(\xi)$ are determined in the following manner:

$$U_2^{(+)}(z)=\frac{(k_3-k_4)(z-m_x)^2+[(k_3-k_4)m_x-l_2](z-m_x)+2\eta_x^2(k_3-k_4)}{2\eta_x\sqrt{2\pi}} \tag{8.32}$$

$$V_2^{(+)}(z)=k_3\frac{(z-m_x)^2+(z-m_x)m_x+2\eta_x^2}{2\eta_x\sqrt{2\pi}} \tag{8.33}$$

When $z=c_2$, $z=a_2$, $z=b_2$ and $z=d_2$ the quantities

$$U_2^{(+)}(c_2),\ V_2^{(+)}(a_2),\ U_2^{(+)}(d_2)$$

and $V_2^{(+)}(b_2)$ in (8.31) are obtained from (8.32).

To determine the coefficient a_3 we have from (8.12)

$$a_3=-\frac{1}{\sqrt{3!}}[I^{(-)}(\xi^3)+I^{(+)}(\xi^3)+3a_1] \tag{8.34}$$

where

$$I^{(-)}(\xi^3) = -U_3^{(-)}(c_1)\,e^{-\frac{(c_1+m_x)^2}{2\eta_x^2}} + V_3^{(-)}(a_1)\,e^{-\frac{(a_1+m_x)^2}{2\eta_x^2}}$$

$$- U_3^{(-)}(d_1)\,e^{-\frac{(d_1+m_x)^2}{2\eta_x^2}} + V_3^{(-)}(b_1)\,e^{-\frac{(b_1+m_x)^2}{2\eta_x^2}} \tag{8.35}$$

$$+ 3\eta_x \frac{k_1 - k_2}{2}\left[\Phi\left(\frac{c_1 + m_x}{\eta_x}\right) + \Phi\left(\frac{d_1 + m_x}{\eta_x}\right)\right]$$

$$- \frac{3}{2}\eta_x k_1\left[\Phi\left(\frac{a_1 + m_x}{\eta_x}\right) + \Phi\left(\frac{b_1 + m_x}{\eta_x}\right)\right] + 3k_2\eta_x$$

$$U_3^{(-)}(z)$$

$$= \frac{(k_1-k_2)(z+m_x)^3 - [(k_1-k_2)m_x+l_1](z+m_x)^2 - 3(k_1-k_2)(z+m_x)\eta_x^2}{2\eta_x^2\sqrt{2\pi}} \tag{8.36}$$

$$- \frac{2\eta_x^2[(k_1 - k_2)m_x + l_1]}{2\eta_x^2\sqrt{2\pi}}$$

$$V_3^{(-)}(z) = k_1 \frac{(z + m_x)^3 - (z + m_x)^2 m_x + 3(z + m_x)\eta_x^2 - 2\eta_x^2 m_x}{2\eta_x^2\sqrt{2\pi}} \tag{8.37}$$

When $z = c_1$, $z = a_1$, $z = d_1$, and $z = b_1$ we obtain from (8.36) and (8.37) the expressions for $U_3^{(-)}(c_1)$, $V_3^{(-)}(a_1)$, $U_3^{(-)}(d_1)$ and $V_3^{(-)}(b_1)$ in (8.35).

Further

$$I^{(+)}(\xi^3) = -U_3^{(+)}(c_2)\,e^{-\frac{(c_2-m_x)^2}{2\eta_x^2}} + V_3^{(+)}(a_2)\,e^{-\frac{(a_2-m_x)^2}{2\eta_x^2}}$$

$$-U_3^{(+)}(d_2)\,e^{-\frac{(d_2-m_x)^2}{2\eta_x^2}} + V_3^{(+)}(b_2)\,e^{-\frac{(b_2-m_x)^2}{2\eta_x^2}} \tag{8.38}$$

$$+ 3\eta_x^2 \frac{k_3 - k_4}{2}\left[\Phi\left(\frac{c_2 - m_x}{\eta_x}\right) + \Phi\left(\frac{d_2 - m_x}{\eta_x}\right)\right]$$

$$- \frac{3}{2}\eta_x k_3\left[\Phi\left(\frac{a_2 - m_x}{\eta_x}\right) + \Phi\left(\frac{b_2 - m_x}{\eta_x}\right)\right] + 3\eta_x k_4$$

where

$$U_3^{(+)}(z)$$

$$=\frac{(k_3-k_4)(z-m_x)^3+[(k_3-k_4)m_x-l_2](z-m_x)^2+3\eta_x^2(k_3-k_4)(z-m_x)}{2\eta_x^2\sqrt{2\pi}}$$

$$+\frac{2\eta_x^2[(k_3-k_4)m_x-l_2]}{2\eta_x^2\sqrt{2\pi}} \tag{8.39}$$

$$V_3^{(+)}(z)=k_3\frac{(z-m_x)^3+(z-m_x)^2m_x+3\eta_x^2(z-m_x)+2\eta_x^2m_x}{2\eta_x^2\sqrt{2\pi}} \tag{8.40}$$

The expressions $U_3^{(+)}(c_2)$, $V_3^{(+)}(a_2)$, $U_3^{(+)}(d_2)$ and $V_3^{(+)}(b_2)$ are obtained from (8.39) and (8.40) by putting $z=c_2$, $z=a_2$, $z=d_2$ and $z=b_2$.

The coefficient a_4 given by (8.13) is determined as follows

$$a_4=\frac{1}{\sqrt{4!}}[I^{(-)}(\xi^4)+I^{(+)}(\xi^4)-6\sqrt{2!}a_2-3a_0] \tag{8.41}$$

where

$$I^{(-)}(\xi^4)=U_4^{(-)}(c_1)e^{-\frac{(c_1+m_x)^2}{2\eta_x^2}}-V_4^{(-)}(a_1)e^{-\frac{(a_1+m_x)^2}{2\eta_x^2}}$$

$$+U_4^{(-)}(d_1)e^{-\frac{(d_1+m_x)^2}{2\eta_x^2}}-V_4^{(-)}(b_1)e^{-\frac{(b_1+m_x)^2}{2\eta_x^2}} \tag{8.42}$$

$$+3\frac{(k_1-k_2)m_x+l_1}{2}\left[\Phi\left(\frac{c_1+m_x}{\eta_x}\right)+\Phi\left(\frac{d_1+m_x}{\eta_x}\right)\right]$$

$$-3\frac{k_1m_x}{2}\left[\Phi\left(\frac{a_1+m_x}{\eta_x}\right)+\Phi\left(\frac{b_1+m_x}{\eta_x}\right)\right]+3(k_1m_x-l_1)$$

The quantities $U_4^{(-)}$ and $V_4^{(-)}$ entering (8.42) are determined in the following manner:

$$U_4^{(-)}(z)$$

$$=\frac{(k_1-k_2)(z+m_x)^4-[(k_1-k_2)m_x+l_1](z+m_x)^3+4\eta_x^2(k_1-k_2)(z+m_x)^2}{2\eta_x^3\sqrt{2\pi}}$$

$$-\frac{3\eta_x^2[(k_1-k_2)m_x+l_1](z+m_x)+8\eta_x^4(k_1-k_2)}{2\eta_x^3\sqrt{2\pi}} \tag{8.43}$$

$$V_4^{(-)}(z) = k_1$$

$$\times \frac{(z+m_x)^4 - (z+m_x)^3 m_x + 4\eta_x^2 (z+m_x)^2 - 3\eta_x^2 (z+m_x) m_x + 8\eta_x^2}{2\eta_x^3 \sqrt{2\pi}} \tag{8.44}$$

Further, the term $I^{(+)}(\xi^4)$ in (8.41) is given by

$$\begin{aligned} I^{(+)}(\xi^4) = & -U_4^{(+)}(c_2) e^{-\frac{(c_2-m_x)^2}{2\eta_x^2}} + V_4^{(+)}(a_2) e^{-\frac{(a_2-m_x)^2}{2\eta_x^2}} \\ & - U_4^{(+)}(d_2) e^{-\frac{(d_2-m_x)^2}{2\eta_x^2}} + V_4^{(+)}(b_2) e^{-\frac{(b_2-m_x)^2}{2\eta_x^2}} \\ & - 3\frac{(k_3-k_4) m_x + l_2}{2}\left[\Phi\left(\frac{c_2-m_x}{\eta_x}\right) + \Phi\left(\frac{d_2-m_x}{\eta_x}\right)\right] \\ & + 3\frac{k_3 m_x}{2}\left[\Phi\left(\frac{a_2-m_x}{\eta_x}\right) + \Phi\left(\frac{b_2-m_x}{\eta_x}\right)\right] + 3(k_4 m_x + l_2) \end{aligned} \tag{8.45}$$

where $U_4^{(+)}(z)$ and $V_4^{(+)}(z)$ are determined by

$$\begin{aligned} U_4^{+}(z) = & \frac{(k_3-k_4)(z-m_x)^4 + [(k_3-k_4)m_x - l_2](z-m_x)^3 + 4\eta_x^2 (k_3-k_4)(z-m_x)^2}{2\eta_x^3\sqrt{2\pi}} \\ & + \frac{3\eta_x^2 [(k_3-k_4) m_x - l_2](z-m_x) + 8\eta_x^4 (k_3-k_4)}{2\eta_x^3\sqrt{2\pi}} \end{aligned} \tag{8.46}$$

$$V_4^{(+)}(z) = k_3$$

$$\times \frac{(z-m_x)^4 + (z-m_x)^3 m_x + 4\eta_x^2 (z-m_x)^2 + 3\eta_x^2 (z-m_x)m_x + 8\eta_x^4}{2\eta_x^3\sqrt{2\pi}} \tag{8.47}$$

The factors $U_4^{(-)}(c_1)$, $U_4^{(-)}(d_1)$, $V_4^{(-)}(b_1)$, $U^{(+)}(c_2)$, $U^{(+)}(d_2)$ and others in (8.42) and (8.45) are obtained from (8.43), (8.44), (8.46) and (8.47) for the corresponding values of the argument z.

From (8.14) the coefficient a_5 is

$$\begin{aligned} a_5 = & -\frac{1}{\sqrt{5!}}[I^{(-)}(\xi^5) + I^{(+)}(\xi^5) \\ & + 10\sqrt{3!}a_3 + 15a_1] \end{aligned} \tag{8.48}$$

where

$$I^{(-)}(\xi^5) = -U_5^{(-)}(c_1)\,e^{-\frac{(c_1+m_x)^2}{2\eta_x^2}} + V_5^{(-)}(a_1)\,e^{-\frac{(a_1+m_x)^2}{2\eta_x^2}}$$
$$-U_5^{(-)}(d_1)\,e^{-\frac{(d_1+m_x)^2}{2\eta_x^2}} + V_5^{(-)}(b_1)\,e^{-\frac{(b_1+m_x)^2}{2\eta_x^2}}$$
$$+15\frac{k_1-k_2}{2}\eta_x\left[\Phi\left(\frac{c_1+m_x}{\eta_x}\right)+\Phi\left(\frac{d_1+m_x}{\eta_x}\right)\right]$$
$$-15\frac{k_1\eta_x}{2}\left[\Phi\left(\frac{a_1+m_x}{\eta_x}\right)+\Phi\left(\frac{b_1+m_x}{\eta_x}\right)\right]+15k_2\eta_x \tag{8.49}$$

$$U_5^{(-)}(z)$$
$$=\frac{(k_1-k_2)(z+m_x)^5+[(k_1-k_2)m_x-l_1](z+m_x)^4+5\eta_x^2(k_1-k_2)(z+m_x)^3}{2\eta_x^4\sqrt{2\pi}}$$
$$-\frac{4\eta_x^2[(k_1-k_2)m_x+l_1](z+m_x)^2-15\eta_x^4(k_1-k_2)(z+m_x)}{2\eta_x^4\sqrt{2\pi}}$$
$$+\frac{8\eta_x^4[(k_1-k_2)m_x+l_1]}{2\eta_x^4\sqrt{2\pi}} \tag{8.50}$$

$$V_5^{(-)}(z)=k_1$$
$$\times\frac{(z+m_x)^5-(z+m_x)^4m_x+5\eta_x^2(z+m_x)^3+4\eta_x^2(z+m_x)^2m_x}{2\eta_x^4\sqrt{2\pi}}$$
$$+k_1\frac{15\eta_x^4(z+m_x)-8\eta_x^4m_x}{2\eta_x^4\sqrt{2\pi}} \tag{8.51}$$

Further, $I^{(+)}(\xi^5)$ in (8.49) is given by

$$I^{(+)}(\xi^5) = -U_5^{(+)}(c_2)\,e^{-\frac{(c_2-m_x)^2}{2\eta_x^2}} + V_5^{(+)}(a_2)\,e^{-\frac{(a_2-m_x)^2}{2\eta_x^2}}$$
$$-U_5^{(+)}(d_2)\,e^{-\frac{(d_2-m_x)^2}{2\eta_x^2}} + V_5^{(+)}(b_2)\,e^{-\frac{(b_2-m_x)^2}{2\eta_x^2}}$$
$$+15\eta_x\frac{k_3-k_4}{2}\left[\Phi\left(\frac{c_2-m_x}{\eta_x}\right)+\Phi\left(\frac{d_2-m_x}{\eta_x}\right)\right] \tag{8.52}$$

$$-15\frac{k_3\eta_x}{2}\left[\Phi\left(\frac{a_2-m_x}{\eta_x}\right)+\Phi\left(\frac{b_2-m_x}{\eta_x}\right)\right]+k_4\eta_x \qquad (8.52)$$

where

$$U_5^{\;+}(z)=$$

$$-\frac{(k_3-k_4)(z-m_x)^5+[(k_3-k_4)\,m_x-l_2]\,(z-m_x)^4+5\eta_x^2\,(k_3-k_4)(z-m_x)^3}{2\eta_x^4\sqrt{2\pi}}$$

$$-\frac{4\eta_x^2\,[(k_3-k_4)\,m_x-l_2]\,(z-m_x)^2+15\eta_x^4\,(k_3-k_4)\,(z-m_x)+}{2\eta_x^4\sqrt{2\pi}} \qquad (8.53)$$

$$\frac{+8\eta_x^4\,[(k_3-k_4)\,m_x-l_2]}{2\eta_x^4\sqrt{2\pi}}$$

$$V_5^{(+)}(z)=k_3$$

$$\times\frac{(z-m_x)^5+(z-m_x)^4\,m_x+5\eta_x^2\,(z-m_x)^3+4\eta_x^2\,(z-m_x)^2}{2\eta_x^4\sqrt{2\pi}}$$

$$+k_3\,\frac{15\eta_x^4\,(z-m_x)+8\eta_x^4\,m_x}{2\eta_x^4\sqrt{2\pi}} \qquad (8.54)$$

The factors $U_5^{(+)}(c_1)$, $U_5^{(-)}(c_2)$, $V_5^{(-)}(a_1)$, $V_5^{(+)}(a_2)$ and others in (8.49) and (8.52) are obtained from (8.50), (8.51), (8.53) and (8.54) for the corresponding values of the argument z.

4.9 REPLACEMENT OF A NON-LINEAR SYSTEM BY AN INFINITE NUMBER OF LINEAR SYSTEMS

The approximate method of investigating the accuracy of non-linear systems with the aid of statistical linearisation is based on replacing the non-linear system by two corresponding linear systems. Such a replacement is fairly approximate, so that the accuracy of determining the mathematical expectation and dispersion at the output of the non-linear system is low, since the transformation of the spectrum by the non-linear element is not considered.

It will be shown in this section that a non-linear system can be replaced by an infinite number of linear systems. Such a replacement is based on determining the correlation function of the random process at the output of the non-linear element.

Let us consider the non-linear system shown in Fig. 4.12.

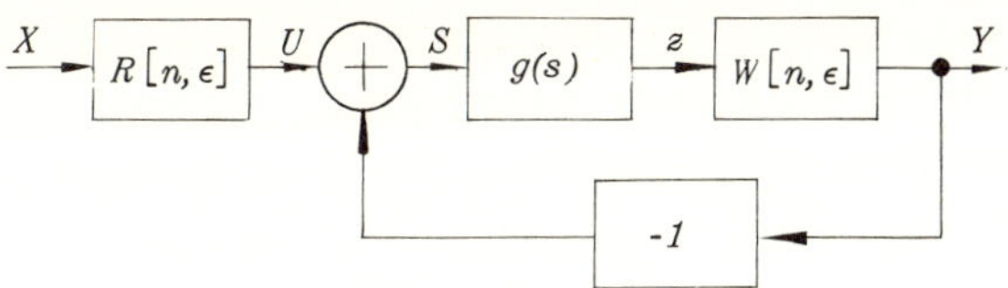

Fig. 4.12 A non-linear system with a non-linear element in the forward loop

As follows from the block diagram, the system is formed by connecting in series a linear system with weighting function $R[n, \varepsilon]$ and a non-linear closed-loop system, the weighting function of the open-loop linear part of which is $W[n, \varepsilon]$. A non-linear element with characteristic $g(s)$ is included in the loop of the second system. Without loss of generality we shall assume that the samplers of both systems work in unison and in phase. This restriction is not a fundamental one, and has been introduced to simplify calculations.

Suppose that a stationary random function X with mathematical expectation m_x and correlation function $K_x[n]$ acts at the input of the system considered. In conformity with the symbols adopted in Fig. 4.12, we can write for the centralised random functions

$$\overset{\circ}{U}[n] = \overset{\circ}{Y}[n] + \overset{\circ}{S}[n] \tag{9.1}$$

Denoting the correlation and cross-correlation functions by

$$M[\overset{\circ}{U}[n]\,\overset{\circ}{U}[n_1]] = K_u[m]$$

$$M[\overset{\circ}{S}[n]\overset{\circ}{S}[n_1]] = K_s[m]$$

$$M[\overset{\circ}{Y}[n]\overset{\circ}{S}[n_1]] = K_{ys}[m]$$

$$M[\overset{\circ}{S}[n]\,Y[n_1]] = K_{sy}[m]$$

and using (9.1) we obtain

$$K_u[m] = K_y[m] + K_s[m] + K_{ys}[m] + K_{sy}[m] \tag{9.2}$$

We next express the correlation functions $K_{ys}[m]$ and $K_{sy}[m]$, in the right-hand side of (9.2), as well as the correlation function $K_y[m]$, in terms of the correlation function $K_s[m]$ of the error of the system. The corresponding expressions can be easily obtained if we assume that the random process $S[n]$ follows a normal distribution law. In fact, on the basis of (8.15) we can then write the following expression for $K_{ys}[m]$:

$$K_{ys}[m] = -\eta_s a_1 \sum_{l=0}^{\infty} W[l] \rho_s [m-l] \tag{9.3}$$

where a_1 is determined from (8.22);

η_s^2 is the dispersion of S;

$\rho_s[m]$ is the normalised correlation function of S.

Hence

$$\rho_s[m] = \frac{K_s[m]}{\eta_s^2} \tag{9.4}$$

Taking into account the properties of the cross-correlation function given by

$$K_{ys}[n-l] = K_{sy}[l-n]$$

we have from (9.3) the following expression for function $K_{sy}[m]$

$$K_{sy}[m] = -\eta_s a_1 \sum_{l=0}^{\infty} W[l] \rho_s [m+l] \tag{9.5}$$

To determine $K_y[m]$ we use (8.1). Thus we have

$$K_y[m] = \sum_{i=1}^{\infty} \frac{a_i^2}{i!} \sum_{l=0}^{\infty} \sum_{\lambda=0}^{\infty} W[l] W[\lambda] \rho_s^i [m-l+\lambda] \tag{9.6}$$

The correlation function $K_u[m]$ can be expressed in terms of $K_x[m]$ of the input action of the system under consideration. Following the theory of linear systems we can write

$$K_u[m] = \sum_{l=0}^{\infty} \sum_{\lambda=0}^{\infty} R[l] R[\lambda] K_x [m-l+\lambda] \tag{9.7}$$

or, if we take into account that

$$K_x[m] = \eta_x^2 \rho_x[m]$$

where $\rho_x[m]$ is the normalised correlation function, we can write the expression for $K_u[m]$ also in the following form:

$$K_u[m] = \eta_x^2 \sum_{l=0}^{\infty} \sum_{\lambda=0}^{\infty} R[l] R[\lambda] \rho_x [m-l+\lambda] \tag{9.8}$$

Representing $K_s[m]$ for the error in the form

$$K_s[m] = \eta_x^2 \rho_s[m]$$

and using (9.3), (9.5), (9.6) and (9.8), we can write (9.2) as follows

$$\eta_x^2 \sum_{l=0}^{\infty} \sum_{\lambda=0}^{\infty} R[l] R[\lambda] \rho_x [m - l + \lambda] = \eta_s^2 \rho_s [m]$$

$$+ \sum_{i=1}^{\infty} \frac{a_i^2}{i!} \sum_{l=0}^{\infty} \sum_{\lambda=0}^{\infty} W[l] W[\lambda] \rho_s^i [m - l + \lambda] \qquad (9.9)$$

$$- \eta_s a_1 \sum_{l=0}^{\infty} W[l] \rho_s [m - l] - \eta_s a_1 \sum_{i=0}^{\infty} W[l] \rho_s [m + l]$$

Equation (9.9) contains the normalised correlation function $\rho_s[m]$ and its powers $\rho_s^l[m]$, $l = 1, 2, \ldots$ It is non-linear and of infinite order. To analyse it we use the procedure [48] enabling non-linear equation of infinite order to be replaced by a system of linear equations.

We write (9.9) in another form picking out the non-linear terms individually:

$$\eta_x^2 \sum_{l=0}^{\infty} \sum_{\lambda=0}^{\infty} R[l] R[\lambda] \rho_x [m - l + \lambda]$$

$$- \sum_{i=2}^{\infty} \frac{a_i^2}{i!} \sum_{l=0}^{\infty} \sum_{\lambda=0}^{\infty} W[l] W[\lambda] \rho_s^i [m - l + \lambda]$$

$$(9.10)$$

$$= \eta_s^2 \rho_s [m] - \eta_s a_1 \sum_{l=0}^{\infty} W[l] \{\rho_s [m - l] + \rho_s [m + l]\}$$

$$+ a_1^2 \sum_{l=0}^{\infty} \sum_{\lambda=0}^{\infty} W[l] W[\lambda] \rho_s [m - l + \lambda]$$

The equality (9.10) is not violated if a factor $h^i (h = 1)$ is placed in front of the terms containing the factor $\frac{a_i^2}{i!}$. We then obtain

$$\eta_x^2 \sum_{l=0}^{\infty} \sum_{\lambda=0}^{\infty} R[l] R[\lambda] \rho_x [m - l + \lambda]$$

$$- \sum_{i=2}^{\infty} h^i \frac{a_i^2}{i!} \sum_{l=0}^{\infty} \sum_{\lambda=0}^{\infty} W[l] W[\lambda] \rho_s^i [m - l + \lambda]$$

$$= \eta_s^2 \rho_s [m] - \eta_s a_1 \sum_{l=0}^{\infty} W[l] \{\rho_s [m - l] + \rho_s [m + l]\}$$

$$+ a_1^2 \sum_{l=0}^{\infty} \sum_{\lambda=0}^{\infty} W[l] W[\lambda] \rho_s [m - l + \lambda] \tag{9.11}$$

The solution of (9.11) is given by the normalised correlation function $\rho_s [m]$. Let us seek this solution in the form of an infinite series

$$\rho_s [m] = \sum_{\alpha=0}^{\infty} h^{\alpha} \rho_{s\alpha} [m] \tag{9.12}$$

Substituting (9.12) into (9.11) we have

$$\sum_{\alpha=0}^{\infty} h^{\alpha} \Big[\eta_s^2 \rho_{s\alpha} [m]$$

$$- \eta_s a_1 \sum_{l=0}^{\infty} W[l] \{\rho_{s\alpha} [m - l] + \rho_{s\alpha} [m + l]\}$$

$$+ a_1^2 \sum_{l=0}^{\infty} \sum_{\lambda=0}^{\infty} W[l] W[\lambda] \rho_{s\alpha} [m - l + \lambda] \Big]$$

$$= \eta_x^2 \sum_{l=0}^{\infty} \sum_{\lambda=0}^{\infty} R[l] R[\lambda] \rho_x [m - l + \lambda]$$

$$- \sum_{i=2}^{\infty} \sum_{\alpha_1 = \alpha_2 = \dots = \alpha_i = 0}^{\infty} h^{i + \alpha_1 + \alpha_2 + \dots + \alpha_i} \frac{a_i^2}{i!} \sum_{l=0}^{\infty} \sum_{\lambda=0}^{\infty} W[l] W[\lambda]$$

$$\times \rho_{s\alpha_1} [m - l + \lambda] \rho_{s\alpha_2} [m - l + \lambda] \dots \rho_{s\alpha_i} [m - l + \lambda] \tag{9.13}$$

Considering (9.13) as an identity for h^i, and equating the terms with the same powers of h, we obtain

$$\eta_s^2 \rho_{s0}[m] - \eta_s a_1 \sum_{l=0}^{\infty} W[l]\{\rho_{s0}[m-l] + \rho_{s0}[m+l]\}$$

$$+ a_1^2 \sum_{l=0}^{\infty} \sum_{\lambda=0}^{\infty} W[l] W[\lambda] \rho_{s0}[m-l+\lambda]$$

$$= \eta_x^2 \sum_{l=0}^{\infty} \sum_{\lambda=0}^{\infty} R[l] R[\lambda] \rho_x[m-l+\lambda]. \tag{9.14}$$

$$\eta_s^2 \rho_{s1}[m] - \eta_s a_1 \sum_{l=0}^{\infty} W[l]\{\rho_{s1}[m-l] + \rho_{s1}[m+l]\}$$

$$+ a_1^2 \sum_{l=0}^{\infty} \sum_{\lambda=0}^{\infty} W[l] W[\lambda] \rho_{s1}[m-l+\lambda] = 0 \tag{9.15}$$

$$\eta_s^2 \rho_{s2}[m] - \eta_s a_1 \sum_{l=0}^{\infty} W[l]\{\rho_{s2}[m-l] + \rho_{s2}[m+l]\}$$

$$+ a_1^2 \sum_{l=0}^{\infty} \sum_{\lambda=0}^{\infty} W[l] W[\lambda] \rho_{s2}[m-l+\lambda]$$

$$= -\frac{a_2^2}{2!} \sum_{l=0}^{\infty} \sum_{\lambda=0}^{\infty} W[l] W[\lambda] \rho_{s0}^2[m-l+\lambda] \tag{9.16}$$

$$\eta_s^2 \rho_{s3}[m] - \eta_s a_1 \sum_{l=0}^{\infty} W[l]\{\rho_{s3}[m-l] + \rho_{s3}[m+l]\}$$

$$+ a_1^2 \sum_{l=0}^{\infty} \sum_{\lambda=0}^{\infty} W[l] W[\lambda] \rho_{s3}[m-l+\lambda]$$

$$= -\frac{a_3^2}{3!} \sum_{l=0}^{\infty} \sum_{\lambda=0}^{\infty} W[l] W[\lambda] \rho_{s0}^3[m-l+\lambda] \tag{9.17}$$

$$\eta_s^2 \rho_{s4}[m] - \eta_s a_1 \sum_{l=0}^{\infty} W[l]\{\rho_{s4}[m-l] + \rho_{s4}[m+l]\}$$

$$+ a_1^2 \sum_{l=0}^{\infty} \sum_{\lambda=0}^{\infty} W[l] W[\lambda] \rho_{s4}[m-l+\lambda]$$

$$= -\frac{a_4^2}{4!} \sum_{l=0}^{\infty} \sum_{\lambda=0}^{\infty} W[l] W[\lambda] \rho_{s0}^4 [m-l+\lambda]$$

$$-\frac{a_2^2}{2!} \sum_{l=0}^{\infty} \sum_{\lambda=0}^{\infty} W[l] W[\lambda] 2\rho_{s0}[m-l+\lambda] \rho_{s2}[m-l+\lambda] \quad (9.18)$$

$$\eta_s^2 \rho_{s5}[m] - \eta_s a_1 \sum_{l=0}^{\infty} W[l]\{\rho_{s5}[m-l] + \rho_{s4}[m+l]\}$$

$$+ a_1^2 \sum_{l=0}^{\infty} \sum_{\lambda=0}^{\infty} W[l] W[\lambda] \rho_{s5}[m-l+\lambda]$$

$$= -\frac{a_5^2}{5!} \sum_{l=0}^{\infty} \sum_{\lambda=0}^{\infty} W[l] W[\lambda] \rho_{s0}^5 [m-l+\lambda]$$

$$-\frac{a_3^2}{3!} \sum_{l=0}^{\infty} \sum_{\lambda=0}^{\infty} W[l] W[\lambda] \rho_{s0}^2 [m-l+\lambda] \rho_{s2}[m-l+\lambda]$$

$$-\frac{a_2^2}{2!} \sum_{l=0}^{\infty} \sum_{\lambda=0}^{\infty} W[l] W[\lambda] \rho_{s0}[m-l+\lambda] \rho_{s3}[m-l+\lambda] \quad (9.19)$$

and so on.

Equations (9.14) to (9.19) constitute a system in $\rho_{s0}[m]$, $\rho_{s1}[m], \ldots$ Here the mutual independence of the equations is important, since each component $\rho_{s\alpha}[m]$ of the correlation function $\rho_s[m]$ satisfies its own equation. At the same time it turns out that each subsequent component $\rho_{s\alpha}[m]$ with higher number can be expressed in terms of components with lower numbers.

We shall give a physical interpretation of Equations (9.14) to (9.19). Let us consider (9.14). Introducing the expression

$$\eta_s^2 \rho_{s0}[m] = K_{s0}[m] \quad (9.20)$$

we write this in the following form:

$$K_{s0}[m] - \frac{a_1}{\eta_s} \sum_{l=0}^{\infty} W[l] \{K_{s0}[m - l] + K_{s0}[m + l]\}$$

$$+ \frac{a_1^2}{\eta_s^2} \sum_{l=0}^{\infty} \sum_{\lambda=0}^{\infty} W[l] W[\lambda] K_{s0}[m - l + \lambda] \tag{9.21}$$

$$= \sum_{l=0}^{\infty} \sum_{\lambda=0}^{\infty} R[l] R[\lambda] K_x[m - l + \lambda]$$

Let us now consider the sampled data system shown in Fig. 4.13(a). Its system is linear, and is formed from the original non-linear system by changing the non-linear element whose characteristic is $g(s)$ by an inertia-free amplifying element with the gain factor $-\frac{a_1}{\eta_s}$. The input action of this linear system is given by the centralised random function $\overset{\circ}{X} = X - m_x$.

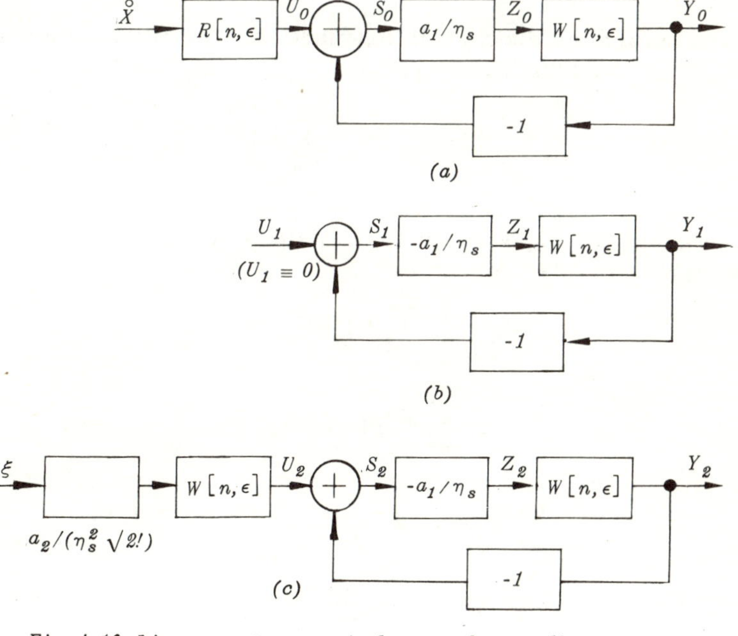

Fig. 4.13 Linear systems equivalent to the non-linear system shown in Fig. 4.12

We shall find the expression for the correlation function S_0 of the error of the system under consideration. In accordance with the symbols used in Fig. 4.13(a), we have the following equations:

$$S_0[n] = U_0[n] - Y_0[n] \tag{9.22}$$

$$U_0[n] = \sum_{l=0}^{\infty} R[l]\,\mathring{X}[n-l] \tag{9.23}$$

$$Y_0[n] = -\frac{a_1}{\eta_s}\sum_{l=0}^{\infty} W[l]\,S_0[n-l] \tag{9.24}$$

Solving (9.22), (9.23) and (9.24) simultaneously, we have the following equation for the error S_0 of the system:

$$S_0[n] - \frac{a_1}{\eta_s}\sum_{l=0}^{\infty} W[l]\,S_0[n-l] = \sum_{l=0}^{\infty} R[l]\,\mathring{X}[n-l] \tag{9.25}$$

To determine the correlation function of the error we write (9.25) for another instant $\bar{t} = n_1$:

$$S_0[n_1] - \frac{a_1}{\eta_s}\sum_{\lambda=0}^{\infty} W[\lambda]\,S_0[n_1-\lambda] = \sum_{\lambda=0}^{\infty} R[\lambda]\,\mathring{X}[n_1-\lambda] \tag{9.26}$$

Multiplying (9.25) by (9.26) term by term and then applying mathematical expectation to both sides of the equation thus obtained, we have

$$\begin{aligned} K_{s0}[m] &- \frac{a_1}{\eta_s}\sum_{l=0}^{\infty} W[l]\,K_{s0}[m-l] \\ &- \frac{a_1}{\eta_s}\sum_{\lambda=0}^{\infty} W[\lambda]\,K_{s0}[m+\lambda] \\ &+ \frac{a_1^2}{\eta_s^2}\sum_{l=0}^{\infty}\sum_{\lambda=0}^{\infty} W[l]\,W[\lambda]\,K_{s0}[m-l+\lambda] \\ &= \sum_{l=0}^{\infty}\sum_{\lambda=0}^{\infty} R[l]\,R[\lambda]\,K_x[m-l+\lambda] \end{aligned} \tag{9.27}$$

where

$$K_{s0}[m] = M[S_0[n]S_0[n_1]]$$

Replacing λ by l in the third term on the left-hand side of (9.27), we finally have

$$K_{s0}[m] - \frac{a_1}{\eta_s}\sum_{l=0}^{\infty} W[l]\{K_{s0}[m-l] + K_{s0}[m+l]\}$$

$$+\frac{a_1^2}{\eta_s^2}\sum_{l=0}^{\infty}\sum_{\lambda=0}^{\infty} W[l]W[\lambda]K_{s0}[m-l+\lambda] \tag{9.28}$$

$$=\sum_{l=0}^{\infty}\sum_{\lambda=0}^{\infty} R[l]R[\lambda]K_x[m-l+\lambda]$$

We note that (9.21) and (9.28) coincide, and therefore conclude that (9.14) gives the correlation function of the error of the linear system shown in Fig. 4.13(a). At the same time it should be borne in mind that the input action of this system is given by the centralised random function $\mathring{X} = X - m_x$.

Let us now analyse (9.15). Introducing the symbol

$$\eta_s^2 \rho_{s1}[m] = K_{s1}[m] \tag{9.29}$$

we write this equation in the following form:

$$K_{s1}[m] - \frac{a_1}{\eta_s}\sum_{l=0}^{\infty} W[l]\{K_{s1}[m-l] + K_{s1}[m+l]\}$$

$$+\frac{a_1^2}{\eta_s^2}\sum_{l=0}^{\infty}\sum_{\lambda=0}^{\infty} W[l]W[\lambda]K_{s1}[m-l+\lambda] = 0 \tag{9.30}$$

We shall now consider the system shown in Fig. 4.13(b). Let us find the expression for the correlation function of the error S_1 of this system. With the symbols used in the figure, we have

$$S_1[n] = U_1[n] - Y_1[n] \tag{9.31}$$

$$Y_1[n] = -\frac{a_1}{\eta_s}\sum_{l=0}^{\infty} W[l]S_1[n-l] \tag{9.32}$$

The simultaneous solution of (9.31) and (9.32) gives the following equation for the error $S_1[n]$:

$$S_1[n] - \frac{a_1}{\eta_s}\sum_{l=0}^{\infty} W[l]\, S_1[n-l] = U_1[n] \tag{9.33}$$

Considering (9.33) for the instant $\bar{t} = n_1$

$$S_1[n_1] - \frac{a_1}{\eta_s}\sum_{\lambda=0}^{\infty} W[\lambda]\, S_1[n_1-\lambda] = U_1[n_1] \tag{9.34}$$

and multiplying (9.33) by (9.34) term by term we find, after applying mathematical expectation to both sides of the equation obtained,

$$K_{s1}[m] - \frac{a_1}{\eta_s}\sum_{l=0}^{\infty} W[l]\{K_{s1}[m-l] + K_{s2}[m+l]\}$$
$$+\frac{a_1^2}{\eta_s^2}\sum_{l=0}^{\infty}\sum_{\lambda=0}^{\infty} W[l]\, W[\lambda]\, K_{s1}[m-l+\lambda] = M[U_1[n]\, U_1[n_1]] \tag{9.35}$$

where

$$K_{s1}[m] = M[S_1[n]\, S_1[n_1]]$$

Putting $U_1 = 0$ in (9.35) we have

$$K_{s1}[m] - \frac{a_1}{\eta_s}\sum_{l=0}^{\infty} W[l]\{K_{s1}[m-l] + K_{s1}[m+l]\}$$
$$+\frac{a_1^2}{\eta_s^2}\sum_{l=0}^{\infty}\sum_{\lambda=0}^{\infty} W[l]\, W[\lambda]\, K_{s1}[m-l+\lambda] = 0 \tag{9.36}$$

We note that (9.30) and (9.36) coincide, and consequently, (9.15) gives the correlation function of the error of the linear system shown in Fig. 4.13(b). At the same time the input action of this system is identically zero ($U_1 \equiv 0$).

We now analyse (9.16), writing it in the form

$$K_{s2}[m] - \frac{a_1}{\eta_s}\sum_{l=0}^{\infty} W[l]\{K_{s2}[m-l] + K_{s2}[m+l]\}$$
$$+\frac{a_1^2}{\eta_s^2}\sum_{l=0}^{\infty}\sum_{\lambda=0}^{\infty} W[l]\, W[\lambda]\, K_{s2}[m-l+\lambda] \tag{9.37}$$

$$= -\frac{a_2^2}{\eta_s^4 \, 2!} \sum_{l=0}^{\infty} \sum_{\lambda=0}^{\infty} W[l] W[\lambda] K_{s0}^2 [m - l + \lambda] \tag{9.37}$$

where

$$\eta_s^2 \rho_{s2} [m] = K_{s2} [m] \tag{9.38}$$

$$\eta_s^4 \rho_{s0}^2 [m] = K_{s0}^2 [m.] \tag{9.39}$$

To give a physical interpretation of (9.37) we shall consider the system shown in Fig. 4.13(c). Let us determine the correlation function of the error S_2 of this system. Using the symbols adopted above we have the following equations:

$$S_2 [n] = U_2 [n] - Y_2 [n] \tag{9.40}$$

$$U_2 [n] = \frac{a_2}{\eta_s^2 \sqrt{2!}} \sum_{l=0}^{\infty} W[l] \xi [n - l] \tag{9.41}$$

$$Y_2 [n] = -\frac{a_1}{\eta_s} \sum_{l=0}^{\infty} W[l] S_2 [n - l] \tag{9.42}$$

Solving (9.40), (9.41) and (9.42) simultaneously, we find an equation for the error S_2 of the system under consideration. This equation has the form:

$$S_2 [n] - \frac{a_1}{\eta_s} \sum_{l=0}^{\infty} W[l] S_2 [n - l] = \frac{a_2}{\eta_x^2 \sqrt{2!}} \sum_{l=0}^{\infty} W[l] \xi [n - l] \tag{9.43}$$

Introducing the symbols*

$$\begin{aligned} K_{s2} [m] &= M [S_2 [n] S_2 [n_1]] \\ K_\xi [m] &= M [\xi [n] \xi [n_1]] \end{aligned} \tag{9.44}$$

we obtain the following expression for the correlation function $K_{s2}[m]$ of the error, proceeding from (9.43):

* The second equation (9.44) means that the mathematical expectation of the random function ξ is zero.

$$K_{s2}[m]-\frac{a_1}{\eta_s}\sum_{l=0}^{\infty}W[l]\{K_{s2}[m-l]+K_{s2}[m+l]\}$$

$$+\frac{a_1^2}{\eta_s^2}\sum_{l=0}^{\infty}\sum_{\lambda=0}^{\infty}W[l]W[\lambda]K_{s2}[m-l+\lambda] \tag{9.45}$$

$$=\frac{a_2^2}{\eta_s^4 2!}\sum_{l=0}^{\infty}\sum_{\lambda=0}^{\infty}W[l]W[\lambda]K_{\xi}[m-l+\lambda]$$

Equation (9.45) coincides with (9.37) up to the sign in the right-hand side, provided the correlation function $K_{\xi}[m]$ of the input action ξ is equal to $K^2_{s0}[m]$, i.e.*

$$K_{\xi}[m]=\eta_s^4\rho^2_{s0}[m]=K^2_{s0}[m] \tag{9.46}$$

Consequently (9.16) gives the correlation function $K_{s2}[m]$ of the error of the linear sampled data system shown in Fig. 4.13(c). At the same time the input action of this system should be ξ, whose correlation function is equal to the square of the correlation function of the error of the system shown in Fig. 4.13(a), i.e. (9.46) must hold.

Analogously, for other equations from (9.14) to (9.19) we can find the linear systems with block diagrams and input actions in which the equations for the correlation functions of the errors coincide with the corresponding equations in the group (9.14) to (9.19).

The foregoing demonstrates that a non-linear system can be replaced for the random terms by an infinite number of linear systems with corresponding input disturbances. These results follow from the fact that the correlation function at the output of the non-linear element is given by a series with an infinite number of terms.

Let us now apply the results obtained above to non-linear sampled data systems.

4.10 ACCURACY OF NON-LINEAR SYSTEMS AND SPECTRUM TRANSFORMATION OF INPUT SIGNAL

We now use results from the preceding section for the approximate analysis of the accuracy of non-linear systems.

* The exact equality of (9.45) and (9.37) is possible only when the gain factor of the system is $\sqrt{-1}\; a_2/(\eta_s^2\sqrt{2!})$, i.e. purely imaginary.

Depending on the accuracy required in determining the probability characteristics, the non-linear system can be replaced by a finite number of linear systems with appropriate input actions. In practical problems, non-linear systems can often be replaced with sufficient accuracy by three or four linear systems, and sometimes by five. In addition, to determine the correlation function of the random process at the output of the non-linear element, we can also confine ourselves to a finite number of terms in the corresponding series. The number of terms taken can be four or five, since the coefficients $a_l \frac{1}{l!}$ decrease rapidly.

Let us now consider the determination of the correlation function and mathematical expectation of the random processes at the output of the non-linear system shown in Fig. 4.12. We shall assume that a random disturbance X with mathematical expectation m_x and correlation function $K_x[n]$ acts at the input of the system. We also assume that the system is at rest up to the instant at which the input disturbance is applied.

To determine the correlation function of the random process at the output we shall use (9.6) from the preceding section. For a limited number of terms of the series this has the form

$$K_y[m] = \sum_{i=1}^{M} \frac{a_i^2}{i!} \sum_{l=0}^{\infty} \sum_{\lambda=0}^{\infty} W[l] W[\lambda] \rho_s^i [m - l + \lambda] \tag{10.1}$$

where $\rho_s[m]$ is the normalised correlation function of the error of this system.

If the system is replaced by N linear systems, then the normalised correlation function $\rho_s[m]$ of the error can be determined by

$$\rho_s[m] = \sum_{\alpha=0}^{N} \rho_{s\alpha}[m] \tag{10.2}$$

which is obtained from (9.12) when $h = 1$. Equation (10.2) approximately gives $\rho_s[m]$ in terms of the normalised error correlation functions $\rho_{s\alpha}[m]$ of the linear systems which replace the original non-linear system.

Substituting the expression for $\rho_s[m]$ given by (10.2) into

(10.1) we have

$$K_y[m] = \sum_{i=1}^{M} \frac{a_i^2}{i!} \sum_{l=0}^{\infty} \sum_{\lambda=0}^{\infty} W[l] W[\lambda] \sum_{\alpha=0}^{N} \rho_{s\alpha}[m-l+\lambda] \quad (10.3)$$

Inverting the order of summation in (10.3) we obtain

$$K_y[m] = \sum_{\alpha=0}^{N} K_{y\alpha}[m] \quad (10.4)$$

where

$$K_{y\alpha}[m] = \sum_{i=1}^{M} \frac{a_i^2}{i!} \sum_{l=0}^{\infty} \sum_{\lambda=0}^{\infty} W[l] W[\lambda] \rho_{s\alpha}[m-l+\lambda] \quad (10.5)$$

$$\alpha = 0, 1, \ldots, N$$

It follows from (10.5) that $K_{y\alpha}[m]$, $\alpha = 0, 1, \ldots. N$, represent the correlation functions of the random processes at the output of the linear systems which replace the non-linear system under consideration. The correlation function $K_y[m]$ of the random process at the output of this system is determined, as follows from (10.4), by the sum of the correlation functions $K_{y\alpha}[m]$ of the random processes at the outputs of the corresponding linear systems.

Thus, to determine the correlation function $K_y[m]$ we must know the error correlation functions $\rho_{s\alpha}[m]$. These can be found by solving equations of the type (9.14) to (9.19), but this is a cumbersome procedure. We shall give here another method based on the weighting function for the error.

We shall consider the linear sampled data system shown in Fig. 4.14. This system is a normal closed-loop system whose output is $S[n]$, the error of the system. We assume that the weighting function $W[n, \varepsilon]$ of the open-loop system is known, and find the weighting function of the closed-loop system whose output is given by the error. This is called the weighting function of error, and is denoted by $W_s[n, \varepsilon]$.

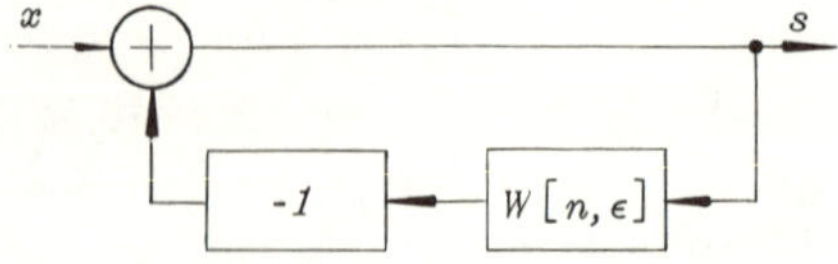

Fig. 4.14 A linear sampled data system

In accordance with the symbols used in the diagram we can write

$$\begin{aligned} s[n, \varepsilon] &= x[n, \varepsilon] - y[n, \varepsilon] \\ y[n, \varepsilon] &= \sum_{m=-\infty}^{\infty} W[n-m, \varepsilon]\, s[m-1,1] \end{aligned} \tag{10.6}$$

Solving (10.6) we have

$$s[n, \varepsilon] = x[n, \varepsilon] - \sum_{m=-\infty}^{\infty} W[n-m, \varepsilon]\, s[m-1,1] \tag{10.7}$$

Since the weighting function of a sampled data system is its response to the σ-function, then, putting $x[n, \varepsilon]|_{\varepsilon=0} = \sigma[n, \varepsilon]_{\varepsilon=0}$, in (10.7), we have

$$W_s[n, \varepsilon] = \sigma[n, 0] - \sum_{m=-\infty}^{\infty} W[n-m, \varepsilon]\, W_s[m-1,1] \tag{10.8}$$

When the condition of physical realisability of the system is used, (10.8) can be written as follows:

$$W_s[n, \varepsilon] = \sigma[n, 0] - \sum_{m=0}^{n} W[n-m, \varepsilon]\, W_s[m-1,1] \tag{10.9}$$

Noting that

$$W_s[m-1,1]|_{m=0} \equiv 0$$

we can finally write the expression for the weighting function $W_s[n, \varepsilon]$,

$$W_s[n, \varepsilon] = \begin{cases} 1, & n = 0 \\ -\sum_{m=0}^{n} W[n-m, \varepsilon]\, W_s[m-1,1], & n \geqslant 1 \end{cases} \tag{10.10}$$

whence it follows that the weighting function of the closed-loop system for the error is easily determined in terms of the weighting function $W[n, \varepsilon]$ of the corresponding open-loop system.

If the random function X with the correlation function $K_x[n]$ acts at the input of the system (Fig. 4.14), we can then write the following expression, from the theory of linear systems, for the correlation function of error:

$$K_s[m]=\sum_{l=0}^{\infty}\sum_{\lambda=0}^{\infty}W_s[l]W_s[\lambda]K_x[m-l+\lambda] \quad (10.11)$$

It is easy to see that the determination of this function by (10.11) is considerably simpler than by the corresponding equations of the type (9.14) to (9.19).

Let us return to the problem of determining $\rho_{s\alpha}[m]$. In conformity with Fig. 4.13 we have

$$K_{s0}[m]=\sum_{l=0}^{\infty}\sum_{\lambda=0}^{\infty}W_s[l]W_s[\lambda]K_{u_0}[m-l+\lambda] \quad (10.12)$$

where the weighting function $W_s[l]$ of the closed-loop system for the error is determined by

$$W_s[n]=\begin{cases} 1 & , \quad n=0 \\ \dfrac{a_1}{\eta_s}\sum\limits_{m=0}^{n}W[n-m]W_s[m], & n\geqslant 1 \end{cases} \quad (10.13)$$

The correlation function $K_{u_0}[m]$, in (10.12) is given in terms of the correlation function of the input action X by

$$K_{u_0}[m]=\sum_{l=0}^{\infty}\sum_{\lambda=0}^{\infty}R[l]R[\lambda]K_x[m-l+\lambda]$$

We now consider the system shown in Fig. 4.13(b). Since it is at rest up to the instant at which the input is applied, then we have

$$K_{s1}[m]\equiv 0 \quad (10.14)$$

as the input of the system is identically zero $(U_1\equiv 0)$.

Next, analogously to (10.12), the correlation function $K_{s2}[m]$ can be found from

$$K_{s2}[m]=\sum_{l=0}^{\infty}\sum_{\lambda=0}^{\infty}W_s[l]W_s[\lambda]K_{u_2}[m-l+\lambda] \quad (10.15)$$

where in accordance with (9.45) and (9.46) the correlation function $K_{u_2}[m]$ is defined as follows:

$$K_{u_2}[m]=-\frac{a_2^2}{2!}\sum_{l=0}^{\infty}\sum_{\lambda=0}^{\infty}W[l]W[\lambda]\rho_{\alpha 0}^2[m-l+\lambda] \quad (10.16)$$

Analogously, for $K_{s3}[m]$ we have in accordance with Fig. 4.13(c)

$$K_{s3}[m]=\sum_{l=0}^{\infty}\sum_{\lambda=0}^{\infty}W_s[l]W_s[\lambda]K_{u_3}[m-l+\lambda] \tag{10.17}$$

where

$$K_{u_3}[m]=-\frac{a_3^2}{3!}\sum_{l=0}^{\infty}\sum_{\lambda=0}^{\infty}W[l]W[\lambda]\rho_{s0}^3[m-l+\lambda] \tag{10.18}$$

In the same way we can write the corresponding expressions for the correlation functions $\rho_{sl}[m]$, $l\geqslant 4$.

Thus, using the theory of linear systems given in Chapter 3, we can find the correlation functions $\rho_{s\alpha}[m]$ of error of the linear systems, and then determine the correlation functions $K_{y\alpha}[m]$ of the random output processes for the same systems. The correlation function $K_y[m]$ at the output of the system is determined by (10.4).

It follows that we must know the coefficients a_α to determine the correlation functions $\rho_{s\alpha}[m]$ of error. It has been shown that these coefficients are determined by the characteristic of the non-linear element, and that they depend on the mathematical expectation and dispersion of the random process at the input of the non-linear element. Thus, we must know the dispersion η_s^2 of the error and its mathematical expectation m_s, in order to determine a_α.

The dispersion η_s^2 of the error can be found by

$$\eta_s^2=K_s[0]=\sum_{a=0}^{N}K_{s\alpha}[0] \tag{10.19}$$

where $K_{s\alpha}[0]=K_{s\alpha}[m]|_{m=0}$ are the error correlation functions.

To determine the mathematical expectation m_s of the error we can use the method of statistical linearisation, according to which the non-linear element is replaced by an inertia-free element with transfer coefficient $k_0(\eta_s, m_s)$. For simplicity we shall consider a case where the characteristic of the non-linear element is skew-symmetric.

In this way, replacing the non-linear element by the corresponding linear one, we can obtain an expression for the mathematical expectation m_s of the error in the steady-state, in the absence of self-excited oscillations:

$$m_s = \frac{\sum_{m=0}^{\infty} R[m]}{1 + k_0(\eta_s, m_s)\sum_{m=0}^{\infty} W[m]} \tag{10.20}$$

where $\sum_{m=0}^{\infty} R[m]$ is the gain factor of the system with weighting function $R[n]$;

$\sum_{m=0}^{\infty} W[m]$ is the gain factor of the system with weighting function $W[n]$.

The mathematical expectation m_y at the output of the non-linear system can be found from

$$m_y = \frac{k_0(\eta_s, m_s)\sum_{m=0}^{\infty} R[m] \times \sum_{m=0}^{\infty} W[m]}{1 + k_0(\eta_s, m_s)\sum_{m=0}^{\infty} W[m]} \tag{10.21}$$

Since to determine $K_y[m]$ we must know the coefficients a_l which are functions of the dispersion and the mathematical expectation of the error of the system, i.e. $a_l = a_l(\eta_s, m_s)$, and since to determine m_s we must know η_s^2, then it is clear that we must use the method of successive approximations to solve the corresponding equations and formulae for $\eta_s, m_s, K_{s\alpha}[m]$ and $K_y[m]$. Digital computers can be used for this purpose.

The determination of the correlation function at the output of a non-linear system by this method is more complicated than that for a linear system. This is because in the corresponding linear systems, we need to carry out the numerical solution of the corresponding differential-difference equations only once to determine the weighting functions $R[m]$ and $W[m]$. All other operations connected with the determination of $K_y[m]$ reduce to the solution of simple formulae.

Note also that programming such problems for digital computers is more laborious than for a linear sampled data system. This is because the error correlation functions $K_{s\alpha}[m]$, necessary for the determination of $K_y[m]$, are computed by the same programmes.

In many practical problems, however, it is sufficient to know the dispersion and mathematical expectation at the output of the system.

To determine the dispersion we can use (10.4) which, when $m=0$, gives

$$D_y=\sum_{\alpha=0}^{N} D_{y\alpha} \tag{10.22}$$

where $D_{y\alpha}$ are the dispersions at the outputs of the linear systems which replace the non-linear system. The dispersion $D_{y\alpha}$ can be found from (10.5) when $m=0$.

Accurate results are obtained when the dispersions $D_{y2}, D_{y3}, \dots, D_{yN}$ are calculated with the assumption that the input disturbances are represented by discrete white noise.

The loss of accuracy in determining the dispersion and the mathematical expectation are negligible when the sampling period of the system with function $W[n]$ is equated with the correlation time.

It is possible to replace the input actions of the linear systems, beginning with the third system, by discrete white noise in the following circumstances. Input disturbance of the system shown in Fig. 4.13(c) (the third system) is the centralised random function ξ whose correlation function is

$$K_\xi[m]=\rho_{s0}^2[m]$$

where $\rho_{s0}[m]$ is the normalised correlation function of the error of the first system (Fig. 4.13(a)). In many cases it turns out, especially for long sampling periods, that $\rho_{s0}^2[m] \ll 1$ when $m \geqslant 1$. Consequently we may assume that ξ is discrete white noise.

The input disturbance of the following system is also given by the centralised random function $\chi[m]$ whose correlation function in accordance with (9.17) must be equal to

$$K_\chi[m]=\rho_{s0}^3[m]$$

In this case there are even more grounds for the assumption that $\rho_{s0}^3[m] \ll 1$ when $m \geqslant 1$, and, consequently, the input disturbance x can also be replaced by discrete white noise. The same is true for other systems.

By replacing the input disturbances of the corresponding systems by white noise, we can simplify the determination of the dispersion and mathematical expectation at the output of a non-linear system.

Finally, we point out that the above procedure for assessing the accuracy of non-linear systems is based on replacing the non-linear system by the corresponding linear system. The method of statistical linearisation, based on replacing the non-linear system by a single linear system (for random terms), follows from this method as a first approximation. At the same time it is important to note that investigation of non-linear systems by the method described is more expensive than the method of statistical linearisation.

Chapter 5

OPTIMAL STATIONARY SYSTEMS WITH INFINITE OBSERVATION TIME

Every sampled data system transforms a set of input actions $\{\varphi_i\}$ into a set of output signals $\{h_j\}$ in accordance with a certain transformation operator H, so that

$$\{h_j[n]\} = H\{\varphi_i[m]\}$$

If $\{\varphi_i\}$ consists solely of deterministic (non-random) functions of time, then the choice of H does not represent any difficulties. In this case the system must usually transform or reproduce (when $H = 1$) the set of actions without dynamic errors or with a given dynamic accuracy.

But in addition to the deterministic actions, $\{\varphi_i\}$ as a rule contains random actions given by their probability characteristics. At the same time the random signals can be either useful signals or noise. In this case the problem of choosing H is made considerably more difficult, since usually it is required to find a transformation of $\{\varphi_i\}$ which will guarantee a sufficient dynamic accuracy of transformation (reproduction) when small random errors are present.

The problem of choosing H for the given requirements of transformation (reproduction) is called the synthesis problem. The requirements on the transformation are given in the form of certain criteria, such as can be found in the books by Pugachev [56] and Solodovnikov [63]. Then, if H is chosen from the conditions of obtaining the extremal values of these criteria, the problem of choosing the transformation is called the synthesis problem of an optimal system. This is most simply solved when $\{\varphi_i\}$ contains only stationary random functions, whilst the optimality criterion is the minimum mean-square error of transformation (reproduction).

In this chapter we shall give a number of methods for determining a stationary optimal sampled data system which guarantees the minimum mean-square error of transformation for a random signal consisting of a useful signal and noise.

5.1 FORMULATION OF THE PROBLEM

Let the input signal φ of a system be the sum of two terms

$$\varphi[n] = s[n] + \xi[n] \tag{1.1}$$

where s is a useful random signal and ξ is a random noise signal. We assume that both terms of the input signal φ are stationary random functions with the mathematical expectations

$$M[s[n]] \equiv 0, \quad M[\xi[n]] \equiv 0 \tag{1.2}$$

and correlation functions

$$\begin{aligned} K_s[n] &= M[s[n_1]\, s[n_2]] \\ K_\xi[n] &= M[\xi[n_1]\, \xi[n_2]] \end{aligned} \tag{1.3}$$

In addition, we assume that the useful random signal s is correlated with the noise signal ξ, since the following cross-correlation functions are non-zero:

$$\begin{aligned} K_{s\xi}[n] &= M[s[n_1]\, \xi[n_2]] \\ K_{\xi s}[n] &= M[\xi[n_1]\, s[n_2]] \end{aligned} \tag{1.4}$$

We formulate the problem of determining the optimal system which is to transform the useful signal s into the signal h according to the formula

$$h[n] = Hs[m] \tag{1.5}$$

where H is the given transformation operator.

Further, we assume that the system having the operator H is an ideal transforming system. We shall give some examples of such systems.

If a system is to pick out the useful signal s, then the weighting function $W_{id}[n]$ of the ideal transforming system is

$$W_{id}[n] = \sigma[n] \tag{1.6}$$

and in accordance with (1.5)

$$h[n] \equiv s[n] \tag{1.7}$$

The weighting function of an ideal predicting device, which computes the value of the signal s at the instant $\bar{t} = n + l$, is given by

$$W_{id}[n] = \sigma[n + l] \tag{1.8}$$

In this case (1.5) gives

$$h[n] \equiv s[n + l] \tag{1.9}$$

Systems with the weighting functions

$$W_{id}[n] = \sigma[n - l] \tag{1.10}$$

$$W_{id}[n] = \Delta^p \sigma[n - m] \tag{1.11}$$

compute the value of the signal $s[n-l]$ and the difference of order p of the signal s at the instant $\bar{t} = n$.

We now return to the formulation of the problem of determining the optimal system. The system which carries out the given transformation belongs to the class of linear systems. This means that the transforming system must carry out only linear operations on the signal s. In addition, we require the system to be physically realisable. This requirement reduces to the fact that the weighting function of the system is to be zero for all negative values of the argument.

As a criterion for the best possible system we use the minimum mean-square error of the transformation, i.e. the difference between the actual signal $s^*[n]$ at the output of the actual system and the desired signal h at the output of the ideal system.

The difference

$$\mathcal{E}[n] = h[n] - s^*[n] \tag{1.12}$$

will be called the error of transformation (reproduction).

The assumptions made in formulating the problem of

determining the optimal sampled data systems coincide exactly with the assumptions which underlie the theory of Wiener [71]. Accordingly the problem is called the discrete analogue of Wiener's problem.

We shall now derive the equations which must be satisfied by the weighting function of the optimal system.

5.2 WEIGHTING FUNCTION OF AN OPTIMAL SYSTEM

According to the assumptions made above, the statistical characteristics of the components of the signal φ do not depend on time. Consequently the system which carries out the transformation of the signal φ must be stationary. Let us denote the weighting function of this system by $W[n]$. Then the equation which determines the value of the signal s^* at the output can be written in the following form:

$$s^*[n] = \sum_{m=0}^{\infty} W[m]\varphi[n-m] \tag{2.1}$$

The lower limit of summation in (2.1) must be zero, since, proceeding from the formulation of the problem (see Section 1), it is required that the system satisfies the condition of physical realisability, that is,

$$W[n] \equiv 0, \quad n < 0 \tag{2.2}$$

Since

$$\varphi[n] = s[n] + \xi[n]$$

the signal s^* in accordance with (2.1) equals

$$s^*[n] = \sum_{m=0}^{\infty} W[m](s[n-m] + \xi[n-m]) \tag{2.3}$$

The error of transformation, $\mathcal{E}$, given by (1.12) and (2.3), can be written in the following form

$$\mathcal{E}[n] = h[n] - \sum_{m=0}^{\infty} W[m](s[n-m] + \xi[n-m]) \tag{2.4}$$

Let us now find the expression for the mean square value error $\mathcal{E}$. To do this we consider the error value for the two instants $\overline{t}_1 = n_1$ and $\overline{t}_2 = n_2$:

$$\mathcal{E}[n_1] = h[n_1] - \sum_{m_1=0}^{\infty} W[m_1](s[n_1 - m_1] + \xi[n_1 - m_1])$$

$$\mathcal{E}[n_2] = h[n_2] - \sum_{m_2=0}^{\infty} W[m_2](s[n_2 - m_2] + \xi[n_2 - m_2])$$

Multiplying these expressions term by term and then applying mathematical expectation to both sides of the equations thus obtained, we find

$$K_e[n_1 - n_2] = K_h[n_1 - n_2] - 2\sum_{m=0}^{\infty} W[m] K_{h\varphi}[n_1 - n_2 + m]$$
$$+ \sum_{m_1=0}^{\infty} W[m_1] \sum_{m_2=0}^{\infty} W[m_2] K_{\varphi}[n_1 - n_2 + m_1 - m_2] \quad (2.5)$$

where $K_e[n_1 - n_2]$ is the correlation function of error,

$$K_{h\varphi}[n] = K_{hs}[n] + K_{h\xi}[n] \quad (2.6)$$

$$K_{\varphi}[n] = K_s[n] + K_{\xi}[n] + K_{s\xi}[n] + K_{\xi s}[n] \quad (2.7)$$

in which $K_{hs}[n]$ and $K_{h\xi}[n]$ denote the cross-correlation functions

$$K_{hs}[n] = M[h[n_1]\, s[n_2]]$$
$$K_{h\xi}[n] = M[h[n_1]\, \xi[n_2]]$$

To determine the mean square value of the error we put $n_1 = n_2$ in (2.5). We then obtain

$$D_e = K_e[0] = K_h[0] - 2\sum_{m=0}^{\infty} W[m] K_{h\varphi}[m]$$
$$+ \sum_{m_1=0}^{\infty} W[m_1] \sum_{m_2=0}^{\infty} W[m_2] K_{\varphi}[m_1 - m_2] \quad (2.8)$$

We shall now determine the necessary and sufficient condition for the weighting function W to transform (2.8) into a minimum, i.e. for the dispersion of the transformation error φ by means of a system with the weighting function W to be a minimum. We then assume that W does in fact ensure the minimum dispersion

$$D_e[W]=\min$$

Then any variation of the weighting function $W[n]+\delta w[n]$, where $w[n]$ is an arbitrary function of n, satisfying the condition

$$w[n]\equiv 0 \text{ when } n<0$$

must satisfy the dispersion D_e, i.e. we have

$$D_e[W+\delta w]>D_e[W]$$

The necessary condition of a minimum of $D_e[W]$ is given by the condition

$$\frac{\partial}{\partial\delta}D_e[W+\delta w]|_{\delta=0}=0 \tag{2.9}$$

To determine $D_e[W+\delta w]$ we replace W in (2.8) by its value and the variation δw. We then have

$$D_e[W+\delta w]=K_h[0]-2\sum_{m=0}^{\infty}(W[m]+\delta w[m])K_{h\varphi}[m]$$

$$+\sum_{m_1=0}^{\infty}(W[m_1]+\delta w[m_1])\sum_{m_2=0}^{\infty}(W[m_2]+\delta w[m_2])K_{\varphi}[m_1-m_2]$$

$$=K_h[0]-2\sum_{m=0}^{\infty}W[m]K_{h\varphi}[m]+\sum_{m_1=0}^{\infty}W[m_1]\sum_{m_2=0}^{\infty}W[m_2]K_{\varphi}[m_1$$

$$-m_2]-2\sum_{m=0}^{\infty}\delta w[m]K_{h\varphi}[m]$$

$$+\sum_{m_1=0}^{\infty}\delta w[m_1]\sum_{m_2=0}^{\infty}W[m_2]K_{\varphi}[m_1-m_2]$$

$$+\sum_{m_1=0}^{\infty}W[m_1]\sum_{m_2=0}^{m_2}\delta w[m_2]K_{\varphi}[m_1-m_2]$$

$$+\delta^2\sum_{m_1=0}^{\infty}w[m_1]\sum_{m_2=0}^{\infty}w[m_2]K_{\varphi}[m_1-m_2]$$

whence it follows that

$$D_e[W+\delta w]=D_e[W]-2\delta D_1+\delta^2 D_2 \tag{2.10}$$

where

$$D_1=\sum_{m_1=0}^{\infty} w[m_1]\left\{K_{h\varphi}[m_1]-\sum_{m_2=0}^{\infty} W[m_2]K_\varphi[m_1-m_2]\right\} \tag{2.11}$$

$$D_2=\sum_{m_1=0}^{\infty} w[m_1]\sum_{m_2=0}^{\infty} w[m_2]K_\varphi[m_1-m_2] \tag{2.12}$$

Applying now (2.9) to (2.10) we have

$$\left.\frac{\partial D_e[W+\delta w]}{\partial \delta}\right|_{\delta=0}=-2D_1=0$$

whence it follows that W, transforming (2.8) into a minimum, must satisfy

$$D_1=0$$

or, when (2.11) is taken into account,

$$K_{h\varphi}[n]=\sum_{m=0}^{\infty} W[m]K_\varphi[n-m], \quad n\geqslant 0 \tag{2.13}$$

It is not difficult to show that (2.13) is not only a necessary, but also a sufficient condition for (2.8) to be a minimum. For this it is sufficient to prove that for (2.13) the following holds

$$D_e[W+\delta w]>D_e[W]$$

where W is determined by (2.13).

We write (2.10) in the following form:

$$D_e[W+\delta w]=D_e[W]-2\delta D_1$$

$$+\delta^2\sum_{m_1=0}^{\infty} w[m_1]\sum_{m_2=0}^{\infty} w[m_2]K_\varphi[m_1-m_2]$$

$$=D_e[W]-2\delta D_1+\delta^2 M\left[\left\{\sum_{m=0}^{\infty} w[m]\varphi[n-m]\right\}^2\right]$$

whence, when (2.13) is taken into account, we obtain

$$D_e[W+\delta w]>D_e[W]$$

since

$$\delta^2 M\left[\left\{\sum_{m=0}^{\infty} w[m]\varphi[n-m]\right\}^2\right] > 0$$

It has thus been shown that the weighting function $W[n]$ of a sampled data transforming system, optimal in the sense that it gives the minimum dispersion of the transformation error, must satisfy (2.13). This is a discrete analogue of the Wiener-Hopf equation which defines the weighting function of an optimal (i.e. giving the minimum mean-square error) continuous system, which carries out an optimal transformation of the input signal in accordance with the assumptions made in Section 1.

To determine the weighting function of the optimal system we must solve (2.13). The solution in the general case is known (see, for example, the book by Solodovnikov [63]). Later other methods will be given which in many practical cases enable us to obtain the required results more quickly.

5.3 OPTIMAL PREDICTION IN THE ABSENCE OF NOISE

We shall consider the problem of optimal prediction of a random signal $\varphi \equiv s$, i.e. the case where the noise ξ is absent. In spite of this drastic idealisation, such problems are encountered in practice, as the following example shows. The trajectory of a certain point moving in a random manner is observed over a prolonged time interval. The observation of the trajectory is carried out by measuring devices whose output constitutes the realisation of the random function (trajectory) and the error of measurement. If then for some reason or other the observation is broken off, its position at a certain future instant of time remains of interest to the observer, and the problem arises of determining the future position of the point. If at the same time the level of the measurement error is not too high, so that the probable position of the point at a future instant of time is determined primarily by the character of its motion at the time of observation, then the problem can be considered as a problem of predicting the random signal when noise is absent.

Accordingly, let us formulate the problem of determining the weighting function of a system which optimally (i.e. with the minimum error dispersion) predicts a random signal s, given by the correlation function $K_s[n]$, over the time $\bar{t}=l$. To solve this problem we must find the equation which is satisfied by the weighting function of this system.

In the given case we have

$$Hs \equiv s[n+l] \tag{3.1}$$

i.e. the signal at the output of the ideal transforming system represents the value of s at the instant $\bar{t}=n+l$.

Furthermore, since in the given case the noise ξ is absent, then from (2.6) we find

$$K_{h\varphi}[n]=K_{hs}[n] \tag{3.2}$$

Using (3.1) we obtain the expression for the cross-correlation function $K_{hs}[n]$

$$K_{hs}[n]=M\left[s[n_1+l]\,s[n_2]\right]=K_s[n+l] \tag{3.3}$$

Furthermore, from (2.7) we have

$$K_{\varphi}[n]=K_s[n] \tag{3.4}$$

We can now write the equation for the optimal weighting function $W[n]$. This is obtained from the general equation (2.13), if the corresponding values of the correlation functions $K_{h\varphi}$ and K_{φ} given by (3.3) and (3.4) are substituted. We then have

$$K_s[n+l]=\sum_{m=0}^{\infty} W[m]\,K_s[n-m],\quad n\geqslant 0 \tag{3.5}$$

Equation (3.5) in fact gives the weighting function of the ideal predictor. As has been mentioned already the solution of this equation can be obtained from the general solution of (2.13). Another method, based on the use of discrete shaping filters, is in use.

It has been shown in Chapter 3 that if the correlation function $K_x[n]$ of the random function X is given, then we can determine a filter which forms X with the given correlation function out of the discrete white noise V (noncorrelated sequence of random numbers). If X is stationary, the discrete shaping filter is a stationary sampled data system characterised by the weighting function $W_f[n]$.

If in series with this system we connect its inverse with the weighting function $W_f^{-}[n]$, then the resulting combination is a system for identical transformation of the input action. Consequently, when white noise is applied to the input of such

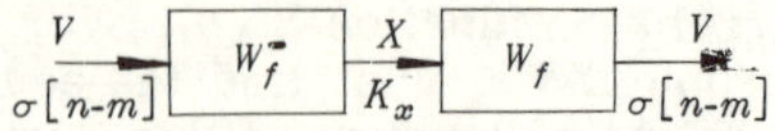

Fig. 5.1 A system which identically transforms the input action

a system, white noise also appears at the output (Fig. 5.1).

From the given correlation function $K_s[n]$ we determine the shaping filter, whose weighting function is denoted by $W_1[n]$. It is clear that, since the filter with the weighting function W_1 is physically realisable, then a filter with the weighting function W_1^-, inverse to the former, is also physically realisable.

If we apply a signal s to the filter with weighting function W_1, then at its output we have discrete white noise V, i.e.

$$V = A_1 s \tag{3.6}$$

where A_1 is the operator of the system with the weighting function W_1^-. Consequently, the determination of an optimal predictor reduces to the determination of an optimal linear transformation A_2 of white noise. Then the linear transformations A_1 and A_2 applied in series to the signal s give its value at the instant $n+l$. Thus the predicted value of the signal $s^*[n+l]$ is

$$s^*[n+l] = A_2\{A_1 s\} = A_2 V \tag{3.7}$$

The formation of the predicted signal is diagrammatically illustrated by Fig. 5.2.

Let $W_2[n]$ denote the weighting function of a system with the operator A_2, i.e. the system which carries out the optimal transformation of white noise into $s^*[n+l]$. Then we have

$$W[n] = W_2[n] * W_1^-[n] \tag{3.8}$$

where $W[n]$ is the weighting function of the optimal predictor.

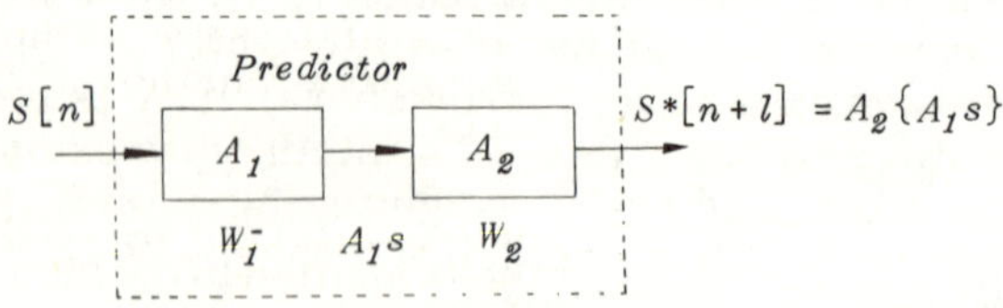

Fig. 5.2 An optimal predictor

It follows from (3.8) that the weighting function $W[n]$ of the predictor is obtained as the result of convolution of the weighting functions W_1^- and W_2.

Let us now determine the optimal linear transformation of white noise, i.e. the weighting function W_2. For this purpose we express the value of the signal s^* at the instant $n+l$ in terms of the white noise V and the weighting function $W_1[n]$ as follows:

$$s^*[n+l] = \sum_{m=-\infty}^{n+l} W_1[n+l-m]\,V[m] \tag{3.9}$$

From (3.9) it follows that the value of $s^*[n+l]$ is determined by V observed up to the instant $\bar{t} \leqslant n$, as well as by the subsequent values of the noise when $\bar{t} > n$. Hence (3.9) can be represented in the form of a sum of the two terms

$$s^*[n+l] = s_1 + s_2 \tag{3.10}$$

where in conformity with (3.9)

$$s_1 = \sum_{m=-\infty}^{n} W_1[n+l-m]\,V[m] \tag{3.11}$$

and

$$s_2 = \sum_{m=n+1}^{n+l} W_1[n+l-m]\,V[m] \tag{3.12}$$

Since in determining the predicted value of the signal we use only the observed values of $V[m]$ over the time interval $[-\infty < m \leqslant n]$, and since the subsequent values of the noise $V[m]$ $(m > n)$ have no connection whatever with the preceding ones, it turns out that the value of the signal s_2 is unknown in principle. By contrast, the signal s_1, determined by the observed values of the noise $V[m]$ $(m \leqslant n)$ can always be found.

We shall next consider another linear transformation of V. This transformation is written in the form

$$f[n] = \sum_{m=-\infty}^{n} W[n-m]\,V[m] \tag{3.13}$$

where W is an arbitrary weighting function. As follows from (3.13), the signal f is formed from the observed values of the white noise $V[m]$ when $m \leqslant n$.

Let us now evaluate the difference of the signals

$$\mathcal{E}=f-s^*$$

where s^* is determined by (3.9). Using (3.9) to (3.12), we have

$$\mathcal{E}^2=(f-s_1-s_2)^2=(f-s_1)^2-2(f-s_1)s_2+s_2^2 \quad (3.14)$$

Applying mathematical expectation to both sides of (3.14) we find

$$M[\mathcal{E}^2]=M[(f-s_1)^2]-2M[(f-s_1)s_2]+M[s_2^2]$$

or

$$M[\mathcal{E}^2]=M[(f-s_1)^2]-2M[fs_2]+2M[s_1s_2]+M[s_2^2] \quad (3.15)$$

Let us determine the mathematical expectation of the products f_1s_2 and s_1s_2 contained in (3.15).

From (3.11) and (3.13) we have

$$M[fs_2]=\sum_{m_1=-\infty}^{n} W[n-m_1]\sum_{m_2=n+1}^{n+l} W[n+l-m_2]\,M\left[V[m_1]V[m_2]\right]$$

Since by the definition of discrete white noise

$$M\left[V[m_1]V[m_2]\right]=K_v[m_1-m_2]=\sigma[m_1-m_2]$$

the last relation can be written in the form

$$M[f\cdot s_2]=\sum_{m_1=-\infty}^{n} W[n-m_1]\sum_{m_2=n+1}^{n+l} W[n+l-m_2]\,\sigma[m_1-m_2]$$

whence it follows that

$$M[fs_2]=0 \quad (3.16)$$

as the limits of summation with respect to the variables m_1 and m_2 do not overlap and the σ-function is nowhere unity.

Furthermore, since s_1 is determined by the values of the noise $V[m]$ when $m\leqslant n$, and s_2 is determined by the subsequent $(m>n)$ values of the noise which have no connection whatever with the preceding values, we can show in the same way that

$$M[s_1s_2]=0 \quad (3.17)$$

Thus, when (3.16) and (3.17) are used, (3.15) can finally be written as follows

$$M[\mathscr{E}^2] = M[(f_1 - s_1)^2] + M[s_2^2] \tag{3.18}$$

whence it follows that the minimum $M[\mathscr{E}^2]$ is attained in the case where the signals s_1 and f coincide identically. This means that the optimal transformation of white noise into the signal $s^*[n+l]$ is given by a linear transformation of the form

$$s^*[n+l] = \sum_{m=0}^{n} W_2[n+l-m]\, V[m] \tag{3.19}$$

where the optimal weighting function $W_2[n]$ is defined as follows

$$W_2[n] = \begin{cases} W_1[n+l] & \text{when } n \geqslant 0 \\ 0 & \text{when } n < 0 \end{cases} \tag{3.20}$$

We recall that $W_1[n]$ is the weighting function of a discrete filter which forms the random signal s with the correlation function $K_s[n]$ out of discrete noise.

The weighting function of the optimal predictor can be determined by (3.8)

$$W[n] = W_1[n+l] * W_1^-[n] \tag{3.21}$$

Applying the discrete Laplace transformation to both sides of (3.21), the transfer function of the optimal predictor is

$$W^*(q) = W_1^*(q)\, W_2^*(q) \tag{3.22}$$

where

$$\begin{aligned} W_1^*(q) &= D\{W_1^-[n]\} \\ W_2^*(q) &= D\{W_1[n+l]\} \end{aligned} \tag{3.23}$$

Hence we can finally formulate the procedure for determining an optimal predictor by means of shaping filters.

1. From the given correlation function $K_s[n]$ of the random signal s we determine the difference equation of the shaping filter and its weighting function $W_f[n] = W_1[n]$, as well as the transfer function $W_2^*(q)$.
2. We set up the difference equation of a filter which is

inverse to the shaping filter. From these difference equations we determine the weighting function $W_1^-[n]$ and the transfer function $W_1^*(q)$.

3. In accordance with (3.22) we determine the transfer function of the optimal predictor.

It is clear that the transfer function $W^*(q)$ of the optimal predictor represents the ratio of polynomials in e^q, i.e.

$$W^*(q) = \frac{S^*_{\text{out}}(q)}{S^*_{\text{in}}(q)} = \frac{\sum_{i=0}^{h} a_i (e^q - 1)^i}{\sum_{i=0}^{k} b_i (e^q - 1)^i} \tag{3.24}$$

Now, reverting to the originals in (3.24) we can immediately obtain the programme for a digital computer for solving the prediction problem. The prediction error can most easily be calculated from (2.8) which is simplified considerably when (2.13), (3.2), (3.3) and (3.4) are used.

Let us apply the results just obtained to determine the transfer function of a system which carries out optimal prediction, over the time $\bar{t} = l$, of a random signal with the correlation function

$$K_s[n] = e^{-\alpha |n|} \tag{3.25}$$

In accordance with the procedure given we must find the difference equation of the shaping filter and the weighting function $W_f[n] = W_1[n]$ of this filter, proceeding from the correlation function $K_s[n]$. It was shown in Chapter 3 that the difference equation of a filter which forms a random process with a correlation function of the form (3.25) out of the discrete white noise V, has the form

$$\Delta s[n] + (1 - e^{-\alpha}) s[n] = \sqrt{1 - e^{-2\alpha}} \cdot \Delta v[n] + \sqrt{1 - e^{-2\alpha}}\, v[n] \tag{3.26}$$

Putting $v[n] = \sigma[n]$ in (3.26) we can find the weighting function of this filter for zero initial conditions

$$W_1[n] = \sqrt{1 - e^{-2\alpha}}\, e^{-\alpha n}, \quad n \geqslant 0 \tag{3.27}$$

Using (3.23) we find

$$W_2[n] = \begin{cases} \sqrt{1 - e^{-2\alpha}}\, e^{-\alpha (n+l)}, & n \geqslant 0 \\ 0, & n < 0 \end{cases} \tag{3.28}$$

Following (3.23) we find the transfer function $W_2^*(q)$, having taken (3.28) into consideration,

$$W_2^*(q) = D\{W_1[n+l]\} = \sum_{n=0}^{\infty} W_1[n+l]\,\mathrm{e}^{-qn}$$

$$= \sqrt{1-\mathrm{e}^{-2\alpha}}\,\mathrm{e}^{-\alpha l}\sum_{n=0}^{\infty}\mathrm{e}^{-\alpha n}\,\mathrm{e}^{-qn} \tag{3.29}$$

$$= \sqrt{1-\mathrm{e}^{-2\alpha}}\,\frac{\mathrm{e}^{-\alpha l}}{\mathrm{e}^{q}-\mathrm{e}^{-\alpha}}\,\mathrm{e}^{q}$$

Replacing the operator Δ in (3.26) by $(\mathrm{e}^{q}-1)$, we obtain the transfer function of the shaping filter

$$W_{\mathrm{f}}^*(q) = \frac{\mathrm{e}^{q}\sqrt{1-\mathrm{e}^{-2\alpha}}}{\mathrm{e}^{y}-\mathrm{e}^{-\alpha}} \tag{3.30}$$

whence we obtain the following expression for the transfer function of a system which is inverse to the shaping filter:

$$W_{\mathrm{I}}^*(q) = D\{W_{\mathrm{I}}^{-}[n]\} = \frac{\mathrm{e}^{q}-\mathrm{e}^{-\alpha}}{\mathrm{e}^{q}\sqrt{1-\mathrm{e}^{-2\alpha}}} \tag{3.31}$$

Substituting the expression for $W_{\mathrm{I}}^*(q)$ and $W_2^*(q)$ given by (3.31) and (3.29) into (3.22) we find the transfer function of the optimal predictor in question

$$W^*(q) = \mathrm{e}^{-\alpha l} \tag{3.32}$$

5.4 OPTIMAL PREDICTION IN THE PRESENCE OF NOISE

Let us find the optimal sampled data system for predicting the random signal s over the time $\bar{t} = l$, if the total signal is a sum of the useful signal s, with correlation function $K_s[n]$, and the noise ξ, with correlation function $K_\xi[n]$. In this case we shall assume that the useful signal s and the noise ξ are correlated, so that $K_{s\xi}[n] \neq 0$ and $K_{\xi s} \neq 0$. This problem is sometimes called the problem of optimal prediction combined with the problem of smoothing (filtering), since the optimal system must determine the value of the useful signal $s^*[n+l]$ at the instant $\bar{t} = n + l$. The same system must simultaneously filter out the noise ξ.

We have

$$Hs = s[n+l] \tag{4.1}$$

i.e. the signal at the output of the ideal sampled data system must coincide with the signal s at the instant $\bar{t} = n + l$. Since $h[n] = s[n+l]$, then

$$K_{hs}[n] = M\left[s[n_1+l]\, s[n_2]\right] = K_s[n+l] \tag{4.2}$$

In addition,

$$K_{h\xi}[n] = M\left[s[n_1+l]\, \xi[n_2]\right] = K_{s\xi}[n+l] \tag{4.3}$$

Consequently, when (4.2) and (4.3) are used we obtain from (2.6) the following expression for the cross-correlation function

$$K_{h\varphi}[n] = K_s[n+l] + K_{s\xi}[n+l] = K_{s\varphi}[n+l] \tag{4.4}$$

Taking into consideration (4.4) and (2.7) we write the equation for the optimal weighting function of the predictor as follows:

$$K_{s\varphi}[n+l] = \sum_{m=0}^{\infty} W[m]\, K_{\varphi}[n-m], \quad n \geqslant 0 \tag{4.5}$$

where the correlation function $K_{\varphi}[n]$ is given by (2.7). The solution of (4.5) gives the weighting function $W[n]$ of the optimal predictor. To solve (4.5) we use the method of discrete shaping filters. In this case we proceed from the equation for the error $\mathcal{E}[n] = h[n] - s^*[n]$, written in the domain of transforms.

Let us consider the sampled data system shown in Fig. 5.3. This system is a parallel combination of the ideal system with the weighting function $W_{id}[n]$ and the optimal system (which is to be determined) with the weighting function $W[n]$. The system has two inputs. The useful random signal s acts

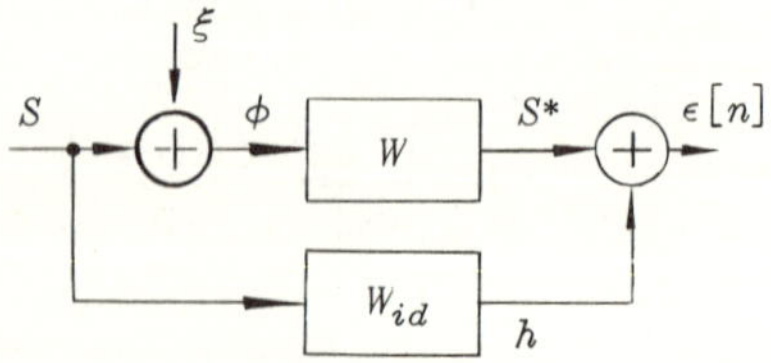

Fig. 5.3 Transformation of the system error

at one input, and the noise ξ at the other. We shall find the spectral density of the error $\mathscr{E}[n]$ on the assumption that the signals s and ξ are correlated with each other.

Let us first find the error correlation function. Following the notation used in the diagram, we have

$$h[n]=\sum_{m=-\infty}^{\infty} W_{\mathrm{id}}[m]\, s[n-m] \tag{4.6}$$

$$\begin{aligned} s^{*}[n] &= \sum_{m=0}^{\infty} W[m]\,(s[n-m]+\xi[n-m]) \\ &= \sum_{m=-\infty}^{\infty} W[m]\,(s[n-m]+\xi[n-m]) \end{aligned} \tag{4.7}$$

The limits of summation in (4.6) are taken as infinite, since the weighting function $W_{\mathrm{id}}[n]$ will not necessarily become zero for $n<0$. The limits of summation in (4.7) are also taken to be infinite, since the weighting function $W[n]$ must satisfy the condition of physical realisability, and hence $W[n]\equiv 0$ for $n<0$.

Using (4.6) and (4.7) we can write the expression for the transformation error

$$\begin{aligned} \mathscr{E}[n] &= \sum_{m=-\infty}^{\infty} W[m]\,(s[n-m]+\xi[n-m]) \\ &\quad - \sum_{m=-\infty}^{\infty} W_{\mathrm{id}}[m]\, s[n-m] \end{aligned} \tag{4.8}$$

Using (4.8) we find the error correlation function

$$\begin{aligned} K_{\mathrm{e}}[m] &= \sum_{l=-\infty}^{\infty} W[l] \sum_{\alpha=-\infty}^{\infty} W[\alpha]\,\{K_s[m+l-\alpha] \\ &\quad + K_{s\xi}[m+l-\alpha]+K_{\xi s}[m+l-\alpha]+K_{\xi}[m+l-\alpha]\} \\ &\quad + \sum_{l=-\infty}^{\infty} W_{\mathrm{id}}[l] \sum_{\alpha=-\infty}^{\infty} W_{\mathrm{id}}[\alpha]\, K_s[m+l-\alpha] \\ &\quad - \sum_{l=-\infty}^{\infty} W[l] \sum_{\alpha=-\infty}^{\infty} W_{\mathrm{id}}[\alpha]\,\{K_s[m+l-\alpha]+K_{\xi s}[m+l-\alpha]\} \\ &\quad - \sum_{l=-\infty}^{\infty} W_{\mathrm{id}}[l] \sum_{\alpha=-\infty}^{\infty} W[\alpha]\,\{K_s[m+l-\alpha]+K_{s\xi}[m+l-\alpha]\} \end{aligned} \tag{4.9}$$

Multiplying both sides of (4.9) by $e^{-j\bar{\omega}m}$ and then summing with respect to m between the limits $m = -\infty$ and $m = \infty$, we obtain the expression for the spectral density of the error

$$S_e(\bar{\omega}) = W^*(j\bar{\omega})\, W^*(-j\bar{\omega})\, [S_s(\bar{\omega}) + S_{s\xi}(j\bar{\omega}) + S_{\xi s}(j\bar{\omega}) + S_\xi(\bar{\omega})]$$
$$+ W^*_{id}(j\bar{\omega})\, W^*_{id}(-j\bar{\omega})\, S_s(\bar{\omega}) - W^*(-j\bar{\omega})\, W^*_{id}(j\bar{\omega})\, [S_s(\bar{\omega})$$
$$+ S_{\xi s}(j\bar{\omega})] - W^*_{id}(-j\bar{\omega}) W^*(j\bar{\omega})\, [S_s(\bar{\omega}) + S_{s\xi}(j\bar{\omega})] \qquad (4.10)$$

where

$$W^*(j\bar{\omega}) = \sum_{m=-\infty}^{\infty} W[m]\, e^{-j\bar{\omega}m} \qquad (4.11)$$

is the frequency characteristic of the optimal system with the weighting function $W[n]$, and

$$W^*_{id}(j\bar{\omega}) = \sum_{m=-\infty}^{\infty} W_{id}[m]\, e^{-j\bar{\omega}m} \qquad (4.12)$$

is the frequency characteristic of the ideal system with weighting function $W_{id}[n]$. The symbols $S(j\bar{\omega})$ with the corresponding indices in (4.10) denote the spectral and cross-spectral densities of the respective random processes. We have

$$S(j\bar{\omega}) = \sum_{m=-\infty}^{\infty} K[m]\, e^{-j\bar{\omega}m} \qquad (4.13)$$

where $K[m]$ is the correlation function of the random process. The cross-spectral densities $S_{s\xi}(j\bar{\omega})$ and $S_{\xi s}(j\bar{\omega})$ are given by (4.13) where $K[m]$ must be substituted by the corresponding cross-correlation functions $K_{s\xi}[n]$ and $K_{\xi s}[n]$.

After certain transformations (4.10) can be reduced to the form

$$S_e(\bar{\omega}) = |W^*(j\bar{\omega}) - W^*_{id}(j\bar{\omega})|^2 \cdot S_s(\bar{\omega}) + |W^*(j\bar{\omega})|^2 S_\xi(\bar{\omega})$$
$$+ [W^*(j\bar{\omega}) - W^*_{id}(j\bar{\omega})]\, W^*(-j\bar{\omega})\, S_{\xi s}(j\bar{\omega}) \qquad (4.14)$$
$$+ [W^*(-j\bar{\omega}) - W^*_{id}(-j\bar{\omega})]\, W^*(j\bar{\omega})\, S_{s\xi}(j\bar{\omega})$$

If the spectral density $S_e(\bar{\omega})$ is known, the dispersion of the error can be found from

$$D_e = \frac{1}{\pi}\int_0^{\pi} S_e(\bar{\omega})\, d\bar{\omega} \tag{4.15}$$

Expression (4.14) gives the spectral density of the error of a system with an arbitrary transformation operator. For a system which predicts the input signal for the time $\bar{t}=l$, the weighting function of the ideal system must be put equal to

$$W_{id}[n]=\sigma[n+l] \tag{4.16}$$

Consequently, the transfer function of the ideal system, when (4.16) is taken into account, is

$$W^*_{id}(q)=\sum_{m=0}^{\infty}\sigma[m+l]\,e^{-qm}=e^{ql} \tag{4.17}$$

Putting $q=j\bar{\omega}$ in (4.17) we find the frequency characteristic of the ideal system

$$W^*_{id}(j\bar{\omega})=e^{j\bar{\omega}l} \tag{4.18}$$

Using (4.15), (4.18) and (4.14) we have the following expression for the dispersion of the error of the predicting system:

$$D_e=\frac{1}{\pi}\int_0^{\pi}\Big\{|W^*(j\bar{\omega})-e^{j\bar{\omega}l}|^2 S_s(\bar{\omega})+|W^*(j\bar{\omega})|^2 S_{\xi}(\bar{\omega})$$

$$+[W^*(-j\bar{\omega})-e^{-j\bar{\omega}l}]W^*(j\bar{\omega})S_{s\xi}(j\bar{\omega})\Big\}\,d\bar{\omega} \tag{4.19}$$

$$+[W^*(j\bar{\omega})-e^{j\bar{\omega}l}]W^*(-j\bar{\omega})S_{\xi s}(j\bar{\omega})$$

Let us transform (4.19). Before everything else the transfer function $W^*(j\bar{\omega})$ of the required predicting system, being a complex number for all values of $\bar{\omega}$, is represented in the exponential form

$$W^*(j\bar{\omega})=W(\bar{\omega})\,e^{j\psi(\bar{\omega})} \tag{4.20}$$

Substituting the expression given by (4.20) for the transfer function $W^*(j\bar{\omega})$ we find

$$D_e = \frac{1}{\pi}\int_0^\pi \Big\{W^2(\bar{\omega})[S_s(\bar{\omega}) + S_\xi(\bar{\omega}) + S_{s\xi}(j\bar{\omega}) + S_{\xi s}(j\bar{\omega})] + S_s(\bar{\omega}) - W(\bar{\omega})[S_s(\bar{\omega}) + S_{\xi s}(j\bar{\omega})]\, e^{j[\bar{\omega}l - \psi(\bar{\omega})]} - W(\bar{\omega})[S_s(\bar{\omega}) + S_{s\xi}(j\bar{\omega})]\, e^{-j[\bar{\omega}l - \psi(\bar{\omega})]}\Big\}\, d\bar{\omega} \tag{4.21}$$

Since the cross-spectral densities have the following property

$$\begin{aligned} S_{s\xi}(j\bar{\omega}) &= S_{\xi s}(-j\bar{\omega}) \\ S_{\xi s}(j\bar{\omega}) &= S_{s\xi}(-j\bar{\omega}) \end{aligned} \tag{4.22}$$

we can write

$$\begin{aligned} S_s(\bar{\omega}) + S_{\xi s}(j\bar{\omega}) &= S(\bar{\omega})\, e^{-j\beta(\bar{\omega})} \\ S_s(\bar{\omega}) + S_{s\xi}(j\bar{\omega}) &= S(\bar{\omega})\, e^{j\beta(\bar{\omega})} \end{aligned} \tag{4.23}$$

Substituting (4.23) into (4.21) we obtain the following expression for the dispersion of the prediction error:

$$D_e = \frac{1}{\pi}\int_0^\pi \{W^2(\bar{\omega})[S_s(\bar{\omega}) + S_\xi(\bar{\omega}) + S_{s\xi}(j\bar{\omega}) + S_{\xi s}(j\bar{\omega})] + S_s(\bar{\omega}) - 2S(\bar{\omega})W(\bar{\omega})\cos[\bar{\omega}l - \psi(\bar{\omega}) + \beta(\bar{\omega})]\}\, d\bar{\omega} \tag{4.24}$$

For the predicting system to be optimal, its frequency characteristic $W^*(j\bar{\omega})$ must be chosen so that the dispersion of the prediction error defined by (4.24) is a minimum. Thus, the problem of determining the optimal predicting system reduces to the determination of the amplitude-frequency and phase-frequency characteristics, $W(\bar{\omega})$ and $\psi(\bar{\omega})$, from the conditions securing a minimum dispersion of the prediction error.

It follows from (4.24) that the dispersion will be minimal if the expression under the integral sign gives a minimum for the chosen $W(\bar{\omega})$ and $\psi(\bar{\omega})$ for every value of $\bar{\omega}$. To determine $W(\bar{\omega})$ and $\psi(\bar{\omega})$ which minimise the dispersion D_e, we find the derivatives with respect to $W(\bar{\omega})$ and $\psi(\bar{\omega})$ of the expression under the integral sign of (4.24) and equate these derivatives to zero. It can be shown that $W(\bar{\omega})$ and $\psi(\bar{\omega})$, thus found, transform the expression under the integral sign of (4.24) into a minimum, and not into a maximum.

Carrying out the transformations just mentioned we find

$$W(\overline{\omega})[S_s(\overline{\omega})+S_\xi(\overline{\omega})+S_{s\xi}(j\overline{\omega})+S_{\xi s}(j\overline{\omega})] \\ -S(\overline{\omega})\cos[\overline{\omega}l-\psi(\overline{\omega})+\beta(\overline{\omega})]=0 \tag{4.25}$$

$$S(\overline{\omega})W(\overline{\omega})\sin[\overline{\omega}l-\psi(\overline{\omega})+\beta(\overline{\omega})]=0 \tag{4.26}$$

Since $S(\overline{\omega})\neq 0$ and $W(\overline{\omega})\neq 0$ we obtain from (4.26) the following expression

$$\psi(\overline{\omega})=\overline{\omega}l+\beta(\overline{\omega}) \tag{4.27}$$

Substituting (4.27) into (4.25) we have

$$W(\overline{\omega})[S_s(\overline{\omega})+S_\xi(\overline{\omega})+S_{s\xi}(j\overline{\omega})+S_{\xi s}(j\overline{\omega})]-S(\overline{\omega})=0$$

whence we find the amplitude-frequency characteristic of the predicting system

$$W(\overline{\omega})=\frac{S(\overline{\omega})}{S_s(\overline{\omega})+S_\xi(\overline{\omega})+S_{s\xi}(j\overline{\omega})+S_{\xi s}(j\overline{\omega})} \tag{4.28}$$

Expressions (4.27) and (4.28) which give the phase-frequency and amplitude-frequency characteristics of the required predicting system can also be found from (4.5), which is satisfied by the weighting function $W[n]$ of the optimal system. Thus, multiplying both sides of (4.5) by $e^{-j\overline{\omega}n}$ and summing with respect to n from $-\infty$ to $+\infty$, we have

$$\sum_{n=-\infty}^{\infty} K_{s\varphi}[n+l]e^{-j\overline{\omega}n}=\sum_{n=-\infty}^{\infty} e^{-j\overline{\omega}n}\sum_{m=0}^{\infty} W[m]K_\varphi[n-m] \tag{4.29}$$

Equation (4.29) can also be written in the following form:

$$\sum_{n=-\infty}^{\infty} K_{s\varphi}[n+l]e^{-j\overline{\omega}(n+l)}e^{j\overline{\omega}l} \\ =\sum_{n=-\infty}^{\infty} e^{-j\overline{\omega}(n-m)}K_\varphi[n-m]\times\sum_{m=0}^{\infty} W[m]e^{-j\overline{\omega}m} \tag{4.30}$$

Introducing the symbols

$$S_{s\varphi}(j\bar{\omega}) = \sum_{m=-\infty}^{\infty} K_{s\varphi}[m]\, e^{-j\bar{\omega}m} \tag{4.31}$$

$$S_{\varphi}(\bar{\omega}) = \sum_{m=-\infty}^{\infty} K_{\varphi}[m]\, e^{-j\bar{\omega}m} \tag{4.32}$$

we have from (4.30)

$$S_{s\varphi}(j\bar{\omega})\, e^{j\bar{\omega}l} = W^*(j\bar{\omega})\, S_{\varphi}(\bar{\omega}) \tag{4.33}$$

Since, on the basis of (4.4)

$$S_{s\varphi}(j\bar{\omega}) = S_s(\bar{\omega}) + S_{s\xi}(j\omega)$$

and according to (2.7)

$$S_{\varphi}(\bar{\omega}) = S_s(\bar{\omega}) + S_{\xi}(\bar{\omega}) + S_{s\xi}(j\bar{\omega}) + S_{\xi s}(j\bar{\omega})$$

(4.33) can be written as follows:

$$\begin{aligned}&[S_s(\bar{\omega}) + S_{s\xi}(j\bar{\omega})]\, e^{j\bar{\omega}l} \\ &= W^*(j\bar{\omega})[S_s(\bar{\omega}) + S_{\xi}(\bar{\omega}) + S_{s\xi}(j\bar{\omega}) + S_{\xi s}(j\bar{\omega})]\end{aligned} \tag{4.34}$$

whence we obtain

$$W^*(j\bar{\omega}) = \frac{S_s(\bar{\omega}) + S_{s\xi}(j\bar{\omega})}{S_s(\bar{\omega}) + S_{\xi}(\bar{\omega}) + S_{s\xi}(j\bar{\omega}) + S_{\xi s}(j\bar{\omega})}\, e^{j\bar{\omega}l} \tag{4.35}$$

Equation (4.35) coincides with the expression for the transfer function (4.20) when (4.27) and (4.28) are used. Thus, substituting (4.27) and (4.28) into (4.20), we have

$$W^*(j\bar{\omega}) = \frac{S(\bar{\omega})}{S_s(\bar{\omega}) + S_{\xi}(\bar{\omega}) + S_{s\xi}(j\bar{\omega}) + S_{\xi s}(j\bar{\omega})}\, e^{j[\bar{\omega}l + \beta(\bar{\omega})]} \tag{4.36}$$

Since in accordance with (4.23)

$$S(\omega)\, e^{j\beta(\bar{\omega})} = S_s(\bar{\omega}) + S_{s\xi}(j\bar{\omega})$$

we find from (4.36) that

$$W^*(j\bar{\omega}) = \frac{S_s(\bar{\omega}) + S_{s\xi}(j\bar{\omega})}{S_s(\bar{\omega}) + S_\xi(\bar{\omega}) + S_{s\xi}(j\bar{\omega}) + S_{\xi s}(j\bar{\omega})} e^{j\bar{\omega}l} \tag{4.37}$$

which coincides with (4.35).

Let us analyse the results just obtained. The expression (4.37) giving the transfer function of an optimal predicting system is obtained without the conditions of physical realisability being taken into consideration. Consequently, the weighting function $W[n]$ of such a system will be non-zero for negative values of the argument, i.e.

$$W[n] \neq 0 \text{ when } n < 0$$

To find the transfer function of an optimal predictor which is physically realisable, we use the method of shaping filters, proceeding from the transfer function $W^*(q)$. For this, using the given correlation function $K_\varphi[n] = K_s[n] + K_\xi[n] + K_{s\xi}[n] + K_{\xi s}[n]$, we find a discrete filter which forms the random signal $\varphi = s + \xi$ with the correlation function $K_\varphi[n]$ from the discrete white noise V. The weighting function of this filter is denoted by $W_f[n]$, whilst its transfer function is denoted by $W_f^*(q)$.

If the random signal φ is applied to a sampled data system with the transfer function $W_f^{*-1}(q)$, which is inverse to the shaping filter, then at the output of this system we obtain discrete white noise V. In this case the problem of choosing the optimal predictor reduces to the problem of determining the optimal linear transformation on discrete white noise.

Taking into account that for optimal prediction the transfer function of the system must be determined by (4.37), we find that the optimal linear transformation on white noise can be carried out by a system which is a series connection of two systems with the transfer functions $W_f^*(q)$ and $W^*(q)$ respectively (Fig. 5.4). Thus, the transfer function of a system which ensures the optimal linear transformation of

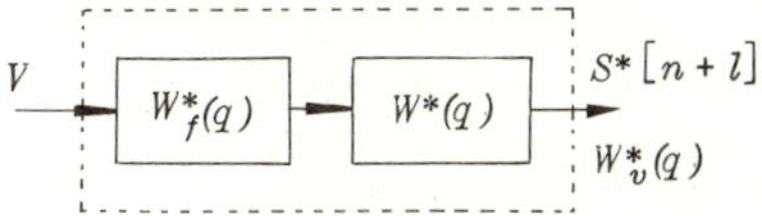

Fig. 5.4 A system of optimal linear transformation of white noise

discrete white noise, in order to obtain the predicted value of the signal $s^*[n+l]$, must be

$$W_v^*(q) = W_f^*(q)\,W^*(q) \tag{4.38}$$

But since $W^*(q)$ is the transfer function of a system which is not physically realisable, the weighting function

$$W_v[n] = D^{-1}\{W_v^*(q)\} \tag{4.39}$$

can be non-zero when $n<0$. But because the predicted value of the signal $s^*[n+l]$ is formed from discrete white noise $V[m]$ when $m \leqslant n$, i.e. in accordance with

$$s^*[n+l] = \sum_{m=0}^{\infty} W_v[m]\,V[n-m]$$

we must stipulate that the weighting function $W_1[n]$ of the system which carried out the optimal operation on white noise is defined as follows:

$$W_1[n] = \begin{cases} W_v[n], & n \geqslant 0 \\ 0, & n<0 \end{cases} \tag{4.40}$$

where $W_v[n]$ is given by (4.39).

Accordingly, the transfer function of an optimal predictor must be equal to

$$W_{\text{pred}}^*(q) = W_f^{*-1}(q)\,W_1^*(q) \tag{4.41}$$

where

$$W_1^*(q) = D\{W_1[n]\} \tag{4.42}$$

In this way, the procedure of determining the transfer function of an optimal sampled data system which predicts the random signal s, when noise ξ is present, reduces to the following:

1. Using the given correlation function $K_\varphi[n]$, determine the transfer function $W_f^*(q)$ of the discrete shaping filter and the transfer function $W_f^{*-1}(q)$ of a filter which is inverse to this shaping filter.
2. Using the given spectral densities $S_s(\bar{\omega}), S_{s\xi}(j\bar{\omega})$ and $S_{\xi s}(j\bar{\omega})$ find the transfer function $W^*(q)$ which is given by (4.37).
3. Find the weighting function $W_v[n]$ of a system whose

transfer function $W_v^*(q)$ is

$$W_v^*(q) = W_f^*(q)\,W^*(q)$$

4. Using (4.42) determine the transfer function $W_1^*(q)$.
5. Using (4.41) find the transfer function $W_{\text{pred}}^*(q)$ of the optimal predictor.

The error of prediction can be found from (4.15), or from the corresponding formulae of the linear theory given in Chapter 3.

An analogous method has been derived by Bode and Shannon [70], for determining the optimal continuous predicting system.

5.5 WEIGHTING FUNCTION OF AN OPTIMAL SYSTEM: METHOD OF INDETERMINATE COEFFICIENTS

In the preceding sections methods have been given for determining optimal sampled data systems by means of discrete shaping filters. These methods are physically straightforward but somewhat cumbersome, since they necessitate the determination of shaping filters. Hence these methods may be difficult if the expressions of the corresponding spectral densities are complicated. In this section we shall present another method for determining optimal sampled data systems based on representing the weighting function of the required optimal system as a series with a finite number of terms which are of exponential form with unknown indices. Because of this the method may be called the method of indeterminate coefficients. In order to determine the unknown coefficients we have to solve comparatively complicated systems of algebraic equations. In spite of this the method is found to be useful in practice, particularly when digital computers are available.

The method will be given as applied to the determination of the weighting function of a sampled data system which predicts a random signal in the presence of noise. This method can also be used to determine the weighting function of a system which carries out any linear operation on a useful random signal, and has been used by Battin to solve the Wiener-Hopf equation [70].

It has been shown that the weighting function of an optimal system which predicts the random signal s over the time

$\bar{t}=l$, when the noise ξ is present, satisfies the following equation:

$$K_{s\varphi}[n+l]=\sum_{m=0}^{\infty} W[m] K_{\varphi}[n-m], \quad n \geqslant 0 \tag{5.1}$$

where the correlation function K_{φ} and the cross-correlation function $K_{s\varphi}$ are given by (2.7) and (4.4) respectively. In what follows we shall assume that the useful random signal s is not correlated with the noise ξ, i.e. $K_{\xi s}[n]=K_{s\xi}[n] \equiv 0$, so that (2.7) and (4.4) give

$$K_{\varphi}[n]=K_s[n]+K_{\xi}[n] \tag{5.2}$$

$$K_{s\varphi}[n]=K_s[n] \tag{5.3}$$

Using (5.2) and (5.3) we write (5.1) for the optimal weighting function as follows:

$$K_s[n+l]=\sum_{m=0}^{\infty} W[m]\left(K_s[n-m]+K_{\xi}[n-m]\right), \quad n \geqslant 0 \tag{5.4}$$

We shall now solve (5.4).

We assume that the correlation functions $K_s[n]$ and $K_{\xi}[n]$ are given in the form of a series with a finite number of terms

$$K_s[n]=\sum_{i=1}^{p} A_i e^{-\alpha_i |n|} \tag{5.5}$$

$$K_{\xi}[n]=\sum_{j=1}^{k} B_j e^{-\beta_j |n|} \tag{5.6}$$

where α_i and β_j can be real or complex.

The correlation functions can be represented in the form (5.5) and (5.6), since the correlation function of any stationary random process can be represented as a sum of exponential terms with any required accuracy.

Noting that correlation functions are even functions, we can write, using (5.5) and (5.6),

$$K_s[n] = \begin{cases} \sum_{i=1}^{p} A_i e^{-\alpha_i n}, & n \geqslant 0 \\ \sum_{i=1}^{p} A_i e^{\alpha_i n}, & n < 0 \end{cases} \tag{5.7}$$

$$K_\xi[n] = \begin{cases} \sum_{j=1}^{k} B_j e^{-\beta_j n}, & n \geqslant 0 \\ \sum_{j=1}^{k} B_j e^{\beta_j n}, & n < 0 \end{cases} \tag{5.8}$$

We now substitute the expressions for the correlation functions (5.7) and (5.8) into (5.4)

$$\sum_{i=1}^{p} A_j e^{-\alpha_i (n+l)}$$

$$= \sum_{m=0}^{\infty} W[m] \left[\sum_{i=1}^{p} A_i e^{-\alpha_i |n-m|} + \sum_{j=1}^{k} B_j e^{-\beta_j |n-m|} \right] \tag{5.9}$$

$$n \geqslant 0$$

Noting that

$$\sum_{m=0}^{\infty} W[m] [A_i e^{-\alpha_i |n-m|} + B_j e^{-\beta_j |n-m|}]$$

$$= \sum_{m=0}^{n} W[m] [A_i e^{-\alpha_i (n-m)} + B_j e^{-\beta_j (n-m)}] \tag{5.10}$$

$$+ \sum_{m=n+1}^{\infty} W[m] [A_i e^{\alpha_i (n-m)} + B_j e^{\beta_j (n-m)}]$$

we transform (5.9) into the following form:

$$\sum_{i=1}^{p} A_i e^{-\alpha_i (n+l)}$$

$$= \sum_{m=0}^{n} W[m] \left[\sum_{i=1}^{p} A_i e^{-\alpha_i (n-m)} + \sum_{j=1}^{k} B_j e^{-\beta_j (n-m)} \right]$$

$$+\sum_{m=n+1}^{\infty} W[m]\left[\sum_{i=1}^{p} A_i e^{\alpha_i (n-m)} + \sum_{j=1}^{k} B_j e^{\beta_j (n-m)}\right] \quad n \geqslant 0 \tag{5.11}$$

Because a sampled data system with constant parameters is described by a difference equation with constant coefficients, the weighting function of such a system as the solution of the corresponding non-homogeneous difference equation, with zero initial conditions and the σ-function instead of the disturbing function, is a linear combination of a finite number of exponential functions with real and complex indices. In addition, the weighting function may in the general case contain also the σ-function.

Consequently, the weighting function $W[n]$ of the required optimal system can be written

$$W[n] = \sum_{\nu=1}^{r} w_\nu e^{-\gamma_\nu n} + w_0 \sigma[n] \quad n \geqslant 0 \tag{5.12}$$

Representation of the weighting function in the form (5.12) is equivalent to representing the sampled data system in the form of a parallel connection of $r+1$ sampled data systems (Fig. 5.5), of which one system is an amplifying element with gain factor w_0, whilst the other r systems are inertial systems whose processes are given by first-order equations of the form

$$\Delta y_\nu[n] + (1 - e^{-\gamma_\nu}) y_\nu[n] = w_\nu x[n] \quad \nu = 1, 2, \ldots, r \tag{5.13}$$

Accordingly, we seek the solution of (5.11) in the form

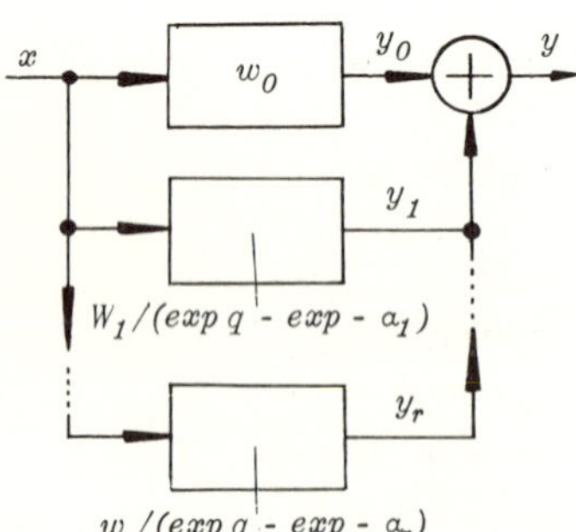

Fig. 5.5 A system with the weighting function (5.12)

(5.12). Substituting (5.12) into (5.11) we find

$$\sum_{i=1}^{p} A_i e^{-\alpha_i (n+l)} = \sum_{m=0}^{n} \left[\sum_{\nu=0}^{r} w_\nu e^{-\gamma_\nu m} + \sigma[m] \right] \times \left[\sum_{i=1}^{p} A_i e^{-\alpha_i (n-m)} + \sum_{j=1}^{k} B_j e^{-\beta_j (n-m)} \right] + \sum_{m=n+1}^{\infty} \left[\sum_{\nu=1}^{r} w_\nu e^{-\gamma_\nu m} + \sigma[m] \right] \times \left[\sum_{i=1}^{p} A_i e^{\alpha_i (n-m)} + \sum_{j=1}^{k} B_j e^{\beta_j (n-m)} \right] \quad n \geqslant 0 \tag{5.14}$$

Inverting the order of summation in (5.14) and carrying out certain transformations we have

$$\begin{aligned} \sum_{i=1}^{p} A_i e^{-\alpha_i (n+l)} &= \sum_{\nu=1}^{r} w_\nu \sum_{i=1}^{p} A_i e^{-\alpha_i n} \sum_{m=0}^{n} e^{-(\gamma_\nu - \alpha_i) m} \\ &+ \sum_{\nu=1}^{r} w_\nu \sum_{j=1}^{k} B_j e^{-\beta_j n} \sum_{m=0}^{n} e^{-(\gamma_\nu - \beta_j) m} \\ &+ w_0 \left[\sum_{i=1}^{p} A_i e^{-\alpha_i n} \sum_{m=0}^{n} e^{\alpha_i m} \sigma[m] + \sum_{j=1}^{k} B_j e^{-\beta_j n} \sum_{m=0}^{n} e^{\beta_j m} \sigma[m] \right] \\ &+ \sum_{\nu=1}^{r} w_\nu \sum_{i=1}^{p} A_i e^{\alpha_i n} \sum_{m=n+1}^{\infty} e^{-(\gamma_\nu + \alpha_i) m} \\ &+ \sum_{\nu=1}^{r} w_\nu \sum_{j=1}^{k} B_j e^{\beta_j n} \sum_{m=n+1}^{\infty} e^{-(\gamma_\nu + \beta_j) m} \\ &+ w_0 \left[\sum_{i=1}^{p} A_i e^{\alpha_i n} \sum_{m=n+1}^{\infty} e^{-\alpha_i m} \sigma[m] + \sum_{j=1}^{k} B_j e^{\beta_j n} \sum_{m=n+1}^{\infty} e^{-\beta_j m} \sigma[m] \right] \\ & n \geqslant 0 \end{aligned} \tag{5.15}$$

Taking into account the properties of the σ-function we have

$$\sum_{i=1}^{p} A_i e^{-\alpha_i n} \sum_{m=0}^{n} e^{\alpha_i m} \sigma[m] = \sum_{i=1}^{p} A_i e^{-\alpha_i n} \qquad (5.16)$$

$$\sum_{j=1}^{k} B_j e^{-\beta_j n} \sum_{m=0}^{n} e^{\beta_j m} \sigma[m] = \sum_{j=1}^{k} B_j e^{-\beta_j n} \qquad (5.17)$$

$$\sum_{i=1}^{p} A_i e^{\alpha_i n} \sum_{m=n+1}^{\infty} e^{-\alpha_i m} \sigma[m] = 0 \qquad (5.18)$$

$$\sum_{j=1}^{k} B_j e^{\beta_j n} \sum_{m=n+1}^{\infty} e^{-\beta_j m} \sigma[m] = 0 \qquad (5.19)$$

After carrying out summation in (5.15) we find the following equations:

$$\sum_{i=1}^{p} A_i e^{-\alpha_i n} \sum_{m=0}^{n} e^{-(\gamma_\nu - \alpha_i) m} = \sum_{i=1}^{p} A_i \frac{e^{-\alpha_i n} - e^{-\gamma_\nu n}}{1 - e^{-(\gamma_\nu - \alpha_i)}} \qquad (5.20)$$

$$\sum_{i=1}^{p} A_i e^{+\alpha_i n} \sum_{m=n+1}^{\infty} e^{-(\gamma_\nu + \alpha_i) m} = \sum_{i=1}^{p} A_i \frac{e^{-\gamma_\nu (n+1)} \cdot e^{-\alpha_i}}{1 - e^{-(\gamma_\nu + \alpha_i)}} \qquad (5.21)$$

Analogously to (5.20) and (5.21), we also have

$$\sum_{j=1}^{k} B_j e^{-\beta_j n} \sum_{m=0}^{n} e^{-(\gamma_\nu - \beta_j) m} = \sum_{j=1}^{k} B_j \frac{e^{-\beta_j n} - e^{-\gamma_\nu n}}{1 - e^{-(\gamma_\nu - \beta_j)}} \qquad (5.22)$$

$$\sum_{j=1}^{k} B_j e^{+\beta_j n} \sum_{m=n+1}^{\infty} e^{-(\gamma_\nu + \beta_j) m} = \sum_{j=1}^{k} B_j \frac{e^{-\gamma_\nu (n+1)} e^{-\beta_j}}{1 - e^{-(\gamma_\nu + \beta_j)}} \qquad (5.23)$$

Using (5.16) to (5.23), we write (5.15) in the form

$$\sum_{i=1}^{p} A_i e^{-\alpha_i (n+l)}$$

$$= \sum_{\nu=1}^{r} w_\nu \sum_{i=1}^{p} A_i \left[\frac{e^{-\alpha_i n} - e^{-\gamma_\nu n}}{1 - e^{-(\gamma_\nu - \alpha_i)}} + \frac{e^{-\gamma_\nu (n+1)} \cdot e^{-\alpha_i}}{1 - e^{-(\gamma_\nu + \alpha_i)}} \right]$$

$$+\sum_{\nu=1}^{r} w_{\nu} \sum_{j=1}^{k} B_j \left[\frac{e^{-\beta_j n} - e^{-\gamma_\nu n}}{1 - e^{-(\gamma_\nu - \beta_j)}} + \frac{e^{-\gamma_\nu (n+1)} \cdot e^{-\beta_j}}{1 - e^{-(\gamma_\nu + \beta_j)}}\right]$$

$$+ w_0 \left[\sum_{i=1}^{p} A_i e^{-\alpha_i n} + \sum_{j=1}^{k} B_j e^{-\beta_j n}\right], \quad n \geqslant 0 \tag{5.24}$$

Equation (5.24) should be regarded as an identity which must satisfy each instant $\bar{t} = n \geqslant 0$. In this case (5.24) gives a system of equations from which the unknowns w_ν, $\nu = 0, 1, \ldots, r$ and γ_ν, $\nu = 0, 1, \ldots, r$ are determined.

We point out that r must be chosen to equal $p + k - 1$, where p and k are the numbers of terms in the series for $K_s[n]$ and $K_\xi[n]$ respectively. The total number of unknowns which must be determined thus equals $2(p + k - 1)$.

Proceeding from (5.24) it is not difficult to obtain the necessary equations for the general case, although there is no need for this, since in each concrete case the required equations are easily obtained.

Chapter 6

OPTIMAL SYSTEMS WITH FINITE OBSERVATION TIME

In the preceding chapter we gave methods for determining optimal sampled data systems where the input actions are stationary random processes with zero mathematical expectations. At the same time it has been assumed that the output signal is formed from an input signal observed over the semi-infinite time interval $(-\infty; n)$. The determination of the optimal system reduces in this case to solving the problem of unconditional minimum of the mean-square error.

In practice we frequently encounter problems where optimal systems must be determined which, in accordance with the given operators, reproduce or transform random signals carrying useful information constituting non-stationary random processes. In this chapter methods will be discussed for determining optimal transforming (reproducing) sampled data systems in the case where besides a stationary random process the input action contains a mathematical expectation which can be represented in the form of a polynomial of finite order with respect to time.

When an optimal sampled data system is determined when the non-stationary part of the signal is present, a more complicated problem must be solved for the conditional minimum of the error of transformation (reproduction). Additional conditions arise from the fact that the mathematical expectation of the signal must be reproduced or transformed with a given accuracy. At the same time, as a rule, it is necessary to satisfy the condition of 'finiteness of memory' of the system, which reduces to the condition that the signal at the output of the transforming (reproducing) system must be formed from values of the input signal observed over a finite time interval.

In the theory of continuous optimal systems it is usual to make a number of assumptions regarding the polynomial which represents the mathematical expectation of the random input signal. These reduce to the following three cases.

1. The coefficients of the polynomial are non-random quantities and can assume arbitrary values. The problem of determining the continuous optimal system in this case was solved by Zadeh and Ragazzini [73], and is called the Zadeh-Ragazzini problem.
2. The coefficients of the polynomial are given. This problem was solved by Solodovnikov [64] and is called Solodovnikov's problem.
3. The coefficients of the polynomial are random quantities. The problem of determining the optimal continuous system in this case is called Semenov's problem [62].

A discrete analogue of the Zadeh-Ragazzini problem has been considered by Perov [68, 69], Tsypkin [74] and Solodovnikov [63].

In this chapter the main attention is focussed on the solution of the problem which has not been considered until very recently, but which is of great importance for practical applications: the discrete analogue of Semenov's problem, of which the discrete analogue of the Zadeh-Ragazzini problem arises as a particular case.

6.1 DISCRETE ANALOGUE OF THE ZADEH-RAGAZZINI PROBLEM

Suppose that the useful signal $X[n]$ observed over the closed time interval $[0; N]$, where N is the time of observation, has the mathematical expectation $m_x[n]$ which is a polynomial of finite degree r in time,

$$m_x[n] = \sum_{k=0}^{r} d_k n^k \tag{1.1}$$

where the coefficients d_k, $k=0, 1, \ldots, r$ are non-random quantities and can assume arbitrary values. In addition, we assume that $\overset{\circ}{X}[n]=X[n]-m_x[n]$, i.e. the centralised random function, is a stationary random function belonging to the class connected with discrete white noise by difference equations of finite order.

We further assume that on the useful signal $X[n]$ is superimposed random noise $\xi[n]$,, about which we assume that it is a stationary random function also belonging to the class of random functions given by linear difference equations of finite order. Without loss of generality we assume that the mathematical expectation of the random function $\xi[n]$ is zero, whilst $\xi[n]$ itself is not correlated with the random function $X[n]$, i.e.

$$M[X[m]\xi[n]]=0$$
$$0\leqslant m\leqslant N,\ 0\leqslant n\leqslant N$$

We shall now formulate the problem of determining a linear sampled data system which guarantees a minimum of the mean-square error of transformation (reproduction) of $X[n]$, with an additional condition, stipulating the absence of transformation (reproduction) errors when the signal

$$m_x[n]=\sum_{k=0}^{r} d_k n^k$$

is transformed (reproduced) by this system, for any value of time $0\leqslant n\leqslant N$ and for any value of the coefficient d_k, $k=0, 1, \ldots, r$.

Let us express mathematically the conditions just formulated which must be satisfied by the sampled data system. We characterise this system by its weighting function $W[n]$. In doing this we stipulate that $W[n]$ is to satisfy the following conditions:

$$W[n]=\begin{cases} W[n], & 0\leqslant n\leqslant N \\ 0, & n<0,\ n>N \end{cases}$$

where N is the time at which the input signal is observed.

The result $X^*[n]$ of the transformation of the input signal by the system with weighting function $W[n]$ can be written as follows:

$$X^*[n]=\sum_{m=0}^{N} W[m]\left(\sum_{k=0}^{r} d_k(n-m)^k+\overset{\circ}{X}[n-m]+\xi[n-m]\right) \tag{1.2}$$
$$0\leqslant n\leqslant N$$

or

$$X^{*}[n]=\sum_{m=0}^{N} W[m] \sum_{k=0}^{r} d_{k}(n-m)^{k}+\sum_{m=0}^{N} W[m] \varphi[n-m] \tag{1.3}$$

$$0 \leqslant n \leqslant N$$

where

$$\varphi[n]=\overset{\circ}{X}[n]+\xi[n] \tag{1.4}$$

In this connection it should be pointed out that, because the random functions $\overset{\circ}{X}$ and ξ belong to the class of random functions related to discrete white noise by linear difference equations of finite order, the random function φ, being the sum of the random functions $\overset{\circ}{X}$ and ξ, also belongs to this class of random functions.

Applying mathematical expectation to both sides of (1.3) and taking into account that

$$M[\varphi[n]]=0, \ 0 \leqslant n \leqslant N$$

we find the expression for the mathematical expectation $m_{x^*}[n]$ of the signal at the output of the transforming system

$$m_{x^*}[n]=\sum_{k=0}^{r} d_{k} \sum_{m=0}^{N} W[m](n-m)^{k}, \ 0 \leqslant n \leqslant N \tag{1.5}$$

It has been mentioned above that the system must accurately transform the mathematical expectation of the input signal. Denoting the weighting function of the ideal transforming system by $W_{\mathrm{id}}[n]$, we can write the condition for this as follows:

$$m_{x^*}[n] \equiv \sum_{m=-\infty}^{\infty} W_{\mathrm{id}}[m]\, m_{x}[n-m], \ 0 \leqslant n \leqslant N \tag{1.6}$$

We point out that the limits of summation in Equation (1.6) are taken to be infinite, since the weighting function of the ideal transforming system need not necessarily satisfy the condition of physical realisability, while also the signal at the output can be formed from values of the signal observed over the time interval $(-\infty; +\infty)$.

When (1.5) is taken into account, the çondition (1.6) can be written as follows:

$$\sum_{k=0}^{r} d_k \sum_{m=0}^{N} W[m](n-m)^k \equiv \sum_{k=0}^{r} d_k \sum_{m=-\infty}^{\infty} W_{\text{id}}[m](n-m)^k \qquad (1.7)$$

$$0 \leqslant n \leqslant N$$

whence

$$\sum_{k=0}^{r} d_k \left[\sum_{m=0}^{N} W[m](n-m)^k - \sum_{m=-\infty}^{\infty} W_{\text{id}}[m](n-m)^k \right] \equiv 0 \qquad (1.8)$$

$$0 \leqslant n \leqslant N$$

Noting that

$$(n-m)^k = \sum_{\nu=0}^{k} (-1)^{\nu} C_k^{\nu} n^{k-\nu} m^{\nu}$$

we can write (1.8) in the following form:

$$\sum_{k=0}^{r} d_k \left[\sum_{m=0}^{N} W[m] \sum_{\nu=0}^{k} (-1)^{\nu} C_k^{\nu} n^{k-\nu} m^{\nu} \right.$$

$$\left. - \sum_{m=-\infty}^{\infty} W_{\text{id}}[m] \sum_{\nu=0}^{k} (-1)^{\nu} C_k^{\nu} n^{k-\nu} m^{\nu} \right] \equiv 0, \quad 0 \leqslant n \leqslant N$$

or, changing the order of summation,

$$\sum_{k=0}^{r} d_k \sum_{\nu=0}^{k} (-1)^{\nu} C_k^{\nu} n^{k-\nu} \left[\sum_{m=0}^{N} W[m] m^{\nu} - \sum_{m=-\infty}^{\infty} W_{\text{id}}[m] m^{\nu} \right] \equiv 0 \qquad (1.9)$$

$$0 \leqslant n \leqslant N$$

Taking into consideration that

$$\sum_{k=0}^{r} \sum_{\nu=0}^{k} = \sum_{\nu=0}^{r} \sum_{k=\nu}^{r}$$

we have from (1.9)

$$\sum_{\nu=0}^{r} \sum_{k=\nu}^{r} d_k C_k^{\nu} n^{k-\nu} \left[\sum_{m=0}^{N} W[m] m^{\nu} - \sum_{m=-\infty}^{\infty} W_{\text{id}}[m] m^{\nu} \right] \equiv 0 \qquad (1.10)$$

$$0 \leqslant n \leqslant N$$

If we introduce the following notation

$$\sum_{k=\nu}^{r} d_k C_k^{\nu} n^{k-\nu} = D_\nu[n] \tag{1.11}$$

$$\nu = 0, 1, \ldots, r; \; 0 \leqslant n \leqslant N$$

(1.10) assumes the form

$$\sum_{\nu=0}^{r} D_\nu[n] \left[\sum_{m=0}^{N} m^\nu W[m] - \sum_{m=-\infty}^{\infty} m^\nu W_{\text{id}}[m] \right] \equiv 0$$

$$0 \leqslant n \leqslant N,$$

whence it follows that, in order to satisfy this equality for every value n belonging to the interval $[0; N]$, we must satisfy the condition

$$\sum_{m=0}^{N} m^\nu W[m] = \sum_{m=-\infty}^{\infty} m^\nu W_{\text{id}}[m]$$

$$\nu = 0, 1, \ldots, r$$

Introducing the notation

$$\mu_\nu = \sum_{m=-\infty}^{\infty} m^\nu W_{\text{id}}[m], \; \nu = 0, 1, \ldots, r \tag{1.12}$$

we finally obtain

$$\sum_{m=0}^{N} m^\nu W[m] = \mu_\nu, \; \nu = 0, 1, \ldots, r \tag{1.13}$$

Thus, for the sampled data system to transform exactly the mathematical expectation of the useful signal, the weighting function $W[n]$ of this system must satisfy the $(r+1)$ conditions (1.13). To illustrate this we shall consider a number of practical cases.

Suppose, for example, that the system to be determined is to predict accurately the useful signal (in absence of noise) over the time $\bar{t} = l$. In this case the weighting function $W_{\text{id}}[n]$ of the ideal system is

$$W_{\text{id}}[n] = \sigma[n + l] \tag{1.14}$$

Substituting (1.14) into (1.12) we have

$$\mu_\nu = \sum_{m=-\infty}^{\infty} m^\nu \sigma[m+l], \quad \nu = 0, 1, \ldots, r$$

Taking into consideration the property of the σ-function

$$\sum_{m=-\infty}^{\infty} f[m]\sigma[m+l] = f[-l]$$

we find from the last relation that

$$\mu_\nu = (-l)^\nu, \quad \nu = 0, 1, \ldots, r$$

Consequently, the weighting function $W[n]$ of the system must satisfy the following conditions:

$$\sum_{m=0}^{N} m^\nu W[m] = (-l)^\nu, \quad \nu = 0, 1, \ldots, r \tag{1.15}$$

If the required system is to form the value of the useful signal at the instant $\bar{t} = n$, with this value corresponding to the preceding instant $\bar{t} = n - l$, then obviously the weighting function of the ideal system must be

$$W_{\text{id}}[n] = \sigma[n-l] \tag{1.16}$$

In this case we obtain from (1.12) the following expression

$$\mu_\nu = \sum_{m=-\infty}^{\infty} m^\nu \sigma[m-l], \quad \nu = 0, 1, \ldots, r$$

whence, taking into account the property of the σ-function, we find that

$$\mu_\nu = l^\nu, \quad \nu = 0, 1, \ldots, r$$

Consequently, the weighting function $W[n]$ must satisfy the following conditions:

$$\sum_{m=0}^{N} m^\nu W[m] = l^\nu, \quad \nu = 0, 1, \ldots, r \tag{1.17}$$

If the sampled data system is to smooth the input signal, we must use the σ-function in the role of the weighting function of the ideal system, i.e.

$$W_{id}[n] = \sigma[n] \tag{1.18}$$

We can then show that the weighting function of the system must satisfy the following conditions:

$$\begin{aligned} &\sum_{m=0}^{N} W[m] = 1 \\ &\sum_{m=0}^{N} mW[m] = 0 \\ &\cdots\cdots\cdots \\ &\sum_{m=0}^{N} m^r W[m] = 0 \end{aligned} \tag{1.19}$$

In a similar manner we can find the corresponding expressions (1.13) in other cases.

We have thus shown that the weighting function of the system must satisfy (1.13). It must also guarantee the minimum dispersion of the transformation error, expressed mathematically as follows:

$$M\{[\sum_{m=-\infty}^{\infty} W_{id}[m] X[n-m] - \sum_{m=0}^{N} W[m] X([n-m] + \xi[n-m])\}^2] = \min, \ 0 \leqslant n \leqslant N \tag{1.20}$$

Taking into consideration that the useful signal $X[n]$ is given by

$$X[n] = m_x[n] + \overset{\circ}{X}[n]$$

and noting also (1.6), we reduce (1.20) to the form

$$M[\{\sum_{m=-\infty}^{\infty} W_{id}[m] \overset{\circ}{X}[n-m] - \sum_{m=0}^{N} W[m] (\overset{\circ}{X}[n-m] + \xi[n-m])\}^2] = \min, \ 0 \leqslant n \leqslant N \tag{1.21}$$

or, when (1.4) is used,

$$M\Big[\Big\{\sum_{m=-\infty}^{\infty} W_{\text{id}}[m]\overset{\circ}{X}[n-m] - \sum_{m=0}^{N} W[m]\varphi[n-m]\Big\}^2\Big] = \min,\quad 0 \leqslant n \leqslant N \tag{1.22}$$

The weighting function $W[n]$ of the system must thus guarantee a minimum of (1.22), and must satisfy the $r+1$ conditions (1.13) at the same time. A system which has such a weighting function is called an optimal system.

Consequently, the determination of an optimal sampled data transforming system reduces to finding a linear system which ensures the conditional minimum of the dispersion of the transformation error.

6.2 EQUATIONS FOR THE WEIGHTING FUNCTION OF AN OPTIMAL SYSTEM

Let us first find the expression for the mean-square error, proceeding from (1.22). Using (1.4) we can write the following expression for the dispersion D_e of the transformation error:

$$\begin{aligned} D_e = {} & M\Big[\sum_{m_1=-\infty}^{\infty} W_{\text{id}}[m_1] \sum_{m_2=-\infty}^{\infty} W_{\text{id}}[m_2]\overset{\circ}{X}[n-m_1]\overset{\circ}{X}[n-m_2]\Big] \\ & - M\Big[\sum_{m_1=-\infty}^{\infty} W_{\text{id}}[m_1] \sum_{m_2=0}^{N} W[m_2]\overset{\circ}{X}[n-m_1] \\ & \times(\overset{\circ}{X}[n-m_2]+\xi[n-m_2])\Big] - M\Big[\sum_{m_1=0}^{N} W[m_1] \\ & \times \sum_{m_2=-\infty}^{\infty} W_{\text{id}}[m_2](\overset{\circ}{X}[n-m_1]+\xi[n-m_1])\overset{\circ}{X}[n-m_2]\Big] \\ & + M\Big[\sum_{m_1=0}^{N} W[m_1] \sum_{m_2=0}^{N} W[m_2](\overset{\circ}{X}[n-m_1]+\xi[n-m_1] \\ & \times(\overset{\circ}{X}[n-m_2]+\xi[n-m_2])\Big] \end{aligned}$$

Carrying out the operation of mathematical expectation on

the right-hand side of this equation we have

$$
\begin{aligned}
D_e = & \sum_{m_1=-\infty}^{\infty} W_{id}[m_1] \sum_{m_2=-\infty}^{\infty} W_{id}[m_2] K_x[m_2 - m_1] \\
& - \sum_{m_1=-\infty}^{\infty} W_{id}[m_1] \sum_{m_2=0}^{N} W[m_2] K_x[m_2 - m_1] \\
& - \sum_{m_1=0}^{N} W[m_1] \sum_{m_2=-\infty}^{\infty} W_{id}[m_2] K_x[m_2 - m_1] \\
& + \sum_{m_1=0}^{N} W[m_1] \sum_{m_2=0}^{N} W[m_2] (K_x[m_2 - m_1] + K_\xi[m_2 - m_1])
\end{aligned}
\tag{2.1}
$$

where

$$K_x[m_2 - m_1] = M[\overset{\circ}{X}[n - m_1] \overset{\circ}{X}[n - m_2]]$$

is the correlation function of the stationary part of the useful signal, whilst

$$K_\xi[m_2 - m_1] = M[\xi[n - m_1] \xi[n - m_2]]$$

is the correlation function of the noise.

Introducing the expression

$$K_\varphi[n] = K_x[n] + K_\xi[n] \tag{2.2}$$

and changing the order of summation in the third term of (2.1) we have the following expression for the dispersion:

$$
\begin{aligned}
D_e = & \sum_{m_1=0}^{N} W[m_1] \sum_{m_2=0}^{N} W[m_2] K_\varphi[m_2 - m_1] \\
& - 2 \sum_{m_1=-\infty}^{\infty} W_{id}[m_1] \sum_{m_2=0}^{N} W[m_2] K_x[m_2 - m_1] \\
& + \sum_{m_1=-\infty}^{\infty} W_{id}[m_1] \sum_{m_2=-\infty}^{\infty} W_{id}[m_2] K_x[m_2 - m_1]
\end{aligned}
\tag{2.3}
$$

Let us now determine the equation to be satisfied by the weighting function $W[n]$ which transforms the expression (2.3) into a minimum and ensures at the same time that the conditions (1.13) are satisfied.

To determine the condition of an extremum (in our case, a minimum) of (2.3) for the given constraints (1.13), the calculus of variations [72] is generally used, according to which the problem just formulated is reduced to finding the minimum of the expression

$$I(W)=D_e-2\sum_{\nu=0}^{r}\gamma_\nu\mu_\nu \tag{2.4}$$

where γ_ν are the Lagrange multiplier, and

$$\mu_\nu=\sum_{n=0}^{N}n^\nu W[n],\quad \nu=0,1,\ldots,r \tag{2.5}$$

By the rules of the calculus of variations, to determine the minimum of (2.4) we must give a variation $\delta W[n]$ (δ is an arbitrary number) to $W[n]$ and find the quantity

$$I(W+\delta W)=I(W)+\varkappa I(W)$$

where $\varkappa I(W)$ is the variation in $I(W)$ caused by δW.

In this case $W[n]$, transforming the expression (2.4) into a minimum, is determined from

$$\frac{\partial}{\partial\delta}I(W+\delta W)\big|_{\delta=0}=0 \tag{2.6}$$

It is important to note here that (2.6) is only a necessary condition for obtaining the minimum of (2.4).

In this way, substituting $W[n]+\delta W[n]$, for $W[n]$ in (2.4) we have, when (2.3) is used,

$$\begin{gathered}I(W+\delta W)=\sum_{m_1=0}^{N}(W[m_1]+\delta W[m_1])\\ \times\sum_{m_2=0}^{N}(W[m_2]+\delta W[m_2])K_\varphi[m_2-m_1]\\ -2\sum_{m_1=-\infty}^{\infty}W_{\text{id}}[m_1]\sum_{m_2=0}^{N}(W[m_2]+\delta W[m_2])K_x[m_2-m_1]\\ -2\sum_{\nu=0}^{r}\gamma_\nu\sum_{n=0}^{N}n^\nu(W[n]+\delta W[n])\end{gathered}$$

We point out that in deriving this expression we have omitted the last term in (2.3), since it does not depend on $W[n]$.

Carrying out certain transformations we arrive at the following equation:

$$I(W+\delta W)=I(W)+2\delta I_1(W)+\delta^2 I_2(W) \tag{2.7}$$

where

$$\begin{aligned} I_1(W)=\sum_{m_1=0}^{N} W[m_1]\sum_{m_2=0}^{N} W[m_2]K_\varphi[m_2-m_1] \\ -\sum_{m_1=-\infty}^{\infty} W_{\mathrm{id}}[m_1]\sum_{m_2=0}^{N} W[m_2]K_x[m_2-m_1] \\ -\sum_{\nu=0}^{r}\gamma_\nu\sum_{n=0}^{N} n^\nu W[n] \end{aligned} \tag{2.8}$$

$$I_2(W)=\sum_{m_1=0}^{N} W[m_1]\sum_{m_2=0}^{N} W[m_2]K_\varphi[m_2-m_1] \tag{2.9}$$

A necessary condition which makes (2.4) a minimum is obtained in accordance with (2.5) by differentiating (2.7) with respect to δ and equating this derivative to zero when $\delta=0$. Accordingly, we find that satisfying the equality

$$I_1(W)=0 \tag{2.10}$$

constitutes a necessary condition for which (2.4) becomes a minimum.

Writing (2.8) in the form

$$\begin{aligned} \sum_{n=0}^{N} W[n]\Big[\sum_{m=0}^{N} W[m]K_\varphi[n-m] \\ -\sum_{m=-\infty}^{\infty} W_{\mathrm{id}}[m]K_x[n-m]-\sum_{\nu=0}^{r}\gamma_\nu n^\nu\Big]=0 \end{aligned} \tag{2.11}$$

we find that (2.10) is satisfied in the case where

$$\begin{aligned} \sum_{m=0}^{N} W[m]K_\varphi[n-m]=\sum_{m=-\infty}^{\infty} W_{\mathrm{id}}[m]K_x[n-m] \\ +\sum_{\nu=0}^{r}\gamma_\nu n^\nu, \quad 0\leqslant n\leqslant N \end{aligned} \tag{2.12}$$

The condition (2.12) is necessary for the dispersion D_e to be a minimum. As in the preceding chapter, we can show that (2.12) is also a sufficient condition.

The weighting function of an optimal sampled data system ensuring the minimum of the mean-square transformation error under the given additional conditions (1.13) must thus satisfy (2.12), which is itself the discrete analogue of the integral equation obtained by Zadeh and Ragazzini [73]. This equation was first obtained by Perov [68].

We shall now solve (2.12), which determines the weighting function of an optimal sampled data system.

6.3 SOLUTION FOR THE WEIGHTING FUNCTION OF AN OPTIMAL SYSTEM

It was shown in the preceding section that the weighting function $W[n]$ of an optimal sampled data system which ensures the minimum dispersion of the transformation error, when the conditions

$$\sum_{n=0}^{N} n^{\nu} W[n] = \sum_{n=-\infty}^{\infty} n^{\nu} W_{id}[n] \qquad \nu = 0, 1, \ldots, r \tag{3.1}$$

of accurate transformation of the mathematical expectation of the useful signal are satisfied, must satisfy

$$\sum_{m=0}^{N} W[m] K_{\varphi}[n-m] = \sum_{m=-\infty}^{\infty} W_{id}[m] K_x[n-m] + \sum_{\nu=0}^{r} \gamma_{\nu} n^{\nu} \qquad 0 \leqslant n \leqslant N \tag{3.2}$$

The solution of (3.2), which gives the weighting function $W[n]$, can be reduced to the solution of two somewhat simpler equations. For this let us write (3.2) in the following form:

$$\sum_{m=0}^{N} W[m] K_{\varphi}[n-m] - \sum_{m=-\infty}^{\infty} W_{id}[m] K_x[n-m] - \sum_{\nu=0}^{r} \gamma_{\nu} n^{\nu} = 0 \qquad 0 \leqslant n \leqslant N \tag{3.3}$$

or

$$\sum_{m=0}^{N} K_{\varphi}[n-m]\left[W[m]-u[m]-\sum_{\nu=0}^{r}\gamma_{\nu}W_{\nu}[m]\right]=0 \quad (3.4)$$

$$0\leqslant n\leqslant N$$

where the functions $u[n]$ and $W_{\nu}[n]$, $\nu=0,1,\dots,r$ are the solutions of equations of the following form:

$$\sum_{m=0}^{N} K_{\varphi}[n-m]u[m]=\sum_{m=-\infty}^{\infty} W_{\mathrm{id}}[m]K_{x}[n-m] \quad (3.5)$$

$$0\leqslant n\leqslant N$$

$$n^{\nu}=\sum_{m=0}^{N} K_{\varphi}[n-m]W_{\nu}[m],\ \nu=0,1,\dots,r \quad (3.6)$$

It follows from (3.4) that for each $0\leqslant n\leqslant N$. the weighting function in question, $W[n]$, is defined by the equation

$$W[n]=u[n]+\sum_{\nu=0}^{r}\gamma_{\nu}W_{\nu}[n] \quad (3.7)$$

$$0\leqslant n\leqslant N$$

It should be noted that $u[n]$ and $W_{\nu}[n], \nu=0,1,\dots,r$, must satisfy the conditions

$$u[n]=\begin{cases}u[n], & 0\leqslant n\leqslant N\\ 0, & n<0,\ n>N\end{cases} \quad (3.8)$$

$$W_{\nu}[n]=\begin{cases}W_{\nu}[n], & 0\leqslant n\leqslant N\\ 0, & n<0,\ n>N\end{cases} \quad (3.9)$$

In this way, to determine $W[n]$ in accordance with (3.7) we must know $u[n]$ and $W_{\nu}[n]$, $\nu=0,1,\dots,r$ which satisfy (3.5) and (3.6) respectively.

The task of determining $W[n]$ is not completed by solving (3.5) and (3.6), i.e. by determining $u[n]$ and $W_{\nu}[n]$. As follows from (3.7), the multipliers γ_{ν} must be known. These must be determined from (3.1) which has to be satisfied by the weighting function of the optimal system.

Substituting the expression for $W[n]$ from (3.7) into (3.1) we have

$$\sum_{n=0}^{N} n^{\mu}\Big\{u[n]+\sum_{\nu=0}^{r}\gamma_{\nu}W_{\nu}[n]\Big\}=\sum_{n=-\infty}^{\infty} n^{\mu} W_{id}[n]$$

$$\mu=0,1,\ldots,r$$

or

$$\sum_{\nu=0}^{r}\gamma_{\nu}\sum_{n=0}^{N} n^{\mu}W_{\nu}[n]=\sum_{n=-\infty}^{\infty} n^{\mu}W_{id}[n]-\sum_{n=0}^{N} n^{\mu}u[n] \quad (3.10)$$

Since $u[n]$, $W_{\nu}[n]$ and $W_{id}[n]$ are known, (3.10) constitute a system of $r+1$ algebraic equations. The $r+1$ unknowns γ_{ν}, $\nu=0,1,\ldots,r$, are given by the solutions of these equations.

Equations (3.5) and (3.6), and (3.7) which determine the weighting function of the optimal system, are analogous to the corresponding integral equations and the corresponding expression which determines the weighting function of the optimal continuous system, obtained by Pugachev. In this way, the task of determining the weighting function is completed by the solution (3.5) and (3.6), which give the functions $u[n]$ and $W_{\nu}[n]$, together with the solution of the system of $r+1$ algebraic equations (3.10) for the unknown Lagrange multipliers $\gamma_{\nu}, \nu=0,1,\ldots,r$.

We shall not give here the method of solving (3.5) and (3.6), because it will be shown in the following sections that the problem just considered (discrete analogue of the Zadeh-Ragazzini problem) is a particular case of the more general problem, called the discrete analogue of Semenov's problem. The method of determining the optimal system in this more general case is fully applicable to the solution of (3.5) and (3.6).

6.4 DISCRETE ANALOGUE OF SEMENOV'S PROBLEM

It has been shown above that when the coefficients d_k, $k=0,1,\ldots,r$, of the polynomial (1.1) are non-random quantities which can assume arbitrary values, the determination of the weighting function of an optimal system is reduced to the determination of the function $W[n]$, which satisfies

$$\sum_{m=0}^{N} W[m]K_{\varphi}[n-m]=\sum_{m=-\infty} W_{\mathrm{id}}[m]K_{x}[n-m]+\sum_{\nu=0}^{r}\gamma_{\nu}n^{\nu} \quad (4.1)$$

$$0\leqslant n\leqslant N$$

and the $r+1$ additional conditions

$$\sum_{m=0}^{N} m^{\nu}W[m]=\mu_{\nu}, \quad \nu=0,1,\ldots,r \qquad (4.2)$$

In this case $W[n]$ is given by

$$W[n]=u[n]+\sum_{\nu=0}^{r}\gamma_{\nu}W_{\nu}[n], \; 0\leqslant n\leqslant N \qquad (4.3)$$

where the functions $u[n]$ and $W_{\nu}[n]$, $\nu=0,1,\ldots,r$, are the solutions of (3.5) and (3.6).

We now assume that the coefficients d_k of (1.1) are random quantities with unbounded dispersions, which are not correlated. Then in order to obtain a finite dispersion of the error of transformation of the input signal by the optimal system, the weighting function of this system must satisfy (4.2), that is, the conditions for exact transformation of the mathematical expectation of the useful signal by the system. Hence it follows that these conditions are not supplementary. Because of this the optimal system ensures an unconditional minimum of dispersion of the transformation error.

Further, since (4.2) applies to the case under consideration, the expression for the dispersion of error also applies, and accordingly (4.1), which gives the weighting function of the optimal systems, remains unaltered. Hence (4.3), expressing the weighting function of the optimal system, also remains unaltered.

Consequently, the solution of the discrete analogue of the Zadeh-Ragazzini problem can immediately be extended to the case where the mathematical expectation of the useful signal is a polynomial whose coefficients are random quantities which are not correlated and have unbounded dispersions. This fact was proved by Semenov [62].

In practice it is not always possible to assume that the random coefficients d_k have unbounded dispersions and are not correlated. On the contrary, the case where the coefficients d_k are correlated with each other and have a finite dispersion is of the greatest interest. The determination of the optimal continuous system in such a case is given by Semenov [62]. The discrete analogue of this problem will be considered below.

6.4.1 Formulation of the problem

We assume that the useful signal X in the neighbourhood of a certain instant $\bar{t}=n$ can be represented as the sum of three terms,

$$X[n+m]=m_x[n+m]+\sum_{k=0}^{r} d_k[n]m^k+\overset{\circ}{X}[n+m] \quad (4.4)$$

where $m_x[n+m]$ is an arbitrary non-random function;

$d_k[n]$, $k=0,1, \ldots, r$ are random quantities (for each value n) with zero mathematical expectations

$$M[d_k]=0, \; k=0,1, \ldots, r \quad (4.5)$$

and the covariance moments

$$\begin{gathered} M[d_i d_j]=D_{ij}[n] \\ i=0,1, \ldots, r, \; j=0,1, \ldots, r \end{gathered} \quad (4.6)$$

$\overset{\circ}{X}[n+m]$ is a stationary random function with the correlation function $K_x[n]$.

We assume that the random function $\overset{\circ}{X}$ belongs to a class of random functions which are connected with discrete white noise by linear difference equations of finite order. In addition, we assume that the values of the random function $\overset{\circ}{X}[n]$ are not correlated with $d_k[n]$, $k=0,1, \ldots, r$ for any value n belonging to the closed interval $[0; N]$, i.e.

$$\begin{gathered} M[\overset{\circ}{X}[n]d_k[\lambda]]\equiv 0, \; k=0,1, \ldots, r \\ 0\leqslant n\leqslant N, \quad 0\leqslant\lambda\leqslant N \end{gathered} \quad (4.7)$$

We further assume that the random noise $X_i[n]$ is superimposed on the useful signal $X[n]$, and assume that it is a stationary random function which also belongs to a class of random functions given by linear difference equations of finite order. Without loss of generality we can take the mathematical expectation of the random function ξ as zero and

$$M[\xi[m]\overset{\circ}{X}[\lambda]]\equiv 0$$
$$0\leqslant m\leqslant N,\ 0\leqslant\lambda\leqslant N \tag{4.8}$$

$$M[\xi[m]d_k[\lambda]]\equiv 0,\ k=0,1,\ldots,r$$
$$0\leqslant m\leqslant N;\quad 0\leqslant\lambda\leqslant N \tag{4.9}$$

i.e. ξ is correlated with neither $\overset{\circ}{X}$ nor with the coefficients d_k, $k=0,1,\ldots,r$.

With these assumptions, let us formulate the problem of determining a linear sampled data transforming system which guarantees a global extremum (in particular, a minimum) of the transformation error dispersion of $X[n]$, with the condition that the output signal $X^*[n]$ of the system is formed from the values, observed over the interval $[0; N]$, of the total signal which is the sum of the useful signal X and the noise ξ.

As before, we characterise the system by the weighting function $W[n]$. At the same time we stipulate that the weighting function is to satisfy the following conditions:

$$W[n]=\begin{cases}W[n], & 0\leqslant n\leqslant N\\ 0, & n<0,\ n>N\end{cases} \tag{4.10}$$

where N is the time at which the input signal is observed.

Proceeding in the same way as with systems we write the signal at the output of the system in the following form:

$$X^*[n]=\sum_{m=0}^{N}W[m]\Big(m_x[n-m]+\sum_{k=0}^{r}d_k[n](-m)^k$$
$$+\overset{\circ}{X}[n-m]+\xi[n-m]\Big)+\psi[n] \tag{4.11}$$

where $\psi[n]$ is a non-random function which is determined by the condition that the mathematical expectation of the useful signal is transformed exactly. Applying mathematical expectation to both sides of (4.11), and noting that $M[X^*[n]]$ must be equal to $Hm_x[n]$, where H is the operator of the ideal system, we have

$$Hm_x[n]=\sum_{m=0}^{N}W[m]m_x[n-m]+\psi[n]$$

whence we obtain the condition for the absence of the transformation error for the mathematical expectation of the useful signal

$$Hm_x[n] - \sum_{m=0}^{N} W[m]\, m_x[n-m] \equiv \psi[n] \qquad (4.12)$$

In this way, if the mathematical expectation m_x of the useful signal is known, the absence of system errors is achieved by forming the signal $X^*[n]$ according to (4.11), where $\psi[n]$ is chosen from (4.12).

It is clear that the $r+1$ additional conditions (4.2) which must be satisfied by $W[n]$ are eliminated, and the latter must be chosen proceeding from the only condition for securing the unconditional minimum of the error when the useful signal is transformed or reproduced by the system.

This problem coincides exactly with the problem formulated by Semenov [62] for determining the corresponding optimal continuous system. For this reason it is called the discrete analogue of Semenov's problem.

6.4.2 Equation determining the weighting function of an optimal system

We shall now determine the equation which must be satisfied by the weighting function of an optimal system. For this we find the expression for the transformation error of the useful signal by the system in question. Denoting the transformation error by $\mathcal{E}$ we have

$$\mathcal{E}[n] = H_m X[n-m] - X^*[n] \qquad (4.13)$$

where $H_m X[n-m]$ is the signal at the output of an ideal transforming system with transformation operator H. Substituting the expression for the useful signal $X[n]$ from (4.4) into (4.13), and bearing in mind that the signal $X^*[n]$ at the output of the required system is formed according to (4.11), we obtain the following expression for $\mathcal{E}$:

$$\mathcal{E}[n] = H_m m_x[n-m] + H_m \sum_{k=0}^{r} d_k[n](-m)^k$$

$$+ H_m \overset{\circ}{X}[n-m] - \sum_{m=0}^{N} W[m]\Big(m_x[n-m] + \sum_{k=0}^{r} d_k[n](-m)^k$$

$$+\overset{\circ}{X}[n-m]+\xi[n-m]\Big)-\psi[n]$$

If the function $\psi[n]$ is chosen from the condition that there is no transformation error in the mathematical expectation m_x of the useful signal, i.e. in accordance with (4.12), then the expression for the error assumes the form

$$\mathscr{E}[n]=H_m\sum_{k=0}^{r}d_k[n](-m)^k+H_m\overset{\circ}{X}[n-m] \\ -\sum_{m=0}^{N}W[m]\Big(\overset{\circ}{X}[n-m]+\xi[n-m]+\sum_{k=0}^{r}d_k[n](-m)^k\Big) \quad (4.14)$$

whence we find the following equation which gives the square of the error:

$$\mathscr{E}^2[n]=H_m\sum_{\nu=0}^{r}d_\nu[n](-m)^\nu\times H_l\sum_{\mu=0}^{\nu}d_\mu[n](-l)^\mu \\ +H_m\overset{\circ}{X}[n-m]\,H_l\overset{\circ}{X}[n-l] \\ +\sum_{l=0}^{N}W[l]\sum_{m=0}^{N}W[m]\Big[\overset{\circ}{X}[n-l]+\xi[n-l]+\sum_{\nu=0}^{r}d_\nu[n](-l)^\nu\Big] \\ \times\Big[\overset{\circ}{X}[n-m]+\xi[n-m]+\sum_{\mu=0}^{r}d_\mu[n](-m)^\mu\Big] \\ -2H_l\overset{\circ}{X}[n-l]\sum_{m=0}^{N}W[m]\Big(\overset{\circ}{X}[n-m]+\xi[n-m] \\ +\sum_{k=0}^{r}d_k[n](-m)^k\Big)+2H_m\sum_{k=0}^{r}d_k[n](-m)^k \\ \times H_l\overset{\circ}{X}[n-l]-2H_l\sum_{\mu=0}^{r}d_\mu[n](-l)^\mu\sum_{m=0}^{N}W[m] \\ \times\Big(\overset{\circ}{X}[n-m]+\xi[n-m]+\sum_{\nu=0}^{r}d_\nu[n](-m)^\nu\Big) \quad (4.15)$$

The dispersion of the transformation error can be found by applying mathematical expectation to both sides of (4.15). By assumptions made about the non-correlation of the random functions $\overset{\circ}{X}[n]$ and $\xi[n]$, and with (4.7) and (4.8) taken into account, we have

$$M\left[H_m\sum_{\nu=0}^{r} d_\nu[n](-m)^\nu H_l\sum_{\mu=0}^{r} d_\mu[n](-l)^\mu\right] = H_m H_l \sum_{\nu=0}^{r}\sum_{\mu=0}^{r} D_{\nu\mu}(-l)^\mu(-m)^\nu \tag{4.16}$$

$$M[H_m\overset{\circ}{X}[n-m]H_l\overset{\circ}{X}[n-l]] = H_m H_l K_x[m-l] \tag{4.17}$$

$$M\left[\sum_{\nu=0}^{r} d_\nu[n](-l)^\nu \overset{\circ}{X}[n-m]\right] = M\left[\sum_{\nu=0}^{r} d_\nu[n](-l)^\nu \xi[n-m]\right] = 0 \tag{4.18}$$

and from (4.15) we find the following expression for the dispersion:

$$\begin{aligned} M[\mathcal{E}^2[n]] = D_e = {} & H_m H_l\left[\sum_{\nu=0}^{r}\sum_{\mu=0}^{r} D_{\nu\mu}(-l)^\mu(-m)^\nu + K_x[m-l]\right] \\ & + \sum_{l=0}^{N} W[l]\sum_{m=0}^{N} W[m]\left[K_x[m-l] + K_\xi[m-l]\right. \\ & \left. + \sum_{\nu=0}^{r}\sum_{\mu=0}^{r} D_{\nu\mu}(-l)^\mu(-m)^\nu\right] - 2H_l\sum_{m=0}^{N} W[m] \\ & \times\left[K_x[m-l] + \sum_{\nu=0}^{r}\sum_{\mu=0}^{r} D_{\nu\mu}(-l)^\mu(-m)^\nu\right] \end{aligned} \tag{4.19}$$

where $K_\xi[m-l]$ is the correlation function of the random function $\xi[n]$ (noise);
$K_x[m-l]$ is the correlation function of the stationary part of the useful signal;
$H_m H_l[\ \ ]$ is the result of applying the operator H with respect to the variables m and l.

To determine the weighting function $W[n]$ which makes the dispersion of the transformation error a minimum we shall use the calculus of variations. In order to to this we add a variation $\delta W[n]$ to the function $W[n]$ in Equation (4.19) (where δ is an arbitrary number).

We then have

$$D_e(W+\delta W)=\sum_{l=0}^{N}\{W[l]+\delta W[l]\}\sum_{m=0}^{N}\{W[m]+\delta W[m]\}$$
$$\times F[m-l]-2H_l\sum_{m=0}^{N}\{W[m]+\delta W[m]\}$$
$$\times G[m-l]+H_mH_lG[m-l] \tag{4.20}$$

where

$$F[m-l]=K_x[m-l]+K_\xi[m-l]+\sum_{\nu=0}^{N}\sum_{\mu=0}^{r}D_{\nu\mu}(-l)^\mu(-m)^\nu \tag{4.21}$$

$$G[m-l]=K_x[m-l]+\sum_{\nu=0}^{r}\sum_{\mu=0}^{r}D_{\nu\mu}(-l)^\mu(-m)^\nu \tag{4.22}$$

$$D_e(W+\delta W)=M[\mathscr{E}^2[n]]|_{W\to W+\delta W} \tag{4.23}$$

Equation (4.20) can be transformed into the following form

$$D_e(W+\delta W)=\sum_{l=0}^{N}W[l]\sum_{m=0}^{N}W[m]F[m-l]$$
$$-2H_l\sum_{m=0}^{N}W[m]G[m-l]+H_mH_lG[m-l]$$
$$+2\delta\sum_{l=0}^{N}W[l]\sum_{m=0}^{N}W[m]F[m-l]-2\delta H_l\sum_{m=0}^{N}W[m]G[m-l]$$
$$+\delta^2\sum_{l=0}^{N}W[l]\sum_{m=0}^{N}W[m]F[m-l]$$

or

$$D_e(W+\delta W)=D_e(W)+2\delta D_1(W)+\delta^2D_2(W) \tag{4.24}$$

where

$$D_e(W)=D_e(W+\delta W)|_{\delta=0} \tag{4.25}$$

$$D_1(W)=\sum_{l=0}^{N}W[l]\sum_{m=0}^{N}W[m]F[m-l]-H_l\sum_{m=0}^{N}W[m]G[m-l] \tag{4.26}$$

$$D_2(W)=\sum_{l=0}^{N} W[l]\sum_{m=0}^{N} W[m]F[m-l] \tag{4.27}$$

The necessary condition for the minimum dispersion of the transformation error, i.e. the condition for the minimum of (4.19), is determined from the following equation:

$$\frac{\partial}{\partial\delta} D_e(W+\delta W)|_{\delta=0}=0 \tag{4.28}$$

Consequently, differentiating (4.24) with respect to δ and then putting $\delta=0$, we have from (4.28)

$$D_1(W)=0$$

or, with (4.26) taken into account,

$$\sum_{l=0}^{N} W[l]\sum_{m=0}^{N} W[m]F[m-l]-H_l\sum_{m=0}^{N} W[m]G[m-l]=0 \tag{4.29}$$

The weighting function $W[n]$ of an optimal system must thus satisfy (4.29).

Bearing in mind that the summation operator and the operator H_l are commutative, we can write (4.29) in the following form:

$$\sum_{m=0}^{N} W[m]\left\{\sum_{l=0}^{N} W[l]F[m-l]-H_lG[m-l]\right\}=0$$

whence it follows that this equality is valid, if

$$\sum_{l=0}^{N} W[l]F[m-l]=H_lG[m-l]$$

or, when (4.21) and (4.22) are used,

$$\sum_{l=0}^{N} W[l]\left(K_x[m-l]+K_\xi[m-l]+\sum_{\nu=0}^{r}\sum_{\mu=0}^{r} D_{\nu\mu}(-l)^{\mu}(-m)^{\nu}\right)$$

$$=H_l\left(K_x[m-l]+\sum_{\nu=0}^{r}\sum_{\mu=0}^{r} D_{\nu}\ (-l)^{\mu}(-m)^{\nu}\right)$$

or, in final form,

$$\sum_{l=0}^{N} W[l](K_x[m-l]+K_\xi[m-l])$$

$$= H_l\Big[K_x[m-l] + \sum_{\nu=0}^{r}\sum_{\mu=0}^{r} D_{\nu\mu}(-l)^{\mu}(-m)^{\nu}\Big]$$
$$-\sum_{l=0}^{N} W[l]\Big[\sum_{\nu=0}^{r}\sum_{\mu=0}^{r} D_{\nu\mu}(-l)^{\mu}(-m)^{\nu}\Big] \tag{4.30}$$

Let us now transform this expression. From the commutative property of linear operators we can write

$$H_l\Big[\sum_{\nu=0}^{r}\sum_{\mu=0}^{r} D_{\nu\mu}(-l)^{\mu}(-m)^{\nu}\Big]$$
$$-\sum_{l=0}^{N} W[l]\Big[\sum_{\nu=0}^{r}\sum_{\mu=0}^{r} D_{\nu\mu}(-l)^{\mu}(-m)^{\nu}\Big]$$
$$=\sum_{\nu=0}^{r}(-m)^{\nu}\sum_{\mu=0}^{r} D_{\nu\mu}\Big[H_l(-l)^{\mu} - \sum_{l=0}^{\mu} W[l](-l)^{\mu}\Big] \tag{4.31}$$
$$=\sum_{\nu=0}^{r}(-1)^{\nu}m^{\nu}\sum_{\mu=0}^{r} D_{\nu\mu}(-1)^{\mu}\Big[H_l l^{\mu} - \sum_{l=0}^{N} W[l]\, l^{\mu}\Big]$$

Introducing the notation

$$\gamma_{\nu}^{*} = (-1)^{\nu}\sum_{\mu=0}^{r} D_{\nu\mu}(-1)^{\mu}\Big[H_l l^{\mu} - \sum_{l=0}^{N} l^{\mu} W[l]\Big] \tag{4.32}$$
$$\nu = 0, 1, \ldots, r$$

we write the right-hand side of (4.31) in the following form

$$\sum_{\nu=0}^{r}(-1)^{\nu}m^{\nu}\sum_{\mu=0}^{r} D_{\nu\mu}(-1)^{\mu}\Big[H_l l^{\mu} - \sum_{l=0}^{N} l^{\mu}W[l]\Big] = \sum_{\nu=0}^{r}\gamma_{\nu}^{*} m^{\nu} \tag{4.33}$$

When the above remarks are taken into consideration, (4.30) assumes the final form

$$\sum_{l=0}^{N} W[l]\,(K_x[m-l] + K_{\xi}[m-l])$$
$$= H_l K_x[m-l] + \sum_{\nu=0}^{r}\gamma_{\nu}^{*} m^{\nu}, \quad 0 \leqslant m \leqslant N \tag{4.34}$$

In this way, the weighting function of the optimal system

must satisfy (4.34). At the same time, by virtue of (4.10), (4.34) gives the weighting function $W[n]$ when $0 \leqslant n \leqslant N$. Outside this time interval $W[n] \equiv 0$.

Let us consider concrete cases of (4.34). Suppose that the weighting function $W_{id}[n]$ of the ideal transforming system coincides with the σ-function, i.e. the system is to filter the useful signal. It is clear that in this case we obtain

$$H_l K_x[m-l] = \sum_{l=-\infty}^{\infty} \sigma[l] K_x[m-l] = K_x[m]$$

and (4.34) assumes the form

$$\sum_{l=0}^{N} W[l](K_x[m-l] + K_\xi[m-l]) = K_x[m] + \sum_{\nu=0}^{r} \gamma_\nu^* m^\nu \qquad 0 \leqslant m \leqslant N \tag{4.35}$$

where γ_ν^* is determined as follows:

$$\gamma_\nu^* = (-1)^\nu \sum_{\mu=0}^{r} D_{\nu\mu}(-1)^{\mu+1} \sum_{l=0}^{N} l^\mu W[l] \qquad \nu = 0, 1, \ldots, r \tag{4.36}$$

In the case where the useful signal is to be predicted over the time $\bar{t} = h$, the weighting function of the ideal system equals $\sigma[n+h]$. For this case we obtain

$$H_l K_x[m-l] = \sum_{l=-\infty}^{\infty} \sigma[l+h] K_x[m-l] = K_x[m+h]$$

and (4.34) assumes the form

$$\sum_{l=0}^{N} W[l](K_x[m-l] + K_\xi[m-l]) = K_x[m+h] + \sum_{\nu=0}^{r} \gamma_\nu^* m^\nu \qquad 0 \leqslant m \leqslant N \tag{4.37}$$

where the coefficients γ_ν^* in accordance with (4.32) are

$$\gamma_\nu^* = (-1)^\nu \sum_{\mu=0}^{r} D_{\nu\mu}(-1)^\mu [(-h)^\mu - \sum_{l=0}^{N} l^\mu W[l]] \qquad \nu = 0, 1, \ldots, r \tag{4.38}$$

In the same way we can obtain the corresponding equations in other concrete cases.

We have thus shown that if the non-stationary part of the useful signal is a polynomial of degree r with respect to time and has random coefficients, then the weighting function of a system which transforms the useful signal in accordance with the operator H with the minimum error dispersion must satisfy an equation of the form

$$\sum_{l=0}^{N} W[l](K_x[m-l]+K_\xi[m-l])=H_l K_x[m-l] + \sum_{\nu=0}^{r} \gamma_\nu^* m^\nu \qquad (4.39)$$

$$0 \leqslant m \leqslant N$$

where the coefficients γ_ν^* are determined from (4.32).

We point out that (4.39) has been obtained on the assumption that $\overset{\circ}{X}$ and ξ are not correlated.

When

$$M[\overset{\circ}{X}[n_1]\xi[n_2]]=K_{x\xi}[n_1-n_2]\neq 0$$

(4.39), which determines the weighting function of the optimal system, assumes the form

$$\sum_{l=0}^{N} W[l]K_{\varphi\varphi}[m-l]=H_l K_{x\varphi}[m-l]+\sum_{\nu=0}^{r}\gamma_\nu^* m^\nu \qquad (4.40)$$

$$0 \leqslant m \leqslant N$$

where

$$K_{\varphi\varphi}[m-l]=K_x[m-l]+K_{x\xi}[m-l]+K_{\xi x}[m-l] + K_\xi[m-l] \qquad (4.41)$$

$$K_{x\varphi}[m-l]=K_x[m-l]+K_{x\xi}[m-l]$$

A comparison of (4.39) and (2.12) shows that they have the same form. Consequently, the solution of (4.39) can be written in the same form as the solution of (2.12). From (3.7) we thus obtain

$$W[n]=u[n]+\sum_{\nu=0}^{r}\gamma_\nu^* W_\nu[n],\ 0\leqslant n\leqslant N \qquad (4.42)$$

where the functions $u[n]$ and $W_\nu[n]$ ($\nu=0, 1, \ldots, r$) are the solutions of equations analogous to (3.5) and (3.6). The coefficients γ_ν^* are determined in the given case from a system of algebraic equations analogous to (3.10).

It can be established that (2.12) and the additional $r+1$ conditions (3.1) are a particular case of (4.39). Thus, let us assume at the beginning that the coefficients d_k of the polynomial (4.4) are such that the correlation moments of connection are given by

$$D_{\nu\mu} = \begin{cases} D_{\mu\mu}, & \nu = \mu \\ 0, & \nu \neq \mu \end{cases} \tag{4.43}$$

In this case we have from (4.32)

$$\gamma_\nu^* = (-1)^\nu D_{\nu\nu} (-1)^\nu \left[H_l l^\nu - \sum_{l=0}^{N} l^\nu W[l] \right]$$

or

$$\gamma_\nu^* = D_{\nu\nu} H_l l^\nu - D_{\nu\nu} \sum_{l=0}^{N} l^\nu W[l] \tag{4.44}$$

Assuming further that

$$D_{\nu\nu} = \infty, \quad \nu = 0, 1, \ldots, r$$

we find from (4.44) that

$$H_l l^\nu - \sum_{l=0}^{N} l^\nu W[l] = 0 \tag{4.45}$$

whence we conclude the following: If the coefficients d_k of the polynomial, which is the non-stationary part of the useful signal, are random quantities which are not correlated and have unbounded dispersions, then (4.32) are transformed into additional $r+1$ conditions coinciding with (3.1). Consequently, the problem considered in the preceding sections, i.e. the discrete analogue of the Zadeh-Ragazzini problem, arises as a particular case from the more general case which is the discrete analogue of Semenov's problem.

It has been shown above that the solution of (4.39) can be obtained in the form (4.42). In fact, the solution of (4.40) can also be found in this form. In what follows, however, we shall give another method for solving these equations. This gives an expression which is more convenient in practice for determining the weighting function of an optimal system.

6.5 SOLUTION FOR THE WEIGHTING FUNCTION OF AN OPTIMAL SYSTEM

6.5.1 Expression for an optimal weighting function

Equation (4.39), or the more general equation (4.40) which determines the weighting function $W[n]$ for $0 \leqslant n \leqslant N$, can be written in the form of a system of linear algebraic equations for $W[0]$, $W[1]$, ..., $W[N]$. Solving this system of equations we can find the values of the weighting function $W[n]$, n, 0, 1, ..., N.

This solution is laborious, particularly for large values of N. In addition, a direct solution of (4.39) does not allow the expression for the weighting function to be found in the closed form necessary when optimal sampled data systems are simulated by means of analogue computers.

In this section a method will be given for solving the equation which determines the weighting function of an optimal system. It is based on the use of the theory of adjoint systems and the theory of discrete shaping filters.

We shall first consider (4.39) for a particular case where the stationary part of the useful signal is missing, i.e. $\overset{\circ}{X}[n] \equiv 0$. In this case the weighting function of the optimal system must satisfy

$$\sum_{l=0}^{N} W[l] K_{\xi}[m-l] = \sum_{\nu=0}^{r} \gamma_{\nu}^{*} m^{\nu}, \quad 0 \leqslant m \leqslant N \tag{5.1}$$

where in conformity with (4.32)

$$\gamma_{\nu}^{*} = (-1)^{\nu} \sum_{\mu=0}^{r} D_{\nu\mu} (-1)^{\mu} \left[H_l l^{\mu} - \sum_{l=0}^{N} l^{\mu} W[l] \right] \tag{5.2}$$

$$\nu = 0, 1, \ldots, r$$

In the preceding section the assumption has been made that ξ belongs to a class of random functions which are connected with discrete white noise by linear difference equations of finite order. This means that the correlation function $K_{\xi}[n-m]$ of the random function ξ can be represented as a series with a finite number of exponential terms

$$K_{\xi}[n-m]=\sum_{i=1}^{k} b_i e^{\beta_i |n-m|} \tag{5.3}$$

where indices β_i of the exponent can be real or complex. At the same time, if β_i are real numbers then they are necessarily negative. If some β_i are complex, then their real parts are also negative. Since $K_{\xi}[n-m]$ is a real function, then for the complex indices β_i the corresponding coefficients b_i must also be complex. Suppose, for example, that $\beta_l = -\alpha_l + j\gamma_l$. Then amongst the β_i, $i=1, 2, \ldots, l-1, l+1, \ldots, k$ there must necessarily be a number β_{l+1} which is conjugate to β_l, that is, $\beta_{l+1} = -\alpha_l - j\gamma_l$. For the sum

$$b_l e^{(-\alpha_l + j\gamma_l)|n-m|} + b_{l+1} e^{(-\alpha_l - j\gamma_l)|n-m|}$$

to be a real number in the present case, b_l and b_{l+1} must in general be complex conjugate numbers.

Since β_i are either negative real numbers or complex numbers with negative real parts, the following equalities hold for $K_{\xi}[n-m]$:

$$\lim_{n-m\to\infty} \Delta_n^r K_{\xi}[n-m] = \lim_{n-m\to-\infty} \Delta_n^r K_{\xi}[n-m] = 0$$
$$r=0, 1, \ldots \tag{5.4}$$

Hence, in what follows, we shall assume that the correlation function $K_{\xi}[n-m]$ of ξ can be represented in the form (5.3).

Substituting the expression for $K_{\xi}[n-m]$ from (5.3) into (5.1), we have

$$\sum_{i=1}^{k} b_i \sum_{l=0}^{N} W[l]\, e^{\beta_i |m-l|} = \sum_{\nu=0}^{r} \gamma_\nu^* m^\nu$$
$$0 \leqslant m \leqslant N \tag{5.5}$$

Since ξ is connected with discrete white noise by linear difference equations of finite order, the correlation function $K_{\xi}[n-m]$ of this random function is connected in a definite manner with Green's function for a certain self-adjoint homogeneous difference boundary-value problem (see

Chapter 3). This result of the theory of discrete random processes, given in Chapter 3, will be used here to solve (5.5).

We assume that ξ can be formed from the white noise V by a shaping filter whose difference equation has the form

$$L(\Delta)\xi[n]=M(\Delta)V[n] \tag{5.6}$$

where $L(\Delta)$ and $M(\Delta)$ are polynomials in the operator Δ, having orders k and h respectively.

We use $G[n-m]$ to denote Green's function for the difference boundary-value problem

$$\begin{gathered} L^{*}(\Delta)L(\Delta)\xi[n]=0 \\ \lim_{n\to\infty}\Delta^{r}\xi[n]=\lim_{n\to-\infty}\Delta^{r}\xi[n]=0 \\ r=0,\ 1,\ \ldots,\ k-1 \end{gathered} \tag{5.7}$$

where the difference operator $L^{*}(\Delta)$ is adjoint to the operator $L(\Delta)$.

The correlation function $K_{\xi}[n-m]$ is related to Green's function $G[n-m]$ by the following equation:

$$K_{\xi}[n-m]=M_{n}(\Delta)M_{n}^{*}(\Delta)G[n-m] \tag{5.8}$$

where the operator $M^{*}(\Delta)$ is adjoint to the difference operator $M(\Delta)$.

Let us now return to (5.1). We replace $K_{\xi}[n-m]$ by its expression from (5.8). We then have

$$\begin{gathered} \sum_{l=0}^{N}W[l]M_{m}(\Delta)M_{m}^{*}(\Delta)G[m-l]=\sum_{\nu=0}^{r}\gamma_{\nu}^{*}m^{\nu} \\ 0\leqslant m\leqslant N \end{gathered}$$

or, by the linearity property of the operators,

$$\begin{gathered} M_{m}(\Delta)M_{m}^{*}(\Delta)\left[\sum_{l=0}^{N}W[l]G[m-l]\right]=\sum_{\nu=0}^{r}\gamma_{\nu}^{*}m^{\nu} \\ 0\leqslant m\leqslant N \end{gathered} \tag{5.9}$$

Since the difference equation (5.6) of a filter forming a stationary random process has constant coefficients, it follows that (5.9) is a non-homogeneous difference equation with constant coefficients. The general solution

$$F[m] = \sum_{l=0}^{N} W[l]\, G[m-l]$$

of this equation can be represented in the form of a sum of the general solution $F_1[m]$ of the homogeneous equation

$$M_m(\Delta)\, M^*_m(\Delta)\, F_1[m] = 0 \tag{5.10}$$

and the particular solution $F_2[m]$ of the non-homogeneous equation

$$M_m(\Delta)\, M^*_m(\Delta)\, F_2[m] = \sum_{\nu=0}^{r} \gamma^*_\nu m^\nu \tag{5.11}$$

In order to find the particular solution of (5.11), we apply to both sides the difference operator

$$[M_m(\Delta)\, M^*_m(\Delta)]^{-1} = M^{-1}_m(\Delta)\, M^{*-1}_m(\Delta)$$

which is inverse to the difference operator $M_m(\Delta)\, M^*_m(\Delta)$. We then have

$$\begin{aligned} F_2[m] &= M^{-1}_m(\Delta)\, M^{*-1}_m(\Delta) \left[\sum_{\nu=0}^{r} \gamma^*_\nu m^\nu \right] \\ &= \sum_{\nu=0}^{r} \gamma^*_\nu M^{-1}_m(\Delta)\, M^{*-1}_m(\Delta)\, m^\nu \end{aligned} \tag{5.12}$$

whence it follows that the particular solution of (5.11) has the form of a polynomial of degree r with respect to time.

Let us now assume that $M(\Delta)$ has the form

$$M(\Delta) = \sum_{i=0}^{h} d_i \Delta^i \tag{5.13}$$

In this case the general solution of (5.10) can be represented as

$$F_1[m] = \sum_{i=1}^{2h} B'_i e^{q_i m} \tag{5.14}$$

where q_i, $i = 1, 2, \ldots, 2h$ are the roots of

$$M(q)\, M^*(q) = 0 \tag{5.15}$$

where the polynomials $M(q)$ and $M^*(q)$ are obtained from $M(\Delta)$ and $M^*(\Delta)$ as follows:

$$M(q) = M(\Delta)|_{\Delta = e^{q} - 1}$$

$$M^*(q) = M^*(\Delta)|_{\Delta = e^{-q} - 1}$$

The general solution of equation (5.9), can thus be written in the following form:

$$\sum_{l=0}^{N} W[l]\, G[m-l] = \sum_{i=1}^{2h} B_i' e^{q_i m} + \sum_{\nu=0}^{r} \gamma_\nu^* M_m^{-1}(\Delta) M_m^{*-1}(\Delta) m^\nu, \quad 0 \leqslant m \leqslant N \tag{5.16}$$

Starting from (5.16), we can obtain an explicit expression for $W[m]$. We use the equation which is satisfied by Green's function $G[n-m]$ for the difference boundary-value problem (5.7):

$$L_n^*(\Delta) L_n(\Delta) G[n-m] = \sigma[n-m] \tag{5.17}$$

We now apply the difference operator $L_n^*(\Delta) L_n(\Delta)$ with respect to m to both sides of (5.16). We obtain

$$\sum_{l=0}^{N} W[l]\, L_m(\Delta) L_m^*(\Delta) G[m-l] = L_m^*(\Delta) L_m(\Delta) \left[\sum_{i=1}^{2h} B_i' e^{q_i m} + \sum_{\nu=0}^{r} \gamma_\nu^* M_m^{-1}(\Delta) M_m^{*-1}(\Delta) m^\nu \right] \tag{5.18}$$

Bearing in mind that Green's function satisfied (5.17), we find from (5.18) that

$$\sum_{l=0}^{N} W[l]\, \sigma[m-l] = L_m^*(\Delta) L_m(\Delta) \left[\sum_{i=1}^{2h} B_i' e^{q_i m} + \sum_{\nu=0}^{r} \gamma_\nu^* M_m^{-1}(\Delta) M_m^{*-1}(\Delta) m^\nu \right] \tag{5.19}$$

or, finally, with the property of the σ-function taken into account,

$$W[m] = L_m^*(\Delta) L_m(\Delta) \left[\sum_{i=1}^{2h} B_i' e^{q_i m} + \sum_{\nu=0}^{r} \gamma_\nu^* M_m^{-1}(\Delta) M_m^{*-1}(\Delta) m^\nu \right], \quad 0 \leq m \leq N \tag{5.20}$$

It follows from (5.20) that the $W[m]$ can contain the σ-function and its differences $\Delta_n^j \sigma[n]$ and $\Delta_n \sigma[N-n]$ when $\bar{t} = 0$ and $\bar{t} = N$. At the same time the maximum order of a difference of the σ-function is $k-h-1$, where k and h are the orders of the difference operators $L(\Delta)$ and $M(\Delta)$ respectively.

The presence of the σ-function and its differences in the general solution of equation is the result of applying the difference operator $L^*(\Delta) L(\Delta) M^{-1}(\Delta) M^{*-1}(\Delta)$ to the functions $f_\nu[n] = \gamma_\nu^* n^\nu$ $(\nu = 0, 1, \ldots, r)$. It is not difficult to establish that if the order of $L(\Delta)$ is k, and the order of $M(\Delta)$ is h, then the operator

$$L^*(\Delta) L(\Delta) M^{-1}(\Delta) M^{*-1}(\Delta) = \frac{\sum_{i=0}^{2k} a_i \Delta^i}{\sum_{i=0}^{2h} d_i \Delta^{i+k-h}} \tag{5.21}$$

Further, bearing in mind that the maximum order of a difference of the σ-function is obtained when $\nu = 0$, we find from (5.21) that this order equals $k-h-1$.

Before writing the solution of (5.9), we note that the result of applying the difference operator $L^*(\Delta) L(\Delta)$ to the function $e^{q_i m}$ can be represented in the form

$$L_m^*(\Delta) L_m(\Delta) e^{q_i m} = B_i'' e^{q_i m}, \quad 0 \leq m \leq N \tag{5.22}$$

where B_i'' is a constant. It is also easy to establish that

$$L_m^*(\Delta) L_m(\Delta) M_m^{-1}(\Delta) M_m^{*-1}(\Delta) m^\nu = A_\nu' m^\nu \qquad 1 \leq m \leq N-1 \tag{5.23}$$

Using (5.22) and (5.23) and assuming that at the ends of the interval $[0; N]$ the weighting function $W[m]$ contains the σ-function and its differences up to and including the order $k-h-1$, we can write the final expression for the weighting function of the optimal system. This expression has the form

$$W[m]=\sum_{i=1}^{2h} B_i e^{q_i m}+\sum_{\nu=0}^{r} A_\nu m^\nu+\sum_{i=0}^{k-h-1} C_i \Delta_m^i \sigma[m]$$

$$+\sum_{i=0}^{k-h-1} D_i \Delta_m^i \sigma[N-m], \quad 0 \leqslant m \leqslant N \tag{5.24}$$

where B_i, A_i, C_i and D_i are constant coefficients, of which C_i, D_i and A_i are real, whilst B_i can be complex.

Expression (5.24), giving the weighting function of an optimal sampled data system, is the discrete analogue of the expression which gives the weighting function of the corresponding continuous optimal system [62, 63].

Expression (5.24) has been obtained for the case where the weighting function of the optimal system satisfies (5.1), i.e. for the case where the stationary part of the continuous signal is absent. It is clear that the solution of (4.40), giving the weighting function, in the general case where the useful input signal is represented in the form (4.4), can be written in the same form.

The coefficients A_i, B_i, C_i and D_i, in (5.24) must be chosen in such a way that this expression when substituted into (5.1) transforms the latter into an identity of every value m within the interval $[0; N]$. The problem of determining the weighting function of an optimal system is not completely solved by determining these coefficients. Indeed, to determine the weighting function completely, we must know q_i, $i=1, 2, \ldots, 2h$, the roots of

$$M(q)\, M^*(q)=0 \tag{5.25}$$

Equation (5.25) can be found by solving the problem which gives the discrete shaping filter for the random process ξ whose correlation function is expressed by (5.3). A method will be given below which allows (5.25) to be found in a sufficiently simple manner without determining the difference equation of the corresponding shaping filter.

6.5.2 Determination of the self-adjoint equation $M(q)\, M^*(q) = 0$

We shall determine a self-adjoint difference boundary-value problem for which Green's function $G[n-m]$ has the form

$$G[n-m] = Ae^{\beta|n-m|} \tag{5.26}$$

where A is a constant.

It is not difficult to show that the random function with correlation functions $K_{\xi}[n-m] = e^{\beta|n-m|}$ can be formed by means of a filter whose difference equation has the form

$$\Delta\xi[n] + (1-e^{\beta})\xi[n] = \sqrt{1-e^{2\beta}}\,\Delta v[n] + \sqrt{1-e^{2\beta}}\,v[n] \tag{5.27}$$

Replacing the operator Δ in (5.27) by the operator $(e^q - 1)$ we find the transfer function $W^*(q)$ of this filter

$$W^*(q) = \frac{e^q\sqrt{1-e^{2\beta}}}{e^q - e^{\beta}} \tag{5.28}$$

The transfer function $R^*(q)$ of the self-adjoint system is determined in accordance with (5.28) as follows:

$$R^*(q) = W^*(q)\, W^*(-q) = \frac{1-e^{2\beta}}{1-e^{\beta}(e^q - e^{-q}) + e^{2\beta}} \tag{5.29}$$

We shall transform the expression for $R^*(q)$ into a more convenient form. From (5.29) we can write

$$R^*(q) = \frac{1-e^{2\beta}}{(1-e^{\beta})^2 - e^{\beta}(e^q - 2 + e^{-q})} \tag{5.30}$$

or

$$R^*(q) = \frac{1-e^{2\beta}}{(1-e^{\beta})^2 - e^{\beta}[(e^q-1) + (e^{-q}-1)]} \tag{5.31}$$

Introducing the notation

$$W^*(q) = \frac{\Psi^*(q)}{\Phi^*(q)} \tag{5.32}$$

(5.31) can be represented in the following form:

$$R^*(q) = \frac{\Psi^*(q)\,\Psi^*(-q)}{\Phi^*(q)\,\Phi^*(-q)} \tag{5.33}$$

where

$$\Phi^*(q) = e^q - e^\beta;\ \Phi^*(-q) = e^{-q} - e^\beta$$

$$\Psi^*(q) = e^q\sqrt{1 - e^{2\beta}};\ \Psi^*(-q) = e^{-q}\sqrt{1 - e^{2\beta}} \tag{5.34}$$

It has been shown in Chapter 1 that we can pass from the transfer function to the difference operator by replacing (e − 1) by Δ; thus, for example,

$$\Phi_n(\Delta) = \Phi^*(q)|_{e^q - 1 = \Delta} \tag{5.35}$$

Replacing $(e^{-q} - 1)$ in the expression for $\Phi^*(-q)$ by the operator Δ^* we obtain the difference operator

$$\Phi^*{}_n(\Delta) = \Phi^*(-q)|_{e^{-q} - 1 = \Delta^*} = \Phi_n(\Delta^*) \tag{5.36}$$

which is adjoint to the difference operator $\Phi_n(\Delta)$. Consequently, in the same way as $(e^{-q} - 1)$ is adjoint to $(e^q - 1)$, the difference operator Δ^* is adjoint to Δ, i.e. we have

$$\Phi^*{}_n(\Delta) = \Phi_n(\Delta^*) \tag{5.37}$$

The operation of Δ on the function $y[n]$ reduces to the calculation of the first-order difference

$$\Delta y[n] = y[n+1] - y[n] \tag{5.38}$$

Equation (5.38) constitutes the original of a transform which is the product of $(e^q - 1)$ and the transform $Y^*(q)$ of $y[n]$, with the condition that $y[0] = 0$, i.e.

$$y[n+1] - y[n] = D^{-1}\{e^q - 1)Y^*(q)\}$$

Using the shift theorem for an independent variable in the domain of originals, we have

$$D^{-1}\{e^{-q} - 1)Y^*(q)\} = y[n-1] - y[n] \tag{5.39}$$

whence it follows that the operation Δ^*, adjoint to Δ, on $y[n]$ reduces to

$$\Delta^* y[n] = y[n-1] - y[n] \tag{5.40}$$

It thus follows from the foregoing that the difference operator $\Phi^*_n(\Delta)$, adjoint to $\Phi_n(\Delta)$, is obtained from $\Phi_n(\Delta)$ by replacing Δ by Δ^*, whose operation on a function of discrete argument is defined by (5.40).

Let us next consider the self-adjoint difference equation

$$\Phi^*_n(\Delta)\Phi_n(\Delta)\xi[n]=0 \tag{5.41}$$

where $\Phi_n(\Delta)$ and its adjoint $\Phi^*_n(\Delta)$ are given by (5.35) and (5.36) respectively. Using (5.37) we can write (5.41) in the form

$$\Phi_n(\Delta^*)\Phi_n(\Delta)\xi[n]=0$$

or, when (5.31) and (5.33) are taken into account,

$$[(1-e^{\beta})^2-e^{\beta}(\Delta+\Delta^*)]\xi[n]=0 \tag{5.42}$$

Let us now define the self-adjoint difference boundary-value problem by the self-adjoint equation (5.42) and the boundary condition

$$\lim_{n\to-\infty}\xi[n]=\lim_{n\to\infty}\xi[n]=0 \tag{5.43}$$

Green's function $G[n-m]$ for this self-adjoint boundary-value is defined as follows:

$$G[n-m]=\frac{1}{1-e^{2\beta}}e^{\beta|n-m|} \tag{5.44}$$

Comparing (5.26) and (5.44) we note that the required problem is defined by (5.42), whilst the boundary conditions are given by (5.43). At the same time the coefficient

$$A=\frac{1}{1-e^{2\beta}}$$

Since by definition Green's function satisfies the non-homogeneous difference equation

$$\Phi_n(\Delta^*)\Phi_n(\Delta)G[n-m]=\sigma[n-m] \tag{5.45}$$

the operator $\Phi_n(\Delta^*)\Phi_n(\Delta)$, when applied to the function $e^{\beta|n-m|}$ with respect to the variable n, gives

$$\Phi_n(\Delta^*)\Phi_n(\Delta)e^{\beta|n-m|}=(1-e^{2\beta})\sigma[n-m] \tag{5.46}$$

Indeed, since

$$\Phi_n(\Delta^*)\Phi_n(\Delta)=(1-e^{\beta})^2-e^{\beta}(\Delta+\Delta^*)$$

we have

$$\Delta_n e^{\beta|n-m|} = e^{\beta|n+1-m|} - e^{\beta|n-m|},\ \Delta_n^* e^{\beta|n-m|} = e^{\beta|n-1-m|} - e^{\beta|n-m|}$$

Taking these equations into consideration we have

$$\Phi_n(\Delta^*)\Phi_n(\Delta)e^{\beta|n-m|} = (1-e^{\beta})e^{\beta|n-m|} - e^{\beta}(e^{\beta|n+1-m|} - 2e^{\beta|n-m|} + e^{\beta|n-1-m|})$$

whence, when $n = m$, we find

$$[\Phi_n(\Delta^*)\Phi_n(\Delta)e^{\beta|n-m|}]\big|_{n=m} = (1-e^{2\beta})$$

which coincides with the right-hand side of Equation (5.46) for $n = m$.

We now express the self-adjoint operator $L^*_n(\Delta)L_n(\Delta)$ by the product of self-adjoint operators of the form

$$\Phi_n(\Delta^*)\Phi_n(\Delta) = (1-e^{\beta_i})^2 - e^{\beta_i}(\Delta+\Delta^*) \tag{5.47}$$

i.e. we put

$$L_n(\Delta^*)L_n(\Delta) = \prod_{i=1}^{k}[(1-e^{\beta_i})^2 - e^{\beta_i}(\Delta+\Delta^*)] \tag{5.48}$$

The operator (5.48) is applied to both sides of (5.5). Then, using (5.45), we have

$$\begin{aligned}
&\sum_{\nu=0}^{r}\gamma_\nu^*\prod_{i=1}^{k}[(1-e^{\beta_i})^2 - e^{\beta_i}(\Delta+\Delta^*)]m^\nu \\
&=\sum_{l=0}^{N}\Big\{b_1(1-e^{2\beta_1})\prod_{i=1}^{k}{}^{(1)}[(1-e^{\beta_i})^2 - e^{\beta_i}(\Delta+\Delta^*)] \\
&\times W[l]\sigma[m-l] + b_2(1-e^{2\beta_2})\prod_{i=1}^{k}{}^{(2)}[(1-e^{\beta_i})^2 - e^{\beta_i}(\Delta+\Delta^*)] \\
&\times W[l]\sigma[m-l] + \ldots + b_j(1-e^{2\beta_j})\prod_{i=1}^{k}{}^{(j)}[(1-e^{\beta_i})^2 \\
&- e^{\beta_i}(\Delta+\Delta^*)]W[l]\sigma[m-l] + \ldots + b_k(1-e^{2\beta k}) \\
&\times\prod_{i=1}^{k}{}^{(k)}[(1-e^{\beta_i})^2 - e^{\beta_i}(\Delta+\Delta^*)]W[l]\sigma[m-l]\Big\}
\end{aligned} \tag{5.49}$$

where the symbol (j) at the product sign $(\Pi^{(j)})$ denotes that when the multiplication is carried out the factor with the number j is omitted.

Carrying out the summation in the right-hand side of (5.49) and taking into account that

$$\sum_{l=0}^{N} b_j(1-e^{2\beta_j})\prod_{i=1}^{k}{}^{(j)}[(1-e^{\beta_i})^2-e^{\beta_i}(\Delta+\Delta^*)]W[l]\sigma[m-l]$$

$$=b_j(1-e^{2\beta_j})\prod_{i=1}^{k}{}^{(j)}[(1-e^{\beta_i})^2-e^{\beta_i}(\Delta+\Delta^*)]W[m]$$

we have

$$\begin{aligned}
&\sum_{\nu=0}^{r}\gamma_\nu^*\prod_{i=1}^{k}[(1-e^{\beta_i})^2-e^{\beta_i}(\Delta+\Delta^*)]m^\nu\\
&=b_1(1-e^{2\beta_1})\prod_{i=1}^{k}{}^{(1)}[(1-e^{\beta_i})^2-e^{\beta_i}(\Delta+\Delta^*)]W[m]\\
&+b_2(1-e^{2\beta_2})\prod_{i=1}^{k}{}^{(2)}[(1-e^{\beta_i})^2-e^{\beta_i}(\Delta+\Delta^*)]W[m]\\
&\cdots\cdots\cdots\cdots\cdots\cdots\cdots\cdots\\
&+b_j(1-e^{2\beta_j})\prod_{i=1}^{k}{}^{(j)}[(1-e^{\beta_j})^2-e^{\beta_i}(\Delta+\Delta^*)]W[m]\\
&\cdots\cdots\cdots\cdots\cdots\cdots\cdots\cdots\\
&+b_k(1-e^{2\beta_k})\prod_{i=1}^{k}{}^{(k)}[(1-e^{\beta_i})^2-e^{\beta_i}(\Delta+\Delta^*)]W[m]
\end{aligned}\tag{5.50}$$

Equation (5.50) is a non-homogeneous difference equation in the unknown function $W[m]$. We write this equation in the compact form:

$$\begin{aligned}
&\sum_{\nu=0}^{r}\gamma_\nu^*\prod_{i=1}^{k}[(1-e^{\beta_i})^2-e^{\beta_i}(\Delta+\Delta^*)]m^\nu\\
&=\sum_{\mu=1}^{k}b_\mu(1-e^{2\beta_\mu})\prod_{i=1}^{k}{}^{(\mu)}[(1-e^{\beta_i})^2-e^{\beta_i}(\Delta+\Delta^*)]W[m]
\end{aligned}\tag{5.51}$$

Taking into account that

$$\frac{\prod_{i=1}^{k}{}^{(\mu)}[(1-e^{\beta_i})^2-e^{\beta_i}(\Delta+\Delta^*)]}{\prod_{i=1}^{k}[(1-e^{\beta_i})^2-e^{\beta_i}(\Delta+\Delta^*)]}=\frac{1}{(1-e^{\beta_\mu})^2-e^{\beta_\mu}(\Delta+\Delta^*)}$$

(5.51) can be written as

$$\sum_{\nu=0}^{r}\gamma_\nu^* m^\nu=\sum_{i=1}^{k}b_i(1-e^{2\beta_i})\frac{1}{(1-e^{\beta_i})^2-e^{\beta_i}(\Delta+\Delta^*)}W[m] \tag{5.52}$$

Comparing (5.11) and (5.12) we find that (5.25) has the form

$$\sum_{i=1}^{k}b_i(1-e^{2\beta_i})\frac{1}{(1-e^{\beta_i})^2-e^{\beta_i}(\Delta+\Delta^*)}\Bigg|_{\substack{\Delta=e^q-1\\ \Delta^*=e^{-q}-1}}=0 \tag{5.53}$$

Replacing Δ and Δ^* in (5.53) by (e^q-1) and $(e^{-q}-1)$ respectively we find

$$\sum_{i=1}^{k}b_i(1-e^{2\beta_i})\frac{1}{(1-e^{\beta_i})^2-e^{\beta_i}(\Delta+\Delta^*)}\Bigg|_{\substack{\Delta=e^q-1\\ \Delta^*=e^{-q}-1}}$$

$$=\sum_{i=1}^{k}\frac{b_i(1-e^{2\beta_i})}{(1-e^{\beta_i})^2-e^{\beta_i}(e^q-2+e^{-q})}$$

or, after certain transformations,

$$\sum_{i=1}^{k}\frac{b_i(1-e^{\beta_i})}{(1-e^{\beta_i})^2-e^{\beta_i}(e^q-2+e^{-q})} \tag{5.54}$$

$$=\frac{\sum_{\mu=1}^{k}b_\mu(1-e^{2\beta_\mu})\prod_{i=1}^{k}{}^{(\mu)}[(1-e^{\beta_i})^2-e^{\beta_i}(e^q-2+e^{-q})}{\prod_{i=1}^{k}[(1-e^{\beta_i})^2-e^{\beta_i}(e^q-2+e^{-q})}$$

Bearing in mind (5.54) and also using (5.53), we have the following equation:

$$M(q)M^*(q)=\sum_{\mu=1}^{k} b_\mu(1-e^{2\beta_\mu})\prod_{i=1}^{k}{}^{(\mu)}[(1-e^{\beta_i})^2 - e^{\beta_i}(e^q-2+e^{-q})]=0 \tag{5.55}$$

It can be seen that this equation has the order $2(k-1)$, and thus, after certain transformations, we obtain from it

$$\sum_{i=0}^{k-1} a_i e^{iq}+\sum_{i=0}^{k-1} a_i e^{-iq}=0 \tag{5.56}$$

where the coefficients $a_i\,(i=0,\ 1,\dots,\ k-1)$ are functions of b_μ and e^{β_μ}. Multiplying both sides of (5.56) by $e^{(k-1)q}$ we have

$$\sum_{i=0}^{2(k-1)} d_i e^{iq}=0 \tag{5.57}$$

whence it follows that the order of (5.55) equals $2(k-1)$. The order of $M(\Delta)$ is $h=k-1$, which is due to the fact that $M(q)|_{e^{q}-1=\Delta}=M(\Delta)$ is the operator in the right-hand side of (5.6), the equation of a discrete filter which forms the random process ξ with correlation function (5.3). Hence we draw the important conclusion, that the highest order of the difference of the σ-function in the general solution of (5.5), giving the weighting function of the optimal system, must be zero.

Thus, on the basis of (5.24) and with these remarks taken into consideration, we can write the expression for the weighting in the following form:

$$W[m]=\sum_{i=1}^{2(k-1)} B_i e^{q_i m}+\sum_{\nu=0}^{r} A_\nu m^\nu+C\sigma[m]+D\sigma[N-m] \qquad 0\leqslant m\leqslant N \tag{5.58}$$

where q_i are the roots of (5.55).

It should be pointed out that (5.55) is transcendental, and accordingly its solution for q_i cannot be found in explicit form. But it is easy to see that there is no need to know the roots of (5.55) in order to determine $W[m]$; it is sufficient to know the values

$$z_i = e^{q_i}, \ i = 1, 2, \ldots, 2(k-1) \tag{5.59}$$

Introducing the substitution (5.59) we can write the expression for the weighting function as follows:

$$W[m] = \sum_{i=1}^{2(k-1)} B_i z_i^m + \sum_{\nu=0}^{r} A_\nu m^\nu + C\sigma[m] + D\sigma[N-m] \tag{5.60}$$

$$0 \leqslant m \leqslant N$$

where z_i $(i = 1, 2, \ldots, 2(k-1)$ are the roots of

$$\sum_{\mu=1}^{k} b_\mu (1 - e^{\beta_\mu}) \prod_{i=1}^{k}{}^{(\mu)} [(1 - e^{\beta_i})^2 z - e^{\beta_i}(z-1)^2] = 0 \tag{5.61}$$

which is obtained from (5.55) by the substitutions $e^q = z$ and $e^{-q} = z^{-1}$.

In concluding this section, we point out that the expression for the weighting function of the optimal system can be written in the form (5.60) in the general case also, where the useful input signal is represented in the form (4.4), i.e. when it contains a stationary term. It follows that the determination of the weighting function of an optimal system reduces to the determination of the coefficients B_i, A_i, C and D and the roots z_i of an equation of the type (5.61). This equation can be obtained if the correlation function K_ξ, defined by (5.3), is given (in a more general case the correlation functions K_x, $K_{x\xi}$ and $K_{\xi x}$ must also be known).

To determine B_i, A_i, C and D we must substitute (5.60) into an equation which is satisfied by the weighting function; we then have to consider this equation as an identity for each instant of time $0 \leqslant m \leqslant N$. This gives the required system of equations whose solution gives the unknown coefficients.

The method just given for the solution of (5.1) is found to be fully applicable for the solution of (3.2), which is in fact a particular case of (5.1). In this case the required system of equations, from which the coefficients A_i, B_i, C and D

are determined, is obtained by substituting (5.60) into (3.2), and also into the $r+1$ additional conditions (3.1). This system of equations constitutes a particular case of the discrete analogue of Semenov's problem.

The method given for the determination of the weighting function of an optimal system is thus sufficiently general. It is, in addition, suitable for solving the discrete analogue of Semenov's problem, as well as the discrete analogue of the Zadeh-Ragazzini problem.

Chapter 7

OPTIMAL NON-STATIONARY SYSTEMS

In this chapter methods are given for determining optimal (in the sense of a minimum dispersion) non-stationary transforming sampled data systems in cases where the useful signal over a finite time interval $[n-N;\ n]$ can be represented in the form

$$S[n-m] = \sum_{\nu=0}^{r} d_\nu[n]\, m^\nu + X[n-m], \quad m=0,\ 1, \dots,\ N$$

where $d_\nu[n]$ are random (for the given n) quantities correlated with each other;

$X[n-m]$ is a non-stationary random function with correlation function $K_x[n;\ m]$.

At the same it is assumed that noise ξ, which is a non-stationary random function with correlation function $K_\xi[n;\ m]$, is superimposed on the useful signal.

The most general methods for determining optimal continuous systems are found in the work of Pugachev [56, 57, 58, 59, 60, 61]. They are based on the use of integral canonical expansions; with the help of discrete shaping filters they can also be used to determine optimal sampled data systems. The methods which will be given in this chapter are based on the ideas of the general theory of optimal systems developed by Pugachev.

The determination of optimal sampled data systems, difficult even in stationary cases, becomes still more complicated in the case of non-stationary actions, particularly when long intervals of observation are used. For this reason methods will be given to enable the determination of the optimal weighting function to be reduced to the solution of certain recurrence formulae instead of the corresponding system of algebraic equations. These methods are based on representing the correlation functions $K_x[n; m]$ and $K_\xi[n; m]$ in the form of polynomials of finite degree in the second variable, i.e.

$$K_x[n; m] = \sum_{\nu=0}^{r_1} a_\nu[n] m^\nu, \quad n \geqslant m$$

$$K_\xi[n; m] = \sum_{\nu=0}^{r_2} g_\nu[n] m^\nu, \quad n \geqslant m$$

where $a_\nu[n]$ and $g_\nu[n]$ are known functions of arbitrary form.

These methods for determining optimal systems are applicable also for the case where the useful signal has the form

$$S[n-m] = \sum_{\nu=0}^{r} d_\nu[n] \varphi_\nu[m] + X[n-m]$$

$$m = 0, 1, \ldots, N$$

where $\varphi_\nu[m]$ are non-random functions of time, whilst the correlation functions K_x and K_ξ can be represented in the form

$$K_x[n; m] = \sum_{\nu=0}^{r_1} a_\nu[n] \varphi_\nu[m], \quad n \geqslant m$$

$$K_\xi[n; m] = \sum_{\nu=0}^{r_2} g_\nu[n] \varphi_\nu[m], \quad n \geqslant m$$

The purpose of giving methods which give optimal systems in the case where $\varphi_\nu[m] = m^\nu$ is to simplify the necessary calculations.

Analogous methods for determining optimal continuous systems are given in the work of Pugachev [75-78].

7.1 FORMULATION OF THE PROBLEM

We assume that the useful signal S which is to be transformed (or reproduced) is the sum of two components

$$S[n]=g[n]+X[n] \tag{1.1}$$

of which $g[n]$ is a polynomial of finite degree with respect to time

$$g[n-m]=\sum_{\nu=0}^{r} d_\nu[n]\,m^\nu,\quad m=0,1,\ldots,\ N \tag{1.2}$$

whose coefficients d_ν $(\nu=0, 1,\ldots, r)$ are correlated with each other and have finite dispersion and zero mathematical expectation, i.e.

$$M[d_\nu d_\mu]=D_{\nu\mu} \tag{1.3}$$

$$M[d_\nu]\equiv 0,\quad \nu,\ \mu=0,\ 1,\ldots,\ r \tag{1.4}$$

We assume that $X[n]$ is a non-stationary random function with zero mathematical expectation and correlation function

$$M[X[n]\,X[m]]=K_x[n;\ m] \tag{1.5}$$

In addition, we assume that

$$M[d_\nu X[n]]\equiv 0,\quad \nu=0,\ 1,\ldots,\ r \tag{1.6}$$

Let us assume further that the noise ξ, a non-stationary random function with zero mathematical expectation, is superimposed on ξ; this function is not correlated with the random function X or with the coefficients d_ν. The correlation function of X is denoted by $K_\xi[n;\ m]$. The assumption that X and ξ have zero mathematical expectations is not vital, and has been introduced only to shorten the calculations.

Suppose that the total signal $Y[n]=S[n]+\xi[n]$ is observed over the interval $[n-N;\ n]$, and that from the observed values we must find the best approximation $S^*[n]$ to the signal $h[n]$, where the latter is obtained by applying the operator H to S,

$$h[n] = HS \tag{1.7}$$

Since X and ξ are non-stationary, the parameters of the system which carries out the best transformation of the signal will depend on time. Denoting the weighting function of such a system by $W[n;\ m]$, we can write the following expression for the signal S^*:

$$S^*[n] = \sum_{m=n-N}^{n} W[n;\ m]\{g[m] + X[m] + \xi[m]\}$$

or, introducing the new variable $n - k = m$, and denoting the independent variable as before,

$$S^*[n] = \sum_{m=0}^{N} W[n;\ n-m]\{g[n-m] + X[n-m] + \xi[n-m]\} \tag{1.8}$$

To assess the system giving the best transformation of the input signal, we apply the criterion of the minimum dispersion of the transformation error for any instant when systematic errors are absent. Since d_ν, $\nu = 0, 1, \ldots, r$ are random quantities, the required system must guarantee an unconditional minimum of the error dispersion, i.e. the determination of an optimal system reduces to the determination of the function $W[n;\ m]$ which will give an unconditional minimum of the expression

$$M\left[\left\{\sum_{m=-\infty}^{\infty} W_{\text{id}}[n;\ m]\, S[m] - \sum_{m=n-N}^{n} W[n;\ m]\, Y[m]\right\}^2\right] \tag{1.9}$$

where $W_{\text{id}}[n;\ m]$ is the weighting function of the ideal transforming system.

We shall now determine the equation which is to be satisfied by $W[n;\ m]$.

7.2 EQUATION FOR THE OPTIMAL WEIGHTING FUNCTION

Since the problem of finding $W[n;\ m]$ is a problem of determining an unconditional extremum (1.9), we can find the required equation by the calculus of variations, as in Chapters 5 and 6. Here, however, we shall use another method, based on the fact that the condition for minimum dispersion of the transformation error reduces to the condition that the transformation error be non-correlated with all the values of the signal $Y[m]=g[m]+X[m]+\xi[m]$, observed over the interval $n-N\leqslant m\leqslant n$. This can be expressed as follows:

$$M\left[\left\{\sum_{m=-\infty}^{\infty} W_{\mathrm{id}}[n;\ n-m]S[n-m]\right.\right.$$
$$\left.\left.-\sum_{m=0}^{N} W[n;\ n-m]Y[n-m]\right\}Y[n-l]\right]=0 \qquad (2.1)$$
$$0\leqslant l\leqslant N$$

Substituting (1.1) into (2.1) and noting that $Y=S+\xi$, we have, using (1.2),

$$M\left[\left\{\sum_{m=-\infty}^{\infty} W_{\mathrm{id}}[n;\ n-m]\left(\sum_{\nu=0}^{r} d_{\nu}[n]m^{\nu}+X[n-m]\right)\right.\right.$$
$$-\sum_{m=0}^{N} W[n;\ n-m]\left(\sum_{\nu=0}^{r} d_{\nu}[n]m^{\nu}+X[n-m]+\xi[n-m]\right) \qquad (2.2)$$
$$\left.\left.\times\left\{\sum_{\mu=0}^{r} d_{\mu}[n]l^{\mu}+X[n-l]+\xi[n-l]\right\}\right]=0$$
$$l=0,\ 1,\ldots,\ N$$

Applying mathematical expectation and taking into account (1.3) to (1.6) we have

$$\sum_{m=-\infty}^{\infty} W_{\mathrm{id}}[n;\ n-m]\left(\sum_{\nu,\ \mu=0}^{r} D_{\nu\mu}m^{\nu}l^{\mu}+K_{x}[n-l;\ n-m]\right)$$
$$-\sum_{m=0}^{N} W[n;\ n-m]\left(\sum_{\nu,\ \mu=0}^{r} D_{\nu\mu}m^{\nu}l^{\mu}\right.$$
$$\left.+K_{x}[n-l;\ n-m]+K_{\xi}[n-l;\ n-m]\right)=0$$
$$l=0,\ 1,\ldots,\ N$$

or

$$\sum_{m=-\infty}^{\infty} W_{id}[n;\ n-m] \sum_{\nu,\ \mu=0}^{r} D_{\nu\mu} m^{\nu} l^{\mu}$$
$$+\sum_{m=-\infty}^{\infty} W_{id}[n;\ n-m] K_x[n-l;\ n-m] -$$
$$-\sum_{m=0}^{N} W[n;\ n-m] \sum_{\nu,\ \mu=0}^{r} D_{\nu\mu} m^{\nu} l^{\mu}$$
$$-\sum_{m=0}^{N} W[n;\ n-m](K_x[n-l;\ n-m]$$
$$+K_{\xi}[n-l;\ n-m])=0$$
$$l=0,\ 1, \ldots,\ N$$

whence

$$\sum_{m=0}^{N} W[n;\ n-m](K_x[n-l;\ n-m]+K_{\xi}[n-l;\ n-m])$$
$$=\sum_{m=-\infty}^{\infty} W_{id}[n;\ n-m] K_x[n-l;\ n-m]$$
$$+\sum_{\nu=0}^{r}\sum_{\mu=0}^{r} D_{\nu\mu} l^{\mu}\left\{\sum_{m=-\infty}^{\infty} m^{\nu}\cdot W_{и}[n;\ n-m]\right.$$
$$\left.-\sum_{m=0}^{N} W[n;\ n-m]\, m^{\nu}\right\}$$
$$l=0,\ 1, \ldots,\ N$$

Putting

$$\sum_{\nu=0}^{r} D_{\nu\mu}\left\{\sum_{m=-\infty}^{\infty} W_{id}[n;\ n-m]\, m^{\nu}\right.$$
$$\left.-\sum_{m=0}^{N} W[n;\ n-m]\, m^{\nu}\right\}=\gamma_{\mu}^{*}[n] \qquad (2.3)$$
$$\mu=0,\ 1, \ldots,\ r$$

in the last expression, we find the equation

$$\sum_{m=0}^{N} W[n;\ n-m](K_x[n-l;\ n-m]+K_\xi[n-l;\ n-m])$$
$$=\sum_{m=-\infty}^{\infty} W_{id}[n;\ n-m]\,K_x[n-l;\ n-m]+\sum_{\mu=0}^{r}\gamma_\mu^*[n]\,l^\mu \tag{2.4}$$
$$0\leqslant l\leqslant N$$

which must be satisfied by the weighting function $W[n;\ m]$ of the optimal system.

The assumptions made in formulating the problem show it to be the discrete analogue of Semenov's problem [62] for non-stationary random actions. At the same time it is not difficult to show that the discrete analogue of the Zadeh-Ragazzini [73] problem for non-stationary random actions arises as a particular case of the problem just considered. Indeed, solving the corresponding problem for a conditional extremum of the transformation error, we can show that $W[n;\ m]$ must satisfy

$$\sum_{m=0}^{N} W[n;\ n-m](K_x[n-l;\ n-m]+K_\xi[n-l;\ n-m])$$
$$=\sum_{m=-\infty}^{\infty} W_{id}[n;\ n-m]\,K_x[n-l;\ n-m]+\sum_{\mu=0}^{r}\gamma_\mu[n]\,l^\mu \tag{2.5}$$
$$0\leqslant l\leqslant N$$

and the additional $r+1$ conditions

$$\sum_{m=-\infty}^{\infty} W_{id}[n;\ n-m]\,m^\mu=\sum_{m=0}^{N} W[n;\ n-m]\,m^\nu \tag{2.6}$$
$$\mu=0,\ 1,\dots,\ r$$

Equation (2.5) coincides with (2.4) when $\gamma_\mu^*[n]$, $\mu=0,\ 1,\ \dots,\ r$, is replaced by the undetermined Lagrange multipliers. Conditions (2.6), on the other hand, are obtained from (2.3), where it must be assumed that the coefficients d_ν are not correlated with each other, i.e.

$$D_{\nu\mu} = \begin{cases} 0, & \nu \neq \mu \\ D_{\nu\nu}, & \mu = \nu \end{cases} \quad (2.7)$$

and that

$$D_{\nu\nu} \to \infty, \ \nu = 0, 1, \ldots, r \quad (2.8)$$

Thus, when the additional conditions (2.6) are used the solution of (2.5) is obtained from the solution of the more general equation (2.4). To do this we must put the dispersion $D_{\nu\nu}$ equal to infinity in the solution of (2.4), whilst the following applies to the connecting moments

$$M[d_\nu d_\mu] = 0, \ \nu \neq \mu$$

In the following sections methods will be given for solving the more general equation (2.4). Before solving this equation we shall write it in vector form.

7.3 VECTOR FORM OF EQUATION (2.4)

As already mentioned, in this chapter we shall consider methods for determining optimal sampled data systems in the case where the correlation functions $K_x[n;\ m]$ and $K_\xi[n;\ m]$ are polynomials in the second variable m:

$$K_x[n;\ m] = \sum_{\nu=0}^{r_1} a_\nu[n]\, m^\nu, \ n \geqslant m \quad (3.1)$$

$$K_\xi[n;\ m] = \sum_{\nu=0}^{r_2} g_\nu[n]\, m^\nu, \ n \geqslant m \quad (3.2)$$

For the general case, denoting

$$m^\nu = b_\nu[m], \quad \nu = 0, 1, \ldots, \quad (3.3)$$

we can write (3.1) and (3.2) in the following form,

$$K_x[n;\ m] = \sum_{\nu=0}^{r_1} a_\nu[n]\, b_\nu[m], \ n \geqslant m \quad (3.4)$$

$$K_\xi[n;\ m]=\sum_{\nu=0}^{r_2} g_\nu[n]\, b_\nu[m],\quad n\geqslant m \tag{3.5}$$

Owing to the fact that the weighting function $W_{\mathrm{id}}[n;\ m]$ of the ideal transforming system usually has the form

$$W_{\mathrm{id}}[n;\ m]=\Delta_n^q\,\sigma[n-m]$$

$$W_{\mathrm{id}}[n;\ m]=\sigma[n+l],\ (l\geqslant 0,\ l<0)$$

or

$$W_{\mathrm{id}}[n;\ m]=\sum_{l=-\infty}^{\infty}\sigma[n-m-l]$$

the expression

$$\sum_{m=-\infty}^{\infty} W_{\mathrm{id}}[n;\ m]\,K_x[l;\ m]$$

can also be represented in a form analogous to (3.4). Assuming, for example that $W_{\mathrm{id}}[n;\ m]$ is equal to $\Delta_n^q\,\sigma[n-m]$, we have

$$\sum_{m=-\infty}^{\infty} W_{\mathrm{id}}[n;\ n-m]\,K_x[n-l;\ n-m]$$

$$=\sum_{m=-\infty}^{\infty}\Delta^q\sigma[m]\,K_x[n-l;\ n-m]$$

whence, using the σ-function, we find that

$$\sum_{m=-\infty}^{\infty}\Delta^q\sigma[m]\,K_x[n-l;\ n-m]=\Delta_n^q K_x[l;\ n]$$

Substituting the expression for the correlation function $K_x[l;\ n]$ from (3.4) into the last equation, we obtain

$$\Delta_n^q K_x[l;\ n]=\Delta_n^q\sum_{\nu=0}^{r_1} a_\nu[l]\,b_\nu[n]=\sum_{\nu=0}^{r_1} a_\nu[l]\,\Delta_n^q b_\nu[n]$$

whence, with (3.3) taken into account, it follows that

$$\sum_{m=-\infty}^{\infty} \Delta_n^q \sigma[n-m] K_x[l;\ m] = \sum_{\nu=0}^{r_1-q} f_\nu[l] b_\nu[n], \quad l \geqslant n \tag{3.6}$$

where

$$f_\nu[n] = \begin{cases} 0, & \nu < q \\ \nu(\nu-1)\dots(\nu-q+1)\, a_\nu[n], & \nu \geqslant q \end{cases} \tag{3.7}$$

In a similar way we could find corresponding expansions of

$$\sum_{m=-\infty}^{\infty} W_{\mathrm{id}}[n;\ m] K_x[l;\ m]$$

in other cases. Thus, in the general case we can write

$$\sum_{m=-\infty}^{\infty} W_{\mathrm{id}}[n;\ m] K_x[l;\ m] = \sum_{\nu=0}^{r_3} f_\nu[n] b_\nu[l] \tag{3.8}$$

For generality we shall assume in the sequel that the expansions (3.4), (3.5) and (3.8) contain q terms, where

$$q = \max\{r_1,\ r_2,\ r_3\} \tag{3.9}$$

so that, for example,

$$K_x[n;\ m] = \sum_{\nu=0}^{q} a_\nu[n] b_\nu[m], \quad n \geqslant m$$

where $a_\nu[n] \equiv 0$ when $\nu > r_1$.

If $a_\nu[n]$, $b_\nu[n]$, $g_\nu[n]$ and $f_\nu[n]$, $\nu = 0,\ 1, \dots,\ q$ are considered as the components of the vectors $\boldsymbol{A}[n]$, $\boldsymbol{B}[n]$, $\boldsymbol{G}[n]$ and $\boldsymbol{F}[n]$ of a $(q+1)$-dimensional space, then the expansions (3.4), (3.5), and (3.8) can be represented by the scalar product of the corresponding vectors,

$$K_x[n;\ m] = \sum_{\nu=0}^{q} a_\nu[n] b_\nu[m] = \boldsymbol{A}[n] \cdot \boldsymbol{B}[m] \quad n \geqslant m \tag{3.10}$$

$$K_\xi[n;\ m] = \sum_{\nu=0}^{q} g_\nu[n] b_\nu[m] = \boldsymbol{G}[n] \cdot \boldsymbol{B}[m], \quad n \geqslant m \tag{3.11}$$

$$\sum_{m=-\infty}^{\infty} W_{\mathrm{id}}[n;\ m] K_x[l;\ m] = \sum_{\nu=0}^{q} f_\nu[n] b_\nu[l] = \boldsymbol{F}[n] \cdot \boldsymbol{B}[l] \quad (3.12)$$

By the symmetry property of correlation functions we have

$$K_x[n;\ m] = \sum_{\nu=0}^{q} a_\nu[m] b_\nu[n],\ n < m$$

$$K_\xi[n;\ m] = \sum_{\nu=0}^{q} g_\nu[m] b_\nu[n],\ n < m$$

whence, when (3.10) and (3.11) are used, we obtain

$$K_x[n;\ m] = \boldsymbol{A}[m] \cdot \boldsymbol{B}[n],\ n < m \quad (3.13)$$

$$K_\xi[n;\ m] = \boldsymbol{G}[m] \cdot \boldsymbol{B}[n],\ n < m \quad (3.14)$$

Assuming further that

$$\begin{gathered} \sum_{\nu=0}^{r} \gamma_\nu^*[n] l^\nu = \sum_{\nu=0}^{q} \gamma_\nu^*[n] l^\nu = \sum_{\nu=0}^{q} \gamma_\nu^*[n] b_\nu[l] \\ \gamma_\nu^*[n] \equiv 0,\ \nu > r \end{gathered} \quad (3.15)$$

and assuming that $\gamma_\nu^*[n]$, $\nu = 0, 1, \ldots, q$ ($q = \max\{r_1, r_2, r_3, r\}$) are the components of the vector $\boldsymbol{\Gamma}^*[n]$ of a $q+1$ dimensional space, we can write

$$\sum_{\nu=0}^{r} \gamma_\nu^*[n] b_\nu[l] = \sum_{\nu=0}^{q} \gamma^*[n] b_\nu[l] = \boldsymbol{\Gamma}^*[n] \cdot \boldsymbol{B}[l] \quad (3.16)$$

We shall next substitute the expression for the correlation functions from (3.10) and (3.11) into (2.4). We then obtain, when (3.12) to (3.16) are taken into consideration,

$$\begin{aligned} &\sum_{m=n-N}^{l} W[n;\ m] (\boldsymbol{A}[l] \cdot \boldsymbol{B}[m] + \boldsymbol{G}[l] \cdot \boldsymbol{B}[m]) \\ &+ \sum_{m=l+1}^{n} W[n;\ m] (\boldsymbol{A}[m] \cdot \boldsymbol{B}[l] + \boldsymbol{G}[m] \cdot \boldsymbol{B}[l]) \quad (3.17) \\ &= \boldsymbol{F}[n] \cdot \boldsymbol{B}[l] + \boldsymbol{\Gamma}^*[n] \cdot \boldsymbol{B}[l],\ n - N \leqslant l \leqslant n \end{aligned}$$

or

$$\sum_{m=n-N}^{l} W[n;\ m](\boldsymbol{A}[l]+\boldsymbol{G}[l])\cdot\boldsymbol{B}[m]$$

$$+\sum_{m=l+1}^{n} W[n;\ m](\boldsymbol{A}[m]+\boldsymbol{G}[m])\cdot\boldsymbol{B}[l] \tag{3.18}$$

$$=(\boldsymbol{F}[n]+\boldsymbol{\Gamma}^{*}[n])\cdot\boldsymbol{B}[l],\ n-N\leqslant l\leqslant n$$

Let us denote the vector formed by the sum of $\boldsymbol{A}[l]$ and $\boldsymbol{G}[l]$ by $\boldsymbol{M}[l]$, and that formed by the sum of $\boldsymbol{F}[n]$ and $\boldsymbol{\Gamma}^{*}[n]$ by $\boldsymbol{I}[n]$, i.e.

$$\boldsymbol{M}[l]=\boldsymbol{A}[l]+\boldsymbol{G}[l]=\{m_0[l];\ m_1[l];\ldots,m_q[l]\} \tag{3.19}$$

$$\boldsymbol{I}[n]=\boldsymbol{F}[n]+\boldsymbol{\Gamma}^{*}[n]=\{i_0[n];\ i_1[n];\ldots,i_q[n]\} \tag{3.20}$$

where the components m_ν and i_ν are defined as follows:

$$\left.\begin{aligned} m_\nu[l]&=a_\nu[l]+g_\nu[l],\ \nu=0,\ 1,\ldots,q\\ i_\nu[n]&=f_\nu[n]+\gamma_\nu^{*}[n],\ \nu=0,\ 1,\ldots,q \end{aligned}\right\} \tag{3.21}$$

Using (3.19) and (3.20) we can write (3.19) in the form

$$\sum_{m=n-N}^{l} W[n;\ m]\,\boldsymbol{M}[l]\cdot\boldsymbol{B}[m]+\sum_{m=l+1}^{n} W[n;\ m]\,\boldsymbol{M}[m]\cdot\boldsymbol{B}[l]$$

$$=\boldsymbol{I}[n]\cdot\boldsymbol{B}[l],\ n-N\leqslant l\leqslant n \tag{3.22}$$

Equations (3.18) and (3.22) are vector equations corresponding to (2.4). Vector equations can easily be transformed into scalar ones, so that, noting

$$\boldsymbol{M}[l]\cdot\boldsymbol{B}[m]=\sum_{\nu=0}^{q} m_\nu[l]\,b_\nu[m]$$

$$\boldsymbol{I}[n]\cdot\boldsymbol{B}[l]=\sum_{\nu=0}^{q} i_\nu[n]\,b_\nu[l]$$

we obtain from (3.22) the scalar equation

$$\sum_{\nu=0}^{q} m_\nu[l] \sum_{m=n-N}^{l} W[n;\ m] b_\nu[m] + \sum_{\nu=0}^{q} b_\nu[l] \sum_{m=l+1}^{n} W[n;\ m] m_\nu[m] = \sum_{\nu=0}^{q} i_\nu[n] b_\nu[l] \quad n-N \leqslant l \leqslant n \tag{3.23}$$

We shall use both vector and scalar forms of these equations.

We shall now solve (3.17) which gives the weighting function of an optimal system, considering first the case where the noise ξ is discrete white noise.

7.4 OPTIMAL SYSTEM WITH DISCRETE WHITE NOISE

We assume that the noise ξ has following correlation function

$$K_\xi[n;\ m] = D_\xi \sigma[n-m] \tag{4.1}$$

In this case (2.4), which gives the weighting function of the optimal system, assumes the form

$$\sum_{m=n-N}^{n} W[n;\ m] K_x[l;\ m] + D_\xi W[n;\ l] = \sum_{m=-\infty}^{\infty} W_{\text{id}}[n;\ m] K_x[l;\ m] + \sum_{\mu=0}^{r} \gamma_\mu^*[n] l^\mu \quad n-N \leqslant l \leqslant n \tag{4.2}$$

Substituting the expression for the correlation function K_x from (3.10) into (4.2), and using (3.12), (3.13) and (3.16), we obtain the vector equation

$$\sum_{m=n-N}^{l} W[n;\ m] \boldsymbol{A}[l] \cdot \boldsymbol{B}[m] + \sum_{m=l+1}^{n} W[n;\ m] \boldsymbol{A}[m] \cdot \boldsymbol{B}[l] + D_\xi W[n;\ l] = (\boldsymbol{F}[n] + \boldsymbol{\Gamma}^*[n]) \cdot \boldsymbol{B}[l] \quad n-N \leqslant l \leqslant n \tag{4.3}$$

which corresponds to the scalar equation (4.2). Equation (4.3) is solved most easily in the case where $\boldsymbol{A}[n]$ and $\boldsymbol{B}[n]$ are equal.

7.4.1 Solution of Equation (4.3) when $\mathbf{A}[n] = \mathbf{B}[n]$

Using (3.20) we write (4.3) in the following form:

$$\boldsymbol{A}[l]\cdot \sum_{m=n-N}^{l} W[n;\ m]\boldsymbol{B}[m] + \boldsymbol{B}[l]\cdot \sum_{m=l+1}^{n} W[n;\ m]\boldsymbol{A}[m] + D_{\xi}W[n;\ l] = \boldsymbol{I}[n]\cdot\boldsymbol{B}[l],\quad n-N\leqslant l\leqslant n \tag{4.4}$$

Since

$$\sum_{m=l+1}^{n} W[n;\ m]\boldsymbol{A}[m] = \sum_{m=n-N}^{n} W[n;\ m]\boldsymbol{A}[m] - \sum_{m=n-N}^{l} W[n;\ m]\boldsymbol{A}[m]$$

(4.4) can be transformed as follows:

$$\boldsymbol{A}[l]\cdot \sum_{m=n-N}^{l} W[n;\ m]\boldsymbol{B}[m] + \boldsymbol{B}[l]\cdot \sum_{m=n-N}^{n} W[n;\ m]\boldsymbol{A}[m] - \boldsymbol{B}[l]\cdot \sum_{m=n-N}^{l} W[n;\ m]\boldsymbol{A}[m] + D_{\xi}W[n;\ l] = \boldsymbol{I}[n]\cdot\boldsymbol{B}[l] \tag{4.5}$$

$$n-N\leqslant l\leqslant n$$

Further, since the vectors $\boldsymbol{A}$ and $\boldsymbol{B}$ are equal, we obtain from (4.5)

$$\boldsymbol{A}[l]\cdot \sum_{m=n-N}^{n} W[n;\ m]\boldsymbol{A}[m] + D_{\xi}W[n;\ l] = \boldsymbol{I}[n]\cdot\boldsymbol{A}[l] \tag{4.6}$$

$$n-N\leqslant l\leqslant n$$

It is clear that

$$\sum_{m=n-N}^{n} W[n;\ m]\boldsymbol{A}[m]$$

is a $q+1$-dimensional vector. Denoting this by

$$\boldsymbol{K}[n]=\sum_{m=n-N}^{n} W[n;\ m]\,\boldsymbol{A}[m] \tag{4.7}$$

we write (4.6) in the form

$$\begin{gathered}\boldsymbol{A}[l]\cdot\boldsymbol{K}[n]+D_{\xi}W[n;\ l]=\boldsymbol{I}[n]\cdot\boldsymbol{A}[l]\\ n-N\leqslant l\leqslant n\end{gathered} \tag{4.8}$$

whence we obtain the following expression for the weighting function:

$$\begin{gathered}W[n;\ l]=\frac{1}{D_{\xi}}(\boldsymbol{I}[n]\cdot\boldsymbol{A}[l]-\boldsymbol{A}[l]\cdot\boldsymbol{K}[n])\\ n-N\leqslant l\leqslant n\end{gathered} \tag{4.9}$$

The unknown in (4.9) is K, defined by (4.7). In order to find this vector we multiply both sides of (4.8) by $\boldsymbol{A}[l]$ and sum with respect to l between the limits $[n-N;\ n]$. We then obtain

$$\begin{aligned}\sum_{l=n-N}^{n}(\boldsymbol{A}[l]\cdot\boldsymbol{K}[n])\cdot\boldsymbol{A}[l]+D_{\xi}\sum_{l=n-N}^{n}W[n;\ l]\,\boldsymbol{A}[l]\\ =\sum_{l=n-N}^{n}(\boldsymbol{I}[n]\cdot\boldsymbol{A}[l])\cdot\boldsymbol{A}[l],\ n-N\leqslant l\leqslant n\end{aligned} \tag{4.10}$$

or, when (4.7) is used,

$$\begin{aligned}\sum_{l=n-N}^{n}(\boldsymbol{A}[l]\cdot\boldsymbol{K}[n])\cdot\boldsymbol{A}[l]+D_{\xi}\boldsymbol{K}[n]\\ =\sum_{l=n-N}^{n}(\boldsymbol{I}[n]\cdot\boldsymbol{A}[l])\cdot\boldsymbol{A}[l]\end{aligned} \tag{4.11}$$

Equation (4.11) allows K to be determined in terms of the given vectors $\boldsymbol{A}$ and $\boldsymbol{I}$. To determine the components $k_{\nu}[n]$, $\nu=0,\ 1,\ldots,q$ of $\boldsymbol{K}[n]$, we write (4.11) in scalar form, using (3.21) and (4.7):

$$\sum_{\nu=0}^{q}k_{\nu}[n]\sum_{l=n-N}^{n}a_{\nu}[l]\,a_{\mu}[l]+D_{\xi}k_{\mu}[n] \tag{4.12}$$

$$=\sum_{\nu=0}^{q} i_\nu[n] \sum_{l=n-N}^{n} a_\nu[l]\, a_\mu[l],\ \mu=0, 1, \ldots, q \qquad (4.12)$$

whence, introducing the notation,

$$\beta_{\nu\mu}=\sum_{l=n-N}^{n} a_\nu[l]\, a_\mu[l],\ \nu,\ \mu=0,\ 1, \ldots, q \qquad (4.13)$$

we obtain the system of algebraic equations

$$\sum_{\nu=0}^{q} k_\nu[n]\,\beta_{\nu\mu}+D_\xi k_\mu[n]=\sum_{\nu=0}^{q} i_\nu[n]\,\beta_{\nu\mu}$$
$$\mu=0,\ 1, \ldots, q \qquad (4.14)$$

for the required $k_\nu[n]$, $\nu=0,\ 1, \ldots, q$.

The system of equations (4.14) can also be written in the form

$$(\beta_{00}+D_\xi)\,k_0[n]+\beta_{10}k_1[n]+\ldots+\beta_{q0}k_q[n]=\sum_{\nu=0}^{q} i_\nu[n]\beta_{\nu 0}$$
$$\beta_{01}k_0[n]+(\beta_{11}+D_\xi)\,k_1[n]+\ldots+\beta_{q1}k_q[n]=\sum_{\nu=0}^{q} i_\nu[n]\,\beta_{\nu 1} \qquad (4.15)$$
$$\cdots\cdots\cdots\cdots\cdots\cdots\cdots\cdots\cdots\cdots$$
$$\beta_{0q}k_0[n]+\beta_{1q}k_1[n]+\ldots+(\beta_{qq}+D_\xi)\,k_q[n]=\sum_{\nu=0}^{q} i_\nu[n]\,\beta_{\nu q}$$

The components $k_\nu[n]$ of the $(q+1)$-dimensional vector $\boldsymbol{K}[n]$ are thus determined from the corresponding system of algebraic equations.

Using (4.9) we can write the expression for the required weighting function in scalar form

$$W[n;\ l]=\frac{1}{D_\xi}(\boldsymbol{I}[n]-\boldsymbol{K}[n])\cdot\boldsymbol{A}[l]$$
$$=\frac{1}{D_\xi}\sum_{\nu=0}^{q}[f_\nu[n]+\gamma_\nu^*[n]-k_\nu[n]]a_\nu[l],\ n-N\leqslant l\leqslant n \qquad (4.16)$$

whence it follows that the weighting function is not yet found by determining $\boldsymbol{K}$, since the expression (4.16) still contains the unknown components $\gamma_\nu^*[n]$ of the $(q+1)$-dimensional vector $\Gamma^*[n]$.

The components $\gamma_\nu^*[n]$, $\nu=0, 1, \ldots, q$ can be found from the system of algebraic equations obtained from (2.3) by substituting (4.16) into this equation.

Using the notation

$$s_\nu = \sum_{m=-\infty}^{\infty} W_{\text{id}}[n;\ n-m]\, m^\nu,\quad \nu = 0,\ 1, \ldots, r \tag{4.17}$$

in (2.3), we write

$$\sum_{\nu=0}^{r} D_{\nu\mu} s_\nu - \sum_{\nu=0}^{r} D_{\nu\mu} \sum_{m=0}^{N} W[n;\ n-m]\, m^\nu = \gamma_\mu^*[n] \qquad \mu = 0,\ 1, \ldots, r \tag{4.18}$$

Substituting the expression of the weighting function from (4.16) into this equation we have

$$\sum_{\nu=0}^{r} D_{\nu\mu} s_\nu - \frac{1}{D_\xi} \sum_{\nu=0}^{r} D_{\nu\mu} \sum_{\alpha=0}^{q} \{f_\alpha[n] + \gamma_\alpha^*[n] - k_\alpha[n]\} \times \sum_{m=0}^{N} a_\alpha[n-m]\, m^\nu = \gamma_\mu^*[n],\ \mu = 0,\ 1, \ldots, q \tag{4.19}$$

Denoting

$$\sum_{m=0}^{N} a_\alpha[n-m]\, m^\nu = \varphi_{\nu\alpha},\ \nu,\ \alpha = 0,\ 1, \ldots, q \tag{4.20}$$

in (4.19), we write

$$\sum_{\nu=0}^{r} D_{\nu\mu} s_\nu - \frac{1}{D_\xi} \sum_{\nu=0}^{r} D_{\nu\mu} \sum_{\alpha=0}^{q} \{f_\alpha[n] - k_\alpha[n]\}\, \varphi_{\nu\alpha}$$

$$- \frac{1}{D_\xi} \sum_{\nu=0}^{r} D_{\nu\mu} \sum_{\alpha=0}^{q} \gamma_\alpha^*[n]\, \varphi_{\nu\alpha} = \gamma_\mu^*[n],\ \mu = 0,\ 1, \ldots, r$$

whence, noting that $\gamma_\alpha^*[n] = 0,\ \alpha > r$, we obtain

$$\gamma_\mu^*[n] + \frac{1}{D_\xi}\sum_{\nu=0}^{r} D_{\nu\mu}\sum_{\alpha=0}^{r}\gamma_\alpha^*[n]\varphi_{\nu\alpha} = \sum_{\nu=0}^{r} D_{\nu\mu}s_\nu$$
$$-\frac{1}{D_\xi}\sum_{\nu=0}^{r} D_{\nu\mu}\sum_{\alpha=0}^{q}\{f_\alpha[n] - k_\alpha[n]\}\varphi_{\nu\alpha} \tag{4.21}$$
$$\mu = 0,\ 1, \ldots, r$$

After suitable transformations (4.21) can be written

$$\gamma_0^*[n]\sum_{\nu=0}^{r} D_{\nu\mu}\varphi_{\nu 0} + \gamma_1^*[n]\sum_{\nu=0}^{r} D_{\nu\mu}\varphi_{\nu 1} + \ldots$$
$$+\gamma_r^*[n]\sum_{\nu=0}^{r} D_{\nu\mu}\varphi_{\nu r} + \gamma_\mu^*[n]\ D_\xi = D_\xi\sum_{\nu=0}^{r} D_{\nu\mu}s_\nu \tag{4.22}$$
$$-\sum_{\nu=0}^{r} D_{\nu\mu}\sum_{\alpha=0}^{q}\{f_\alpha[n] - k_\alpha[n]\}\varphi_{\nu\alpha}$$
$$\mu = 0,\ 1, \ldots, r$$

Equations (4.22) constitute a system of algebraic equations for the unknowns $\gamma_\nu^*[n], \nu = 0, 1, \ldots. r$. The solution of the system thus completes the determination of the optimal weighting function.

In the case where the coefficients d_ν, $\nu = 0, 1, \ldots, r$ of the polynomial (1.2) are non-random quantities which can assume arbitrary values, the weighting function of the optimal system is given by the expression

$$W[n;\ l] = \frac{1}{D_\xi}\sum_{\nu=0}^{q}[f_\nu[n] + \gamma_\nu[n] - k_\nu[n]]\,a_\nu[l]$$
$$n - N \leqslant l \leqslant n \tag{4.23}$$

which formally coincides with (4.16). At the same time the components $k_\nu[n]$, $\nu = 0, 1, \ldots, q$ of $K[n]$ are determined by solving (4.15). The system of equations determining the components $\gamma_\nu[n]$, $\nu=0, 1, \ldots, r$ of $\Gamma[n]$ can be obtained from (4.22) with the condition that the coefficients of the polynomial (1.2) are not correlated with each other and have unbounded dispersions.

Accordingly, putting

$$D_{\nu\mu} = \begin{cases} 0\ , & \nu \neq \mu \\ D_{\mu\mu}, & \nu = \mu \end{cases}$$

in (4.22), we have

$$\gamma_0 [n] D_{\mu\mu}\varphi_{\mu 0} + \gamma_1 [n] D_{\mu\mu}\varphi_{\mu 1} + \ldots + \gamma_r [n] D_{\mu\mu}\varphi_{\mu r} + \gamma_\mu [n] D_\xi$$
$$= D_\xi s_\mu D_{\mu\mu} - D_{\mu\mu} \sum_{\alpha=0}^{q} \{f_\alpha [n] - k_\alpha [n]\} \varphi_{\mu\alpha}$$
$$\mu = 0,\ 1, \ldots, r$$

Multiplying both sides of this equation by

$$\frac{1}{D_{\mu\mu}},\ \mu = 0,\ 1, \ldots, q$$

and then going to the limit with $D_{\mu\mu} \to \infty$, we find the required system of equations for $\gamma_\nu [n],\ \nu = 0,\ 1, \ldots, q$

$$\gamma_0 [n] \varphi_{\mu 0} + \gamma_1 [n] \varphi_{\mu 1} + \cdots + \gamma_r [n] \varphi_{\mu r}$$
$$= D_\xi s_\mu - \sum_{\alpha=0}^{q} \{f_\alpha [n] - k_\alpha [n]\} \varphi_{\mu\alpha} \qquad (4.24)$$
$$\mu = 0,\ 1. \ldots, r$$

Thus a method has been given for solving (4.3) when $\boldsymbol{A}$ and $\boldsymbol{B}$ coincide identically.

The case of the greatest practical interest is that where $\boldsymbol{A} \neq \boldsymbol{B}$.

7.4.2 Solution of Equation (4.3) in the general case

It was shown in Section 7.4.1 that (4.3) can be transformed into the following form:

$$\boldsymbol{A}[l] \cdot \sum_{m=n-N}^{l} W[n;\ m] \boldsymbol{B}[m] + \boldsymbol{B}[l] \cdot \sum_{m=n-N}^{n} W[n;\ m] \boldsymbol{A}[m]$$
$$- \boldsymbol{B}[l] \cdot \sum_{m=n-N}^{l} W[n;\ m] \boldsymbol{A}[m] + D_\xi W[n;\ l] = \boldsymbol{I}[n] \cdot \boldsymbol{B}[l]$$
$$n - N \leqslant l \leqslant n \qquad (4.25)$$

whence it follows that this equation can also be written

$$\left(\boldsymbol{I}[n]-\sum_{m=n-N}^{n} W[n;\ m]\,\boldsymbol{A}[m]\right)\cdot\boldsymbol{B}[l]$$

$$=\sum_{m=n-N}^{l} W[n;\ m](\boldsymbol{A}[l]\cdot\boldsymbol{B}[m]-\boldsymbol{B}[l]\cdot\boldsymbol{A}[m])+D_{\xi}W[n;\ l]$$

$$n-N\leqslant l\leqslant n$$

or

$$\left(\boldsymbol{I}[n]-\sum_{m=n-N}^{n} W[n;\ m]\,\boldsymbol{A}[m]\right)\cdot\boldsymbol{B}[l]$$

$$=\sum_{m=n-N}^{l} W[n;\ m]\,\boldsymbol{H}[l;\ m]+D_{\xi}W[n;\ l],\quad n-N\leqslant l\leqslant n \tag{4.26}$$

where the vector $\boldsymbol{H}$ is defined as

$$\boldsymbol{H}[l;\ m]=\boldsymbol{A}[l]\cdot\boldsymbol{B}[m]-\boldsymbol{B}[l]\cdot\boldsymbol{A}[m] \tag{4.27}$$

Accordingly, in the general case of $\boldsymbol{A}[n]\neq\boldsymbol{B}[n]$ the weighting function $W[n;\ l]$ of the optimal system satisfies (4.26). We shall now show that the solution of (4.26) can be represented as the scalar product of two $(q+1)$-dimensional vectors, i.e.

$$W[n;\ m]=\boldsymbol{U}[n]\cdot\boldsymbol{P}[m],\quad n-N\leqslant m\leqslant n \tag{4.28}$$

or, in scalar form,

$$W[n;\ m]=\sum_{\nu=0}^{q} u_{\nu}[n]\,p_{\nu}[m],\quad n-N\leqslant m\leqslant n \tag{4.29}$$

where $u_{\nu}[n]$ and $p_{\nu}[m]$ are the components of $\boldsymbol{U}[n]$ and $\boldsymbol{P}[m]$ respectively.

Substituting the expression for the weighting function $W[n;\ m]$ from (4.28) into (4.26) we obtain

$$\left(\boldsymbol{I}[n]-\sum_{m=n-N}^{n}(\boldsymbol{U}[n]\cdot\boldsymbol{P}[m])\cdot\boldsymbol{A}[m]\right)\cdot\boldsymbol{B}[l]$$

$$=\sum_{m=n-N}^{l}(\boldsymbol{U}[n]\cdot\boldsymbol{P}[m])\cdot\boldsymbol{H}[l;\ m]+D_{\xi}\boldsymbol{U}[n]\cdot\boldsymbol{P}[l]$$

$$n-N\leqslant l\leqslant n$$

or

$$\left(\boldsymbol{I}[n]-\sum_{m=n-N}^{n}(\boldsymbol{U}[n]\cdot\boldsymbol{P}[m])\cdot\boldsymbol{A}[m]\right)\cdot\boldsymbol{B}[l]$$
$$=\left(\sum_{m=n-N}^{l}\boldsymbol{P}[m]\cdot\boldsymbol{H}[l;\ m]+D_{\xi}\boldsymbol{P}[l]\right)\cdot\boldsymbol{U}[n] \qquad (4.30)$$
$$n-N\leqslant l\leqslant n$$

whence it follows that (4.30) holds when

$$\boldsymbol{B}[l]=\sum_{m=n-N}^{l}\boldsymbol{P}[m]\cdot\boldsymbol{H}[l;\ m]+D_{\xi}\boldsymbol{P}[l] \qquad (4.31)$$
$$n-N\leqslant l\leqslant n$$

$$\boldsymbol{U}[n]=\boldsymbol{I}[n]-\sum_{m=n-N}^{n}(\boldsymbol{U}[n]\cdot\boldsymbol{P}[m])\,\boldsymbol{A}[m] \qquad (4.32)$$

Consequently, (4.30) with the two unknown vectors $\boldsymbol{U}[n]$ and $\boldsymbol{P}[m]$ is decomposed into two equations, each containing only a single unknown. The determination of the weighting function thus reduces to the solution of (4.31) and (4.32). It is first necessary to determine $\boldsymbol{P}$ from (4.31) and then, substituting this vector into (4.32), to find $\boldsymbol{U}$.

To determine the components $p_{\alpha}[n]$, $\alpha=0, 1, \ldots, q$ of $\boldsymbol{P}[n]$, we write (4.31) in scalar form. From (4.27) we obviously have

$$\boldsymbol{H}[l;\ m]=\boldsymbol{A}[l]\cdot\boldsymbol{B}[m]-\boldsymbol{B}[l]\cdot\boldsymbol{A}[m]$$
$$=\sum_{\nu=0}^{q}a_{\nu}[l]\,b_{\nu}[m]-\sum_{\nu=0}^{q}b_{\nu}[l]\,a_{\nu}[m] \qquad (4.33)$$

Using (4.33) we write (4.31) as

$$\boldsymbol{B}[l]=\sum_{m=n-N}^{l}\boldsymbol{P}[m]\sum_{\nu=0}^{q}a_{\nu}[l]\,b_{\nu}[m] \qquad (4.34)$$
$$-\sum_{m=n-N}^{l}\boldsymbol{P}[m]\sum_{\nu=0}^{q}b_{\nu}[l]\,a_{\nu}[m]+D_{\xi}\boldsymbol{P}[l],\ n-N\leqslant l\leqslant n$$

or, in scalar form,

$$b_\alpha[l] = \sum_{m=n-N}^{l} p_\alpha[m] \sum_{\nu=0}^{q} a_\nu[l] b_\nu[m]$$

$$- \sum_{m=n-N}^{l} p_\alpha[m] \sum_{\nu=0}^{q} b_\nu[l] a_\nu[m] + D_\xi p_\alpha[l], \quad \alpha = 0, 1, \ldots q \tag{4.35}$$

$$n - N \leqslant l \leqslant n$$

It is not difficult to see that (4.35) allows the recurrence formulae to be obtained, whence we can determine the components $p_\alpha[m]$, $n - N \leqslant m \leqslant n$. Indeed, putting $l = n - N$ in (4.35), we have for any value $\alpha = 0, 1, \ldots, q$

$$b_\alpha[n-N] = p_\alpha[n-N] \sum_{\nu=0}^{q} a_\nu[n-N] b_\nu[n-N]$$

$$- p_\alpha[n-N] \sum_{\nu=0}^{q} b_\nu[n-N] a_\nu[n-N] + D_\xi p_\alpha[n-N]$$

whence we obtain

$$p_\alpha[n-N] = \frac{b_\alpha[n-N]}{D_\xi} \tag{4.36}$$

Now putting $l = n - N + 1$ in (4.35) we find

$$b_\alpha[n-N+1] = p_\alpha[n-N+1] \sum_{\nu=0}^{q} a_\nu[n-N+1] b_\nu[n-N]$$

$$+ p_\alpha[n-N+1] \sum_{\nu=0}^{q} a_\nu[n-N+1] b_\nu[n-N+1]$$

$$- p_\alpha[n-N] \sum_{\nu=0}^{q} b_\nu[n-N+1] a_\nu[n-N]$$

$$- p_\alpha[n-N+1] \sum_{\nu=0}^{q} b_\nu[n-N+1] a_\nu[n-N+1]$$

$$+ D_\xi p_\alpha[n-N+1]$$

or

$$b_\alpha[n-N+1]=p_\alpha[n-N]\sum_{\nu=0}^{q}(a_\nu[n-N+1]b_\nu[n-N]$$
$$-b_\nu[n-N+1]a_\nu[n-N])+D_\xi p_\alpha[n-N+1]$$

whence

$$p_\alpha[n-N+1]=$$
$$\frac{b_\alpha[n-N+1]-p_\alpha[n-N]\times\sum_{\nu=0}^{q}\{a_\nu[n-N+1]b_\nu[n-N]-b_\nu[n-N+1]a_\nu[n-N]\}}{D_\xi} \quad (4.37)$$

In the general case we have, from (4.35),

$$b_\alpha[m]=\sum_{l=n-N}^{m-1}p_\alpha[l]\sum_{\nu=0}^{q}(a_\nu[m]b_\nu[l]-b_\nu[m]a_\nu[l])+D_\xi p_\alpha[m]$$
$$\alpha=0,1,\ldots,q,\ n-N\leqslant m\leqslant n$$

whence

$$p_\alpha[m]=\frac{b_\alpha[m]-\sum_{l=n-N}^{m-1}p_\alpha[l]\sum_{\nu=0}^{q}(a_\nu[m]b_\nu[l]-b_\nu[m]a_\nu[l])}{D_\xi} \quad (4.38)$$
$$\alpha=0,1,\ldots,q,\ n-N\leqslant m\leqslant n$$

Thus, the components $p_\alpha[m]$, $\alpha=0,1,\ldots,q$ of $\boldsymbol{P}[m]$ are simply determined by means of the recurrence formula (4.38) in terms of the components of $\boldsymbol{A}[l]$ and $\boldsymbol{B}[l]$. At the same time it should be remembered that (4.31) is valid for $n-N\leqslant m\leqslant n$. Consequently, $\boldsymbol{P}[m]\equiv 0$ when $m<n-N$.

Let us now determine $\boldsymbol{U}[n]$. For this we return to (4.32), which we write in scalar form:

$$u_\alpha[n]=i_\alpha[n]-\sum_{m=n-N}^{n}a_\alpha[m]\sum_{\nu=0}^{q}u_\nu[n]p^\nu[m]$$

or, changing the order of summation,

$$u_\alpha[n]=i_\alpha[n]-\sum_{\nu=0}^{q}u_\nu[n]\sum_{m=n-N}^{n}a_\alpha[m]p_\alpha[m] \quad (4.39)$$

Since the components $p_\nu[n] (\nu = 0, 1, \ldots, q)$ of $\mathbf{P}[n]$ and the components $a_\alpha[m] (\alpha = 0, 1, \ldots, q)$ of $\boldsymbol{A}[m]$ are known, the quantities

$$\varphi_{\nu\alpha}[N] = \sum_{m=n-N}^{n} a_\alpha[m]\, p_\nu[m]$$
$$\alpha, \nu = 0, 1, \ldots, q \tag{4.40}$$

are not difficult to calculate. Accordingly, using (4.40), we obtain from (4.39) the following system of equations whence we can determine the components $u_\alpha[n]$ of $\boldsymbol{U}[n]$:

$$\begin{aligned} &(1 + \varphi_{00}) u_0[n] + \varphi_{10} u_1[n] + \ldots + \varphi_{q0} u_q[n] = i_0[n] \\ &\varphi_{01} u_0[n] + (1 + \varphi_{11}) u_1[n] + \ldots + \varphi_{q1} u_q[n] = i_1[n] \\ &\ldots\ldots\ldots\ldots\ldots\ldots\ldots\ldots\ldots \\ &\varphi_{0q} u_0[n] + \varphi_{1q} u_1[n] + \ldots + (1 + \varphi_{qq}) u_q[n] = i_q[n] \end{aligned} \tag{4.41}$$

The problem of finding the weighting function is not solved completely by determining $u_\nu[n]$ from (4.41). In fact, the $u_\nu[n]$ depend on the functions

$$i_\nu[n] = f_\nu[n] + \gamma_\nu^*[n]$$

where $f_\nu[n]$ are the components of $\boldsymbol{F}[n]$, and $\gamma_\nu^*[n]$ are the unknown components of $\Gamma^*[n]$.

Equation (2.3) must be used to determine $\gamma_\nu^*[n]$; to do this (4.28) must be substituted into it. This yields a system of algebraic equations, by solving which the unknowns $\gamma_\nu^*[n]$, $\nu = 0, 1, \ldots, q$, are determined. This in fact completes the solution.

It has thus been shown that the optimal weighting function which satisfies (4.26) is a scalar product of two vectors, i.e.

$$W[n; m] = \boldsymbol{U}[n] \cdot \boldsymbol{P}[m]$$
$$n - N \leqslant m \leqslant n$$

where the components of $\boldsymbol{P}[m]$ are determined from the recurrence formula (4.38), whilst the components of $\boldsymbol{U}[n]$ are determined by solving (4.41).

We point out that the determination of the optimal system when the coefficients d_ν, $\nu = 0,1,\ldots, r$ of the polynomial (1.2) are non-random quantities, with unknown arbitrary values, arises as a particular case from the problem just considered. In this case we must assume that

$$D_{\nu\mu} = \begin{cases} 0, & \nu \neq \mu \\ D_{\mu\mu}, & \nu = \mu \end{cases} \tag{4.42}$$

and

$$D_{\mu\mu} \to \infty, \ \mu = 0, 1, \ldots, r \tag{4.43}$$

i.e. d_ν are random quantities not correlated with each other, which have infinitely large dispersions.

7.5 EXPRESSION FOR THE MINIMUM ERROR DISPERSION

From (1.9) we have

$$D = M\left[\left\{\sum_{m=-\infty}^{\infty} W_{id}[n;\ n-m]\,S[n-m] - \sum_{m=0}^{N} W[n;\ n-m]\,Y[n-m]\right\}^2\right] \tag{5.1}$$

where in accordance with (1.1) and (1.2)

$$S[n-m] = \sum_{\nu=0}^{r} d_\nu m^\nu + X[n-m] \tag{5.2}$$

$$Y[n-m] = S[n-m] + \xi[n-m] \tag{5.3}$$

Bearing in mind the condition for minimum dispersion D expressed by (2.1) we find from (5.1), when (5.2) and (5.3) are taken into account, that

$$\begin{aligned} D_{\min} = M\Bigg[\Bigg\{&\sum_{m=-\infty}^{\infty} W_{id}[n;\ n-m]\left(\sum_{\nu=0}^{r} d_\nu m^\nu + X[n-m]\right) \\ &- \sum_{m=0}^{N} W[n;\ n-m]\left(\sum_{\nu=0}^{r} d_\nu m^\nu + X[n-m] + \xi[n-m]\right)\Bigg\} \\ &\times \sum_{l=-\infty}^{\infty} W_{id}[n;\ n-l]\left(\sum_{\mu=0}^{r} d_\mu l^\mu + X[n-l]\right)\Bigg] \end{aligned} \tag{5.4}$$

We shall carry out mathematical expectation on the right-hand side of (5.4). Using (1.3) to (1.6), and assuming at the same time that the random functions X and ξ are not correlated with each other, we obtain

$$D_{\min} = \sum_{m=-\infty}^{\infty} W_{id}[n;\ n-m] \sum_{l=-\infty}^{\infty} W_{id}[n;\ n-l] \times\left(\sum_{\nu,\ \mu=0}^{r} D_{\nu\mu} m^{\nu} l^{\mu} + K_x[n-m;\ n-l]\right) - \sum_{m=0}^{N} W[n;\ n-m] \sum_{l=-\infty}^{\infty} W_{id}[n;\ n-l] \times\left(\sum_{\nu,\ \mu=0}^{r} D_{\nu\mu} m^{\nu} l^{\mu} + K_x[n-m;\ n-l]\right) \quad (5.5)$$

The expression (5.5) thus obtained is rather general, and unsuitable for practical calculations. In each concrete case (i.e. for each particular form of the weighting function $W_{id}[n;\ m]$ of the ideal transforming system) it is advisable to use its own expression for the minimum dispersion. These expressions can be found easily.

For example, suppose that a system must carry out optimal smoothing (filtering) of the input signal. In this case we have

$$\sum_{m=-\infty}^{\infty} W_{id}[n;\ n-m]\, S[n-m] = S[n] = d_0[n] + X[n] \quad (5.6)$$

and the expression (5.4) for the minimum dispersion assumes the following form:

$$D_{s\,\min} = M\left[\left\{ d_0 + X[n] - \sum_{m=0}^{N} W[n;\ n-m]\left(\sum_{\nu=0}^{r} d_{\nu} m^{\nu} + X[n-m] + \xi[n-m]\right)\right\}(d_0 + X[n])\right] \quad (5.7)$$

where the index s ($D_{s\ \min}$) denotes that the system determines the smoothed value of the signal.

Carrying out mathematical expectation on the right-hand side of (5.7), we find the final expression which gives the dispersion at the output of the optimal smoothing system:

$$D_{s\ \min} = D_{00} + K_x[n;\ n] - \sum_{m=0}^{N} W[n;\ n-m] \times (D_{00} + K_x[n-m;\ n]) \tag{5.8}$$

Let us suppose now that the system must carry out optimal prediction of the input signal over the time $\bar{t} = N_0$. In this case the required signal

$$\sum_{m=-\infty}^{\infty} W_{id}[n;\ n-m]\, S[n-m] = S[n+N_0] = \sum_{\nu=0}^{r} d_\nu [n] N_0^\nu + X[n+N_0] = S_y \tag{5.9}$$

and the expression (5.4) for the minimum dispersion assumes the form

$$D_{s_y \min} = M\Bigg[\Bigg\{\sum_{\nu=0}^{r} d_\nu N_0^\nu + X[n+N_0] - \sum_{m=0}^{N} W[n;\ n-m]\left(\sum_{\nu=0}^{r} d_\nu m^\nu + X[n-m] + \xi[n-m]\right)\Bigg\} \times \left(\sum_{\mu=0}^{r} d_\mu N_0^\mu + X[n+N_0]\right)\Bigg]$$

whence we find that

$$D_{s_y \min} = \sum_{\nu,\mu=0}^{r} D_{\nu\mu} N_0^{\nu+\mu} + K_x[n+N_0;\ n+N_0] - \sum_{m=0}^{N} W[n;\ n-m]\left(\sum_{\nu,\mu=0}^{r} D_{\nu\mu} m^\nu N_0^\mu + K_x[n-m;\ n+N_0]\right) \tag{5.10}$$

Expression (5.10) thus gives the minimum dispersion of the predicting system. Putting $N_0 = 0$ in this expression, we find the formula for the minimum dispersion of the smoothing system, i.e. for $N_0=0$ the expression (5.10) coincides with (5.8), obtained in another way.

If the system is to determine the rate of change of the input signal at the instant $\bar{t}=n$, then

$$\sum_{m=-\infty}^{\infty} W_{\text{id}}[n;\ n-m]\, S[n-m]=d_1[n]+\dot{X}[n]$$

and (5.4) assumes the form

$$D_{\dot{s}\,\min}=M\Bigg[\Bigg\{d_1[n]+\dot{X}[n] -\sum_{m=0}^{N} W[n;\ n-m]\Bigg(\sum_{\nu=0}^{r} d_\nu[n]\, m^\nu + X[n-m]+\xi[n-m]\Bigg)\Bigg\}(d_1[n]+\dot{X}[n])\Bigg]$$

whence

$$D_{\dot{s}\,\min}=D_{11}+K_{\dot{x}}[n;\ n] -\sum_{m=0}^{N} W[n;\ n-m]\,(D_{11}\, m+K_{x\dot{x}}[n-m;\ n]) \tag{5.11}$$

$$K_{\dot{x}}[n;\ n]=M[\dot{X}[n]\,\dot{X}[n]]$$

$$K_{x\dot{x}}[n-m;\ n]=M[X[n-m]\,\dot{X}[n]]$$

Similarly we could find the corresponding expressions giving the minimum error dispersion in other concrete cases.

The expressions obtained above allow the minimum dispersion to be found when the input signal $g[n]$ has random coefficients d_ν, $\nu=0,1,\ldots,r$. Let us now consider another case, when d_ν, $\nu=0,1,\ldots,r$ are not random quantities. We proceed from (5.1), which we write

$$D=M\Bigg[\Bigg\{\sum_{m=-\infty}^{\infty} W_{\text{id}}[n;\ n-m](g[n-m]+X[n-m]) -\sum_{m=0}^{N} W[n;\ n-m](g[n-m]+X[n-m]$$

$$+\xi[n-m])\Bigg\}\Bigg\{\sum_{l=-\infty}^{\infty} W_{\text{id}}[n;\ n-l](g[n-l]+X[n-l])$$

$$-\sum_{l=0}^{N} W[n;\ n-l](g[n-l]+X[n-l]+\xi[n-l])\Bigg\}\Bigg] \tag{5.12}$$

Since in this case the weighting function of the optimal system must satisfy the $r+1$ conditions (2.6), i.e.

$$\sum_{m=-\infty}^{\infty} W_{\text{id}}[n; n-m]m^{\nu} = \sum_{m=n-N}^{n} W[n;\ m]m^{\nu},\ \nu=0,\ 1,\ldots,\ r$$

or, which amounts to the same thing,

$$\sum_{m=-\infty}^{\infty} W_{\text{id}}[n;\ n-m]\,g[n-m] = \sum_{m=0}^{N} W[n;\ n-m]\,g[n-m]$$

we obtain from (5.12) the following expression:

$$D=M\Bigg[\Bigg\{\sum_{m=-\infty}^{\infty} W_{\text{id}}[n;\ n-m]\,X[n-m]$$

$$-\sum_{m=0}^{N} W[n;\ n-m](X[n-m]+\xi[n-m])\Bigg\}$$

$$\times\Bigg\{\sum_{l=-\infty}^{\infty} W_{\text{id}}[n;\ n-l]\,X[n-l]- \tag{5.13}$$

$$-\sum_{l=0}^{N} W[n,\ n-l](X[n-l]+\xi[n-l])\Bigg\}\Bigg]$$

Carrying out mathematical expectation on the right-hand side of (5.13) we have after certain transformations

$$D=\sum_{l=0}^{N} W[n;\ n-l]\Bigg[\sum_{m=0}^{N} W[n;\ n-m](K_x[n-m;\ n-l]$$

$$+K_{\xi}[n-m;\ n-l])-\sum_{m=-\infty}^{\infty} W_{\text{id}}[n; n-m]\,K_x[n-m; n-l]$$

$$-\sum_{m=-\infty}^{\infty} W_{\text{id}}[n;\ n-m]\,K_x[n-m;\ n-l]\Bigg]$$

$$+\sum_{m=-\infty}^{\infty} W_{\text{id}}[n;\ n-m] \sum_{l=-\infty}^{\infty} W_{\text{id}}[n;\ n-l]\, K_x[n-m;\ n-l]$$

Since on the basis of (2.5)

$$\sum_{m=0}^{N} W[n;\ n-m]\,(K_x[n-m;\ n-l]+K_\xi[n-m;\ n-l])$$

$$-\sum_{m=-\infty}^{\infty} W_{\text{id}}[n;\ n-m]\, K_x[n-m;\ n-l]$$

$$=\sum_{\nu=0}^{r} \gamma_\nu[n]\, l^\nu,\quad 0\leqslant l\leqslant N$$

we find from (5.14) the expression for the minimum dispersion

$$D_{\min}=\sum_{l=-\infty}^{\infty} W_{\text{id}}[n;\ n-l] \sum_{m=-\infty}^{\infty} W_{\text{id}}[n;\ n-m]\, K_x[n-m;\ n-l]-\sum_{l=0}^{N} W[n;\ n-l] \sum_{m=-\infty}^{\infty} W_{\text{id}}[n;\ n-m]\, K_x[n-m;\ n-l]+\sum_{\nu=0}^{r}\gamma_\nu[n]\sum_{l=0}^{N} W[n;\ n-l]\, l^\nu \tag{5.14}$$

This expression gives the error dispersion of an optimal system having an arbitrary transformation operator. Like (5.5), however, it is of little use for practical calculations. Accordingly, it is advisable to find the expressions for the minimum dispersion in each individual case.

For example, suppose that the required system is to carry out optimal smoothing of the input signal. In this case

$$\sum_{m=-\infty}^{\infty} W_{\text{id}}[n;\ m]\, X[m]=X[n]$$

and the expression for the dispersion assumes the form

$$D_s=M\left[\left\{X[n]-\sum_{m=0}^{N} W[n;\ n-m]\,(X[n-m]\right.\right.$$

$$+\xi[n-m])\Big\}\times\Big\{X[n]-\sum_{l=0}^{N}W[n;\ n-l](X[n-l]$$
$$+\xi[n-l])\Big\}\Big] \tag{5.15}$$

whence, after carrying out mathematical expectation, we obtain

$$D_s=K_x[n;n]-2\sum_{m=0}^{N}W[n;\ n-m]K_x[n;\ n-m]$$
$$+\sum_{m=0}^{N}W[n;\ n-m]\sum_{l=0}^{N}W[n;\ n-l](K_x[n-l;\ n-m] \tag{5.16}$$
$$+K_\xi[n-l;\ n-m])$$

Since in the given case

$$\sum_{m=0}^{N}W[n;\ n-m](K_x[n-l;\ n-m]+K_\xi[n-l;\ n-m])$$
$$-K_x[n-l;\ n]=\sum_{\nu=0}^{r}\gamma_\nu[n]l^\nu$$

which follows from (2.5) when $W_{id}[n;\ m]=\sigma[n-m]$, we find from (5.16) the expression for the minimum error dispersion of the smoothing system

$$D_{s\,\min}=K_x[n;\ n]-\sum_{l=0}^{N}W[n;\ n-l]K_x[n;\ n-l]$$
$$+\sum_{\nu=0}^{r}\gamma_\nu[n]\sum_{l=0}^{N}W[n;\ n-l]l^\nu \tag{5.17}$$

This expression can be obtained directly from the general expression (5.14), by replacing the weighting function $W_{id}[n;\ m]$of the ideal transforming system by the σ-function. Thus, putting $W_{id}[n;\ m]=\sigma[n-m]$ in (5.14), we have

$$D_{s\,\min}=\sum_{l=-\infty}^{\infty}\sigma[l]\sum_{m=-\infty}^{\infty}\sigma[m]K_x[n-m;\ n-l]$$

$$-\sum_{l=0}^{N} W[n;\, n-l] \sum_{m=-\infty}^{\infty} \sigma[m] K_x[n-m;\, n-l]$$

$$+\sum_{\nu=0}^{r} \gamma_\nu[n] \sum_{l=0}^{N} W[n;\, n-l]\, l^\nu$$

whence, when the property of the σ-function is taken into account,

$$D_{s\,\min} = K_x[n;\, n] - \sum_{l=0}^{N} W[n;\, n-l] K_x[n;\, n-l]$$

$$+\sum_{\nu=0}^{r} \gamma_\nu[n] \sum_{l=0}^{N} W[n;\, n-l]\, l^\nu$$

which coincides with (5.17).

Putting

$$W_{\mathrm{id}}[n;\, m] = \sigma[n-m+N_0] \tag{5.18}$$

in (5.14) we obtain the expression for the error dispersion of an optimal system which predicts the input signal over the time $t=N_0$. Thus, substituting the expression of $W_{\mathrm{id}}[n;\, m]$ from (5.18) into (5.14), we have

$$D_{s_y\min} = \sum_{l=-\infty}^{\infty} \sigma[l+N_0] \sum_{m=-\infty}^{\infty} \sigma[m+N_0] K_x[n-m;\, n-l]$$

$$-\sum_{l=0}^{N} W[n;\, n-l] \sum_{m=-\infty}^{\infty} \sigma[m+N_0] K_x[n-m;\, n-l]$$

$$+\sum_{\nu=0}^{r} \gamma_\nu[n] \sum_{l=0}^{N} W[n;\, n-l]\, l^\nu$$

whence, when the property of the σ-function is used, we obtain

$$D_{s_y\min} = K_x[n+N_0;\, n+N_0] - \sum_{l=0}^{N} W[n;\, n-l] K_x[n+N_0;\, n$$

$$-l] + \sum_{\nu=0}^{r} \gamma_\nu[n] \sum_{l=0}^{N} W[n;\, n-l]\, l^\nu \tag{5.19}$$

Putting the time of prediction $\overline{t}=N_0=0$ in (5.19) we obtain an expression coinciding with (5.17).

Let us find now the expression for the minimum error dispersion of a system determining the rate of change of the input signal at the instant $\overline{t}=n$. We then have

$$\sum_{m=-\infty}^{\infty} W_{1d}[n;\ m]\, X[m] = \dot{X}[n] \tag{5.20}$$

and the expression (5.13) for the error dispersion of determining the rate assumes the form

$$D_{\dot{s}} = M\Bigg[\Bigg\{\dot{X}[n] - \sum_{m=0}^{N} W[n;\ n-m]\,(X[n-m] + \xi[n-m])\Bigg\} \times \Bigg\{\dot{X}[n] - \sum_{l=0}^{N} W[n;\ n-l]\,(X[n-l] + \xi[n-l])\Bigg\} \tag{5.21}$$

whence, after carrying out mathematical expectation, we obtain

$$\begin{aligned} D_{\dot{s}} = {} & K_{\dot{x}}[n;\ n] - \sum_{l=0}^{N} W[n;\ n-l]\, K_{\dot{x}x}[n;\ n-l] \\ & - \sum_{m=0}^{N} W[n;\ n-m]\, K_{x\dot{x}}[n-m;\ n] \\ & + \sum_{l=0}^{N} W[n;\ n-l] \sum_{m=0}^{N} W[n;\ n-m]\,(K_x[n-m;\ n-l] \\ & + K_\xi[n-m;\ n-l]) \end{aligned} \tag{5.22}$$

It is not difficult to show that the weighting function of a system computing the rate of change of the input signal satisfies the following equation

$$\begin{aligned} & \sum_{m=0}^{N} W[n;\ n-m]\,(K_x[n-m;\ n-l] + K_\xi[n-m;\ n-l]) \\ & = K_{\dot{x}x}[n;\ n-l] + \sum_{\nu=0}^{r} \gamma_\nu[n]\, l^\nu,\ 0 \leqslant l \leqslant N \end{aligned} \tag{5.23}$$

and the $r+1$ additional conditions

$$\sum_{m=0}^{N} W[n;\ n-m]=0$$

$$\sum_{m=0}^{N} mW[n;\ n-m]=-1 \tag{5.24}$$

$$\sum_{m=0}^{N} m^{\nu}W[n;\ n-m]=0;\ \nu=2,3,\dots,r$$

Equation (5.23) is easily obtained from the condition

$$M\left[\left\{\dot{S}[n]-\sum_{m=0}^{N} W[n;\ n-m]Y[n-m]\right\}Y[n-l]\right]=0$$
$$0\leqslant l\leqslant N \tag{5.25}$$

which is necessary and sufficient for the minimum dispersion of determining the rate of change of the input signal.

The additional conditions (5.24) are determined from the condition that there are no systematic errors in measuring the rate. This condition is expressed as follows:

$$g[n]\equiv \sum_{m=n-N}^{n} W[n;\ m]g[m] \tag{5.26}$$

Putting $n-l=m$ in (5.26) we have

$$\sum_{m=n-N}^{n} W[n;m]g[m]=\sum_{l=0}^{N} W[n;\ n-l]g[n-l]$$
$$=g[n]\sum_{l=0}^{N} W[n;\ n-l]-\dot{g}[n]\sum_{l=0}^{N} W[n:\ n-l]l$$
$$+\frac{1}{2}\ddot{g}[n]\sum_{l=0}^{N} W[n;\ n-l]l^2-\dots \tag{5.27}$$
$$\dots+(-1)^r\frac{1}{r!}g^{(r)}[n]\sum_{l=0}^{N} W[n;\ n-l]l^r$$

Comparing (5.26) and (5.27) we obtain (5.24).

Let us now return to (5.22), which we write

$$D_{\dot{s}} = K_{\dot{x}}[n;n] - \sum_{l=0}^{N} W[n;\, n-l]\, K_{\dot{x}x}[n;\, n-l]$$
$$-\sum_{l=0}^{N} W[n;\, n-l]\left\{\sum_{m=0}^{N} W[n;\, n-m]\,(K_x[n-m;\, n-l] \right. \tag{5.28}$$
$$\left. + K_{\xi}[n-m;\, n-l]) - K_{\dot{x}x}[n;\, n-l]\right\}$$

Using (5.27) we obtain from (5.28) the expression for the minimum dispersion

$$D_{\dot{s}\,\min} = K_{\dot{x}}[n;\, n] - \sum_{l=0}^{N} W[n;\, n-l]\, K_{\dot{x}x}[n;\, n-l]$$
$$-\sum_{\nu=0}^{r} \gamma_{\nu}[n] \sum_{l=0}^{N} W[n;\, n-l]\, l^{\nu} \tag{5.29}$$

Similarly we can find the corresponding expressions for the minimum error dispersion in other cases.

We shall now solve a number of concrete problems of determining optimal systems by these methods.

7.6 WEIGHTING FUNCTION OF AN OPTIMAL REPRODUCING SYSTEM

In this section we shall determine the weighting function of an optimal sampled data system which reproduces the useful signal:

$$S[n] = d_0 + d_1 n + X[n] \tag{6.1}$$

where d_0 and d_1 are random coefficients, such that

$$M[d_0] = M[d_1] = 0$$
$$M[d_i^2] = D_{ii}, \quad i = 0,1 \tag{6.2}$$
$$M[d_{\nu} d_{\mu}] = 0, \quad \nu \neq \mu$$

whilst X is a non-stationary random function with zero mathematical expectation, having a correlation function of the form

$$K_x[n;m]=1+\alpha^2nm \tag{6.3}$$

At the same time it is assumed that white noise ξ having the correlation function

$$K_\xi[n;m]=D_\xi\sigma[n-m] \tag{6.4}$$

is superimposed on the useful signal (6.1).

Following the method given in the preceding sections we must first introduce a number of vectors. Let us now define these vectors.

1. **The vectors $\boldsymbol{A}[n]$ and $\boldsymbol{B}[n]$.** From (3.10) we have

$$K_x[n;m]=\boldsymbol{A}[n]\cdot\boldsymbol{B}[m]=\sum_{\nu=0}^{q}a_\nu[n]b_\nu[m],\ n\geqslant m$$

Using (6.3), we find that

$$1+\alpha^2nm=\sum_{\nu=0}^{1}a_\nu[n]b_\nu[m]$$

whence we obtain

$$\begin{aligned}a_0[n]=1,\quad a_1[n]=\alpha n\\ b_0[m]=1,\quad b_1[m]=\alpha m\end{aligned} \tag{6.5}$$

Accordingly

$$\boldsymbol{A}[n]=\{1;\ \alpha n\},\ \boldsymbol{B}[m]=\{1;\ \alpha m\} \tag{6.6}$$

2. **The vector $\boldsymbol{F}[n]$.** Since in the given case (i.e. reproduction) the weighting function of the ideal system is

$$W_{\text{id}}[n;\ m]=\sigma[n-m]$$

then we have

$$\sum_{m=-\infty}^{\infty}W_{\text{id}}[n;m]K_x[l;m]=\sum_{m=-\infty}^{\infty}\sigma[n-m]K_x[l;m]=K_x[l;n]$$

Consequently, when (3.12) is taken into account, we obtain

$$K_x[n; m] = \boldsymbol{F}[n]\cdot\boldsymbol{B}[m] = 1 + \alpha^2 nm = \sum_{\nu=0}^{1} f_\nu[n]\, b_\nu[m]$$

whence, finally,

$$\boldsymbol{F}[n] = \{f_0[n];\ f_1[n]\} = \{1;\ \alpha n\} \tag{6.7}$$

3. **The vector** $\Gamma^*[n]$. From (3.16), when (6.1) is taken into account, we have

$$\Gamma^*[n]\cdot\boldsymbol{B}[l] = \sum_{\nu=0}^{r} \gamma_\nu^*[n]\, b_\nu[l] = \sum_{\nu=0}^{1} \gamma_\nu^*[n]\, b_\nu[l]$$

whence

$$\Gamma^*[n] = \{\gamma_0^*[n];\ \gamma_1^*[n]\} \tag{6.8}$$

4. **The vector** $I[n]$. According to (3.20), $\boldsymbol{I}$ is formed by the sum of $\boldsymbol{F}$ and Γ^*:

$$\boldsymbol{I}[n] = \boldsymbol{F}[n] + \Gamma^*[n]$$

Consequently, with (6.7) and (6.8) taken into account, we find

$$\boldsymbol{I}[n] = \{i_0[n];\ i_1[n]\} = \{1 + \gamma_0^*[n];\ \alpha n + \gamma_1^*[n]\} \tag{6.9}$$

5. **The vector** $K[n]$. According to (4.7),

$$\boldsymbol{K}[n] = \sum_{m=n-N}^{n} W[n; m]\,\boldsymbol{A}[m]$$

Consequently, when (6.6) is used,

$$\boldsymbol{K}[n] = \{k_0[n];\ k_1[n]\} \tag{6.10}$$

Since, in the case under consideration, $\boldsymbol{A}[n]$ and $\boldsymbol{B}[n]$ coincide, the weighting function of the optimal system is determined in accordance with (4.16) by the following expression:

$$W[n;l] = \frac{1}{D_\xi} \sum_{\nu=0}^{1} \{f_\nu[n] + \gamma_\nu^*[n] - k_\nu[n]\} a_\nu[l] \qquad (6.11)$$

$$n - N \leqslant l \leqslant n$$

where the components $k_\nu[n]$ $(\nu = 0,1)$ of $\boldsymbol{K}[n]$ are found by solving the system of algebraic equations (4.15), which in this case has the form

$$\begin{aligned} (\beta_{00} + D_\xi) k_0[n] + \beta_{10} k_1[n] &= i_0[n]\beta_{00} + i_1[n]\beta_{10} \\ \beta_{01} k_0[n] + (\beta_{11} + D_\xi) k_1[n] &= i_0[n]\beta_{01} + i_1[n]\beta_{11} \end{aligned} \qquad (6.12)$$

The coefficients $\beta_{\nu\mu}$ $(\nu, \mu = 0,1)$ in (6.12) are determined, in accordance with (4.13), from

$$\beta_{\nu\mu} = \sum_{l=n-N}^{n} a_\nu[l] a_\mu[l], \quad (\nu,\mu = 0,1) \qquad (6.13)$$

Thus, substituting the values of $a_\nu[n]$ from (6.6) into (6.13) we have

$$\beta_{00} = \sum_{l=n-N}^{n} 1 = N+1$$

$$\beta_{01} = \sum_{l=n-N}^{n} \alpha l = \alpha \frac{2n(N+1) - N(N+1)}{2}$$

$$\beta_{10} = \sum_{l=n-N}^{n} \alpha l = \alpha \frac{2n(N+1) - N(N+1)}{2} \qquad (6.14)$$

$$\beta_{11} = \sum_{l=n-N}^{n} \alpha^2 l^2 = \alpha^2 \frac{6n^2(N+1) - 6nN(N+1) + N(N+1)(2N+1)}{6}.$$

Let us now determine the components of $\boldsymbol{K}[n]$. When (6.9) is used the solution of (6.12) gives

$$k_0[n] = \frac{\begin{vmatrix} (1+\gamma_0^*[n])\beta_{00} + (\alpha n + \gamma_1^*[n])\beta_{10} & \beta_{11} \\ (1+\gamma_0^*[n])\beta_0 + (\alpha n + \gamma_1^*[n])\beta_{11} & \beta_{11} + D_\xi \end{vmatrix}}{\Delta} \qquad (6.15)$$

$$k_1[n] = \frac{\begin{vmatrix} \beta_{00} + D_\xi & (1+\gamma_0^*[n])\beta_{00} + (\alpha n + \gamma_1^*[n])\beta_{10} \\ \beta_{01} & (1+\gamma_0^*[n])\beta_{01} + (\alpha n + \gamma_1^*[n])\beta_{11} \end{vmatrix}}{\Delta} \tag{6.16}$$

where

$$\Delta = \begin{vmatrix} \beta_{00} + D_\xi & \beta_{10} \\ \beta_{01} & \beta_{11} + D_\xi \end{vmatrix} \tag{6.17}$$

If for brevity we introduce the following notation:

$$\begin{aligned} \omega_{00} &= \frac{\beta_{00}(\beta_{11} + D_\xi) - \beta_{11}\beta_{01}}{\Delta} \\ \omega_{10} &= \frac{\beta_{10}(\beta_{11} + D_\xi) - \beta_{11}^2}{\Delta} \\ \omega_{01} &= \frac{\beta_{01}(\beta_{00} + D_\xi) - \beta_{00}\beta_{01}}{\Delta} \\ \omega_{11} &= \frac{\beta_{11}(\beta_{00} + D_\xi) - \beta_{01}\beta_{10}}{\Delta} \end{aligned} \tag{6.18}$$

then from (6.15) and (6.16) we have

$$\begin{aligned} k_0[n] &= \omega_{00}(1+\gamma_0^*[n]) + \omega_{10}(\alpha n + \gamma_1^*[n]) \\ k_1[n] &= \omega_{01}(1+\gamma_0^*[n]) + \omega_{11}(\alpha n + \gamma_1^*[n]) \end{aligned} \tag{6.19}$$

Substituting (6.19) into the expression for the weighting function (6.11), we have when (6.7) and (6.5) are used

$$\begin{gathered} W[n;l] = \frac{1}{D_\xi}\{(1+\gamma_0^*[n]) - \omega_{00}(1+\gamma_0^*[n]) \\ - \omega_{10}(\alpha n + \gamma_1^*[n])\} + \frac{1}{D_\xi}\{(\alpha n + \gamma_1^*[n]) - \omega_{01}(1+\gamma_0^*[n]) \\ - \omega_{11}(\alpha n + \gamma_1^*[n])\}\alpha l \\ n - N \leqslant l \leqslant n \end{gathered} \tag{6.20}$$

As follows from (6.20), we must find the components $\gamma_\nu^*[n]$ ($\nu = 0,1$) of $\Gamma^*[n]$, in order to complete the determination of the weighting function. It was shown in Section 7.4 that the components $\gamma_\nu^*[n]$ are determined from (4.22). The latter is obtained from (2.3) after substituting the expression for the weighting function into it. In the present case (4.22) assumes the form

$$\gamma_0^*[n]\sum_{\nu=0}^{1} D_{\mu\nu}\varphi_{\nu 0}+\gamma_1^*[n]\sum_{\nu=0}^{1} D_{\nu\mu}\varphi_{\nu 1}+\gamma_\mu^*[n]D_\xi$$
$$=D_\xi\sum_{\nu=0}^{1} D_{\nu\mu}s_\nu-\sum_{\nu=0}^{1} D_{\nu\mu}\sum_{\alpha=0}^{1}\{f_\alpha[n]-k_\alpha[n]\}\varphi_{\nu\alpha} \tag{6.21}$$
$$\mu=0,1$$

where on the basis of (4.17)

$$s_\nu=\sum_{m=-\infty}^{\infty} W_{\mathrm{id}}[n;m]m^\nu=\sum_{m=-\infty}^{\infty}\sigma[n-m]\,m^\nu=n^\nu \tag{6.22}$$

and in accordance with (4.20)

$$\varphi_{\nu\alpha}=\sum_{m=n-N}^{n} a_\alpha[m](n-m)^\nu,\quad (\nu,\alpha=0,1) \tag{6.23}$$

Further, assuming that in this case

$$D_{10}=D_{01}=0 \tag{6.24}$$

and substituting the values of f_ν and k_ν from (6.7) and (6.19) respectively into (6.21), we obtain for $\gamma_\nu^*[n]$

$$\Omega_{00}\gamma_0^*[n]+\Omega_{10}\gamma_1^*[n]=R_0[n]$$
$$\Omega_{01}\gamma_0^*[n]+\Omega_{11}\gamma_1^*[n]=R_1[n] \tag{6.25}$$

where

$$\Omega_{00}=D_{00}(\varphi_{00}-\omega_{00}\varphi_{00}-\omega_{01}\varphi_{01})+D_\xi$$
$$\Omega_{10}=D_{00}(\varphi_{01}-\omega_{10}\varphi_{00}-\omega_{11}\varphi_{01})$$
$$\Omega_{01}=D_{11}(\varphi_{10}-\omega_{00}\varphi_{01}-\omega_{01}\varphi_{11})$$
$$\Omega_{11}=D_{11}(\varphi_{11}-\omega_{10}\varphi_{01}-\omega_{11}\varphi_{11})+D_\xi$$
$$R_0[n]=D_{00}s_0D_\xi-D_{00}[\varphi_{00}(1-\omega_{00}-\omega_{10}an)+\varphi_{01}(an-\omega_{01}-\omega_{11}an)] \tag{6.26}$$
$$R_1[n]=D_{11}s_1D_\xi-D_{11}[\varphi_{01}(1-\omega_{00}-\omega_{10}an)+\varphi_{11}(an-\omega_{01}-\omega_{11}an)]$$

Solving (6.25) and then substituting the values $\overset{*}{\gamma}_0[n]$ and $\overset{*}{\gamma}_1[n]$ thus found into (6.20), which gives the weighting function, we have

$$W[n;l]=w_0[n]+w_1[n]\cdot l, \quad n-N\leqslant l\leqslant n \tag{6.27}$$

where the functions $w_0[n]$ and $w_1[n]$ are determined as follows:

$$w_0[n]=\frac{1}{D_\xi}\left[(1-\omega_{00}-\omega_{10}\,\alpha n)+(1-\omega_{00})\frac{\Delta_0}{\Delta}-\omega_{10}\frac{\Delta_1}{\Delta}\right] \tag{6.28}$$

$$w_1[n]=\frac{\alpha}{D_\xi}\left[(\alpha n-\omega_{01}-\omega_{11}\alpha n)-\omega_{01}\frac{\Delta_0}{\Delta}+(1-\omega_{11})\frac{\Delta_1}{\Delta}\right] \tag{6.29}$$

$$\Delta=\begin{vmatrix}\Omega_{00} & \Omega_{10}\\ \Omega_{01} & \Omega_{11}\end{vmatrix} \tag{6.30}$$

$$\Delta_0=\begin{vmatrix}R_0[n] & \Omega_{10}\\ R_1[n] & \Omega_{11}\end{vmatrix},\quad \Delta_1=\begin{vmatrix}\Omega_{00} & R_0[n]\\ \Omega_{01} & R_1[n]\end{vmatrix} \tag{6.31}$$

The minimum dispersion can be found from (5.8), which in this case gives

$$D_{s\,\min}=D_{00}+1+\alpha^2n^2-\sum_{m=0}^{N}W[n;\,n-m](1+D_{00}+\alpha(n-m)n) \tag{6.32}$$

Substituting the expression of the weighting function from (6.27) into (6.32), we find that

$$\begin{aligned}D_{s\,\min}&=D_{00}+1+\alpha^2n^2-w_0[n](1+D_{00})\sum_{m=0}^{N}1\\&-w_0[n]\,\alpha n\sum_{m=0}^{N}(n-m)-w_1[n](1+D_{00})\sum_{m=0}^{N}(n-m)\\&-w_1[n]\,\alpha n\sum_{m=0}^{N}(n-m)^2\end{aligned} \tag{6.33}$$

After carrying out the summation in (6.33) we find

$$D_{s\,\min}=D_{00}+1+\alpha^2n^2-[w_0[n]\,an$$
$$+w_1[n](1+D_{00})]\,\frac{2n(N+1)-N(N+1)}{2} \qquad (6.34)$$
$$-w_1[n]\,an\,\frac{6n^2(N+1)-6n(N+1)N+N(N+1)(2N+1)}{6}$$

The above solution of the problem is valid when the coefficients of the polynomial (6.1) are not random quantities, and can assume arbitrary values. In this case the weighting function of the optimal smoothing system is determined from (6.27) to (6.31) with the condition that Ω_{ij} and R_i $(i,\ j=0,1)$ are given by

$$\begin{aligned}
\Omega_{00}&=\varphi_{00}-\omega_{00}\varphi_{00}-\omega_{01}\varphi_{01}\\
\Omega_{10}&=\varphi_{01}-\omega_{10}\varphi_{00}-\omega_{11}\varphi_{01}\\
\Omega_{01}&=\varphi_{10}-\omega_{00}\varphi_{01}-\omega_{01}\varphi_{11}\\
\Omega_{11}&=\varphi_{11}-\omega_{10}\varphi_{01}-\omega_{11}\varphi_{11}\\
R_0[n]&=s_0D_\xi-\varphi_{00}(1-\omega_{00}-\omega_{10}an)\\
&\quad-\varphi_{01}(an-\omega_{01}-\omega_{11}an)\\
R_1[n]&=s_1D_\xi-\varphi_{01}(1-\omega_{00}-\omega_{10}an)\\
&\quad-\varphi_{11}(an-\omega_{01}-\omega_{11}an)
\end{aligned} \qquad (6.35)$$

Equations (6.35) are obtained from (6.21) with the condition that

$$D_{\nu\mu}=\begin{cases}0, & \nu\neq\mu\\ \infty, & \nu=\mu\end{cases}$$

The minimum dispersion is determined from (5.17).

The above example demonstrates that, in accordance with the method developed in Section 7.4, the determination of the weighting function reduces to the solution of two algebraic equations (since the degree of the polynomial (6.1) is unity), regardless of the value of the interval N of observation. The solution thus obtained is therefore valid for any N. But if the method of direct solution of (2.4) is used, the problem then reduces to the solution of a system of N-th-order algebraic equations, which is valid only for a given value of N. For other N the solution must be repeated from beginning to end. This circumstance emphasises the advantage of the method given above.

7.7 WEIGHTING FUNCTION OF A SYSTEM OF OPTIMAL PREDICTION

We shall consider the determination of the optimal weighting function of a system predicting, over the time $\bar{t}=N_0$, the useful signal

$$S[n]=d_0+d_1 n+X[n] \tag{7.1}$$

where d_0 and d_1 are non-random (arbitrary) coefficients, and $X[n]$ is a non-stationary random function with the following correlation function:

$$K_x[n;\ m]=e^{-\alpha n|m|} \tag{7.2}$$

At the same time we assume that noise ξ with the correlation function

$$K_\xi[n;\ m]=D_\xi\sigma[n-m] \tag{7.3}$$

is superimposed on the useful signal (7.1).

The correlation function (7.2) can be represented in the form of an infinite series

$$K_x[n;\ m]=\sum_{\nu=0}^{\infty}(-1)^\nu\alpha^\nu\frac{1}{\nu!}(nm)^\nu,\quad m\geqslant 0 \tag{7.4}$$

Depending on the values of α over the interval $0\leqslant m\leqslant N$, (7.4) may in practice be confined to a finite number of terms. Further, to simplify representation, we put

$$K_x[n;\ m]=1-\alpha nm,\ 0\leqslant m\leqslant N \tag{7.5}$$

and, for rigour, we assume that $0\leqslant n\leqslant\frac{1}{\alpha N}$.

Following the method given in Section 7.4, we now introduce the vectors necessary for the solution of this problem.

1. **The vectors $\mathbf{A}[n]$ and $\mathbf{B}[n]$.** Using (3.10) and taking into consideration (7.5) we obtain

$$K_x[n;\ m]=\boldsymbol{A}[n]\cdot\boldsymbol{B}[m]$$
$$=\sum_{\nu=0}^{1}a_\nu[n]\,b_\nu[m],\ n\geqslant m$$

where

$$a_0[n] = 1, \quad a_1[n] = -\alpha n$$
$$b_1[m] = 1, \quad b_1[m] = m \tag{7.6}$$

Consequently,

$$\boldsymbol{A}[n] = \{1; \ -\alpha n\}, \ \boldsymbol{B}[m] = \{1; \ m\} \tag{7.7}$$

2. **The vector $\boldsymbol{F}[n]$.** Since in the case of prediction the weighting function of the ideal system is

$$W_{\text{id}}[n; \ m] = \sigma[n - m + N_0]$$

we have

$$\sum_{m=-\infty}^{\infty} W_{\text{id}}[n; \ m] K_x[l; \ m]$$

$$= \sum_{m=-\infty}^{\infty} \sigma[n - m + N_0] K_x[l; \ m] = K_x[l; \ n + N_0]$$

Thus, with (3.12) taken into consideration, we can write

$$\sum_{m=-\infty}^{\infty} W_{\text{id}}[n; \ m] K_x[l; \ m]$$

$$= \boldsymbol{F}[n] \cdot \boldsymbol{B}[l] = \sum_{\nu=0}^{1} f_\nu[n] b_\nu[l] = 1 - \alpha(n + N_0) l$$

whence

$$\boldsymbol{F}[n] = \{f_0[n]; \ f_1[n]\} = \{1; \ -\alpha(n + N_0)\} \tag{7.8}$$

3. **The vector $\Gamma[n]$.** With the notation adopted, the components of $\Gamma[n]$ are given by the undetermined Lagrange multipliers $\gamma_\nu[n]$ $(\nu = 0, 1, \ldots, r)$ in (2.5). The latter is satisfied by the weighting function of the optimal system, provided d_ν $(\nu = 0, 1, \ldots, r)$ of the polynomial $g[n]$ are not random quantities. Thus, when (2.5) and (7.1) are used, we have from (3.16)

$$\Gamma[n] \cdot \boldsymbol{B}[l] = \sum_{\nu=0}^{r} \gamma_\nu[n] b_\nu[l] = \sum_{\nu=0}^{1} \gamma_\nu[n] b_\nu[l]$$

whence

$$\Gamma[n] = \{\gamma_0[n];\ \gamma_1[n]\} \tag{7.9}$$

4. **The vector** $I[n]$. From (3.20), substituting for $\Gamma^*[n]$ the vector $\Gamma[n]$, we have

$$I[n] = F[n] + \Gamma[n] = \{i_0[n];\ i_1[n]\}$$

and, with (7.8) and (7.9) taken into account, we obtain

$$I[n] = \{1 + \gamma_0[n];\ -\alpha(n + N_0) + \gamma_1[n]\} \tag{7.10}$$

Since $A[n]$ and $B[n]$ are not equal, the weighting function is determined in accordance with (4.29),

$$W[n;\ m] = \sum_{\nu=0}^{1} u_\nu[n]\, p_\nu[m], \quad n - N \leqslant m \leqslant n \tag{7.11}$$

where the components $u_\nu[n]$ of $U[n]$ are determined from the recurrence formula (4.38), whilst the components $p_\nu[m]$ of $P[m]$ are determined from (4.41).

It can easily be verified that in the given case

$$\sum_{\nu=0}^{q=1} (a_\nu[m]\, b_\nu[l] - b_\nu[m]\, a_\nu[l]) = 0$$

Accordingly, (4.38) has the form

$$p_\alpha[m] = \frac{b_\alpha[m]}{D_\xi}, \quad n - N \leqslant m \leqslant n \qquad \alpha = 0,\ 1 \tag{7.12}$$

whence, using (7.6), we have

$$\begin{aligned} p_0[m] &= \frac{1}{D_\xi} = \text{const}, \quad n - N \leqslant m \leqslant n \\ p_1[m] &= \frac{m}{D_\xi}, \quad n - N \leqslant m \leqslant n \end{aligned} \tag{7.13}$$

To determine the components $u_\nu[n]$ of $U[n]$ we use (4.41), which in this case assumes the form

$$\begin{aligned} (1 + \varphi_{00})\, u_0[n] + \varphi_{10}\, u_1[n] &= i_0[n] \\ \varphi_{01} u_0[n] + (1 + \varphi_{11}) u_1[n] &= i_1[n] \end{aligned} \tag{7.14}$$

where from (4.40)

$$\varphi_{ij} = \sum_{m=n-N}^{n} a_j[m]\, p_i[m], \; (i, j = 0, 1) \tag{7.15}$$

Substituting the expressions for a_j and p_i from (7.6) and (7.13) into (7.15), we have:

$$\varphi_{00} = \sum_{m=n-N}^{n} 1 \cdot \frac{1}{D_\xi} = \frac{N+1}{D_\xi}$$

$$\varphi_{10} = \sum_{m=n-N}^{n} 1 \cdot \frac{m}{D_\xi} = \frac{2n(N+1) - N(N+1)}{2D_\xi} \tag{7.16}$$

$$\varphi_{01} = -\sum_{m=n-N}^{n} \alpha m \cdot \frac{1}{D_\xi} = -\alpha\, \frac{2n(N+1) - N(N+1)}{2D_\xi}$$

$$\varphi_{11} = -\sum_{m=n-N}^{n} \alpha \frac{m^2}{D_\xi}$$

$$= -\alpha\, \frac{6n^2(N+1) - 6nN(N+1) + N(N+1)(2N+1)}{6}$$

Solving (7.14) and using (7.10), we find the required components u_ν of $\boldsymbol{U}$

$$u_0[n] = \frac{1 + \varphi_{11} + \varphi_{10}\alpha(n + N_0)}{\Delta} + \frac{1 + \varphi_{11}}{\Delta}\gamma_0[n] - \frac{\varphi_{10}}{\Delta}\gamma_1[n]$$

$$u_1[n] = -\frac{\varphi_{01} + \alpha(1 + \varphi_{00})(n + N_0)}{\Delta} \tag{7.17}$$

$$-\frac{\varphi_{01}}{\Delta}\gamma_0[n] + \frac{1 + \varphi_{00}}{\Delta}\gamma_1[n]$$

where

$$\Delta = \begin{vmatrix} 1 + \varphi_{00} & \varphi_{10} \\ \varphi_{01} & 1 + \varphi_{11} \end{vmatrix} \tag{7.18}$$

Introducing the symbols

$$\omega_{00} = \frac{1 + \varphi_{11} + \alpha(n + N_0)\varphi_{10}}{\Delta} \tag{7.19}$$

$$\omega_{01} = \frac{1+\varphi_{11}}{\Delta}, \quad \omega_{02} = -\frac{\varphi_{10}}{\Delta}$$

and

$$\begin{aligned} \omega_{10} &= -\frac{\varphi_{01} + \alpha(1+\varphi_{00})(n+N_0)}{\Delta} \\ \omega_{11} &= -\frac{\varphi_{01}}{\Delta}, \quad \omega_{12} = \frac{1+\varphi_{00}}{\Delta} \end{aligned} \tag{7.20}$$

we write (7.17) in the form

$$\begin{aligned} u_0[n] &= \omega_{00} + \omega_{01}\gamma_0[n] + \omega_{02}\gamma_1[n] \\ u_1[n] &= \omega_{10} + \omega_{11}\gamma_0[n] + \omega_{12}\gamma_1[n] \end{aligned} \tag{7.21}$$

When (7.21) and (7.13) are used we have from (7.11) the following expression for the weighting function:

$$\begin{aligned} W[n;m] = {} & \frac{\omega_{00} + \omega_{01}\gamma_0[n] + \omega_{02}\gamma_1[n]}{D_\xi} \\ & + \frac{\omega_{10} + \omega_{11}\gamma_0[n] + \omega_{12}\gamma_1[n]}{D_\xi} m \\ & n - N \leqslant m \leqslant n \end{aligned} \tag{7.22}$$

whence it follows that $\gamma_0[n]$ and $\gamma_1[n]$, the undetermined Lagrange multipliers, must be found in order to determine the weighting function in the final form.

Since the weighting function must satisfy the additional conditions (2.6), we have the following equations for γ_0 and γ_1:

$$\begin{aligned} \sum_{m=0}^{N} W[n;\, n-m]m &= -N_0 \\ \sum_{m=0}^{N} W[n;\, n-m] &= 1 \end{aligned} \tag{7.23}$$

which are obtained from (2.6) when

$$W_{id}[n;\, n-m] = \sigma[m+N_0]$$

Substituting the expression for the weighting function from (7.22) into (7.23), we find that

$$\begin{aligned} \Omega_{00}\gamma_0[n] + \Omega_{10}\gamma_1[n] &= R_0 \\ \Omega_{10}\gamma_0[n] + \Omega_{11}\gamma_1[n] &= R_1 \end{aligned} \tag{7.24}$$

where

$$\Omega_{00}=\frac{\omega_{01}}{D_\xi}(N+1)+\frac{\omega_{11}}{D_\xi}\frac{2n(N+1)-N(N+1)}{2}$$

$$\Omega_{10}=\frac{\omega_{02}}{D_\xi}(N+1)+\frac{\omega_{12}}{D_\xi}\frac{2n(N+1)-N(N+1)}{2}$$

$$\Omega_{01}=\frac{\omega_{01}}{D_\xi}\frac{2n(N+1)-N(N+1)}{2}$$

$$+\frac{\omega_{11}}{D_\xi}\frac{3n(N+1)N-N(N+1)(2N+1)}{6}$$

$$\Omega_{11}=\frac{\omega_{02}}{D_\xi}\frac{2n(N+1)-N(N+1)}{2}+\frac{\omega_{12}}{D_\xi}\frac{3n(N+1)N-N(N+1)(2N+1)}{6}$$

$$R_0=1-\frac{\omega_{00}}{D_\xi}(N+1)-\frac{\omega_{10}}{D_\xi}\frac{2n(N+1)-N(N+1)}{2}$$

$$R_1=-N_0-\frac{\omega_{00}}{D_\xi}\frac{2n(N+1)-N(N+1)}{2}$$

$$-\frac{\omega_{10}}{D_\xi}\frac{3nN(N+1)-N(N+1)(2N+1)}{6} \qquad (7.25)$$

The unknowns $\gamma_0[n]$ and $\gamma_1[n]$ are easily determined by solving (7.24), thus completing the determination of the system. The dispersion of the prediction error can be found from (5.19) and when $N_0=0$ the formulae just derived determine the weighting function of an optimal smoothing system.

It follows that in the general case ($\boldsymbol{A}[n]\neq \boldsymbol{B}[n]$) determination of the optimal weighting function is fairly simple. For any interval of observation N it reduces to the solution of the corresponding recurrence formulae instead of the solution of systems of N-th-order algebraic equations. In addition, the expressions obtained are found to be suitable for any value of N, but when (2.4) is solved by the direct method the corresponding system of algebraic equations must be solved anew for each new N.

7.8 OPTIMAL SYSTEM WITH NON-DISCRETE WHITE NOISE

7.8.1 Equation for the weighting function of an optimal system

It was shown above that the weighting function of an optimal system, which transforms a useful signal of the form (1.1) with superimposed noise ξ having the correlation function $K_\xi[n;\ m]$, satisfies the equation

$$\sum_{m=0}^{N} W[n;\ n-m](K_x[n-l;\ n-m]+K_\xi[n-l;n-m])$$

$$=\sum_{m=-\infty}^{\infty} W_{\text{id}}[n;\ n-m]\,K_x[n-l;\ n-m] \qquad (8.1)$$

$$+\sum_{\mu=0}^{r}\gamma_\mu^*[n]\,l^\mu$$

$$n-N\leqslant l\leqslant n$$

where in accordance with (2.3)

$$\gamma_\mu^*[n]=\sum_{\nu=0}^{r} D_{\nu\mu}\left\{\sum_{m=-\infty}^{\infty} W_{\text{id}}[n;\ n-m]\,m^\nu\right.$$

$$\left.-\sum_{m=0}^{N} W[n;\ n-m]\,m^\nu\right\} \qquad (8.1a)$$

$$\mu=0,1,\ldots r$$

In Section 7.3, Equation (8.1) was written in vector form

$$\sum_{m=n-N}^{l} W[n;\ m](\boldsymbol{A}[l]\cdot\boldsymbol{B}[m]+\boldsymbol{G}[l]\cdot\boldsymbol{B}[m])$$

$$+\sum_{m=l+1}^{n} W[n;\ m](\boldsymbol{A}[m]\cdot\boldsymbol{B}[l]+\boldsymbol{G}[m]\cdot\mathrm{B}[l]) \qquad (8.2)$$

$$=\boldsymbol{F}[n]\cdot\boldsymbol{B}[l]+\boldsymbol{\Gamma}^*[n]\cdot\boldsymbol{B}[l]$$

$$n-N\leqslant l\leqslant n$$

where the vectors and their scalar products are defined as follows:

$$\boldsymbol{A}[n]\cdot\boldsymbol{B}[m]=\sum_{\nu=0}^{q} a_\nu[n]\,b_\nu[m]=K_x[n;\ m]$$

$$\boldsymbol{A}[n]=\{a_0[n];\ a_1[n];\ldots;\ a_q[n]\} \qquad (8.3)$$

$$\boldsymbol{B}[m]=\{b_0[m];\ b_1[m];\ldots;b_q[m]\}$$

$$G[n]\cdot B[m] = \sum_{\nu=0}^{q} g_\nu[n]\, b_\nu[m] = K_\xi[n;\ m]$$
$$G[n] = \{g_0[n];\ g_1[n];\ldots;\ g_q[n]\} \tag{8.4}$$

$$\Gamma^*[n]\cdot B[m] = \sum_{\nu=0}^{q} \gamma_\nu^*[n]\, b_\nu[m] = \sum_{\nu=0}^{q} \gamma_\nu^*[n] m^\nu$$
$$\Gamma^*[n] = \{\gamma_0^*[n];\ \gamma_1^*[n];\ldots;\ \gamma_q^*\} \tag{8.5}$$

$$F[n]\cdot B[l] = \sum_{\nu=0}^{q} f_\nu[n]\, b_\nu[l] =$$
$$= \sum_{m=-\infty}^{\infty} W_{\text{id}}[n;\ n-m]\, K_x[n-l;\ n-m]$$
$$F[n] = \{f_0[n];\ f_1[n];\ldots;f_q[n]\} \tag{8.6}$$

In Section 7.4 methods were given for solving (8.1) when the noise ξ is discrete white noise. In this section a method will be given for solving (8.1) in the general case, when the correlation function $K_\xi[n;\ m]$ of the noise can be represented in the form (8.4).

Defining the vectors

$$M[l] = A[l] + G[l] = \{m_0[l];\ m_1[l];\ldots;\ m_q[l]\}$$
$$I[n] = F[n] + \Gamma^*[n] = \{i_0[n];\ i_1[n];\ldots;i_q[n]\} \tag{8.7}$$

where the components m_ν and i_ν are

$$m_\nu[l] = a_\nu[l] + g_\nu[l],\ \nu = 0, 1, \ldots, q$$

$$i_\nu[n] = f_\nu[n] + \gamma_\nu^*[n],\ \nu = 0, 1, \ldots, q$$

we write (8.2) in the form

$$\sum_{m=n-N}^{l} W[n;\ m]\, M[l]\cdot B[m] + \sum_{m=l+1}^{n} W[n;\ m]\, M[m]\cdot B[l]$$
$$= I[n]\cdot B[l],\ n-N \leqslant l \leqslant n$$

Further, taking into consideration that

$$\sum_{m=l+1}^{n} W[n;\ m]\, \boldsymbol{M}[m]\cdot \boldsymbol{B}[l]$$

$$= \sum_{m=n-N}^{n} W[n;\ m]\, \boldsymbol{M}[m]\cdot \boldsymbol{B}[l] \tag{8.8}$$

$$- \sum_{m=n-N}^{l} W[n;\ m]\, \boldsymbol{M}[m]\cdot \boldsymbol{B}[l]$$

we transform (8.8) into the form

$$\sum_{m=n-N}^{l} W[n;\ m]\,(\boldsymbol{M}[l]\cdot \boldsymbol{B}[m] - \boldsymbol{M}[m]\cdot \boldsymbol{B}[l])$$

$$= (I[n] - \sum_{m=n-N}^{n} W[n;\ m]\, \boldsymbol{M}[m])\cdot \boldsymbol{B}[l], \quad n-N \leqslant l \leqslant n \tag{8.9}$$

Thus, the weighting function $W[n;\ m]$ of the required system, which transforms the signal

$$Y[n-m] = \sum_{\nu=0}^{r} d_{\nu}[n]\, m^{\nu} + X[n-m] + \xi[n-m]$$

with zero systematic error and a minimum random error, satisfies (8.9). When $l = n-N,\ n-N+1, \ldots, n$ this equation yields a system of N-th-order algebraic equations whose solution gives the optimal weighting function. It will be shown below that the determination of the weighting function $W[n;\ m]$ (mathematically the solution of a system of linear algebraic equations) can be reduced to the solution of the corresponding recurrence formula. Before giving the method for solving (8.1), let us find the general expression for the weighting function.

7.8.2 General expression for the weighting function of an optimal system

Let us consider (8.1) which is satisfied by the weighting function of the optimal system. Changing the variables we can write this equation in the following form:

$$\sum_{m=n-N}^{n} W[n;\ m](K_x[l;\ m]+K_\xi[l;\ m])$$
$$=\sum_{m=-\infty}^{\infty} W_{\text{id}}[n;\ m]K_x[l;\ m] \tag{8.10}$$
$$+\sum_{\mu=0}^{r}\gamma_\mu^*[n]\,l^\mu,\ n-N\leqslant l\leqslant n$$

whence, introducing the notation

$$K_x[l;\ m]+K_\xi[l;\ m]=K_\varphi[l;\ m], \tag{8.11}$$

we have

$$\sum_{m=n-N}^{n} W[n;\ m]K_\varphi[l;\ m]=\sum_{m=-\infty}^{\infty} W_{\text{id}}[n;\ m]K_x[l;\ m]$$
$$+\sum_{\mu=0}^{r}\gamma_\mu^*[n]\,l^\mu,\ n-N\leqslant l\leqslant n \tag{8.12}$$

Let us now assume that the random process

$$\varphi[n]=X[n]+\xi[n]$$

with the correlation function $K_\varphi[l;\ m]$, which can be determined from (8.11), is connected with the discrete white noise V by linear difference equations of finite order having the following form:

$$L_n(\Delta, n)\varphi[n]=\Phi_n(\Delta, n)V[n] \tag{8.13}$$

where $L_n(\Delta, n)$ and $\Phi_n(\Delta, n)$ are difference operators with variable coefficients, of orders k and h respectively.

We use $G[n;\ m]$ to denote Green's function for the following difference boundary-value problem

$$L_n^*(\Delta,\ n)L_n(\Delta,\ n)\varphi[n]=0$$
$$\lim_{n\to-\infty}\Delta^r\varphi[n]=\lim_{n\to\infty}\Delta^r\varphi[n]=0 \tag{8.14}$$
$$r=0, 1, \ldots, k-1$$

where $L^*_n(\Delta, n)$ is an operator adjoint to the operator $L_n(\Delta, n)$.

It has been shown in Chapter 3 that the correlation function $K_\varphi[n; m]$ of the random process φ given by (8.13) is connected with Green's function by the relation

$$K_\varphi[n; m] = \Phi_n(\Delta, n)\Phi^*_n(\Delta, n)G[n; m] \tag{8.15}$$

where $\Phi^*_n(\Delta, n)$ is an operator, adjoint to $\Phi_n(\Delta, n)$.

Furthermore, it should be noted that by definition Green's function satisfies the equation

$$L^*_n(\Delta, n)L_n(\Delta, n)G[n; m] = \sigma[n - m] \tag{8.16}$$

We can now, using (8.15), write (8.12) in the form

$$\sum_{m=n-N}^{n} W[n; m]\Phi^*_l(\Delta, l)\Phi^*_l(\Delta, l)G[l; m]$$
$$= \sum_{m=-\infty}^{\infty} W_{\text{id}}[n; m]K_x[l; m]$$
$$+ \sum_{\mu=0}^{r} \gamma^*_\mu[n]l^\mu, \quad n - N \leqslant l \leqslant n$$

or, by linearity of the operators,

$$\Phi_l(\Delta, l)\Phi^*_l(\Delta, l)\left[\sum_{m=n-N}^{n} W[n; m]G[l; m]\right]$$
$$= \sum_{m=-\infty}^{\infty} W_{\text{id}}[n; m]K_x[l; m] + \sum_{\mu=0}^{r} \gamma^*_\mu[n]l^\mu, \quad n - N \leqslant l \leqslant n \tag{8.17}$$

whence we obtain a non-homogeneous difference equation with variable coefficients.

To obtain the general solution of this equation we apply the operator $\Phi^{-1}_l(\Delta, l)\Phi^{*-1}_l(\Delta, l)$, which is inverse to the operator $\Phi_l(\Delta, l)\Phi^*_l(\Delta, l)$, to both sides. We then have

$$\sum_{m=n-N}^{n} W[n; m]G[l; m] = \Phi^{-1}_l(\Delta, l)\Phi^{*-1}_l(\Delta, l)$$

$$\times\left[\sum_{m=-\infty}^{\infty} W_{\mathrm{id}}[n;\ m]\, K_x[l;\ m]\right] + \Phi_l^{-1}(\Delta,\ l)\,\Phi_l^{*-1}(\Delta,\ l)$$
$$\times\left[\sum_{\mu=0}^{r} \gamma_\mu^*[n]\, l^\mu\right],\quad n-N \leqslant l \leqslant n \tag{8.18}$$

To find the explicit expression for the weighting function we apply the operator $L_l^*(\Delta,\ l)\, L_l(\Delta,\ l)$ with respect to the variable l to both sides of (8.18). Using the linearity property of the operators we find

$$\sum_{m=n-N}^{n} W[n;\ m]\, L_l^*(\Delta,\ l)\, L_l(\Delta,\ l)\, G[l;\ m]$$
$$= L_l^*(\Delta,\ l)\, L_l(\Delta,\ l)\, \Phi_l^{-1}(\Delta,\ l)\, \Phi_l^{*-1}(\Delta,\ l)$$
$$\times\left[\sum_{m=-\infty}^{\infty} W_{\mathrm{id}}[n;\ m]\, K_x[l;\ m] \right. \tag{8.19}$$
$$\left. + \sum_{\mu=0}^{r} \gamma_\mu^*[n]\, l^\mu\right],\quad n-N \leqslant l \leqslant n$$

Taking into account that according to (8.16)

$$L_l^*(\Delta,\ l)\, L_l(\Delta,\ l)\, G[l;\ m] = \sigma[l-m]$$

we have

$$\sum_{m=n-N}^{n} W[n;\ m] L_l^*(\Delta,\ l)\, L_l(\Delta, l) G[l; m]$$
$$= \sum_{m=n-N}^{n} W[n;\ m]\, \sigma[l-m] = W[n;\ l]$$

and, accordingly, (8.19) assumes the form

$$W[n;\ l] = L_l^*(\Delta,\ l)\, L_l(\Delta,\ l) \Phi_l^{-1}(\Delta,\ l)\, \Phi_l^{*-1}(\Delta,\ l)$$
$$\times\left[\sum_{m=-\infty}^{\infty} W_{\mathrm{id}}[n;\ m]\, K_x[l;\ m] \right. \tag{8.20}$$
$$\left. + \sum_{\mu=0}^{r} \gamma_\mu^*[n]\, l^\mu\right],\quad n-N \leqslant l \leqslant n$$

Expression (8.20) is the solution of (8.10), giving the weighting function of the system.

Since the order of the difference operator $\Phi(\Delta, n)$ is $h = k - 1$ (which follows from Chapter 6), (8.20) can be written

$$W[n;\ l] = g[n;\ l] + \varkappa_0[n]\sigma[n-l] + \varkappa_1[n]\sigma[n-N-l]$$
$$n - N \leqslant l \leqslant n \tag{8.21}$$

where the functions $\varkappa_0[n]$ and $\varkappa_1[n]$ are defined as follows:

$$L_l^*(\Delta, l)L_l(\Delta, l)\Phi_l^{-1}(\Delta, l)\Phi_l^{*-1}(\Delta, l)\gamma_0^*[n] = \begin{cases} \varkappa_0[n], & l = n \\ \varkappa_1[n], & l = n - N \end{cases} \tag{8.22}$$

The presence of the σ-function in (8.21) is the result of applying the operator $L_l^*(\Delta,\ l)L_l(\Delta,\ l)\Phi_l^{-1}(\Delta,\ l)\times\Phi_l^{*-1}(\Delta,\ l)$ to the function $\gamma_\nu^*[n]l^\nu$ when $l = n$ and $l = n - N$ (i.e. at the ends of the interval of observation). At the same time the σ-function appears for $\nu = 0$, which in fact confirms the validity of (8.22). This clearly reveals the meaning of $g[n;\ l]$. It remains to be pointed out that this function satisfies the condition

$$g[n;\ l] = \begin{cases} g[n;\ l], & n - N \leqslant l \leqslant n \\ 0, & l < n - N,\ l > n \end{cases} \tag{8.23}$$

Equation (8.23) ensures that the weighting function $W[n;\ m]$ is identically zero outside the interval of observation $[n - N;\ n]$.

The weighting function of the optimal system is thus given by (8.20). Consequently, if we know the difference operators $L(\Delta,\ n)$ and $\Phi(\Delta,\ n)$ of the discrete filter which forms the random process $\varphi[n] = X[n] + \xi[n]$, from white noise, the problem of determining the optimal system reduces to finding the result of applying the operator

$$L^*(\Delta,\ l)L(\Delta,\ l)\Phi^{-1}(\Delta,\ l)\Phi^{*-1}(\Delta,\ l)$$

to the known function

$$\sum_{m=-\infty}^{\infty} W_{\text{id}}[n;\ m]K_x[l;\ m] + \sum_{\nu=0}^{r} \gamma_\nu^*[n]l^\nu$$

But, as was mentioned earlier, the determination of the difference equations of discrete shaping filters for non-stationary random processes is in itself complicated. Hence it may not always be practicable to determine the weighting function from (8.20).

We shall now give another method for finding the weighting function, which enables the required function to be found by solving certain recurrence formulae.

7.8.3 Determination of the weighting function of the optimal system

We shall proceed from the condition that (8.3) to (8.6) are valid for the random processes under consideration. In this case, as was demonstrated in Section 7.8.1, the weighting function satisfies (8.9). Consequently, the problem to the solution of (8.9).

Let us seek the solution of this equation in the form (8.21). However, if we substitute the expression of the weighting function from (8.21) into (8.9), we obtain an equation of the same form as the initial one, but with respect to the function $g[n;\ l]$. In addition the equation thus obtained also contains two other unknown functions $\varkappa_0[n]$ and $\varkappa_1[n]$.

Thus a direct substitution (8.21) into the initial equation (8.21) does not simplify the problem. We shall therefore proceed as follows. We consider the function $g[n;\ l]$ as a scalar product of the two $(q+1)$-dimensional vectors $\boldsymbol{U}[n]$ and $\boldsymbol{P}[l]$, i.e.

$$
\begin{gathered}
g[n; l] = \boldsymbol{U}[n] \cdot \boldsymbol{P}[l], \quad n - N \leqslant l \leqslant n \\
\boldsymbol{U}[n] = \{u_0[n]; u_1[n]; \ldots; u_q[n]\} \\
\boldsymbol{P}[l] = \{p_0[l]; p_1[l]; \ldots; p_q[l]\}
\end{gathered} \tag{8.24}
$$

Using (8.24) we represent the expression for the weighting function in the following form

$$
\begin{gathered}
W[n; l] = \boldsymbol{U}[n] \cdot \boldsymbol{P}[l] + \varkappa_0[n]\, \sigma[n-l] \\
+ \varkappa_1[n]\, \sigma[n-N-l] \\
n - N \leqslant l \leqslant n
\end{gathered} \tag{8.25}
$$

or, in scalar form,

$$W[n;l]=\sum_{\nu=0}^{q}u_{\nu}[n]\,p_{\nu}[l]+\varkappa_0[n]\,\sigma[n-l]+\varkappa_1[n] \times\sigma[n-N-l] \quad n-N\leqslant l\leqslant n \tag{8.26}$$

By representing the weighting function $W[n;\ l]$ in the form (8.25) we obtain simpler equations with fewer unknowns from (8.9), which greatly facilitates the task of finding $W[n;\ l]$. Accordingly, we shall assume that $W[n;\ l]$ is expressed by (8.25).

To obtain the necessary relationships, from which $\boldsymbol{U}[n]$, $\boldsymbol{P}[l]$ and the functions $\varkappa_0[n]$ and $\varkappa_1[n]$ entering (8.25) may be determined, we substitute (8.25) into (8.8). We have

$$\begin{aligned}&\sum_{m=n-N}^{l}\boldsymbol{U}[n]\cdot\boldsymbol{P}[m]\,\boldsymbol{M}[l]\cdot\boldsymbol{B}[m]\\&+\sum_{m=n-N}^{l}(\varkappa_0[n]\,\sigma[n-m]+\varkappa_1[n]\,\sigma[n-N-m])\,\boldsymbol{M}[l]\\&\times\boldsymbol{B}[m]+\sum_{m=l+1}^{n}\boldsymbol{U}[n]\cdot\boldsymbol{P}[m]\,\boldsymbol{M}[m]\cdot\boldsymbol{B}[l]\\&+\sum_{m=l+1}^{n}(\varkappa_0[n]\,\sigma[n-m]+\varkappa_1[n]\,\sigma[n-N-m]\,\boldsymbol{M}[m]\\&\times\boldsymbol{B}[l]=\boldsymbol{I}[n]\cdot\boldsymbol{B}[l],\ n-N\leqslant l\leqslant n\end{aligned}\tag{8.27}$$

Making use of the property of the σ-function we write the second and last terms of the left-hand side of (8.27) as follows:

$$\begin{aligned}&\sum_{m=n-N}^{l}(\varkappa_0[n]\,\sigma[n-m]+\varkappa_1[n]\,\sigma[n-N-m])\,\boldsymbol{M}[l]\cdot\boldsymbol{B}[m]\\&\qquad=\varkappa_1[n]\,\boldsymbol{M}[l]\cdot\boldsymbol{B}[n-N]\\&\sum_{m=l+1}^{n}(\varkappa_0[n]\,\sigma[n-m]+\varkappa_1[n]\,\sigma[n-N-m])\,\boldsymbol{M}[m]\cdot\boldsymbol{B}[l]\\&\qquad=\varkappa_0[n]\,\boldsymbol{M}[n]\cdot\boldsymbol{B}[l]\end{aligned}\tag{8.28}$$

Making use of (8.28) and noting also that

$$\sum_{m=l+1}^{n} U[n]\cdot P[m]\, M[m]\cdot B[l]$$

$$= \sum_{m=n-N}^{n} U[n]\cdot P[m] M[m]\cdot B[l]$$

$$- \sum_{m=n-N}^{l} U[n]\cdot P[m]\, M[m]\cdot B[l]$$

we transform (8.27) into the form

$$\sum_{m=n-N}^{l} U[n]\cdot P[m]\,(M[l]\cdot B[m] - M[m]\cdot B[l])$$

$$+\varkappa_1[n]\, M[l]\cdot B[n-N] + \sum_{m=n-N}^{n} U[n]\cdot P[m]\, M[m]\cdot B[l] \tag{8.29}$$

$$+\varkappa_0[n]\, M[n]\cdot B[l] = I[n]\cdot B[l]$$

$$n-N \leqslant l \leqslant n$$

This equation contains four unknowns, the vectors U and P and the functions $\varkappa_0$ and $\varkappa_1$. In order to find the independent equations from which these may be determined we proceed as follows. Considering (8.29) over the time interval $n-N+1 \leqslant l \leqslant n-1$, then over the time interval $[n-N+1;\ n-1]$ the weighting function $W[n;l]$, given by (8.25), equals

$$W[n;l] = U[n]\cdot P[l],\quad n-N+1 \leqslant l \leqslant n-1 \tag{8.30}$$

Substituting (8.30) into (8.8), and carrying out the necessary transformations, we find

$$\sum_{m=n-N}^{l} U[n]\cdot P[m]\,(M[l]\cdot B[m] - M[m]\cdot B[l])$$

$$= I[n]\cdot B[l] - \sum_{m=n-N}^{n} U[n]\cdot P[m]\, M[m]\cdot B[l] \tag{8.31}$$

$$n-N+1 \leqslant l \leqslant n-1$$

Since

$$\sum_{m=n-N}^{l} \boldsymbol{U}[n]\cdot\boldsymbol{P}[m]=\sum_{m=n-N}^{l}\sum_{\nu=0}^{q} u_{\nu}[n]\,p_{\nu}[m]$$

$$=\sum_{\nu=0}^{q} u_{\nu}[n]\sum_{m=n-N}^{l} p_{\nu}[m]=\boldsymbol{U}[n]\cdot\sum_{m=n-N}^{l}\boldsymbol{P}[m]$$

(8.31) can also be written in the form

$$\boldsymbol{U}[n]\cdot\sum_{m=n-N}^{l}\boldsymbol{P}[m](\boldsymbol{M}[l]\cdot\boldsymbol{B}[m]-\boldsymbol{M}[m]\cdot\boldsymbol{B}[l])$$

$$=(\boldsymbol{I}[n]-\boldsymbol{U}[n]\cdot\sum_{m=n-N}^{n}\boldsymbol{P}[m]\cdot\boldsymbol{M}[m])\cdot\boldsymbol{B}[l] \qquad (8.32)$$

$$n-N+1\leqslant l\leqslant n-1$$

Equation (8.32) contains the two unknown vectors $\boldsymbol{U}$ and $\boldsymbol{P}$. But, as may easily be verified, this equation can be decomposed into two independent equations, each with only one unknown:

$$\boldsymbol{B}[l]=\sum_{m=n-N}^{l}\boldsymbol{P}[m](\boldsymbol{M}[l]\cdot\boldsymbol{B}[m]-\boldsymbol{M}[m]\cdot\boldsymbol{B}[l]) \qquad (8.33)$$

$$n-N+1\leqslant l\leqslant n-1$$

$$\boldsymbol{U}[n]=\boldsymbol{I}[n]-\boldsymbol{U}[n]\cdot\sum_{m=n-N}^{n}\boldsymbol{P}[m]\cdot\boldsymbol{M}[m] \qquad (8.34)$$

Thus, (8.33) contains only one unknown, the vector $\boldsymbol{P}[m]$. Determining $\boldsymbol{P}[m]$ and substituting this vector into (8.34), we obtain a new equation with the single unknown vector $\boldsymbol{U}[n]$.

To determine the functions $\varkappa_0[n]$ and $\varkappa_1[n]$ we can use (8.29) when $l=n-N$ and $l=n$, which in this case gives two algebraic equations for the unknowns $\varkappa_0$ and $\varkappa_1$. Let us now derive these equations. Putting $l=n-N$ in (8.29) we have

$$\varkappa_0[n]\,\boldsymbol{M}[n]\cdot\boldsymbol{B}[n-N]+\varkappa_1[n]\,\boldsymbol{M}[n-N]\cdot\boldsymbol{B}[n-N]$$

$$=\boldsymbol{I}[n]\cdot\boldsymbol{B}[n-N]-\sum_{m=n-N}^{n}\boldsymbol{U}[n]\cdot\boldsymbol{P}[m]\,\boldsymbol{M}[m]\cdot\boldsymbol{B}[n-N] \qquad (8.35)$$

Similarly, for $l=n$, we can obtain from (8.29) the other equation for $\varkappa_0$ and $\varkappa_1$

$$\varkappa_0[n]\,\boldsymbol{M}[n]\cdot\boldsymbol{B}[n]+\varkappa_1[n]\,\boldsymbol{M}[n]\cdot\boldsymbol{B}[n-N] \\ =\boldsymbol{I}[n]\cdot\boldsymbol{B}[n]-\sum_{m=n-N}^{n}\boldsymbol{U}[n]\cdot\boldsymbol{P}[m]\,\boldsymbol{M}[n]\cdot\boldsymbol{B}[m] \tag{8.36}$$

Consequently, we have four equations. From them we can determine the four unknowns entering (8.25): $\boldsymbol{P}$ and $\boldsymbol{U}$ from (8.33) and (8.34), and $\varkappa_0[n]$ and $\varkappa_1[n]$ from (8.35) and (8.36).

The problem then consists of determining the vectors $P[n]$ and $U[n]$ and the functions $\varkappa_0[n]$ and $\varkappa_1[n]$.

To solve this problem, let us consider (8.33), which we write in scalar form as

$$b_\alpha[l]=\sum_{m=n-N}^{l}p_\alpha[m]\sum_{\nu=0}^{q}(m_\nu[l]\,b_\nu[m]-m_\nu[m]\,b_\nu[l]) \\ \alpha=0,1,\ldots,q,\quad n-N+1\leqslant l\leqslant n-1 \tag{8.37}$$

It is not difficult to see that the components $p_\alpha[m]$, $\alpha=0,1,\ldots,q$ of the vector $\boldsymbol{P}[m]$ can be determined from (8.37) when $n-N\leqslant m\leqslant n-2$. Thus putting $l=n-N+1$ in (8.37), we find

$$b_\alpha[n-N+1]=\sum_{m=n-N}^{n-N+1}p_\alpha[m] \\ \times\sum_{\nu=0}^{q}(m_\nu[n-N+1]\,b_\nu[m]-m_\nu[m]\,b_\nu[n-N+1]) \\ =p_\nu[n-N]\sum_{\nu=0}^{q}(m_\nu[n-N+1]\,b_\nu[n-N] \\ -m_\nu[n-N]\,b_\nu[n-N+1])$$

whence we obtain

$$p_\alpha[n-N]=\frac{b_\alpha[n-N+1]}{\sum_{\nu=0}^{q}(m_\nu[n-N+1]\,b_\nu[n-N]-m_\nu[n-N]\,b_\nu[n-N+1])} \\ \alpha=0,1,\ldots,q \tag{8.38}$$

When $l = n - 1$, we find from (8.37)

$$b_\alpha[n-1] = \sum_{m=n-N}^{n-1} p_\alpha[m]$$

$$\times \sum_{\nu=0}^{q} (m_\nu[n-1]\, b_\nu[m] - m_\nu[m]\, b_\nu[n-1])$$

$$= p_\alpha[n-2] \sum_{\nu=0}^{q} (m_\nu[n-1]\, b_\nu[n-2] - m_\nu[n-2]\, b_\nu[n-1])$$

$$+ \sum_{m=n-N}^{n-3} p_\alpha[m] \sum_{\nu=0}^{q} (m_\nu[n-1]\, b_\nu[m] - b_\nu[n-1]\, m_\nu[m])$$

whence it follows that

$$p_\alpha[n-2] = \frac{b_\alpha[n-1] - \sum\limits_{m=n-N}^{n-3} p_\alpha[m] \sum\limits_{\nu=0}^{q} (m_\nu[n-1]\, b_\nu[m] - m_\nu[m]\, b_\nu[n-1])}{\sum\limits_{\nu=0}^{q} (m_\nu[n-1]\, b_\nu[n-2] - m_\nu[n-2]\, b_\nu[n-1])} \tag{8.39}$$

$$\alpha = 0, 1, \ldots, q$$

Thus, when $l = n - N + 1,\ n - N + 2, \ldots, n - 1$, (8.37) allows the required components $p_\alpha[m]$ to be determined for the values $m = n - N,\ n - N + 1, \ldots, n - 2$. It should be noted that the functions $p_\alpha[m]$ with the larger values of m are expressed in terms of those with smaller values of m.

For the general case $(n - N + 1 \leqslant l \leqslant n - 2)$, we can write, proceeding from (8.37),

$$p_\nu[l-1] = \frac{b_\alpha[l] - \sum\limits_{m=n-N}^{l-2} p_\alpha[m] \sum\limits_{\nu=0}^{q} (m_\nu[l]\, b_\nu[m] - m_\nu[m]\, b_\nu[l])}{\sum\limits_{\nu=0}^{q} (m_\nu[l]\, b_\nu[l-1] - m_\nu[l-1]\, b_\nu[l])} \tag{8.40}$$

$$\alpha = 0, 1, \ldots, q,\ n - N + 1 \leqslant l \leqslant n - 1$$

Thus, the components $p_\alpha[m]$, $\alpha=0, 1, \ldots, q$ of the vector $\boldsymbol{P}[m]$ are determined from (8.40) for $m=n-N$, $n-N+1$, $\ldots$, $n-2$. This formula is essentially a recurrence relationship which establishes a connection between the quantities $p_\alpha[m]$ with smaller and larger values of m.

In order to determine the functions $p_\alpha[m]$ for $m=n-1$ and $m=n$ we calculate the first- and second-order differences of both sides of (8.37) with respect to the variable l. Putting then $l=n-1$ in the equations thus obtained, we find the required expressions which define $p_\alpha[n-1]$ and $p_\alpha[n]$.

To determine these expressions, we first write (8.37) in the following form:

$$b_\alpha[l]=\sum_{\nu=0}^{l} m_\nu[l] \sum_{m=n-N}^{l} p_\alpha[m]\, b_\nu[m]$$
$$-\sum_{\nu=0}^{q} b_\nu[l] \sum_{m=n-N}^{l} p_\alpha[m]\, m_\nu[m], \quad \alpha=0, 1, \ldots, q \tag{8.41}$$
$$n-N+1 \leqslant l \leqslant n-1$$

We then use the following easily-proved formula to calculate the first-order difference of both sides of (8.41) with respect to

$$\Delta_l\left\{\sum_{\nu=0}^{q} m_\nu[l] \sum_{m=k}^{l} p_\alpha[m]\, b_\nu[m]\right\}$$
$$=\sum_{\nu=0}^{q} \Delta_l m_\nu[l] \sum_{m=k}^{l} p_\alpha[m]\, b_\nu[m] \tag{8.42}$$
$$+\sum_{\nu=0}^{q} m_\nu[l+1]\, p_\alpha[l+1]\, b_\nu[l+1]$$

It is also easy to show that

$$\Delta_l\left\{\sum_{\nu=0}^{q} m_\nu[l] \sum_{m=k}^{l} p_\alpha[m]\, b_\nu[m]\right\}$$
$$=\sum_{\nu=0}^{q} m_\nu[l]\, p_\alpha[l+1]\, b_\nu[l+1] \tag{8.43}$$
$$+\sum_{\nu=0}^{q} \Delta m_\nu[l] \sum_{m=k}^{l+1} p_\alpha[m]\, b_\nu[m]$$

Using either (8.42) or (8.43), we obtain from (8.41),

$$\Delta b_\alpha[l]=\sum_{\nu=0}^{q}\Delta m_\nu[l]\sum_{m=n-N}^{l}p_\alpha[m]\,b_\nu[m]$$

$$-\sum_{\nu=0}^{q}\Delta b_\nu[l]\sum_{m=n-N}^{l}p_\alpha[m]\,m_\nu[m] \tag{8.44}$$

$$\alpha=0,1,\ldots,q,\quad n-N+1\leqslant l\leqslant n-1$$

or, after certain transformations,

$$\Delta b_\alpha[l]=\sum_{m=n-N}^{l}p_\alpha[m]\sum_{\nu=0}^{q}(\Delta m_\nu[l]\,b_\nu[m]-\Delta b_\nu[l]\,m_\nu[m]) \tag{8.45}$$

$$\alpha=0,1,\ldots,q,\quad n-N+1\leqslant l\leqslant n-1$$

Putting $l=n-1$ in (8.44) we find

$$\Delta b_\alpha[n-1]=p_\alpha[n-1]$$

$$\times\sum_{\nu=0}^{q}m_\nu[n]\,b_\nu[n-1]-b_\nu[n]\,m_\nu[n-1])$$

$$+\sum_{m=n-N}^{n-2}p_\alpha[m]\sum_{\nu=0}^{q}(\Delta m_\nu[n-1]\,b_\nu[m]-\Delta b_\nu[n-1]\,m_\nu[n-1])$$

$$\alpha=0,1,\ldots,q$$

whence it follows that

$$p_\alpha[n-1]$$

$$=\frac{\Delta b_\alpha[n-1]-\sum\limits_{m=n-N}^{n-2}p_\alpha[m]\sum\limits_{\nu=0}^{q}(\Delta m_\nu[n-1]b_\nu[m]-\Delta b_\nu[n-1]m_\nu[m])}{\sum\limits_{\nu=0}^{q}(m_\nu[n]\,b_\nu[n-1]-b_\nu[n]\,m_\nu[n-1])} \tag{8.46}$$

$$\alpha=0,1,\ldots,q$$

Thus, the components $p_\alpha[m]$, $\alpha=0,1,\ldots,q$ of $\boldsymbol{P}[m]$, when $m=n-1$, are determined from (8.46). We shall now find the corresponding expression for $p_\alpha[m]$ when $m=n$. To do this we calculate the first-order difference of both sides of (8.44) with respect to l. Using (8.42) we obtain

$$\Delta^2 b_\alpha[l] = \sum_{\nu=0}^{q} \Delta^2 m_\nu[l] \sum_{m=n-N}^{l} p_\alpha[m] b_\nu[m]$$

$$+ \sum_{\nu=0}^{q} \Delta m_\nu[l+1] p_\alpha[l+1] b_\nu[l+1]$$

$$- \sum_{\nu=0}^{q} \Delta^2 b_\nu[l] \sum_{m=n-N}^{l} p_\alpha[m] m_\nu[m]$$

$$- \sum_{\nu=0}^{q} \Delta b_\nu[l+1] p_\alpha[l+1] m_\nu[l+1]$$

or, after certain transformations,

$$\Delta^2 b_\alpha[l] = \sum_{m=n-N}^{l} p_\alpha[m] \sum_{\nu=0}^{q} (\Delta^2 m_\nu[l] b_\nu[m] - \Delta^2 b_\nu[l] m_\nu[m])$$
$$+ p_\alpha[l+1] \sum_{\nu=0}^{q} (m_\nu[l+2] b_\nu[l+1] - m_\nu[l+1] b_\nu[l+2]) \tag{8.47}$$

When $l = n-1$ we find from (8.47) the expression

$$p_\alpha[n] = \frac{\Delta^2 b_\alpha[n-1] - \sum_{m=n-N}^{n-1} p_\alpha[m] \sum_{\nu=0}^{q} (\Delta^2 m_\nu[n-1] b_\nu[m] - \Delta^2 b_\nu[n-1] m_\nu[m])}{\sum_{\nu=0}^{q} (m_\nu[n+1] b_\nu[n] - m_\nu[n] b_\nu[n+1])} \tag{8.48}$$

$$\alpha = 0, 1, \ldots, q$$

giving the components $p_\alpha[m]$ of $\boldsymbol{P}[m]$ for $m=n$.

We have thus obtained the corresponding formulae for determining $\boldsymbol{P}$. The next problem consists of determining $\boldsymbol{U}[n]$, from (8.34).

Using (8.7) and (8.24) we write (8.34) in scalar form as

$$u_\alpha[n] = i_\alpha[n] - \sum_{\nu=0}^{q} u_\nu[n] \sum_{m=n-N}^{n} p_\nu[m] m_\alpha[m] \tag{8.49}$$
$$\alpha = 0, 1, \ldots, q$$

whence, introducing the notation

$$\varphi_{\nu\alpha}=\sum_{m=n-N}^{n} p_{\nu}[m]\, m_{\alpha}[m], \quad (\nu, \alpha=0, 1, \ldots, q) \tag{8.50}$$

we have

$$u_{\alpha}[n]=i_{\alpha}[n]-\sum_{\nu=0}^{q} u_{\nu}[n]\varphi_{\nu\alpha}, \quad \alpha=0, 1, \ldots, q \tag{8.51}$$

When $\alpha=0, 1, \ldots, q$, (8.51) forms a system of algebraic $(q+1)$-th order equations for the unknown functions $u_{\alpha}[n]$, $\alpha=0, 1, \ldots, q$. This system can be written in the form

$$\begin{aligned}
&(1+\varphi_{00})\,u_0[n]+\varphi_{10}u_1[n]+\ldots+\varphi_{q0}u_q[n]=i_0[n]\\
&\varphi_{01}u_0[n]+(1+\varphi_{11})\,u_1[n]+\ldots+\varphi_{q1}u_q[n]=i_1[n]\\
&\ldots\ldots\ldots\ldots\ldots\ldots\ldots\ldots\ldots\ldots\\
&\varphi_{0q}u_0[n]+\varphi_{1q}u_1[n]+\ldots+(1+\varphi_{qq})\,u_q=i_q[n]
\end{aligned} \tag{8.52}$$

Since (8.52) is obtained from (8.34), and (8.34) in its turn from (8.1) which has a single solution, the determinant of (8.52) is non-zero. Consequently, (8.52) uniquely defines $u_{\alpha}[n]$, $\alpha=0, 1, \ldots, q$.

Thus the task of determining $\boldsymbol{U}[n]$ is completed by solving a system of algebraic $(q+1)$-th order equations of the form (8.52).

Having determined $\boldsymbol{P}$ and $\boldsymbol{U}$ from (8.35) and (8.36), we can find the functions $\varkappa_0[n]$ and $\varkappa_1[n]$. The task of determining the optimal weighting function is not completed by solving (8.35) and (8.36). In fact, as follows from (8.52), the components of $\boldsymbol{U}[n]$ depend on $\gamma_{\nu}^{*}[n]$, $\nu=0, 1, \ldots, q$. Consequently, $\varkappa_0[n]$ and $\varkappa_1[n]$ also depend on $\gamma_{\nu}^{*}[n]$. Accordingly, we must find the components $\gamma_{\nu}^{*}[n]$, $\nu=0, 1, \ldots, q$ of $\Gamma^{*}[n]$, in order to determine the weighting function in its final form. For this we must use (8.1a), which, when the expression for the weighting function from (8.21) is substituted into it, gives a system of $q+1$ algebraic equations for the unknowns $\gamma_{\nu}^{*}[n]$. The determination of the weighting function is completed by solving this system of equations.

All the results obtained in this section remain valid for the case where the coefficients d_{ν}, $\nu=0, 1, \ldots, r$ of (1.1) are non-random quantities with arbitrary values. Expressions (8.40), (8.46) and (8.48), from which the components $p_\alpha[m]$, $\alpha=0, 1, \ldots, q$, $n-N \leqslant m \leqslant n$, are determined, as well as (8.52), (8.35) and (8.36), from which $u_\alpha[n]$, $\varkappa_0[n]$ and $\varkappa_1[n]$ are determined, remain unaltered in this case. In (8.1a) we must put

$$D_{\nu\mu}=\begin{cases} 0, & \nu\neq\mu \\ \infty, & \nu=\mu \end{cases}$$

thus obtaining a system of $q+1$ algebraic equations from which the components $\gamma_\nu[n]$, $\nu=0, 1, \ldots, q$ of $\mathbf{\Gamma}[n]$ may be determined. Expression (8.21) for the weighting function of the optimal system also remains unaltered.

Accordingly, the solution to the problem of determining the optimal system when the coefficients d_ν in the useful signal are non-random quantities, is obtained as a particular case from the solution of the more general problem.

In concluding we point out that the minimum dispersion of the transformation error of the input signal by the optimal system can be calculated by the formula obtained in Section 7.5.

7.9 WEIGHTING FUNCTION OF AN OPTIMAL FILTERING SYSTEM

The method given in the preceding section will now be illustrated by a concrete example.

Suppose that we have to determine the weighting function of a non-stationary system which ensures optimal isolation of the useful signal

$$S[m]=d_0+d_1 m+X[m] \tag{9.1}$$

from a signal Y observed over the time interval $[n-N;\ n]$, this signal being a sum of the useful signal $S[m]$ and the noise ξ. We assume that d_0 and d_1 are non-random quantities with arbitrary values, whilst $X[n]$ is a non-stationary random function with zero mathematical expectation, having a correlation function of the form

$$K_x[n;\ m]=e^{-\alpha|m|n} \tag{9.2}$$

We further assume that the correlation function of ξ has the form

$$K_\xi[n;m] = e^{-\beta|n-m|} \tag{9.3}$$

Since the method given for determining the weighting functions of non-stationary optimal systems assumes that the correlation functions $K_x[n;m]$ (useful signal) and $K_\xi[n;m]$ (noise) are given in the form of a power series with respect to the second variables, (9.2) and (9.3) can be represented by the following expansions:

$$\begin{aligned} K_x[n;m] &= 1 - anm + \frac{a^2n^2}{2}m^2 - \frac{a^3n^3}{6}m^3 + \dots \\ K_\xi[n;m] &= e^{-\beta n}\left(1 + \beta m + \frac{\beta^2}{2}m^2 + \frac{\beta^3}{6}m^3 + \dots\right) \end{aligned} \tag{9.4}$$

In (9.4) we confine ourselves to two terms and assume for what is to follow that

$$K_x[n;m] = 1 - anm, \quad n - N \leqslant m \leqslant n \tag{9.5}$$

$$K_\xi[n;m] = e^{-\beta n} + \beta e^{-\beta n} m, \quad n \geqslant m \tag{9.6}$$

Let us now introduce the required vectors.

1. **The vectors $\boldsymbol{A}[n]$ and $\boldsymbol{B}[n]$.** In accordance with (8.3), when (9.5) is taken into account, we have

$$\begin{aligned} K_x[n;\ m] &= \boldsymbol{A}[n] \cdot \boldsymbol{B}[m] \\ &= \sum_{\nu=0} a_\nu[n] b_\nu[m], \quad n \geqslant m \end{aligned}$$

where

$$\begin{aligned} a_0[n] &= 1, \qquad a_1[n] = -an \\ b_0[m] &= 1, \qquad b_1[m] = m \end{aligned} \tag{9.7}$$

Consequently,

$$\begin{aligned} \boldsymbol{A}[n] &= \{1; -an\} \\ \boldsymbol{B}[m] &= \{1;\ m\} \end{aligned} \tag{9.8}$$

2. **The vector G[n].** From (8.4), when (9.6) is used

$$K_{\xi}[n;\ m] = \boldsymbol{G}[n] \cdot \boldsymbol{B}[m] = \sum_{\nu=0}^{1} g_{\nu}[n]\, b_{\nu}[m]$$

whence we obtain

$$G[n] = \{g_0[n];\ g_1[n]\} = \{e^{-\beta n};\ \beta e^{-\beta n}\} \tag{9.9}$$

3. **The vector F[n].** In the given case (filtering) the weighting function of the ideal transforming system is $W_{id}[n;\ m] = \sigma[n-m]$ and from (8.6) we obtain

$$\sum_{m=-\infty}^{\infty} W_{id}[n;\ m]\, K_x[l;\ m] = \sum_{m=-\infty}^{\infty} \sigma[n-m]\, K_x[l;\ m]$$

$$= K_x[l;\ n] = 1 - \alpha ln = \boldsymbol{F}[n] \cdot \boldsymbol{B}[l] = \sum_{\nu=0}^{1} f_{\nu}[n]\, b_{\nu}[l]$$

whence

$$\boldsymbol{F}[n] = \{f_0[n];\ f_1[n]\} = \{1;\ -\alpha n\} \tag{9.10}$$

4. **The vector Γ[n].** According to the symbols used, the components of $\boldsymbol{\Gamma}[n]$ are given by the undetermined Lagrange multipliers $\gamma_{\nu}[n]$, $\nu = 0, 1, \ldots, r$. From (8.5), where γ_{ν}^{*} are replaced by γ_{ν} and $\boldsymbol{\Gamma}^{*}$ is replaced by $\boldsymbol{\Gamma}$, we have when (9.1) is taken into account

$$\boldsymbol{\Gamma}[n] \cdot \boldsymbol{B}[m] = \sum_{\nu=0}^{1} \gamma_{\nu}[n]\, m^{\nu}$$

whence

$$\boldsymbol{\Gamma}[n] = \{\gamma_0[n];\ \gamma_1[n]\} \tag{9.11}$$

5. **The vectors M[n] and I[n].** From (8.7), when (9.8) to (9.11) are used, we obtain

$$\begin{aligned} \boldsymbol{M}[n] &= \boldsymbol{A}[n] + \boldsymbol{G}[n] = \{m_0[n];\ m_1[n]\} \\ \boldsymbol{I}[n] &= \boldsymbol{F}[n] + \boldsymbol{\Gamma}[n] = \{i_0[n];\ i_1[n]\} \end{aligned} \tag{9.12}$$

where the components are given by

$$m_0[n]=1+e^{-\beta n},\quad m_1[n]=-\alpha n+\beta e^{-\beta n}$$
$$i_0[n]=1+\gamma_0[n],\; i_1[n]=-\alpha n+\gamma_1[n] \tag{9.13}$$

Let us now determine the required weighting function. In accordance with (8.26) the weighting function of the optimal system is expressed by

$$W[n;\, l]=\sum_{\nu=0}^{1} u_\nu[n]\, p_\nu[l]+\varkappa_0[n]\,\sigma[n-l]+\varkappa_1[n]\,\sigma[n-N-l]$$
$$n-N\leqslant l\leqslant n \tag{9.14}$$

where the functions $u_\nu[n]$, $p_\nu[l]$, $\varkappa_0[n]$ and $\varkappa_1[n]$ are yet to be determined.

To determine $p_\alpha[l]$ when $n-N\leqslant l\leqslant n-2$ we use the recurrence formula (8.40), which in this case has the form

$$p_\alpha[l-1]=\frac{b_\alpha[l]-\sum\limits_{m=n-N}^{l-2} p_\alpha[m]\sum\limits_{\nu=0}^{1}(m_\nu[l]\,b_\nu[m]-m_\nu[m]\,b_\nu[l])}{\sum\limits_{\nu=0}^{1}(m_\nu[l]\,b_\nu[l-1]-m_\nu[l-1]\,b_\nu[l])} \tag{9.15}$$

$$\alpha=0,\ 1,\quad n-N+1\leqslant l\leqslant n-1$$

Substituting the expressions for $m_\nu[l]$ and $b_\nu[l]$ from (9.13) into (9.15) we obtain

$$p_\alpha[l-1]=\frac{l^\alpha-\sum\limits_{m=n-N}^{l-2} p_\alpha[m]\,[(1+\beta m)\,e^{-\beta l}-(1+\beta l)\,e^{-\beta m}]}{e^{-\beta l}(1-e^{\beta}+\beta l-\beta-\beta l\, e^{\beta})} \tag{9.16}$$

$$\alpha=0,\ 1,\quad n-N+1\leqslant l\leqslant n-1$$

To determine the functions $p_\alpha[m]$, when $m=n-1$ and $m=n$, we use (8.46) and (8.48), which here have the form

$$p_\alpha[n-1] = \frac{\Delta b_\alpha[n-1] - \sum_{m=n-N}^{n-2} p_\alpha[m] \sum_{\nu=0}^{1} (\Delta m_\nu[n-1] b_\nu[m] - \Delta b_\nu[n-1] m_\nu[n-1])}{\sum_{\nu=0}^{1} (m_\nu[n] b_\nu[n-1] - b_\nu[n] m_\nu[n-1])} \tag{9.17}$$

$$\alpha = 0, 1$$

$$p_\alpha[n] = \frac{\Delta^2 b_\alpha[n-1] - \sum_{m=n-N}^{n-1} p_\alpha[m] \sum_{\nu=0}^{1} (\Delta^2 m_\nu[n-1] b_\nu[m] - \Delta^2 b_\nu[n-1] m_\nu[m])}{\sum_{\nu=0}^{1} (m_\nu[l+2] b_\nu[l+1] - m_\nu[l+1] b_\nu[l+2])} \tag{9.18}$$

$$\alpha = 0, 1$$

Substituting the expression for m_ν and b_ν, into (9.17), we obtain after certain transformations

$$p_\alpha[n-1] = \frac{\Delta (n-1)^\alpha - \sum_{m=n-N}^{n-2} p_\alpha[m] [e^{-\beta n} (e^{-2\beta} - e^{-\beta} + \beta m - \beta m e^\beta - \beta e^\beta) - \alpha (m-n+1)]}{e^{-\beta n} (1 - e^{-\beta} + \beta n - \beta - \beta n e^\beta)} \tag{9.19}$$

$$\alpha = 0, 1$$

Analogously, from (9.18) we obtain

$$p_\alpha[n] = -\frac{\sum_{m=n-N}^{n-2} p_\alpha[m] [e^{-\beta} - 2 + e^\beta + m\beta e^{-\beta} - 2m\beta + m\beta e^\beta] e^{\beta n}}{e^{-\beta n} [e^\beta - 1 + \beta n e^{-\beta} - \beta (n+1)]} \tag{9.20}$$

$$\alpha = 0, 1$$

The calculation of the functions $p_\alpha[i]$, $l = n-N$, $n-N+1, \ldots, n$ does not present any difficulties. Let us now determine $u_0[n]$ and $u_1[n]$.

To do this we use (8.52), which here has the form

$$\begin{aligned} (1+\varphi_{00}) u_0[n] + \varphi_{10} u_1[n] &= i_0[n] \\ \varphi_{01} u_0[n] + (1+\varphi_{11}) u_1[n] &= i_1[n] \end{aligned} \tag{9.21}$$

where in accordance with (8.50)

$$\varphi_{\nu\alpha} = \sum_{m=n-N}^{n} p_\nu[m]\, m_\alpha[m], \ (\nu,\ \alpha = 0,1) \tag{9.22}$$

Remembering that, according to (9.13), $i_0[n] = 1 + \gamma_0[n]$ and $i_1[n] = -\alpha n + \gamma_1[n]$, we find from (9.21)

$$\begin{aligned} u_0[n] &= \omega_{00} + \omega_{01}\gamma_0[n] + \omega_{02}\gamma_1[n] \\ u_1[n] &= \omega_{10} + \omega_{11}\gamma_1[n] + \omega_{12}\gamma_1[n] \end{aligned} \tag{9.23}$$

where

$$\begin{aligned} \omega_{00} &= \frac{1 + \varphi_{11} + \varphi_{10}\alpha n}{\Delta}, & \omega_{10} &= -\frac{\varphi_{01} + \alpha n(1 + \varphi_{00})}{\Delta} \\ \omega_{01} &= \frac{1 + \varphi_{11}}{\Delta}, & \omega_{11} &= -\frac{\varphi_{01}}{\Delta} \\ \omega_{02} &= -\frac{\varphi_{10}}{\Delta}, & \omega_{12} &= \frac{1 + \varphi_{00}}{\Delta} \end{aligned} \tag{9.24}$$

$$\Delta = \begin{vmatrix} 1 + \varphi_{00} & \varphi_{10} \\ \varphi_{01} & 1 + \varphi_{11} \end{vmatrix}$$

To determine $\varkappa_0[n]$ and $\varkappa_1[n]$, which are contained in the expression for the weighting function, we make use of Equations (8.35) and (8.36). Let us first introduce the following symbols:

$$\begin{aligned} \lambda_{00} &= \boldsymbol{M}[n]\cdot\boldsymbol{B}[n-N], \ \lambda_{01} = \boldsymbol{M}[n]\cdot\boldsymbol{B}[n] \\ \lambda_{10} &= \boldsymbol{M}[n-N]\cdot\boldsymbol{B}[n-N], \ \lambda_{11} = \boldsymbol{M}[n]\cdot\boldsymbol{B}[n-N] \end{aligned} \tag{9.25}$$

and

$$\psi_{\nu\mu} = \sum_{m=n-N}^{n} p_\nu[m]\, b_\mu[m], \ (\nu,\ \mu = 0,1) \tag{9.26}$$

Then, with (9.13) and (9.23) taken into account, (8.35) and (8.36) can be written in the form

$$\begin{aligned} \varkappa_0[n]\lambda_{00} + \varkappa_1[n]\lambda_{10} &= \Omega_{00} + \Omega_{10}\gamma_0[n] + \Omega_{20}\gamma_1[n] \\ \varkappa_0[n]\lambda_{01} + \varkappa_1[n]\lambda_{11} &= \Omega_{01} + \Omega_{11}\gamma_0[n] + \Omega_{21}\lambda_1[n] \end{aligned} \tag{9.27}$$

where the following notation has been used:

$$\Omega_{00}=1-\alpha n(n-N)-\omega_{00}[\varphi_{00}+(n-N)\varphi_{01}]-\omega_{10}[\varphi_{10}+(n-N)\varphi_{11}]$$
$$\Omega_{10}=1-\omega_{01}[\varphi_{00}+(n-N)\varphi_{01}]-\omega_{11}[\varphi_{10}+(n-N)\varphi_{11}]$$
$$\Omega_{20}=(n-N)-\omega_{02}[\varphi_{00}+(n-N)\varphi_{01}]-\omega_{12}[\varphi_{10}+(n-N)\varphi_{11}] \quad (9.28)$$

$$\Omega_{01}=1-\alpha n^2-\omega_{00}[m_0[n]\psi_{00}+m_1[n]\psi_{01}]-\omega_{10}[m_0[n]\psi_{10}+m_1[n]\psi_{11}],$$
$$\Omega_{11}=1-\omega_{01}[m_0[n]\psi_{00}+m_1[n]\psi_{01}]-\omega_{11}[m_0[n]\psi_{10}+m_1[n]\psi_{11}]$$
$$\Omega_{12}=n-\omega_{02}[m_0[n]\psi_{00}+m_1[n]\psi_{01}]-\omega_{12}[m_0[n]\psi_{10}+m_1[n]\psi_{11}]$$

Solving (9.27) we find

$$\varkappa_0[n]=R_{00}+R_{10}\gamma_0[n]+R_{20}\gamma_1[n]$$
$$\varkappa_1[n]=R_{01}+R_{11}\gamma_0[n]+R_{21}\gamma_1[n] \quad (9.29)$$

where for the sake of brevity we have used the notation

$$R_{00}=\frac{\Omega_{00}\lambda_{11}-\Omega_{01}\lambda_{10}}{\Delta}, \quad R_{10}=\frac{\Omega_{10}\lambda_{11}-\Omega_{11}\lambda_{10}}{\Delta}$$
$$R_{20}=\frac{\Omega_{20}\lambda_{11}-\Omega_{21}\lambda_{10}}{\Delta}, \quad R_{01}=\frac{\Omega_{01}\lambda_{00}-\Omega_{00}\lambda_{01}}{\Delta}$$
$$R_{11}=\frac{\Omega_{11}\lambda_{00}-\Omega_{10}\lambda_{01}}{\Delta}, \quad R_{21}=\frac{\Omega_{21}\lambda_{00}-\Omega_{20}\lambda_{01}}{\Delta} \quad (9.30)$$
$$\Delta=\begin{vmatrix}\lambda_{00} & \lambda_{10}\\ \lambda_{01} & \lambda_{11}\end{vmatrix}$$

We can now write the expression for the required weighting function. Using (9.14), (9.23) and (9.29) we have

$$W[n;\ l]=\omega_{00}p_0[l]+\omega_{10}p_1[l]+(\omega_{01}p_0[l]+\omega_{11}p_1[l])\gamma_0[n]$$
$$+(\omega_{02}p_0[l]+\omega_{12}p_1[l])\gamma_1[n]+R_{00}+R_{01}+R_{10}\gamma_0[n]\sigma[n-l]$$
$$+R_{20}\gamma_1[n]\sigma[n-l]+R_{11}\gamma_0[n]\sigma[n-N-l] \quad (9.31)$$
$$+R_{21}\gamma_1[n]\sigma[n-N-l]$$

whence it follows that $\gamma_0[n]$ and $\gamma_1[n]$ must be found in order to determine the weighting function in its final form. To do this we must use (8.1) which, when

$$D_{\nu\mu}=\begin{cases}0, & \nu\neq\mu\\ \infty, & \nu=\mu\end{cases}$$

are transformed into the additional conditions, which we have not yet used, to be satisfied by the weighting function. We shall use these to determine γ_0 and γ_1.

It is not difficult to show that in the given case (filtering) the additional conditions reduce to the following:

$$\sum_{m=n-N}^{n} W[n;\ m]=1$$

$$\sum_{m=n-N}^{n} (n-m)\,W[n;\ m]=0 \tag{9.32}$$

Substituting the expression of the weighting function from (9.31) into (9.32), we find the following equations for $\gamma_0[n]$ and $\gamma_1[n]$:

$$\left.\begin{aligned}\gamma_0[n]\,\mu_{00}+\gamma_1[n]\,\mu_{10}&=\chi_1[n,\ N]\\ \gamma_0[n]\,\mu_{01}+\gamma_1[n]\,\mu_{11}&=\chi_2[n,\ N]\end{aligned}\right\} \tag{9.33}$$

where

$$\chi_1[n,\ N]=1-(R_{00}+R_{01})(N+1)-\sum_{m=n-N}^{n}(\omega_{00}p_0[m]+\omega_{10}p_1[m])$$

$$\mu_{00}=R_{10}+R_{11}+\sum_{m=n-N}^{n}(\omega_{01}p_0[m]+\omega_{11}p_1[m])$$

$$\mu_{10}=R_{20}+R_{21}+\sum_{m=n-N}^{n}(\omega_{02}p_0[m]+\omega_{12}p_1[m]) \tag{9.34}$$

$$\chi_2[n;\ N]=-(R_{00}+R_{01})\frac{N(N+1)}{2}-\sum_{m=n-N}^{n}(n-m)(\omega_{00}p_0[m]+\omega_{10}p_1[m])$$

$$\mu_{01}=NR_{11}+\sum_{m=n-N}^{n}(n-m)(\omega_{01}p_0[m]+\omega_{11}p_1[m])$$

$$\mu_{11}=NR_{21}+\sum_{m=n-N}^{n}(n-m)(\omega_{02}p_0[m]+\omega_{12}p_1[m])$$

The solution of (9.33) completes the determination of the weighting function of the system which ensures optimal filtration of a signal of the form (9.1). The minimum dispersion of the filtration error can be found from (5.8).

REFERENCES

1. TSYPKIN, Ya. Z., *Sampling systems theory and its applications*, Vols 1 & 2, Pergamon Press (1963).
2. TSYPKIN, Ya. Z., 'Frequency method of analysis of discontinuous control systems', *Avtomat. telemekh.* 14, No. 1 (1953).
3. TSYPKIN, Ya. Z., 'Calculation of discontinuous control systems in the case of stationary random actions', *Avtomat. telemekh.*, 14, No. 4 (1953).
4. TSYPKIN, Ya. Z., 'Considering the form of pulses in discontinuous control systems', *Avtomat. telemekh.*, 16, No. 5 (1955).
5. TSYPKIN, Ya. Z., 'Discrete automatic systems, problems of the theory, future developments', *Works of the conference on the theory and application of discrete automatic systems*, Izd. Akad. Nauk SSSR (1960).
6. TSYPKIN, Ya. Z., 'On automatic control systems containing digital computers', *Avtomat. telemekh.*, 17, No. 8 (1956).
7. TSYPKIN, Ya. Z., 'Investigation of steady-state processes in sampled data follow-up systems', *Avtomat. telemekh.*, 17, No. 8 (1956).
8. TSYPKIN, Ya. Z., 'Sampled data systems with extrapolating devices', *Avtomat. telemekh.*, 19, No. 5 (1958).
9. PUGACHEV, V.S., 'Random functions given by ordinary differential equations', Tr. VVIA im. Zhukovskogo, No. 118 (1944).
10. PUGACHEV, V.S., 'Fundamentals of the general theory of random functions', *Trudy Akad. Art. Nauk* (1952).
11. PUGACHEV, V.S., 'General correlation theory of random functions', *Izv. Akad. Nauk SSSR, Ser. Matem.*, 17, No. 5 (1953).
12. PUGACHEV, V.S., 'The general theory of random functions and its use in automatic control theory', *Works of the Second All-Union Conference on Automatic Control Theory*, II, (1955).
13. PUGACHEV, V.S., *Theory of random functions*, Pergamon Press (1965).
14. PEROV, V.P., *Statistical synthesis of sampled data systems*, Sovetskoe Radio (1959).
15. PEROV, V.P., 'Synthesis of sampled data circuits and systems with sampled data feedback', *Avtomat. telemekh.*, 18, No. 12 (1957).
16. SOLODOV, A.V., 'Statistical investigation of non-stationary processes

in linear systems with the use of inverse simulating devices', *Avtomat. telemekh.* 19, No. 4 (1958).
17. SOLODOVNIKOV, V.V., Introduction to the Statistical dynamics of automatic control systems, Constable (1960).
18. GEL'FOND, A.O., *Calculus of finite differences,* Fizmatgiz (1959).
19. POSPELOV, G.S., 'Sampled data automatic control systems', *Avtomat. Upravlenie Vychislitel. Tekhn.,* No. 3, Mashgiz (1960).
20. TARTAKOVSKII, G.P., 'On the theory of linear sampled data systems with variable parameters', *Radiotekh. Elektron.,* No. 11 (1957).
21. TARTAKOVSKII, G.P., 'Time and frequency characteristics of linear sampled data systems with variable parameters', *Radiotekh. Elektron.,* No. 2 (1956).
22. TARTAKOVSKII, G.P., 'On the use of simulation in analysing linear sampled data systems with variable parameters', *Avtomat. telemekh.* 20, No. 5 (1959).
23. LANING, J.H. and BATTIN, R.G., Random Processes in Automatic Control, McGraw-Hill (1956).
24. KRUT'KO, P.D., 'The problem of determining discrete shaping filters', *Avtomat. telemekh.,* 22, No. 11 (1961).
25. KRUT'KO, P.D., 'A method for continuous determination of the dispersion of the output coordinate of a linear sampled data system', *Izv. Akad. Nauk SSSR, Energ. avtom.,* No. 6 (1962).
26. KRUT'KO, P.D., 'Discrete analogue of Dirac's δ-function'. *Avtomat. Telemekh.,* 23, No. 7 (1962).
27. KRUT'KO, P.D., 'The problem of determining the input signal of a linear sampled data system, which is equivalent to initial conditions', *Izv. Akad. Nauk. SSSR, Tekhn. kibern.,* No. 1 (1963).
28. MOZZHUKHIN, N.M., 'On a method of statistical analysis of the dynamics of linear automatic control systems by means of analogue computers', *Izv. Akad. Nauk SSSR, Energ. i avtom.,* No. 4 (1961).
29. BATKOV, A.M., 'Certain problems of the theory of linear systems with variable parameters in the case of random actions', *Avtomat. Upravlenie Vychislitel. Tekhn.,* No. 3, Mashgiz (1960).
30. NAIMARK, M.A., *Linear differential operators,* Gostekhizdat (1954).
31. KUZIN, L.T., 'The application of z-transformation to the analysis of control systems with computers', *Avtomat. Upravlenie Vychislitel. Tekhn.,* No. 1, Mashgiz (1958).
32. JAMES, H.M., NICHOLS, N.B. and PHILLIPS, R.S., Theory of Servomechanisms, MIT Press (Radiation laboratory Series No. 25) (1947).
33. KRAPIVIN, V.K., 'Probability characteristics of linear automatic control system in the case of non-stationary disturbances', *Sb. Nauchn. Tr. VVIA im. Zhukovskogo,* I (1954).
34. ANDREEV, N.I., 'Determination of the correlation function of the error at the output of a system with constant coefficients when a non-stationary disturbance is fed to the input of this system', *Sb. Nauchn. Tr. VVIA im. Zhukovskogo,* I (1954).
35. RAEVSKII, S. Ya., 'Non-stationary random processes in linear systems with constant parameters', *Trudy VVIA im. Zhukovskogo,* No. 495 (1954).
36. KAZAKOV, I.E., 'Approximate probability analysis of the accuracy of operation of essentially non-linear systems', *Avtomat. i telemekh.,* 17, No. 5 (1956).

37. KAZAKOV, I.E., 'Certain problems of the theory of statistical linearisation and its uses', a paper given at the First International Congress of IFAC on Automatic Control, *Akad. Nauk SSSR,* 1960.
38. POPOV, E.P. and PAL'TOV, I.P., *Approximate methods for investigating non-linear automatic systems,* Fizmatgiz (1960).
39. PYATNITSKII, G.I., 'Action of random processes on discontinuous control systems', *Avtomat. i telemekh.,* 21, No. 5 (1960).
40. TSYPKIN, Ya. Z., 'Elements of the theory of digital automatic systems', A paper given at the first International Congress of IFAC on Automatic Control. *Akad. Nauk SSSR,* (1960).
41. TSYPKIN, Ya. Z., *Sampling systems theory and its applications,* Vols 1 & 2, Pergamon Press (1963).
42. KUZNETSOV, P.I., STRATONOVICH, R.L. and TIKHONOV, V.I., 'Passage of random functions through non-linear systems', *Avtomat. i telemekh.,* 14, No. 4 (1953).
43. BOOTON, R.C., 'The analysis of non linear control systems with random inputs', *Proc. Sympos. Non linear Circuit Analysis,* Vol. 2 (1953).
44. AMIANTOV, I.M. and TIKHONOV, V.I., 'Action of normal fluctuations on typical non-linear elements', *Izv. Akad. Nauk SSSR,* No. 4 (1956).
45. BUNIMOVICH, V.I., *Fluctuating processes in radio receivers,* Sovetskoe Radio (1951).
46. CRAMER, H., *Mathematical methods of statistics,* John Wiley (1945).
47. PUPKOV, K.A., 'A method of investigating the accuracy of essentially non-linear automatic control systems by means of an equivalent transfer function', *Avtomat. Telemekh.* 21, No. 2 (1960).
48. PYATNITSKII, G.I., 'Action of stationary random processes on a control system which contains pronounced non-linear elements', *Avtomat. i telemekh.,* 21, No. 4 (1960).
49. PUGACHEV, V.S., *Theory of random functions,* Pergamon Press (1965).
50. RICE, S.O., 'Theory of fluctuating noise', A collection of translated articles, edited by N.A. Zheleznov.
51. KUZNETSOV, P.I., STRATONOVICH, P.L. and TIKHONOV, V.I., 'Quasi-momental functions in the theory of random processes', *Dokl. Akad. Nauk SSSR,* 94, No. 4 (1954).
52. KAZAKOV, I.E., 'On the problem of investigating the dispersion in non-linear systems', *Sb. nauchn. VVIA im. Zhukovskogo* (1954).
53. KAZAKOV, I.E., 'An approximate method of statistical investigation of non-linear systems', *Tr. VVIA im. Zhukovskogo,* No. 394 (1954).
54. CRAMER, H., Random quantities and probability distributions (1947).
55. POPOV, E.P., *The dynamics of automatic control systems,* Pergamon Press (1962).
56. PUGACHEV, V.S., *Theory of random functions,* Pergamon Press (1965).
57. PUGACHEV, V.S., 'The general condition for the minimum mean square error of a dynamical system', *Avtomat. telemekh.,* 17, No. 4 (1956).
58. PUGACHEV, V.S., 'The use of canonical expansions of random functions in determining optimal linear systems', *Avtomat. telemekh.,* 17, No. 6 (1956).
59. PUGACHEV, V.S., 'A possible general solution of the problem of determining an optimal dynamical system', *Avtomat. telemekh.,* 17, No. 7 (1956).

60. PUGACHEV, V.S., 'Integral canonical forms of random functions and their use in determining optimal linear systems', *Avtomat. telemekh.*, 18, No. 11 (1957).
61. PUGACHEV, V.S., 'Determination of an optimal system on an arbitrary criterion', *Avtomat. telemekh.*, 19, No. 6 (1958).
62. SEMENOV, V.M., 'On the extrapolation theory of random processes'. *Sb. nauchn. tr. VVIA im. Zhukovskogo,* I (1954).
63. SOLODOVNIKOV, V.V., *Introduction to the statistical dynamics of automatic control systems,* Constable (1960).
64. SOLODOVNIKOV, V.V. and MATVEEV, P.S., 'Synthesis of the correcting devices of follow-up systems for given requirements and dynamic accuracy when noise is present', *Avtomat. telemekh.*, 16, No. 3 (1955).
65. BATKOV, A.M. and SOLODOVNIKOV, V.V., 'A method of determining the optimal characteristics of a class of adaptive systems', *Avtomat. telemekh.* 18, No. 5 (1957).
66. KURAKIN, K.I., 'An analytical method of synthesis of linear automatic control systems when noise is present', *Avtomat. i telemekh.*, 19, No. 5 (1958).
67. JOHNSON, K., 'Optimum linear discrete filtering of signal containing nonrandom component', *IRE, Trans. Inform. Theory* (1956).
68. PEROV, V.P., 'Synthesis of sampled data circuits and systems with sampled data feedback', *Avtomat. telemekh.*, 18, No. 12 (1957).
69. PEROV, V.P., *Statistical synthesis of sampled data systems,* Sovetskoe Radio (1959).
70. LANING, J.H. and BATTIN, R.G., Random Processes in Automatic Control, McGraw-Hill (1956).
71. WIENER, N., *Extrapolation, interpolation and smoothing of stationary time series,* John Wiley (1949).
72. LAVRENT'EV, M.A. and LYUSTERNK, A.A., *A course in the calculus of variations,* Gostekhizdat (1951).
73. ZADEH, L.A. and RAGAZZINI, I.R., 'An extension of Wiener's theory of prediction', *J. Appl. Phys.*, 2, (July 1950).
74. TSYPKIN, Ya. Z., *Sampling systems theory and its applications,* Vols 1 & 2, Pergamon Press (1963).
75. SHINBROT, M., 'A generalization of a method for the solution of the integral equation arising in optimization of time-varying linear systems with nonstationary inputs', *IRE, Trans. Inform. Theory,* 3, No. 4 (1957).
76. SHINBROT, M., 'Optimization of time-varying linear systems with nonstationary inputs', *Trans. Am. Soc. Mech. Engrs.*, 80, No. 2 (1958).
77. LI CHEN WONG, 'A generalisation of the Shinbrot method for determining the optimum weighting function for non-stationary random actions', *Izv. Akad. Nauk SSSR, Otd. Tekhn. Nauk* No. 3 (1959).
78. LI CHEN WONG, 'A generalisation of the Shinbrot method for determining the optimum weighting function for non-stationary random actions when the disturbance is white noise', *Izv. Akad. Nauk SSSR. Otd. Tekhn. Nauk,* No. 4 (1959).
79. KRUT'KO, P.D., 'A method of determining the optimal weighting functions of discrete filters for a certain class of non-stationary random processes', *Izv. Akad. Nauk SSSR, Tekhn. Kilbern.*, No.1 (1963).

80. KRUT'KO, P.D., 'Determination of an optimal discrete filter for non-stationary random input and white noise', *Izv. Akad. Nauk SSSR, Tekhn. Kilbern.*, No. 2 (1963).

Additional references

81. FLÜGGE-LOTZ, I., *Discontinuous automatic control,* Oxford (1954).
82. JURY, E.I., *Sampled-data control systems,* Wiley (1958).
83. KUZNETSOV, P.I. *el al* (eds.), *Non-linear transformations of stochastic processes,* Pergamon (1965).
84. LINDORFF, D.P., *Theory of sampled-data control systems,* Wiley (1965).
85. PUGACHEV, V.S., *Theory of stochastic processes and its application to automatic control,* Pergamon (1966).
86. SOLODOV, A.N., *Linear automatic control systems with variable parameters,* Blackie (1966).
87. SOLODOVNIKOV, V.V., *Automatic control and computer engineering,* Pergamon (1961-63).
88. TRUXAL, J.G., *Automatic feedback control systems,* McGraw-Hill (1955).

INDEX